Precalculus with Calculator Applications

GEOMETRY FORMULAS

Pythagorean Theorem   For a right triangle

$$a^2 + b^2 = c^2$$

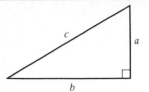

Triangle $\alpha + \beta + \gamma = 180°$ Perimeter $= a + b + c$

$$\text{Area} = \tfrac{1}{2} bh$$

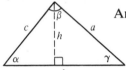 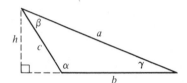

Parallelogram

Area $= bh$

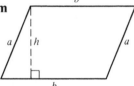

Trapezoid

Area $= \tfrac{1}{2} h (b_1 + b_2)$

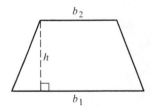

Circle

Circumference $= 2\pi r$
Area $= \pi r^2$

Sphere

Surface area $= 4\pi r^2$
Volume $= \tfrac{4}{3} \pi r^3$

Cone (right circular)

Lateral Surface $= \pi r \ell$
Volume $= \tfrac{1}{3} \pi r^2 h$

Cylinder (right circular)

Lateral Surface $= 2\pi rh$
Volume $= \pi r^2 h$

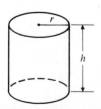

Precalculus with Calculator Applications

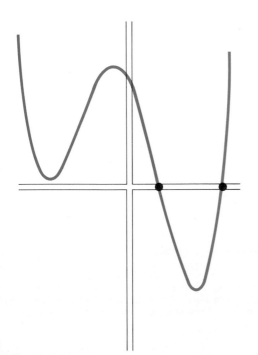

**Joseph
Elich**
Utah State University

**Carletta J.
Elich**
Logan High School

▲▼ ADDISON-WESLEY PUBLISHING COMPANY
Reading, Massachusetts · Menlo Park, California ·
London · Amsterdam · Don Mills, Ontario · Sydney

Sponsoring editor: Pat Mallion
Production editor: Rima Zolina
Designer: Marie McAdam
Illustrator: Parkway Illustrated Press
Art coordinator: Loretta Bailey
Cover design: Richard Hannus
Cover photograph: E. P. Miles, Jr., Miles Color Art

Library of Congress Cataloging in Publication Data

Elich, Joseph, 1918–
  Precalculus with calculator applications.

  Includes index.
  1. Mathematics—1961–   . I. Elich,
Carletta J., 1935–   . II. Title.
QA39.2.E43   512′.1   81-3544
ISBN 0–201–13345–8   AACR2

ISBN 0-201-13345-8
ABCDEFGHIJ-DO-8987654321

Preface

The primary goal of a precalculus course is to provide students with the essential training needed to approach a course in calculus with the confidence necessary for successful completion. This training involves a good understanding of the concepts related to functions in general and to the important special cases (polynomial, exponential, logarithmic, and trigonometric, or circular, functions) in particular. Crucial to the training process is maintenance of good balance between emphasis on understanding fundamental ideas and on developing and expanding the basic skills of arithmetic and algebra. This is the guiding principle adhered to by the authors. The prominent features of the book are the following:

1. Emphasis is on basic ideas throughout the text. Problems in Exercise Sets at the end of each section have been carefully selected with two purposes in mind: to provide the student with an opportunity to apply fundamental concepts, thus leading to a better grasp of ideas, and to offer extensive practice in developing manipulative algebraic skills that are necessary for success in the study of subsequent courses in calculus. Included are many problems in which students are expected to "think through" a solution rather than solve them by rote processes.

2. The use of calculators has been integrated into the material throughout the text, but the primary focus is not on computations. Calculators are used whenever it is felt that they will contribute to the ultimate goals of the course. Lengthy and tedious computations frequently result in distraction from fundamental considerations. Use of high-speed computing devices minimizes such distractions. Although calculators are primarily employed for computational purposes, they can also help to motivate and develop concepts. We attempt to exploit this role whenever possible. For instance, in Section 1.2 we consider a calculator as a function machine to reinforce the definition of a function; then we relate composition of functions to pressing appropriate successive keys on a calculator.

3. Applications are included throughout the text whenever appropriate. The numbers used are realistic since these can be handled by calculators just as easily as the carefully selected simple numbers that commonly appear in mathematics textbooks and that are chosen for the sole purpose of avoiding even slightly cumbersome computations. Calculators can add the dimension of approximate numbers often ignored in mathematics courses.

4. Appendix A contains a relatively complete introduction to the employment of calculators for those students who have had no previous experience with them. Included in separate sections is a discussion of algebraic operating systems (AOS) calculators based on algebraic entry and RPN calculators based on Reverse Polish Notation. Keys for elementary computations are carefully described and followed by several detailed examples and practice problems. In most cases, students can master this material on their own. Additional instruction on the use of special function keys is given throughout the text proper, as needed and when appropriate. In our experience, minimal class time has been required for calculator instruction.

5. Appendix B contains a brief review of concepts and properties of real numbers. This material can be included in courses that emphasize the structure of the real-number system.

6. Appendix C includes a relatively detailed treatment of computation with approximate numbers. This topic is often avoided in mathematics courses, but it is important in applications.

7. Throughout the entire book, the pattern of topic presentation is the following: introduction of basic ideas; illustration of these ideas by several examples worked in detail; Exercise Set of problems carefully designed to practice with the new concepts and reinforce previously encountered ideas. Also included are Review Exercises that utilize the material studied up to that point.

8. A concept, a technique, or a fact can best be learned if it is encountered frequently and in a variety of settings. We exploit this truism by including problems in Exercise Sets that repeatedly use ideas introduced in earlier sections. For instance, the idea of combining functions to get new functions is introduced in Chapter 1 and it is reinforced throughout the remainder of the text in the Exercise Sets. Thus, for example, after completing the study of this book, the student should have a good understanding of composition of functions and be prepared for the introduction to the Chain Rule in a subsequent calculus course.

9. Whenever appropriate, presentation relies heavily on graphs. The reader should find that in many situations an accompanying picture is invaluable for providing insight into the various algebraic techniques for problem solving. One of the important uses of calculators is in drawing accurate graphs.

10. The Exercise Sets contain a large number of problems ranging from simple to challenging. In each section the reader will find several easy-to-follow Examples that illustrate the various types of problems included as exercises. In some cases, the use of calculators allows us to introduce problem-solving methods that are not part of a traditional course.

11. *Looking ahead to calculus.* The fundamentals of calculus are based on concepts related to limits. A thorough understanding of the ideas involved is difficult the first time they are presented. A preliminary introduction based upon numerical examples along with geometrical interpretations gives the students an intuitive feeling for the abstract definitions of $\epsilon - \delta$ studied in calculus courses. Thus we have included a section "Looking Ahead to Calculus" at the end of Chapters 2, 3, 5, and 7, in which various types of limit problems are examined from a numerical point of view. These sections should be considered optional; they are particularly appropriate for those situations in which programmable calculators or microcomputers are available to students.

12. This book is designed for a one-semester or two-quarter course. A prerequisite of high-school geometry and intermediate algebra is assumed.

The authors would like to express their appreciation to reviewers of both books: Ellen E. Casey, Massachusetts Bay Community College; Robert G. Clawson, Brigham Young University; Neville C. Hunsaker, Utah State University; Philip H. Mahler, Henry Ford Community College; Gordon L. Nipp, California State College, Bakersfield; Janet P. Ray, Seattle Central Community College; Joshua H. Rabinowitz, University of Illinois at Chicago Circle.

Logan, Utah   J. E.
November, 1981   C. J. E.

Foreword to the Teacher

The following statements explain some of the prominent pedagogical features of this book.

1. *Order of topics.* Although a sequential ordering is essential in presentation of some of the topics, there is considerable latitude in the order in which chapters can be studied. Chapter 1 contains the basics of functions and should be considered prerequisite for all remaining topics. The material on polynomials (Chapter 2), exponential–logarithmic functions (Chapter 3), and trigonometry (Chapters 4 and 5) can be studied in any order.

Chapters 1 through 5 essentially constitute a study of elementary functions and should be included in any precalculus course. Chapter 7 contains material on sequences of real numbers and provides an important introduction to the topic of sequences and infinite series to be studied in calculus.

Although the topics within a chapter are sequential, there are some portions that can be omitted or easily modified. For instance, in Section 2.7 the Remainder and Factor Theorems are important but the use of synthetic division for evaluating polynomials can be minimized or even omitted since nested-form techniques for polynomial evaluation by using calculators are introduced in the preceding sections.

Although applications are scattered throughout the book, there are some sections that are devoted exclusively to applied problems. These provide excellent experiences with problem-solving techniques and are a desirable feature in any course. These include: Section 3.5 (exponential growth and decay); Sections 4.2, 4.6, and 4.7 (applications in trigonometry); and Section 7.6 (mathematics of finance).

2. *Notation and approximate-number solutions.* As in most mathematics texts, we take liberties in using the "equals" symbol to include "approximately equals"; the context of discussion should make it clear when it is so used. For instance, it is important for the student to realize that $\sqrt{2}$ is a symbol for a number that cannot be written explicitly in decimal form, and when we ask for solutions in *exact* form, such a symbol is the only acceptable answer. On the other hand, if we ask for a solution *rounded off to* or *correct to* five decimal places, we expect the approximate value 1.41421.

Appendix C includes a discussion of the arithmetic of approximate numbers. In general, approximate numbers are studied in numerical-analysis courses. In Exercise Sets we ask students to state answers to a given number of decimal places, rather than always apply rules of approximate-number arithmetic. We prefer that students concentrate on the main ideas under discussion and not be distracted by the need to apply approximate-number rules.

3. *Modern approach.* The approach in this book is consistent with the recommendations of the National Council of Teacher of Mathematics for the curriculum of the 1980's:

> *The use of electronic tools such as calculators and computers should be integrated into the core mathematics curriculum. . . Calculators and computers should be used in imaginative ways for exploring, discovering, and developing mathematical concepts and not merely for checking computational values or for drill and practice.*

4. *Review material.* Most precalculus textbooks include a review chapter or two at the beginning consisting of topics studied in elementary-algebra courses. In general, students find a detailed review boring, particularly at the beginning of a new course of study. Therefore, we have included appropriate review material in the first section of each of the first three chapters. Sections 1.0, 2.0, and 3.0 each contain material prerequisite for that chapter and are intended for review. The review consists of a brief summary of pertinent ideas and facts; this is followed by a set of problems that should be considered the core of the review. The review sections can be included as an integral part of the course, or can be given as out-of-class assignments to be done concurrently with the work of the chapter, or omitted entirely when students have good mastery of elementary algebra.

5. *Supplementary material.* Three separate supplements are available. The *Instructor's Manual* contains suggested class period schedules, a brief overview of material of each section, and lists of problems for each chapter that can be used for tests. *The Answers Booklet* has anwers to *all* of the problems, while the *Solutions Manual* includes a discussion of solutions to most of the problems of the text.

Contents

1 Functions 1

2 Polynomial Functions 103

3 Exponential and Logarithmic Functions 155

4 Trigonometric (Circular) Functions 205

Trigonometric Identities, Inverse Functions, Equations, Graphs 273

Systems of Equations and Inequalities 321

Functions on Natural Numbers 365

8 Analytic Geometry: Conics and Parametric Equations   411

9 Polar Coordinates   455

10 Complex Numbers   471

Appendixes   501

Answers to Odd-Numbered Exercises   545

Index   595

Functions

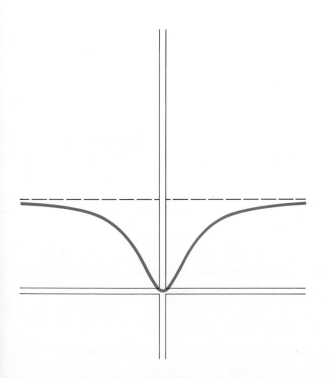

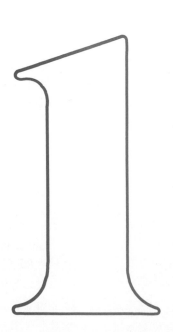

Prerequisite to the study of more advanced topics in mathematics and to applications of mathematics in real-life situations is a sound grasp of concepts related to functions. In this chapter some of the basic notions related to functions in general are explored, and then in later chapters our attention is focused on the study of special classes of functions; these include polynomial, exponential, logarithmic, and trigonometric functions.

Before proceeding with our task, we first present a brief review of topics from elementary algebra that are necessary for continuing with the subsequent material of this chapter. The problems in Exercise 1.0 should provide a good test of how much review work is needed.

1.0 INTRODUCTORY REVIEW

The basis for most of mathematics is the real-number system. Students of this course have already had considerable experience working with numbers without always being aware of the basic properties involved. We begin this section with a brief review of the various types of real numbers and the related terminology. This course does not include a formal discussion of the structure of the real-number system, but a brief description of their properties is given in Appendix B.

Before considering classification of real numbers, we review some notations related to sets. Also included in this section is a brief review of the number line, the rectangular coordinate system, the arithmetic of complex numbers, and algebra skills. (There will be more of this review in Sections 2.0 and 3.0.) We are not interested here in repeating a course in introductory algebra; however, a brief review can be helpful (and in some cases even necessary) in refreshing or reinforcing concepts and skills that may have been forgotten. Our primary interest here is to give each student a working knowledge of the topic, which is essential for successful completion of subsequent sections of this book.

Set Notation

It is customary to use capital letters (such as A, B, and so on) to denote sets. In many cases a given set is indicated by enclosing a *listing of elements* (or members) of the set within braces; the order in which these are listed is of no importance. For example, set A, which consists of the first four positive even integers, can be described by

$$A = \{2, 4, 6, 8\} \qquad \text{or} \qquad A = \{6, 8, 2, 4\}.$$

We say that "2 *is an element of A*" and denote this by $2 \in A$. Similarly, $4 \in A$, $6 \in A$, and $8 \in A$. We also write $5 \notin A$ to denote "5 *is not an element of A.*"

Frequently it is not possible or not convenient to list the elements of a set, and a *set builder* notation is used. For instance, the set B of all positive even integers is given by

$$B = \{x \,|\, x \text{ is a positive even integer}\}.$$

This is read "B is the set of elements x such that x is a positive even integer"; the vertical bar is read "such that." There are occasions when set B will be written by listing a few elements as follows:

$$B = \{2, 4, 6, \ldots\},$$

where the context of discussion will make it clear what numbers follow after 6.

In the examples above we say that A is a *finite set,* while B is an *infinite set.* Note that every element of A is also in B. The notation used to indicate this is $A \subseteq B$; this is read "A is a *subset* of B."

Subsets   In general, $E \subseteq F$ indicates that each element of E is also in F. This does not preclude the possibility that E and F are the same set, and so we could write $E \subseteq E$. If $E \subseteq F$ and there is at least one element of F that is not in E, then we write $E \subset F$. In this case we say that "E is a *proper subset* of F." In the examples above both $A \subseteq B$ and $A \subset B$ are acceptable—the latter simply gives more information.

Equality of sets   If $E \subseteq F$ and $F \subseteq E$, then we say that sets E and F are *equal* and write $E = F$.

Empty set   In many situations it is necessary to talk about a set with no elements. For example, the set C, described by

$$C = \{x \,|\, x \text{ is a real number and } x^2 + 1 = 0\},$$

can have no elements since x^2 is nonnegative; then $x^2 + 1$ is positive and cannot be equal to zero. A set such as C is called the *empty* or *null set* and denoted by the symbol $\varnothing$. Thus $C = \varnothing$.

Combining sets   There are two set operations (with associated notations) that will be helpful in making precise mathematical statements. Suppose E and F are any two sets. The *union* and *intersection* of E and F are sets defined as follows:

$$\textit{Union: } E \cup F = \{x \,|\, x \in E \textit{ OR } x \in F\}$$
$$\textit{Intersection: } E \cap F = \{x \,|\, x \in E \textit{ AND } x \in F\}$$

Fig. 1.1

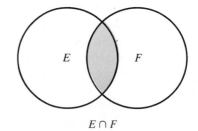

$E \cup F$ $\qquad\qquad\qquad\qquad\qquad\qquad\qquad\qquad$ $E \cap F$

The word *OR* as used in mathematics means x is in E or in F or in both E and F. If $E \cap F = \varnothing$, then sets E and F are said to be *disjoint*. The union and intersection of sets E and F are illustrated schematically in Fig. 1.1, where the shaded regions indicate the resulting sets.

Example 1   Suppose $A = \{1, 3, 5, 7\}$, $B = \{2, 3, 5, 7, 9\}$, $C = \{2, 9\}$. Determine:

a) $A \cup B$ **b)** $A \cap B$ **c)** $A \cap C$ **d)** $B \cup C$

Solution   **a)** $A \cup B = \{1, 2, 3, 5, 7, 9\}$ **b)** $A \cap B = \{3, 5, 7\}$
            c) $A \cap C = \varnothing$ **d)** $B \cup C = B$ ■

Real Numbers

Most people first learn about the counting numbers in childhood; then, as their experience broadens, the world of numbers is expanded to include fractions, negative numbers, square roots, and so on. Students who have reached this course in mathematics have had some exposure to the entire set of real numbers. Here we summarize the hierarchy and terminology associated with the various subsets of the set of real numbers. The notations introduced here will be used throughout this book.

The set of real numbers is denoted by **R**. The following sets are subsets of **R**:

$$
\begin{aligned}
\textit{Natural numbers:} \quad & \mathbf{N} = \{1, 2, 3, \ldots\}^* \\
\textit{Whole numbers:} \quad & \mathbf{W} = \{0, 1, 2, 3, \ldots\} \\
\textit{Integers:} \quad & \mathbf{J} = \{\ldots, -3, -2, -1, 0, 1, 2, 3, \ldots\} \\
\textit{Rational numbers:} \quad & \mathbf{Q} = \{a/b \,|\, a \text{ and } b \text{ are integers and } b \neq 0\}\dagger \\
\textit{Irrational numbers:} \quad & \mathbf{H} = \{x \,|\, x \in \mathbf{R} \text{ and } x \notin \mathbf{Q}\}
\end{aligned}
$$

Thus any real number that is not rational is called an *irrational number*. The following are examples of irrational numbers:

$$\sqrt{2}, \ \sqrt[3]{5}, \ -\sqrt[5]{8}, \ \frac{1 + \sqrt{5}}{2}, \ \pi, \ \pi - 5, \ \ldots$$

This means, for instance, that it is impossible to find integers a and b so that a/b is exactly equal to $\sqrt{2}$; that is, $(a/b)^2$ will never equal 2. This fact can be proved, as we shall see in Chapter 2. Here we are more interested in proper vocabulary and correct use of the properties of real numbers than in a careful development of the real-number system.

---

*   This set will also be referred to as the *set of counting numbers or positive integers*.

†   Here we are assuming that a/b is in lowest terms.

Note: It is conventional to consider, say, $\sqrt{4}$ as a *positive number* whose square is 4; there is only one such number, 2; and so $\sqrt{4} = 2$. In general, if b is a nonnegative number, then there is one and only one nonnegative number, denoted by $\sqrt{b}$, whose square is b.

Real numbers can be expressed in decimal notation, for example $1/4 = 0.25$. However, for many numbers the decimal representation is nonterminating. For instance,

$$\frac{7}{11} = 0.636363\ldots = 0.\overline{63}.$$

We say that $7/11$ has a repeating, nonterminating decimal representation.

Decimal representations can be used to distinguish between rational and irrational numbers:

> A number is *rational* if its decimal representation is terminating or nonterminating and repeating. A number is *irrational* if its decimal representation is nonterminating and nonrepeating.

The subsets of the set of real numbers are illustrated schematically in Fig. 1.2. In Fig. 1.2(a), **R** is shown as the union of two disjoint sets, **Q** and **H**; Fig. 1.2(b) gives the hierachy of the subsets of **Q**.

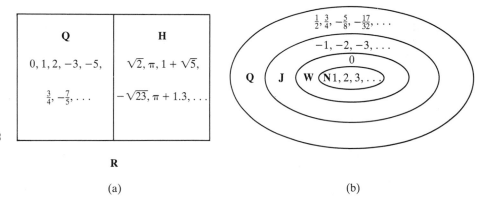

Fig. 1.2

(a)   (b)

Example 2   For each of the following, determine whether the given statement is true or false.

   a) $\mathbf{N} \subset \mathbf{Q}$ b) $\mathbf{J} \cap \mathbf{Q} = \mathbf{J}$ c) $\mathbf{Q} \cap \mathbf{H} = \varnothing$

   d) $\mathbf{J} \cap \mathbf{H} = \mathbf{J}$ e) $\sqrt{16} \in \mathbf{H}$ f) $\sqrt[3]{-8} \in \mathbf{J}$

Solution   a) True   b) True   c) True   d) False

   e) False (since $\sqrt{16} = 4$)   f) True (since $\sqrt[3]{-8} = -2$).

The Number Line

In many situations it is possible to describe properties of real numbers in terms of points on a line. In geometry we learn that a line L consists of infinitely many points. We assume that a one-to-one correspondence can be established between the set of real numbers and the set of points on line L so that each real number x is associated with a unique point of L, and conversely, this same correspondence associates each point of L with a unique real number.* This correspondence, illustrated in Fig. 1.3, gives the *number line*.

Fig. 1.3
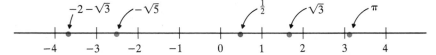

To avoid cumbersome language in reference to a number line, we take some liberties; for instance, we say "the point 2" rather than "the point associated with the real number 2."

Suppose u and v are two given real numbers. The *less than* property of real numbers can be described by referring to the number line, as shown in Fig. 1.4:

Fig. 1.4

u is less than v, denoted by $u < v$, means that point u is to the left of point v on the number line.

In later sections of this book, we will make extensive use of number line ideas in problems related to graphing.

Example 3   Show each of the following subsets of **R** on a number line.

a) $\{-1, 4\}$ b) $\{x \mid x < -2 \text{ or } x \geq 3\}$

c) $\{x \mid x \geq 1 \text{ and } x < 4\}$ d) $\{x \mid -2 \leq x \leq 4\}$

Solution   a)    b)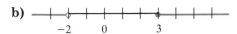

Fig. 1.5   c) 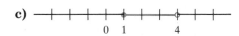   d)

---

*   In advanced mathematics texts this is referred to as the Cantor–Dedekind axiom.

Each set is shown as the colored portion of the number line in Fig. 1.5. The set given in (a) consists of only two points. The open circles enclosing -2 in (b) and 4 in (c) indicate that those two points do not belong to the given sets. Part (d) illustrates the use of shortened notation: $-2 \leq x \leq 4$ means $-2 \leq x \, and \, x \leq 4$. The notation $a < x < b$ can be used only if x is *between* a and b. ▪

Rectangular Coordinate System*

The idea of a number line can be expanded to establish a one-to-one correspondence between the set of points in a plane and the set of ordered pairs of real numbers. This is illustrated in Fig. 1.6, where the horizontal number line is called the *x-axis,* the vertical line is called the *y-axis,* and the point of their intersection is called the *origin.* Any point P in the plane is associated with an ordered pair of real numbers (u, v), as shown .Thus every point in the plane has a first name and a second name; *the first is always the horizontal coordinate, and the second is the vertical coordinate.*

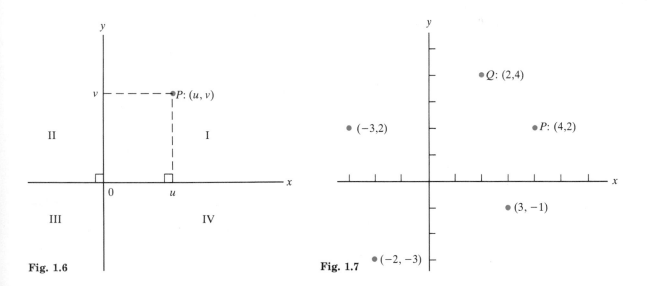

Fig. 1.6   Fig. 1.7

The coordinate axes divide the plane into four regions called *quadrants;* they are numbered I, II, III, IV, as shown in Fig. 1.6. The ordered pairs associated with several points are shown in Fig. 1.7. Note that the points $P{:}(4, 2)$ and $Q{:}(2, 4)$ are different; this illustrates the reason for calling these pairs *ordered* and leads to the following definition.

---

*   Also referred to as the *cartesian coordinate system,* named for the great French mathematician and philosopher René Descartes (1596–1650).

Definition   The ordered pairs (a, b) and (c, d) are the same, or *are equal*, if and only if $a = c$ and $b = d$.

Equality of ordered pairs is indicated by writing $(a, b) = (c, d)$.

Pythagorean Theorem and Distance Formula
First recall the following important theorem from geometry (Fig. 1.8):

Pythagorean theorem   If a and b are lengths of the sides (or legs) of a *right triangle* and c is the length of the hypotenuse, then

$$c^2 = a^2 + b^2. \tag{1.1}$$

The converse is also true; suppose a, b, and c are lengths of the three sides of a triangle; if $c^2 = a^2 + b^2$, then the triangle is a right triangle.

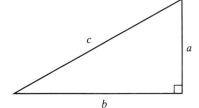

Fig. 1.8

Distance formula   The Pythagorean theorem provides us with a formula for finding the distance between two points in a plane. Let P and Q be the two points (as illustrated in Fig. 1.9), and let d represent the distance between P and Q. Then d is given by

$$d = \overline{PQ} = \sqrt{(x_1 - x_2)^2 + (y_1 - y_2)^2}. \tag{1.2}$$

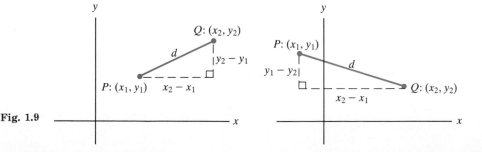

Fig. 1.9

Example 4   Suppose the lengths of the three sides of a triangle are $a = 4, b = 5, c = 7$. Is the triangle a right triangle?

Solution $$a^2 + b^2 = 16 + 25 = 41, \qquad c^2 = 49, \qquad a^2 + b^2 \neq c^2;$$

therefore the triangle is not a right triangle.   ■

Example 5   Find the distance between the given points.

a) $A:(3, -4)$ and $B:(2, 3)$ **b)** $C:(-1, -3)$ and $D:(1, -2)$

Solution   **a)** $\overline{AB} = \sqrt{(3 - 2)^2 + (-4 - 3)^2} = \sqrt{1^2 + (-7)^2} = \sqrt{50} = 5\sqrt{2}$
b) $\overline{CD} = \sqrt{(-1 - 1)^2 + (-3 + 2)^2} = \sqrt{(-2)^2 + (-1)^2} = \sqrt{5}$ ■

Arithmetic of Complex Numbers

Most of our discussion will deal with real numbers; however, there will be a few problems in this text for which there is no solution within the system of real numbers. For example, if we wish to find the values of x that satisfy the equation $x^2 + 1 = 0$, it can easily be argued that there are no real-number solutions. Thus we introduce a "new number" i, also denoted by $\sqrt{-1}$, with the property that $i^2 = -1$. Hence the numbers i and $-i$ are solutions of $x^2 + 1 = 0$.

The set of complex numbers

Definition   The set **C** of complex numbers is given by

$$\mathbf{C} = \{u + vi \,|\, u \text{ and } v \text{ are real numbers}\}.$$

If $v = 0$, then $u + vi = u$; since u is a real number, we see that any real number is also a complex number. That is, $\mathbf{R} \subset \mathbf{C}$.

If $u = 0$, then $u + vi = vi$; for any given real number v, the value vi is called a *pure imaginary number,* or simply an *imaginary number.* For instance, we have the following:

3 is a real number and also a complex number,

$2i$ is an imaginary number and also a complex number,

$3 - 2i$ is a complex number.

The set of complex numbers is the union of three disjoint sets: the real numbers, the imaginary numbers, and numbers of the form $u + vi$, where $u \neq 0$ and $v \neq 0$.

The arithmetic of complex numbers follows familiar rules of algebra of binomial expressions, where i is treated as if it were a variable; whenever i^2 occurs, replace it by -1. The following example illustrates how we add, subtract, multiply, and divide complex numbers and write the result in the form $u + vi$, with u and v real numbers.

Example 6   Perform the indicated operations and simplify.

   a) $(2 - 3i) + (-1 + 5i)$ **b)** $(1 + i) - (-4 + 3i)$

   c) $(1 + 2i) \cdot (3 - i)$ **d)** $(1 - 3i) \div (1 + i)$

Solution   **a)** $(2 - 3i) + (-1 + 5i) = 2 - 3i - 1 + 5i = (2 - 1) + (-3 + 5)i = 1 + 2i$

   b) $(1 + i) - (-4 + 3i) = 1 + i + 4 - 3i = (1 + 4) + (1 - 3)i = 5 - 2i$

   c) $(1 + 2i)(3 - i) = 3 - i + 6i - 2i^2 = 3 + 5i - 2(-1) = 5 + 5i$

   d) $\dfrac{1 - 3i}{1 + i} = \dfrac{(1 - 3i)(1 - i)}{(1 + i)(1 - i)} = \dfrac{1 - 4i + 3i^2}{1 - i^2}$

$$= \dfrac{1 - 4i - 3}{1 - (-1)} = \dfrac{-2 - 4i}{2} = -1 - 2i \qquad \blacksquare$$

Algebraic Operations

We assume that the student is familiar with some of the basic properties of algebra. Here we give several examples as a review.

Example 7   Perform the indicated operations and simplify.

   a) $(2x - 3)(x + 1)$ **b)** $(2y - 1)(4y^2 + 2y + 1)$

   c) $(x^3 - 2x + 1) \div (x + 2)$ **d)** $\dfrac{2}{x - 1} - \dfrac{3}{x + 2}$

Solution   **a)** $(2x - 3)(x + 1) = 2x(x + 1) - 3(x + 1)$

$$= 2x^2 + 2x - 3x - 3 = 2x^2 - x - 3$$

   b) $(2y - 1)(4y^2 + 2y + 1) = 2y(4y^2 + 2y + 1) - 1(4y^2 + 2y + 1)$

$$= 8y^3 + 4y^2 + 2y - 4y^2 - 2y - 1 = 8y^3 - 1$$

   c)

$$
\begin{array}{r}
x^2 - 2x\ \ + 2 \\
x + 2\,\overline{)\,x^3 - 2x + 1} \\
\underline{x^3 + 2x^2 } \\
-2x^2 - 2x + 1 \\
\underline{-2x^2 - 4x } \\
2x + 1 \\
\underline{2x + 4} \\
-3
\end{array}
$$

Thus $\dfrac{x^3 - 2x + 1}{x + 2} = x^2 - 2x + 2 - \dfrac{3}{x + 2}$.

   d) $\dfrac{2}{x - 1} - \dfrac{3}{x + 2} = \dfrac{2(x + 2)}{(x - 1)(x + 2)} - \dfrac{3(x - 1)}{(x - 1)(x + 2)}$

$$= \dfrac{2x + 4 - (3x - 3)}{(x - 1)(x + 2)} = \dfrac{-x + 7}{(x - 1)(x + 2)} \qquad \blacksquare$$

Factoring   First let us illustrate what it means to factor a given algebraic expression. We say that $x^2 - 4$ factors into $(x + 2)(x - 2)$ because expanding the product $(x - 2)(x + 2)$ and simplifying, we get $x^2 - 4$. Thus we write $x^2 - 4 = (x + 2)(x - 2)$ and say that $x + 2$ and $x - 2$ are *factors* of $x^2 - 4$. In elementary algebra it is understood that factoring a given expression means writing the expression in terms of a product involving only variables and integers. Hence we would say that the expression $x^2 - 3$ cannot be factored since it cannot be written as a product of two factors involving x and integers. However, $x^2 - 3$ can be expressed as a product as follows: $x^2 - 3 = (x - \sqrt{3})(x + \sqrt{3})$.

The following formulas are useful in factoring algebraic expressions. Each can be verified by expanding the right side and simplifying.

$$
\begin{aligned}
\textit{Common factor:} \quad & au + av = a(u + v) & (1.3) \\
\textit{Perfect square:} \quad & u^2 + 2uv + v^2 = (u + v)^2 & (1.4) \\
& u^2 - 2uv + v^2 = (u - v)^2 & (1.5) \\
\textit{Difference of squares:} \quad & u^2 - v^2 = (u + v)(u - v) & (1.6)
\end{aligned}
$$

Example 8   Factor each of the following expressions.

a) $2x^2 - 8y^2$ **b)** $4x^2 - 12x + 9$ **c)** $2x^2 + 5x - 3$

Solution   **a)** By using formulas (1.3) and (1.6), we obtain

$$2x^2 - 8y^2 = 2(x^2 - 4y^2) = 2[x^2 - (2y)^2] = 2(x + 2y)(x - 2y).$$

b) Here formula (1.5) applies:

$$4x^2 - 12x + 9 = (2x)^2 - 12x + (3)^2 = (2x - 3)^2.$$

c) None of the given formulas apply here, but we can try various possibilities involving factors of the type $(ax + b)(cx + d)$, where $a \cdot c = 2$ and $b \cdot d = -3$. We find that $(2x - 1)(x + 3)$ yields the middle term of $5x$. Hence

$$2x^2 + 5x - 3 = (2x - 1)(x + 3).$$ ∎

Solving Equations and Inequalities
Here we illustrate solution of linear equations and linear inequalities.

Example 9   Solve each of the equations.

a) $2x - 3 = 5$ **b)** $5 - 3x = x + 4$

Solution   **a)** Adding 3 to each side of $2x - 3 = 5$, we get $2x = 8$. Dividing both sides by 2 gives $x = 4$. Hence 4 is a solution of $2x - 3 = 5$. This means that if x is replaced by 4, the result, $2 \cdot 4 - 3 = 5$, is a true statement.

b) Adding $(-x - 5)$ to each side of the given equation yields an equation with all terms involving x on one side of the equals sign and the remaining terms on the other side. Collecting like terms and simplifying gives the solution:

$$5 - 3x + (-x - 5) = x + 4 + (-x - 5),$$

$$-4x = -1, \qquad x = \frac{1}{4}.$$

Therefore $\frac{1}{4}$ is a solution of the given equation.   ■

Example 10   Solve the given inequalities and show the solutions on a number line.

a) $2x - 3 < 7$ **b)** $5 - 3x \geq 8$

Solution   The procedure for solving linear inequalities is similar to that used in Example 9 to solve linear equations, *except when we multiply or divide both sides by a negative number, the direction of the inequality must be reversed.*

a) Add 3 to both sides of the given inequality and divide the result by 2; this gives

$$(2x - 3) + 3 < 7 + 3,$$

$$2x < 10,$$

$$x < 5.$$

Thus any number less than 5 will satisfy the given inequality; this is shown on the number line:

b) Subtracting 5 from both sides of the given inequality yields

$$(5 - 3x) - 5 \geq 8 - 5,$$

$$-3x \geq 3.$$

Now divide both sides by -3 and reverse the direction of the inequality to get

$$\frac{-3x}{-3} \leq \frac{3}{-3},$$

$$x \leq -1.$$

The solution is shown on the number line:

■

Exercises 1.0

The problems of this set are grouped according to topic.

Set Notation

In problems 1 through 6, write each set by listing its elements within braces.

1. The set of counting numbers between 3 and 7
2. The set of natural numbers between 10 and 15
3. $\{2, 3, 4, 5\} \cup \{1, 3, 5\}$
4. $\{2, 3, 5\} \cap \{2, 4, 6\}$
5. $\{1, 0.1, 0.01\} \cap \{1, \frac{1}{10}, \frac{1}{20}\}$
6. $\{1, -1, 4, -4\} \cup \{1, 4, 8\}$

In problems 7 through 16, determine whether each statement is true or false.

7. $5 \in \{2, 3\}$
8. $\{2, 4\} \subset \{1, 2, 3, 4\}$
9. $\{2, 4\} \subseteq \{1, 2, 3, 4\}$
10. $\{3, 4, 8\} = \{4, 8, 3\}$
11. $\{3, 5, 7\} \subseteq \{x \mid x \text{ is an odd number}\}$
12. $5 \notin \{x \mid x \text{ is an even number}\}$
13. $3 \in \{1, 3, 5\} \cap \{2, 4\}$
14. $\varnothing \subseteq \{2, 3\}$
15. $4 \in \{x \mid x^2 - 4x = 0\}$
16. $\{-1, 1\} \subseteq \{x \mid x^2 - 1 = 0\}$

In problems 17 through 20, sets A and B are: $A = \{2, 3, 5, 7\}$, $B = \{2, 3, 6, 8\}$. List within braces the elements of each of the following sets.

17. $\{x \mid x \in A \text{ and } x \in B\}$
18. $\{x \mid x \in A \text{ or } x \in B\}$
19. $\{y \mid y \in A \text{ and } y \notin B\}$
20. $\{y \mid y \in A \text{ and } y \text{ is not an even number}\}$

Subsets of Real Numbers

In addition to the notations described in this section, let **P** denote the set of prime numbers. A *prime number* is a positive integer greater than 1 whose only divisors are 1 and itself (see Appendix D for a table of primes):

$$\mathbf{P} = \{2, 3, 5, 7, 11, 13, 17, 19, 23, 29, 31, \ldots\}.$$

In problems 21 through 26, determine whether each of the given statements is true or false.

21. a) $8 \in \mathbf{N}$
    b) $-4 \in \mathbf{N}$
    c) $-64 \in \mathbf{J}$
    d) $431 \in \mathbf{P}$
22. a) $1 + \sqrt{2} \in \mathbf{Q}$
    b) $3 + \sqrt{5} \in \mathbf{H}$
    c) $\pi \in \mathbf{R}$
    d) $\pi + 3 \in \mathbf{H}$
23. a) $\{0\} \subseteq \mathbf{W}$
    b) $\{\sqrt{4}\} \subset \mathbf{N}$
    c) $\mathbf{Q} \subseteq \mathbf{H}$
    d) $\mathbf{N} \subset \mathbf{W}$
24. a) $\mathbf{Q} \cap \mathbf{H} = \varnothing$
    b) $\mathbf{N} \cap \mathbf{Q} = \mathbf{Q}$
    c) $\mathbf{P} \cup \mathbf{N} = \mathbf{N}$
    d) $\{0\} \cup \mathbf{N} = \mathbf{W}$
25. a) $107 \in \mathbf{P}$
    b) $247 \notin \mathbf{Q}$
    c) $\{-1, 1\} \subset \mathbf{J}$
    d) $\mathbf{P} \subset \mathbf{Q}$
26. $\{y \mid y = n^2 - n + 41, n \in \mathbf{N}\} \subseteq \mathbf{P}$

Number Line

In problems 27 through 30, show the given subsets of **R** on a number line.

27. a) $\{-3, 5\}$
    b) $\{x \mid x \leq 2 \text{ or } x \geq 5\}$
    c) $\{x \mid x < 4 \text{ and } x \geq -2\}$
28. a) $\{1.5, \sqrt{5}\}$
    b) $\{x \mid x \leq -2 \text{ or } x \geq 4\}$
    c) $\{x \mid x \leq -1\} \cup \{x \mid x \geq 2\}$

29. a) $\{-2, \pi, \pi - 2\}$ **b)** $\{x \mid -3 < x \leq 4\}$ **c)** $\{x \mid x \leq \sqrt{2} \text{ or } x > \pi\}$

30. a) $\{\sqrt{2}, \dfrac{1}{\sqrt{2}}, 3 + \sqrt{2}\}$ **b)** $\{x \mid 1 - \sqrt{5} \leq x \leq 3\}$ **c)** $\{x \mid x \leq 2\} \cap \{x \mid x \geq -2\}$

Rectangular Coordinates

In problems 31 and 32, plot the given points and give the quadrant in which each is located.

31. $A:(3, 4)$ $B:(-2, 3)$ $C:(5, -2)$ $D:(-2, -4)$

32. $A:(1, \sqrt{3})$ $B:(1 - \sqrt{5}, 2)$ $C:(\pi, -\pi)$ $D:(1 - \pi, -\sqrt{2})$

In problems 33 and 34, plot the points (x, y) given in tabular form.

33.

x	-2	-1	0	1	2
y	-5	-3	-1	1	3

34.

x	-2	-1	0	1	2
y	3	0	-1	0	3

In problems 35 and 36, determine the distance between points P and Q. In each case give answers in exact form and also as a decimal approximation rounded off to two places.

35. a) $P:(2, 4)$, $Q:(-3, 2)$ **b)** $P:(-1, -3)$, $Q:(2, -5)$

36. a) $P:(3\sqrt{2}, -1)$, $Q:(-\sqrt{2}, 3)$ **b)** $P:(1 - \sqrt{2}, 4)$, $Q:(1 + \sqrt{2}, -1)$

In problems 37 through 42, the two sides of a *right triangle* are labeled a and b, and the hypotenuse is c. Two quantities are given; find the third rounded off to two decimal places.

37. $a = 5$, $b = 8$ **38.** $a = 17$, $c = 33$ **39.** $b = 24.3$, $c = 48.7$

40. $a = 5$, $b = 12$ **41.** $a = 20$, $b = 48$ **42.** $a = 43.73$, $b = 74.56$

In problems 43 through 46, the lengths of three sides of a triangle are given; determine whether or not it is a right triangle.

43. $a = 5$, $b = 12$, $c = 13$ **44.** $a = 12$, $b = 16$, $c = 20$

45. $a = 24208$, $b = 10575$, $c = 26417$ **46.** $a = 3784$, $b = 2730$, $c = 4666$

Arithmetic of Complex Numbers

In problems 47 through 50, express results in the form $x + yi$, where x and y are real numbers.

47. Given $u = -3 + 4i$ and $v = 1 - i$, determine

     a) $u + v$ **b)** $u - v$ **c)** $u \cdot v$ **d)** $u \div v$

48. Perform the indicated operations.

     a) $(3 + 2i)(-1 + i)$ **b)** $i(1 - 2i)$

     c) $(1 - 2i)(1 + 2i)$ **d)** $i(1 - i)(1 + i)$

49. Perform the indicated operations.

     a) $(1 - i) \div (1 + i)$ **b)** $i \div (2 - i)$

     c) $(2 - i) \div i$ **d)** $[i(3 + i)] \div (1 - i)$

50. Given that $u = 1 - i$, evaluate:

    a) $u^2 + 1$ **b)** $2 - u^2$

    c) $(1 + u) \div (1 - u)$ **d)** $u \div (1 + u)$

Linear Equations

In problems 51 through 53, solve the given equations for x.

51. a) $x + 2 = 5$ **b)** $3x - 2 = 5 + x$

52. a) $2x - 3 = 2 - 3x$ **b)** $(x - 1)(x + 1) = x^2$

53. a) $5 - 3x = x - 3$ **b)** $x^2 + 4 = x(x + 4)$

Linear Inequalities

In problems 54 through 56, solve the given inequalities.

54. a) $2x - 3 < -9$ **b)** $x + 4 \geq 3x - 6$

55. a) $3 - 2x \leq x + 6$ **b)** $x(x - 2) > (x + 1)(x + 3)$

56. a) $\dfrac{2x - 3}{4} \leq 1 - x$ **b)** $(1 + x)(1 - x) < x(2 - x)$

Algebraic Operations

In problems 57 through 60, complete the statement by entering the appropriate algebraic expression within the parentheses.

57. $x^2 - y^2 - x + y = x^2 - y^2 - ($ $)$ **58.** $x - 2 - x^3 + 2x^2 = x - 2 - x^2($ $)$

59. $x^2 - 2x - 4x + 8 = x^2 - 2x - 4($ $)$ **60.** $1 - (x - 1)^2 = x($ $)$

In problems 61 through 70, perform the indicated operations and simplify.

61. $(x^2 - 2x + 5) - (3x - 2)$ **62.** $(x + 3)(2x - 5)$ **63.** $(3x - 1)(x + 4)$

64. $(2x - 3)(2x + 3)$ **65.** $(2x^2 + 1)(x - 3)$ **66.** $(2x - \sqrt{3})(2x + \sqrt{3})$

67. $(2x^3 + 3x^2 - 5x - 3) \div (2x + 1)$ **68.** $(x^2 - 4) \div (x + 2)$ **69.** $(x^3 + 2x - 1) \div (x - 3)$

70. $(x^3 - 4x) \div (x - 2)$

In problems 71 through 80, determine which expressions can be factored, and then factor as far as possible.

71. $3x^2 - 12$ **72.** $x^2 + 4$ **73.** $2x^2 - 5x + 2$

74. $5x^2 + 7x - 6$ **75.** $3x^2 - x + 2$ **76.** $4 - (2x - 3)^2$

77. $(x + 4)^2 - 36$ **78.** $4y^2 + 4y + 1$ **79.** $8x^2 - 32x^4$

80. $1 - (x - 2)^2$

1.1   FUNCTIONS, RELATIONS, AND GRAPHS

The concept of a function is basic for most of mathematics. Very often, elements of two sets are associated by some rule of correspondence. For example, consider the set $A = \{1, 2, 3\}$ and the rule "Associate each number in A with a number that is one greater than its square." This rule associates 1 with 2, 2 with 5, and 3 with 10, and can be considered a correspondence in which each element of A corresponds to *exactly one* member of the set $B = \{2, 5, 10\}$. This association can also be described by giving the resulting set of *ordered pairs:* $\{(1, 2), (2, 5), (3, 10)\}$; or it can be thought of as *a mapping* of set A onto set B and shown schematically by

$$
\begin{array}{ccc}
1 & 2 & 3 \\
\downarrow & \downarrow & \downarrow \\
2 & 5 & 10
\end{array}
$$

This leads to the definition of a function in general.

Definition 1.1

Function as a correspondence

A function f from a set $\mathfrak{D}$ onto a set $\mathfrak{R}$ is a rule of *correspondence* that assigns to each element x of $\mathfrak{D}$ a *unique element y* of $\mathfrak{R}$. The element y is called the *image* of x under f. Set $\mathfrak{D}$ is called the *domain* of the function, and set $\mathfrak{R}$ is called the *range* of f.

The correspondence referred to in Definition 1.1 is given by a rule usually stated in equation or formula form, although sometimes it is described verbally or given by a table or a graph. In many situations it will be more convenient to use other definitions of the concept of a function. There are two such definitions, each of which is equivalent to Definition 1.1.

Definition 1.1(a)

Function as ordered pairs

A function is a *set f of ordered pairs* in which no two ordered pairs have the same first component; that is, the set will not include ordered pairs (a, b) and (a, c) where $b \neq c$. The set of first components of the ordered pairs in f is called the *domain of f*, and the set of second components is called the *range of f*.

Definition 1.1(b)

Function as a mapping

A function f is a *mapping* that associates with each element x in a set $\mathfrak{D}$ (called the *domain* of f) a *unique element y* in a set $\mathfrak{R}$ (called the *range* of f).

Function Notation

In most cases the rule describing the correspondence in Definition 1.1 can be written in equation form as

$$y = f(x).$$

This is read "*y* equals *f* at *x*" or "*y* equals *f* of *x*," and it states that *y*, also denoted by $f(x)$, is the element of $\mathcal{R}$ corresponding to *x* in $\mathcal{D}$. Here *x* is called the *independent* variable and *y* the *dependent variable*.

As a set of ordered pairs, *f* can be written

$$f = \{(x, y) \mid x \in \mathcal{D} \text{ and } y = f(x)\}.$$

As a mapping, *f* is described by

$$f : x \longrightarrow f(x) \qquad \text{or} \qquad \mathcal{D} \xrightarrow{\ f\ } \mathcal{R}.$$

In the definitions above the letter *f* was used to denote a function. In some discussions it will be necessary to use other letters, such as *g*, *h*, *F*, *G*, etc., to denote functions. Accordingly, the domain and range of, say, *f* and *g* shall be denoted by $\mathcal{D}(f)$, $\mathcal{R}(f)$ and $\mathcal{D}(g)$, $\mathcal{R}(g)$, respectively.

Let us now consider a few examples.

Example 1   Suppose $\mathcal{D} = \{1, 2, 3\}$ and the rule of correspondence for *f* is: for each $x \in \mathcal{D}$, the image *y* corresponding to *x* is one less than twice *x*. In formula form this rule is given by

$$y = 2x - 1 \text{ for each } x \in \mathcal{D}.$$

a) Show the correspondence by listing the ordered pairs.

b) Determine the range of *f*.

c) Draw a diagram to show *f* as a mapping.

Solution   **a)** $f = \{(1, 1), (2, 3), (3, 5)\}$ **b)** $\mathcal{R}(f) = \{1, 3, 5\}$

c) $\mathcal{D}(f) = \{1, 2, 3\}$ From (a) or (c) we see that *f* is a function.
$$\qquad\quad \downarrow\ \downarrow\ \downarrow$$
$$\mathcal{R}(f) = \{1, 3, 5\}$$

Example 2   A function *g* is described by $g(x) = x^2 - 1$ and $\mathcal{D}(g) = \{-2, -1, 0, 1, 2\}$.

a) Give *g* by listing the set of ordered pairs.

b) Determine $\mathcal{R}(g)$.   **c)** Show *g* as a mapping.

Solution   **a)** $g = \{(-2, 3), (-1, 0), (0, -1), (1, 0), (2, 3)\}$

b) $\mathcal{R}(g) = \{-1, 0, 3\}$

c) $\mathcal{D}(g) = \{\ 0,\ \ -1, 1,\ -2, 2\}$
$$\qquad\qquad\ \downarrow\ \ \ \ \Downarrow\ \ \ \Downarrow$$
$$\mathcal{R}(g) = \{-1,\ \ \ 0,\ \ \ \ 3\ \ \}$$

In Examples 1 and 2, both *f* and *g* are functions. However, there is an important property that *f* has but *g* does not. Looking at part (c) of the solutions, note that each element in the range of *f* has associated with it exactly one element of $\mathcal{D}$, but this is not so for *g*; each of the range elements 0 and 3 has two elements of

$\mathfrak{D}(g)$ associated with it. We say that f is a *one-to-one function,* whereas g is not. This leads to the following definition.

Definition 1.2   Suppose f is a function with domain $\mathfrak{D}$ and range $\mathfrak{R}$. We say that f is a *one-to-one function* (or a *one-to-one mapping*) if to each element of $\mathfrak{R}$ there corresponds one and only one element of $\mathfrak{D}$. That is, if b and c are two different elements of $\mathfrak{D}$, then $f(b) \neq f(c)$.

Example 3   The following table gives a correspondence between values of u and v.

u	-1	0	1	2	3
v	6	3	2	3	6

The rule of correspondence here is given by $v = u^2 - 2u + 3$. Suppose f is the set of ordered pairs (u, v), and g is the set of ordered pairs (v, u) given in this table.

a) List the ordered pairs in f. Is f a function?

b) List the ordered pairs in g. Is g a function?

Solution   **a)** $f = \{(-1, 6), (0, 3), (1, 2), (2, 3), (3, 6)\}$; f is a function.

             b) $g = \{(6, -1), (3, 0), (2, 1), (3, 2), (6, 3)\}$; g is not a function since 6 is associated with two different numbers, and so is 3.   ■

In Example 3(b) we have a set of ordered pairs that is not a function. This occurs frequently in mathematics, so it is convenient to have a term that represents any set of ordered pairs.

Definition 1.3   Any set of ordered pairs is called a *relation.* The set of first components is called the *domain of the relation;* the set of second components is the *range of the relation.*

Thus every function is a relation, but a relation is not necessarily a function since there are sets of ordered pairs, as seen in Example 3(b), that are not functions.

Example 4   Suppose set g is the set of ordered pairs given by

$$g = \{(x, y) \mid x \in A, y \in A, \text{ and } x < y\},$$

where $A = \{1, 2, 3, 4\}$. List the elements in g. Is g a relation? Is g a function?

Solution $g = \{(1, 2), (1, 3), (1, 4), (2, 3), (2, 4), (3, 4)\}$. Hence g is a relation since it is a set of ordered pairs. However, g is not a function since 1 occurs as a first component in more than one ordered pair.   ▪

Accepted Conventions

Each of the Examples 1 through 4 involves finite sets. Most of the problems of interest to us will involve functions with domain **R** or a subset of **R**. When we are discussing any particular function, it is important that we *not only have a rule of correspondence clearly stated but also understand the domain*. In order to avoid the need to state the domain explicitly in most problems involving a function, we adopt the following convention.

> *Whenever the domain is not explicitly stated, it will be assumed that it is the largest subset of real numbers for which the rule of correspondence yields range values that are real numbers.* That is, if f is a function and $x \in \mathfrak{D}(f)$, then $f(x)$ is a real number.

One of the outstanding features of mathematics is precision of language. However, it is frequently too cumbersome to carry this point to the extreme, and it becomes necessary to take some liberties with language without the fear of causing any misunderstanding. For instance, instead of saying, "The function g whose rule of correspondence is given by the formula $g(x) = x/(x - 1)$ and domain $\mathfrak{D} = \{x \mid x \in \mathbf{R} \text{ and } x \neq 1\}$," we shall use the abbreviated form: "The function g given by $g(x) = x/(x - 1)$" or occasionally we shall simply say, "The function $g(x) = x/(x - 1)$."

Example 5   Suppose the rule of correspondence for function f is given by $f(x) = \sqrt{x + 2}$.

a) Determine $\mathfrak{D}(f)$.

b) Find the values of $f(x)$ that correspond to the following values of x: $-2, -1, 0, 1, 2.4$. Express results in exact form; if an answer is not an integer, give it also in decimal form rounded off to two places.

Solution   **a)** For each x in $\mathfrak{D}(f)$ we want $f(x)$ to be a real number; that is, $x + 2$ must be nonnegative, which means that $x \geq -2$. Hence $\mathfrak{D}(f) = \{x \mid x \geq -2\}$.

b) $f(-2) = 0; \qquad f(-1) = 1; \qquad f(0) = \sqrt{2} = 1.41; \qquad f(1) = \sqrt{3} = 1.73;$
$f(2.4) = \sqrt{4.4} = 2.10.$ ▪

Example 6   A function g is given by $g = \left\{ (x, y) \mid y = \dfrac{x}{(x - 1)(x + 2)} \right\}$.

a) Determine $\mathfrak{D}(g)$.

b) Give the ordered pairs in g that correspond to the following values of x: $-1, 0, 2, \sqrt{5}$.

Solution   **a)** We see that the given formula yields a real number for every real-number value of x, except $x = 1$ or $x = -2$. For instance, if $x = 1$, then

$$y = \frac{1}{(1-1)(1+2)} = \frac{1}{0 \cdot 3} = \frac{1}{0},$$

which is not defined; similarly for $x = -2$. Therefore,

$$\mathfrak{D}(g) = \{x \mid x \neq -2, x \neq 1\}.$$

b) $x = -1, y = \dfrac{-1}{(-1-1)(-1+2)} = \dfrac{-1}{-2} = \dfrac{1}{2};$

$x = 0, y = \dfrac{0}{(0-1)(0+2)} = \dfrac{0}{-2} = 0;$

$x = 2, y = \dfrac{2}{(2-1)(2+2)} = \dfrac{2}{4} = \dfrac{1}{2};$

$x = \sqrt{5}, y = \dfrac{\sqrt{5}}{(\sqrt{5}-1)(\sqrt{5}+2)} = 0.43$ (to two decimal places).

Hence the ordered pairs of g include $\left(-1, \frac{1}{2}\right)$, $(0, 0)$, $\left(2, \frac{1}{2}\right)$, $(\sqrt{5}, 0.43)$.   ■

Example 7   Suppose function f is given by $f(x) = x^2 - 2x$. Determine:

a) $f(-4)$ **b)** $f(3x)$ **c)** $f(u - 1)$

Solution   **a)** $f(-4) = (-4)^2 - 2(-4) = 24$ **b)** $f(3x) = (3x)^2 - 2(3x) = 9x^2 - 6x$

c) $f(u - 1) = (u - 1)^2 - 2(u - 1) = u^2 - 4u + 3$ ■

Graphs

Almost all functions and relations in this text will consist of ordered pairs of *real numbers*. Such ordered pairs can be shown graphically as a set of points in a plane.

Suppose we wish to draw a graph of a function f defined by $y = f(x)$. In general the procedure will consist of the following steps.

1. Determine the domain $\mathfrak{D}(f)$.

2. Take several values of x in $\mathfrak{D}(f)$ and find the corresponding values of y given by the rule $y = f(x)$. A prudent choice of x values will be necessary to get the essential features of the graph.

3. Step 2 will yield a table of x, y values; plot the corresponding (x, y) points on a rectangular system of coordinates with an appropriate scale for each of the x and y axes.

4. Draw a smooth curve connecting the points plotted in Step 3. Here we need to be careful in connecting the points, since we want to be certain that the x, y values of any point on the curve satisfy the given rule of correspondence. Using common sense and experience will be helpful. There are exceptions to Step 4 (see Example 10).

Example 8   Draw a graph of the function f given by $f(x) = \sqrt{1 - x}$.

Solution   Since only those values of x are acceptable for which $f(x)$ is a real number, we require $1 - x \geq 0$. Solution of this inequality gives $\mathfrak{D}(f) = \{x \mid x \leq 1\}$.

Suppose we take values of x: 1, 0, -1, -2, -3, -4 and complete the following table in which the values of $f(x)$ are rounded off to one decimal place.

x	1	0	-1	-2	-3	-4
$f(x)$	0	1	1.4	1.7	2	2.2

Fig. 1.10

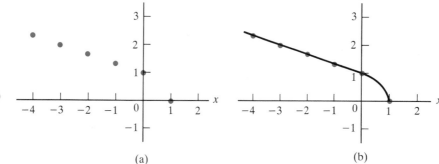

(a)   (b)

The points (x, y), where $y = f(x)$, are now plotted as shown in Fig. 1.10(a). A smooth curve connecting these points is shown in Fig. 1.10(b). This is the graph of the function f.

In a similar manner we can draw a graph of a relation consisting of a set of ordered pairs of real numbers by plotting several points from the given set of ordered pairs and connecting them in an appropriate way.

Example 9   Draw a graph of the relation $\{(x, y) \mid y^2 = x\}$.

Solution   Since $y^2 = x$, the values of x are nonnegative, and so the domain is the set $\mathfrak{D} = \{x \mid x \geq 0\}$. Now make a table of x, y values that satisfy $y^2 = x$, noting that for each $x > 0$ we get two values of y: $y = \sqrt{x}$ and $y = -\sqrt{x}$. That is, the given set of ordered pairs can be written as

$$\{(x, y) \mid y^2 = x\} = \{(x, y) \mid y = \sqrt{x}\} \cup \{(x, y) \mid y = -\sqrt{x}\}.$$

x	0	1	2	3	4	5	6
y	0	± 1	± 1.4	± 1.7	± 2	± 2.2	± 2.4

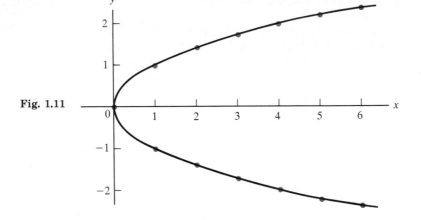

Fig. 1.11

Plotting these points and drawing a smooth curve through them gives the graph shown in Fig. 1.11. Note that the given set of ordered pairs does not define a function since for each $x > 0$, there are two corresponding values of y. ▪

In Example 8, the set of ordered pairs is a function, but in Example 9 the set is not a function. Graphically we see that each vertical line $x = k$, with $k \leq 1$, intersects the graph in Fig. 1.10 exactly once. We cannot say the same for the graph in Fig. 1.11 since vertical lines of the form $x = k$ for $k > 0$ intersect the graph at two points. This leads to the following graphical characterization of a function.

A set of ordered pairs of real numbers with domain $\mathfrak{D}$ represents a function if and only if every vertical line $x = k$, where $k \in \mathfrak{D}$, intersects the graph of f at exactly one point.

Fig. 1.12

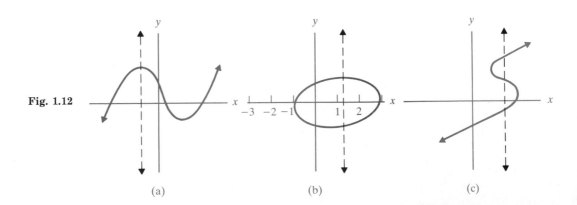

(a)   (b)   (c)

Thus, the graph shown in Fig. 1.12(a) represents a function, but those in Figs. 1.12(b) and (c) do not.

It is also helpful to note from the graphs in Fig. 1.12 that the domain of the relations shown in (a) and (c) is implied to be the set of real numbers, whereas the domain of the relation in (b) is $\{x \,|\, -1 \leq x \leq 3\}$.

The graph of a function need not always be a connected curve. This is illustrated in the following examples.

Example 10   Suppose $\mathcal{D}(f) = \{1, 2, 3 \ldots 7\}$, and the rule of correspondence is given by: $f(x)$ is *the number* of prime numbers less than or equal to x. Draw a graph of f.

Solution   Let $y = f(x)$. First make a table of x, y values. Recall that the set **P** of primes is $\mathbf{P} = \{2, 3, 5, 7, 11, 13, 17 \ldots\}$. To evaluate $f(x)$ we simply count how many prime numbers are less than or equal to x. For example, to evaluate $f(6)$, we note that the prime numbers less than or equal to 6 are 2, 3, 5; there are three of them. Therefore $f(6) = 3$.

x	1	2	3	4	5	6	7
y	0	1	2	2	3	3	4

By plotting the (x, y) points given in this table, we get the graph shown in

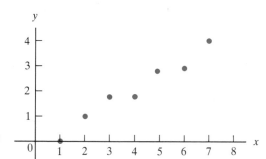

Fig. 1.13

Fig. 1.13. Here we do not connect the plotted points with a smooth curve since to do so would imply that all real numbers between 1 and 7 are in $\mathcal{D}(f)$. That is, the graph consists of seven *isolated points*.

Example 11   Suppose $\mathcal{D}(g) = \{x \,|\, 1 \leq x \leq 7\}$ and the rule of correspondence for g is: $g(x)$ is the number of prime numbers that are less than or equal to x. Draw a graph of g.

Solution   This is similar to Example 10, except that here all real numbers between 1 and 7 are included in the domain. Let $y = g(x)$. In addition to the points given in

Example 10, we must consider other domain points. It is readily apparent that for $1 \le x < 2$, $g(x) = 0$; for $2 \le x < 3$, $g(x) = 1$; and so on.

The graph of the function g is shown in Fig. 1.14. It consists of line segments (which do not include the right-hand endpoints) and one isolated point $(7, 4)$.

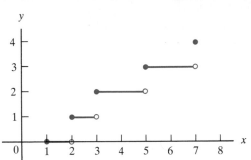

Fig. 1.14

Exercises 1.1

In problems 1 through 8, a set $\mathfrak{D}$ is given and an associated set f of ordered pairs is described.

a) Give f by listing the ordered pairs.

b) Is f a function? Is f a one-to-one function?

c) State the range $\mathfrak{R}$ of f.

d) Draw a diagram that illustrates f as a mapping of $\mathfrak{D}$ onto $\mathfrak{R}$.

1. $\mathfrak{D} = \{-1, 0, 1, 2, 3\}$; $f = \{(x, y) \mid x \in \mathfrak{D} \text{ and } y = 3x + 1\}$

2. $\mathfrak{D} = \{-4, -2, 0, 2, 4\}$; $f = \{(x, y) \mid x \in \mathfrak{D} \text{ and } y = x^2 - 2x\}$

3. $\mathfrak{D} = \{-4, -1, 2, 5\}$; $f = \{(u, v) \mid u \in \mathfrak{D} \text{ and } 2u + 3v = 1\}$

4. $\mathfrak{D} = \{-3, -1, 1, 3\}$; $f = \{(v, u) \mid v \in \mathfrak{D} \text{ and } 2u + 3v = 1\}$

5. $\mathfrak{D} = \{-1, 0, 1, 3\}$ $f = \{(x, y) \mid x \in \mathfrak{D} \text{ and } 3x^2 + 2y = 1\}$

6. $\mathfrak{D} = \{-2, -1, 0, 1, 2\}$;
   $f = \{(x, y) \mid x \in \mathfrak{D} \text{ and the value of } y \text{ is obtained by squaring } x \text{ and then adding 3}\}$.

7. $\mathfrak{D} = \{0, 1, 2, 3\}$;
   $f = \{(u, v) \mid u \in \mathfrak{D} \text{ and the value of } v \text{ is obtained by multiplying } u \text{ by 1 less than } u\}$.

8. $\mathfrak{D} = \{-2, 0, 2, 4\}$;
   $f = \{(x, y) \mid x \in \mathfrak{D} \text{ and the value of } y \text{ is the quotient of the square of } x \text{ and 1 more than } x\}$.

In problems 9 through 12, a set of ordered pairs is given. In each case answer the following.

a) Is the given set a function? Is it a one-to-one function?

b) State the domain and range for the given set of ordered pairs.

c) Give a verbal statement that could be used to describe the rule of correspondence for the given set of ordered pairs.

9. $f = \{(-1, 1), (0, 0), (1, 1), (2, 4), (3, 9), (4, 16)\}$.

10. $g = \{(1, 2), (2, 4), (3, 8), (4, 16), (5, 32)\}$.

11. $h = \{(0, 0), (1, 1), (4, 2), (9, 3), (1, -1), (4, -2), (9, -3)\}$.

12. $f = \{(-1, -2), (0, 1), (1, 4), (2, 7), (3, 10), (4, 13)\}$.

In Problems 13 and 14, use the rule of correspondence between u and v given by the following table.

u	-1	0	1	2
v	1	0	1	4

13. Let f be the set of all ordered pairs (u, v) given by the table.

   a) Give f as a listing of ordered pairs.

   b) Is f a function?

   c) State the domain and range of f.

14. Let g be the set of all ordered pairs (v, u) given by the table.

   a) Give g as a listing of ordered pairs.

   b) Is g a function?

   c) State the domain and range of g.

In problems 15 and 16, a diagram is given showing a mapping of set $\mathcal{D}$ onto set $\mathcal{R}$.

a) List the corresponding set of ordered pairs.

b) Is the mapping one-to-one?

15. $\mathcal{D}$: $\{1, 3, 5\}$

   $\downarrow \downarrow \downarrow$

   $\mathcal{R}$: $\{0, 2, 4\}$

16. $\mathcal{D}$: $\{-2, 2, -1, 1, 0\}$

   $\downarrow\!\!\downarrow \quad \downarrow\!\!\downarrow \quad \downarrow$

   $\mathcal{R}$: $\{\quad 5 \qquad 2 \quad 1\}$

17. A manufacturer advertises the efficiency of a car as "15 kilometers per liter." This describes a rule of correspondence between the amount of gasoline used (x liters) and the distance traveled (y kilometers), thus giving a set of ordered pairs (x, y).

   a) What is the domain implied by this rule?

   b) In each of the following ordered pairs, the first member is given; find the corresponding second member.

   (1L,   ), (2L,   ), (3.6L,   ), (10L,   )

18. In problem 17, the phrase "15 kilometers per liter" can also be considered as a rule of correspondence between the distance traveled (u kilometers) and the amount of gasoline used (v liters), thus yielding a set of ordered pairs (u, v).

   a) What is the domain implied in this set of ordered pairs?

   b) Complete the following ordered pairs by determining the second member in each case.

   (45 km,   ), (175 km,   ), (215 km,   ), (460 km,   )

In each of the problems 19 through 28, a function is described by the given formula.

a) Using the convention stated on page 19, determine the domain of the function.
b) Evaluate the function at the given values of the independent variable; if the result is not a rational number, then give it in approximate decimal form rounded off to two places.

19. $f(x) = x^2 - 2x + 3$, x: $-2, 0, \sqrt{5}$ **20.** $g(x) = \dfrac{1-x}{1+x}$, x: $-2, 0, \pi$

21. $h(x) = 3 - 4x$, x: $-2, 1, 1 + \sqrt{5}$ **22.** $f(x) = 5 - 3x^2$, x: $-1, 2, 1 - \sqrt{3}$

23. $g(x) = \dfrac{x^2}{x-2}$, x: $-2, 0, \sqrt{5}$ **24.** $f(t) = \dfrac{1}{t} - \dfrac{1}{t-1}$, t: $-1, -0.5, \sqrt{3}$

25. $g(t) = \dfrac{1}{t+1} - \dfrac{1}{t}$, t: $-2, 0.5, \sqrt{3}$ **26.** $h(x) = x - 1$, x: $1, 3, \sqrt{7}$

27. $f(t) = \sqrt{1-t}$, t: $1, -3, -\sqrt{7}$ **28.** $g(x) = \sqrt{x^2 + 1}$, x: $-1, 3, \sqrt{2}$

In each of the problems 29 through 33, a function is described. Determine the specified image values. If an answer is an irrational number, give result rounded off to two decimal places.

29. $f(x) = x^2 - 3x + 4$ **a)** $f(0)$ **b)** $f(3)$ **c)** $f(\sqrt{2})$

30. $f(t) = \dfrac{t+3}{t}$ **a)** $f(-1)$ **b)** $f(2)$ **c)** $f(-\sqrt{3})$

31. $f(x) = \sqrt{1 + \sqrt{x}}$ **a)** $f(0)$ **b)** $f(3)$ **c)** $f(9)$

32. $f(x) = 3\sqrt{x^2 - 2x}$ **a)** $f(-2)$ **b)** $f(-1)$ **c)** $f(4)$

33. $f(x) = \sqrt{\sqrt{x} - 2}$ **a)** $f(4)$ **b)** $f(9)$ **c)** $f(12)$

34. Suppose $f(x) = 2x + 3$; determine   **a)** $f(4x)$ **b)** $f(x + 1)$

35. Suppose $f(x) = 3x^2 + 2x - 3$; determine   **a)** $f(1 - u)$ **b)** $f(3 + x)$

36. Given that $g(x) = \dfrac{x}{x-1}$, determine   **a)** $g(u + 1)$ **b)** $g(1 - 3x)$

37. Given that $g(x) = \dfrac{x^2}{x+1}$, determine   **a)** $g(u - 1)$ **b)** $g(2x - 1)$

38. Let f be a function with domain $\mathbf{N}$ (the set of natural numbers), and let the rule of correspondence be given by "$f(x)$ is the *number* of primes less than or equal to x" (see Example 10 of this section). Find:

     a) $f(12)$ **b)** $f(19)$ **c)** $f(100)$

     Hint: Use the table of prime numbers given in Appendix D.

39. $f(x) = \begin{cases} 1 \text{ if } x \text{ is a rational number,} \\ 0 \text{ if } x \text{ is an irrational number.} \end{cases}$ Find:

     a) $f(2)$ **b)** $f(0.73)$ **c)** $f(\sqrt{5})$

40. $f(x) = \begin{cases} 1 \text{ if } x^2 \text{ is a rational number,} \\ x \text{ if } x^2 \text{ is an irrational number.} \end{cases}$ Find:

     a) $f(\sqrt{2})$ **b)** $f(1 + \sqrt{2})$ **c)** $f(-\tfrac{3}{4})$

In problems 41 through 54, draw a graph of the given function. In each case make a table of x, y values, plot the corresponding points, and connect them with a smooth curve when appropriate.

41. $y = 2x + 1$ **42.** $y = 4 - x$ **43.** $y = 4 - x^2$

44. $y = x^2 - 4$ **45.** $y = \sqrt{x}$ **46.** $y = \sqrt{-x}$

47. $f = \{(x, y) \mid 3x + 2y = 0\}$ **48.** $f = \{(x, y) \mid 2x^2 + y = 0\}$

49. $f = \{(x, y) \mid x^2 - 1 - y = 0\}$ **50.** $f = \{(x, y) \mid 2x - y = 3\}$

51. $f = \{(x, y) \mid y$ equals the greatest integer that is less than or equal to $x\}$

52. $\mathfrak{D}(f) = \{-2, -1, 0, 1, 2\}$ and $f(x) = x^2 + 1$

53. $\mathfrak{D}(f) = \{-4, -3, -2, -1, 0\}$ and $f(x) = \sqrt{-x}$

54. $\mathfrak{D}(f) = \{-\sqrt{3}, -1, 0, 1, \sqrt{3}\}$ and $f(x) = \begin{cases} 2x & \text{if } x \text{ is rational} \\ x^2 & \text{if } x \text{ is irrational} \end{cases}$

In each of the problems 55 through 60, a graph of a relation is given.

a) Determine whether or not the graph represents a function.

b) State the domain of the relation.

55.

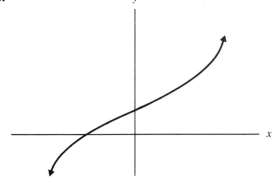

56.

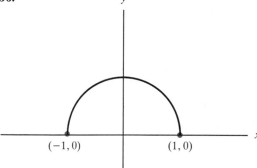

57.

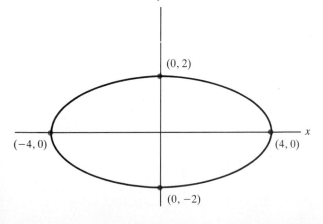

58.

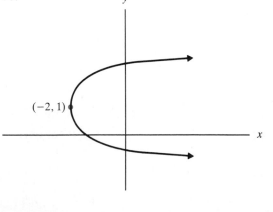

59.

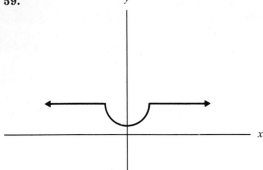

60.

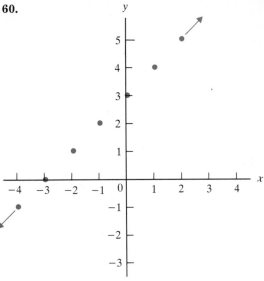

In problems 61 through 64, functions f, g, and h are described by the following graphs. In each case find the value of the function at the given values of x.

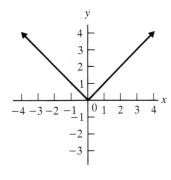

$y = f(x)$

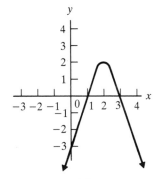

$y = g(x)$

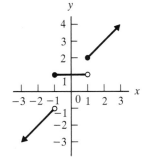

$y = h(x)$

61. a) $f(0)$ **b)** $g(0)$ **c)** $h(1)$

62. a) $f(2)$ **b)** $g(1)$ **c)** $h(-1)$

63. a) $f(-3)$ **b)** $g(2)$ **c)** $h(-2)$

64. a) $f(4)$ **b)** $g(4)$ **c)** $h(3)$

1.2 COMBINING FUNCTIONS

In this section we explore five ways in which given functions can be combined to get new functions. These are referred to as sum, difference, product, quotient and composition functions.

Sum, Difference, Product, and Quotient Functions

Two functions can be combined in four ways to give new functions in a manner analogous to combining two numbers by any of the four arithmetic operations $+$, $-$, $\times$, or $\div$. However, before we discuss the arithmetic of functions, it is necessary to first give a definition of equality of two functions.

Definition 1.4

Equality of functions

Suppose f and g are functions with domains $\mathcal{D}(f)$ and $\mathcal{D}(g)$, respectively. We say that f equals g, denoted by $f = g$, if and only if (1) $\mathcal{D}(f) = \mathcal{D}(g)$ and (2) $f(x) = g(x)$ for each x in their common domain.

The following two examples illustrate this definition.

Example 1   Suppose f, g, and h are functions defined by

$$f(x) = x, \qquad g(x) = \frac{x^2 - x}{x - 1}, \qquad h(x) = \frac{x^3 + x}{x^2 + 1}.$$

a) Is $f = g$?   **b)** Is $f = h$?

Solution   **a)** Since $\mathcal{D}(f) = \mathbf{R}$ and $\mathcal{D}(g) = \{x \mid x \neq 1\}$, $\mathcal{D}(f) \neq \mathcal{D}(g)$, and therefore $f \neq g$ by part (1) of Definition 1.4.

b) First note that $\mathcal{D}(f) = \mathcal{D}(h) = \mathbf{R}$. For each real number x,

$$h(x) = \frac{x^3 + x}{x^2 + 1} = \frac{x(x^2 + 1)}{x^2 + 1} = x.$$

Hence (1) and (2) of Definition 1.4 are satisfied; therefore $f = h$.   ▪

Example 2   The rule of correspondence given by

$$g(x) = \frac{x^2 - 4}{x - 2}$$

defines a function with domain

$$\mathcal{D}(g) = \{x \mid x \neq 2\}.$$

Use algebraic simplification on the rule for g to determine a function f so that $g = f$.

Solution $g(x) = \dfrac{x^2 - 4}{x - 2} = \dfrac{(x + 2)(x - 2)}{(x - 2)} = x + 2$ for every real number x except 2.

Thus a rule for the desired function is given by $f(x) = x + 2$, where $\mathfrak{D}(f) = \{x \mid x \neq 2\}$. ∎

Definition 1.5

> *Sum, difference, product, quotient functions*
>
> Suppose f and g are functions. New functions $f + g, f - g, f \cdot g$ and f/g are given by
>
> **1.** $(f + g)(x) = f(x) + g(x)$, $\mathfrak{D}(f + g) = \mathfrak{D}(f) \cap \mathfrak{D}(g)$
>
> **2.** $(f - g)(x) = f(x) - g(x)$, $\mathfrak{D}(f - g) = \mathfrak{D}(f) \cap \mathfrak{D}(g)$
>
> **3.** $(f \cdot g)(x) = f(x) \cdot g(x)$, $\mathfrak{D}(f \cdot g) = \mathfrak{D}(f) \cap \mathfrak{D}(g)$
>
> **4.** $(f/g)(x) = \dfrac{f(x)}{g(x)}$, $\mathfrak{D}\left(\dfrac{f}{g}\right) = \{x \mid x \in \mathfrak{D}(f) \cap \mathfrak{D}(g)$ and $g(x) \neq 0\}$

It is important to realize that the rules of correspondence stated in Definition 1.5 for the four functions are not merely formal manipulations of symbols. For instance in (1), the plus sign in $f(x) + g(x)$ indicates the sum of two numbers (remember that $f(x)$ and $g(x)$ are numbers), whereas the plus sign in $f + g$ is used to denote a function whose rule of correspondence assigns to each x in $\mathfrak{D}(f) \cap \mathfrak{D}(g)$ the number $f(x) + g(x)$.

Example 3   Suppose $f, g,$ and h are functions given by

$$f(x) = x + 1, \qquad g(x) = \sqrt{x}, \qquad h(x) = \frac{x}{x - 1}.$$

Give formulas and corresponding domains for each of the functions:

a) $f + g$ **b)** $g - h$ **c)** $g \cdot h$ **d)** f/h

Solution   First note that the domains for $f, g,$ and h are, respectively,

$$\mathfrak{D}(f) = \mathbf{R}, \qquad \mathfrak{D}(g) = \{x \mid x \geq 0\}, \qquad \mathfrak{D}(h) = \{x \mid x \neq 1\}.$$

a) $(f + g)x = f(x) + g(x) = (x + 1) + \sqrt{x}$ for $x \in \mathfrak{D}(f) \cap \mathfrak{D}(g)$. Hence,

$$(f + g)(x) = x + 1 + \sqrt{x}; \qquad \mathfrak{D}(f + g) = \{x \mid x \geq 0\}.$$

b) $(g - h)(x) = g(x) - h(x) = \sqrt{x} - \dfrac{x}{x - 1}$ for $x \in \mathfrak{D}(g) \cap \mathfrak{D}(h)$. Thus

$$(g - h)(x) = \sqrt{x} - \frac{x}{x - 1}; \qquad \mathfrak{D}(g - h) = \{x \mid x \geq 0 \text{ and } x \neq 1\}.$$

c) $(g \cdot h)(x) = g(x) \cdot h(x) = \sqrt{x}\left(\dfrac{x}{x-1}\right)$ for $x \in \mathfrak{D}(g) \cap \mathfrak{D}(h)$. Therefore,

$$(g \cdot h)(x) = \dfrac{x\sqrt{x}}{x-1}; \qquad \mathfrak{D}(g \cdot h) = \{x \mid x \geq 0 \text{ and } x \neq 1\}.$$

d) $\left(\dfrac{f}{h}\right)(x) = \dfrac{f(x)}{h(x)} = \dfrac{x+1}{x/(x-1)} = \dfrac{(x+1)(x-1)}{x} = \dfrac{x^2-1}{x},$

where $\mathfrak{D}\left(\dfrac{f}{h}\right) = \{x \mid x \in \mathfrak{D}(f) \cap \mathfrak{D}(h) \text{ and } h(x) \neq 0\}$. Thus

$$\left(\dfrac{f}{h}\right)(x) = \dfrac{x^2-1}{x}; \qquad \mathfrak{D}\left(\dfrac{f}{h}\right) = \{x \mid x \neq 0 \text{ and } x \neq 1\}. \qquad \blacksquare$$

Composition of Functions

Let us consider another way of combining two functions to get a new function. The idea is illustrated first by an example.

Suppose a spherical balloon is being inflated in such a manner that its radius r is given as a function of time t by the formula $r = g(t) = \sqrt{t}$, where t and r are given in appropriate units. Since the volume V of a sphere is given as a function of r by the formula $V = f(r) = (4/3)\pi r^3$, we can also express V as a function of t by replacing r with $\sqrt{t}$. That is,

$$V = f(g(t)) = f(\sqrt{t}) = \tfrac{4}{3}\pi(\sqrt{t})^3.$$

Here we have an example of combining functions g and f to get a new function given by the formula $V = f(g(t))$. We call this function the composition of f and g and denote it by $f \circ g$. Thus $(f \circ g)(t) = f(g(t))$.

The example above leads us to the following definition of composition of functions in general.

Definition 1.6

Suppose f and g are functions. The *composite function of g followed by f* is a function, denoted by $f \circ g$, which assigns to the number x the number $f(g(x))$. That is,

$$(f \circ g)(x) = f(g(x)),$$

where $\mathfrak{D}(f \circ g) = \{x \mid x \in \mathfrak{D}(g) \text{ and } g(x) \in \mathfrak{D}(f)\}$.

Note that the domain of $f \circ g$ as given here includes all of the values of x for which $f(g(x))$ is meaningful. That is, we first evaluate g at x, and so x must be in $\mathfrak{D}(g)$; then f is evaluated at the result $g(x)$, and so we need $g(x)$ to be in $\mathfrak{D}(f)$. This is illustrated schematically in Fig. 1.15, in which b is accepted by $f \circ g$, but c is not since $g(c)$ is not in $\mathfrak{D}(f)$.

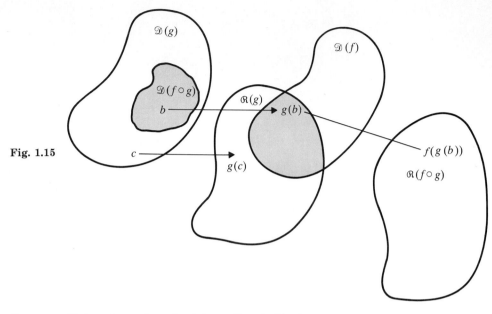

Fig. 1.15

Example 4   Suppose $f(x) = 2x + 3$ and $g(x) = x^2 - 1$. Evaluate

    a) $(f \circ g)(2)$ **b)** $(f \circ f)(-4)$

Solution   **a)** $(f \circ g)(2) = f\big(g(2)\big) = f(3) = 9.$

        b) $(f \circ f)(-4) = f\big(f(-4)\big) = f(-5) = -7.$

Example 5   Suppose $f(x) = \sqrt{x}$ and $g(x) = 1/(x - 1)$. Determine formulas and corresponding domains for the composite functions:

    a) $f \circ g$ **b)** $g \circ f$

Solution   **a)** $(f \circ g)(x) = f\big(g(x)\big) = f\left(\dfrac{1}{x - 1}\right) = \dfrac{1}{\sqrt{x - 1}}.$

        First note that $\mathfrak{D}(f) = \{x \mid x \geq 0\}$ and $\mathfrak{D}(g) = \{x \mid x \neq 1\}$. Hence the do-
main of $f \circ g$ includes any number x for which $x \in \mathfrak{D}(g)$ and $g(x) \in \mathfrak{D}(f)$;
that is, $x \neq 1$ and $1/(x - 1) \geq 0$. Thus $\mathfrak{D}(f \circ g) = \{x \mid x > 1\}$.

        b) $(g \circ f)(x) = g\big(f(x)\big) = g(\sqrt{x}) = \dfrac{1}{\sqrt{x} - 1},$

        where $\mathfrak{D}(g \circ f) = \{x \mid x \geq 0 \text{ and } x \neq 1\}$.

Example 6   Suppose f is a function given by $f(x) = \sqrt{4 - x}$, $\mathfrak{D}(f) = \{x \mid 0 \leq x \leq 4\}$, and g is
a function defined on the set of positive integers **N** by the rule: "$g(x)$ equals *the
number* of prime numbers less than or equal to x." List the ordered pairs of the
functions:

    a) $f \circ g$ **b)** $g \circ f$

Solution   The domains of f and g are $\mathfrak{D}(f) = \{x \mid 0 \le x \le 4\}$ and $\mathfrak{D}(g) = \{1, 2, 3, \ldots\}$. The set **P** of prime numbers is $\mathbf{P} = \{2, 3, 5, 7, 11, 13, 17, \ldots\}$.

It is also helpful to list a few of the ordered pairs in g. For example, to evaluate $g(7)$, the rule for g counts the number of primes less than or equal to 7. These are 2, 3, 5, and 7. Thus $g(7) = 4$. Hence,

$$g = \{(1, 0), (2, 1), (3, 2), (4, 2), (5, 3),(6, 3), (7, 4), (8, 4), (9, 4), (10, 4), (11, 5) \ldots\}.$$

a) $(f \circ g)(x) = f(g(x)) = \sqrt{4 - g(x)}$. Since $g(x)$ must be in $\mathfrak{D}(f)$, $0 \le g(x) \le 4$. The listing of g shows that the acceptable values of x are 1, 2, 3, 4, $\ldots$, 10. Therefore $\mathfrak{D}(f \circ g) = \{1, 2, 3, \ldots, 10\}$.

Let us evaluate $f \circ g$ at a few of these values of x.

$$
\begin{aligned}
(f \circ g)(1) &= \sqrt{4 - g(1)} = \sqrt{4 - 0} = 2, \\
(f \circ g)(2) &= \sqrt{4 - g(2)} = \sqrt{4 - 1} = \sqrt{3}, \\
(f \circ g)(3) &= \sqrt{4 - g(3)} = \sqrt{4 - 2} = \sqrt{2}, \\
&\vdots \\
(f \circ g)(10) &= \sqrt{4 - g(10)} = \sqrt{4 - 4} = 0.
\end{aligned}
$$

Therefore $f \circ g$ consists of the following set of ordered pairs.

$$f \circ g = \{(1, 2), (2\sqrt{3}), (3, \sqrt{2}), (4, \sqrt{2}), (5, 1), (6, 1), (7, 0), (8, 0), (9, 0), (10, 0)\}.$$

b) $(g \circ f)(x) = g(f(x))$. Here we require that $0 \le x \le 4$ and $f(x)$ is a positive integer. Since $f(x) = \sqrt{4 - x}$, we see that there are only two values of x satisfying this requirement, namely, 0 and 3. That is, $f(0) = 2$ and $f(3) = 1$. Therefore $\mathfrak{D}(g \circ f) = \{0, 3\}$ and

$$(g \circ f)(0) = g(f(0)) = g(2) = 1, \qquad (g \circ f)(3) = g(f(3)) = g(1) = 0.$$

Hence the function $g \circ f$ is given by $g \circ f = \{(0, 1), (3, 0)\}$.   ■

Note from Examples 5 and 6 that functions $f \circ g$ and $g \circ f$ are not equal. In general, this is so.

Example 7   Suppose functions f and g are given by $f(x) = 1 - x^2$ and $g(x) = \sqrt{x}$. Determine a formula for the function $f \circ g$.

Solution $(f \circ g)(x) = f(g(x)) = f(\sqrt{x}) = 1 - (\sqrt{x})^2 = 1 - x$. Here we must be careful to state the domain for $f \circ g$. It is necessary that $x \in \mathfrak{D}(g)$ and $g(x) \in \mathfrak{D}(f)$; and so $(f \circ g)(x) = 1 - x$, where $\mathfrak{D}(f \circ g) = \{x \mid x \ge 0\}$.

Note that the function h given by $h(x) = 1 - x$ can be evaluated at any real number, *but such a function is not equal to $f \circ g$ since $f(g(x))$ is not defined* for any value of x less than zero.   ■

Example 8   Suppose f and g are functions given by

$$f(x) = x^2 - 2x - 1 \qquad \text{and} \qquad g(x) = \sqrt{x} + 1.$$

In each of the following, find the roots of the given equation.

a) $(f \circ g)(x) = 3x$ **b)** $(f \circ g)(x) = f(x) + 3x - 2$

Solution   First find $(f \circ g)(x)$:

$$(f \circ g)(x) = f(g(x)) = f(\sqrt{x} + 1) = (\sqrt{x} + 1)^2 - 2(\sqrt{x} + 1) - 1$$
$$= x + 2\sqrt{x} + 1 - 2\sqrt{x} - 2 - 1 = x - 2.$$

Here the acceptable values of x are given by $x \in \mathcal{D}(g)$ and $g(x) \in \mathcal{D}(f)$. Thus $(f \circ g) = \{x \mid x \geq 0\}$. Thus function $f \circ g$ is given by $(f \circ g)(x) = x - 2$ for $x \geq 0$.

a) The equation $(f \circ g)(x) = 3x$ is equivalent to $x - 2 = 3x$ and $x \geq 0$. Therefore x must satisfy $x = -1$ and $x \geq 0$. Hence there is no solution.

b) The equation $(f \circ g)(x) = f(x) + 3x - 2$ is equivalent to the equation $x - 2 = (x^2 - 2x - 1) + 3x - 2$ and $x \geq 0$. Simplifying, we get $x^2 - 1 = 0$ and $x \geq 0$. The only value of x satisfying this is $x = 1$, so 1 is the only solution. ◾

Example 9   Given F as a function defined by $F(x) = (1 + x)^6$, find two functions f and g such that $f \circ g = F$.

Solution   One solution is to take $f(x) = x^6$ and $g(x) = 1 + x$. Then

$$(f \circ g)(x) = f(g(x)) = f(1 + x) = (1 + x)^6.$$

Therefore $(f \circ g)(x) = F(x)$; also $\mathcal{D}(F) = \mathcal{D}(f \circ g) = \mathbf{R}$, and so $f \circ g = F$. Another solution is given by $f(x) = x^3$ and $g(x) = (1 + x)^2$. It should be clear that there are other solutions. ◾

Function Machine

It is instructive to think of a function in terms of an input–output machine. Suppose a machine is built according to the rule that describes a given function f, so that when a number, say c, is taken from the domain $\mathcal{D}(f)$ and entered into the input slot, it is processed by the machine and the corresponding $f(c)$ exits from the output. This is shown schematically in Fig. 1.16.

Consider the following examples.

Example 10   The *square machine* is one that corresponds to the rule $f(x) = x^2$, i.e., it is a machine in which the number entered into the input is multiplied by itself. For example, if -2 is entered, out comes 4, as shown in Fig. 1.17. This corresponds to the ordered pair $(-2, 4)$ of f. ◾

Example 11   The *composition function machine*. For two functions f and g, we can illustrate the $f \circ g$ function machine by combining the f and g machines, as shown in Fig. 1.18. In this type of machine care must be exercised in selecting the x from $\mathcal{D}(g)$

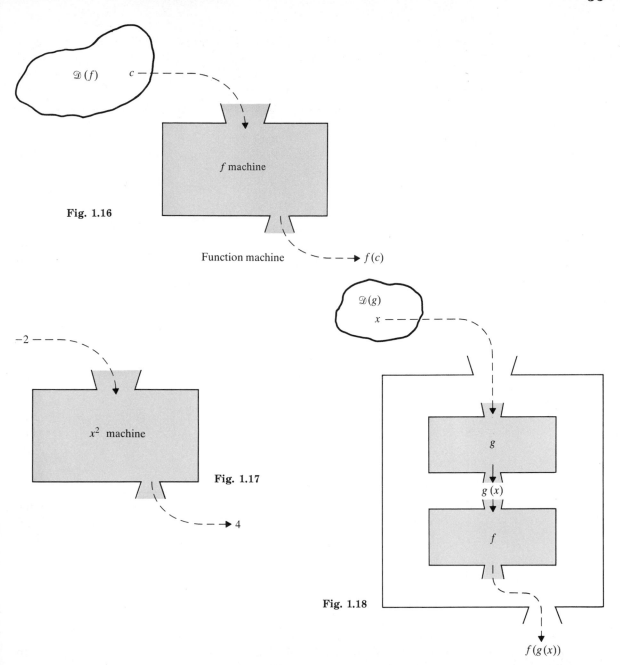

Fig. 1.16

Function machine - - - → $f(c)$

Fig. 1.17

Fig. 1.18

$f(g(x))$

that is to be entered; if the output $g(x)$ from the g component cannot be digested by the f component (that is, if $g(x) \notin \mathfrak{D}(f)$), then the $f \circ g$ machine will reject that value of x and indicate it by *Error*. The user of such a machine should be aware that the only values of x that it will accept are from the set $\{x \mid x \in \mathfrak{D}(g)$ and $g(x) \in \mathfrak{D}(f)\}$.

Example 12   In order to describe the *addition machine* it is necessary to consider a function of two variables; that is, the domain of such a function will itself be a set of ordered pairs of real numbers. Suppose we denote the function we are describing by $\boxed{+}$; the domain of $\boxed{+}$ is the set $\mathcal{D}(\boxed{+}) = \{(u, v) \mid u \text{ and } v \text{ are real numbers}\}$. The rule for the $\boxed{+}$ function is given by $\boxed{+}((a, b)) = $ sum of a and b. The range of $\boxed{+}$ is given by $\mathcal{R}(\boxed{+}) = \mathbf{R}$. For example, $\boxed{+}((3, 5)) = 3 + 5 = 8$. Thus we can think of $\boxed{+}$ *as a function* on the set of ordered pairs of real numbers (Fig. 1.19). ■

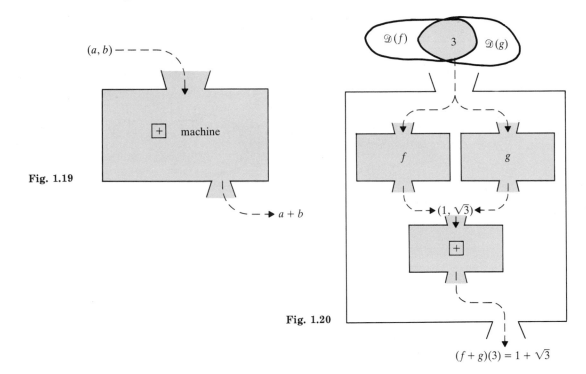

Fig. 1.19

Fig. 1.20

In an analogous manner we can consider the other binary operations $\boxed{-}$, $\boxed{\times}$, $\boxed{\div}$ as functions of two variables.

Example 13   The *sum function machine*. Suppose f and g are functions given by $f(x) = \sqrt{4 - x}$ and $g(x) = \sqrt{x}$. We can "build" a machine for the $f + g$ function by combining the $f, g,$ and $\boxed{+}$ machines. Figure 1.20 shows what happens when the number 3 is entered into such a machine. ■

In a similar manner we could "build" function machines for the various types of functions described in Section 1.2. Although our discussion so far has been in terms of fictitious machines, it does introduce us to a real function machine, the hand-held calculator, which can be considered a truly magnificent multifunction machine.

Scientific Calculator as a Function Machine

In Chapters 2 and 3 polynomial, exponential and logarithmic functions are discussed. These along with trigonometric functions constitute a set of basic functions in the sense that almost any function we consider is one of them or can be described in terms of sum, difference, product, quotient, or composition of these functions.

The scientific calculator has a set of function keys we can press in an appropriate order to "build" any one of these basic functions. Here we discuss some of these keys; we consider others (such as the exponential, logarithmic, and trigonometric keys) in appropriate places throughout the book.

One-variable function keys   We will consider the one-variable function keys
[1/x], [x²], and [√x]. By pressing the [1/x] key, we cause the calculator to become a *reciprocal machine*. For example, if we enter 2 into the display by pressing the [2] key, then instruct the calculator to become a reciprocal machine by pressing the [1/x] key, it will process 2 and give 0.5 as the output in the display. This is illustrated schematically in Fig. 1.21.

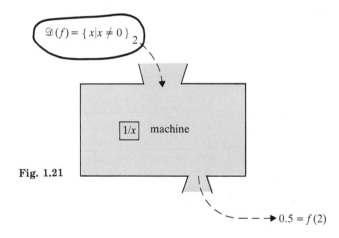

Fig. 1.21

Note that the [1/x] machine will not accept $x = 0$. The reader should try pressing [0] and [1/x] and then observe the response of the calculator.

In a similar manner the [x²] and [√x] keys cause the calculator to behave like $g(x) = x^2$ and $h(x) = \sqrt{x}$ function machines, respectively. Also we can combine the [1/x], [x²], and [√x] keys to get other functions. For example, if $f(x) = 1/x$, $f \circ g$ is the function given by $(f \circ g)(x) = f\big(g(x)\big) = f(x^2) = 1/x^2$. To evaluate $f \circ g$ at 2, that is $(f \circ g)(2)$, we first enter 2 into the display, then press the [x²] and [1/x] keys in that order. The display will then show 0.25; that is, $(f \circ g)(2) = 0.25$.

Example 14   Suppose $f(x) = 1/x$, $g(x) = x^2$, and $h(x) = \sqrt{x}$. Evaluate each of the following by using a calculator. Give answers rounded off to two decimal places.

a) $f(-1.7)$ **b)** $g(3.6)$ **c)** $h(4.3)$

d) $(g \circ f)(-1.3)$ **e)** $(g \circ h)(3.2)$ **f)** $(g \circ h)(-4.1)$

Solution   **a)** Enter 1.7 into the calculator display and then press the change-sign key, usually labeled $\boxed{\text{CHS}}$ or $\boxed{+/-}$. Next press the $\boxed{1/x}$ key, and the calculator will show $-0.5882\ldots$. Thus $f(-1.7) = -0.59$.

b) Entering 3.6 into the display and pressing the $\boxed{x^2}$ key gives $g(3.6) = 12.96$.

c) Enter 4.3, press $\boxed{\sqrt{x}}$, and you get $h(4.3) = 2.07$.

d) $(g \circ f)(-1.3) = g(f(-1.3))$. Enter -1.3, then press the $\boxed{1/x}$ and $\boxed{x^2}$ keys, and you get $(g \circ f)(-1.3) = 0.59$.

e) $(g \circ h)(3.2) = g(h(3.2))$. Enter 3.2, then press the $\boxed{\sqrt{x}}$ and $\boxed{x^2}$ keys, and you get $(g \circ h)(3.2) = 3.20$.

f) $(g \circ h)(-4.1) = g(h(-4.1))$. Enter -4.1, and then press the $\boxed{\sqrt{x}}$ key. At this point the calculator indicates *Error*. The reason is that -4.1 is not in $\mathcal{D}(h)$, and so we conclude that $(g \circ h)(-4.1)$ is undefined.

A formula for $g \circ h$ is given by

$$(g \circ h)(x) = g(h(x)) = g(\sqrt{x}) = (\sqrt{x})^2 = x,$$

where we must restrict the values of x to $x \geq 0$. Note that the calculator evaluation procedure includes this restriction on x, as illustrated in (e) and (f). ▪

Two-variable function keys We now consider the keys $\boxed{+}, \boxed{-}, \boxed{\times}, \boxed{\div}$, and $\boxed{y^x}$. As illustrated in Example 12, each of the binary operations $+$, $-$, $\times$, and $\div$ can be considered as a function on a set of ordered pairs of real numbers. For example, $+$ is a function (which we shall denote by $\boxed{+}$) with domain $\mathcal{D}(\boxed{+}) = \{(x, y) \mid x \in \mathbf{R}, y \in \mathbf{R}\}$ and defined by $\boxed{+}((u, v)) = u + v$. Similarly, the functions $\boxed{-}, \boxed{\times}$, and $\boxed{\div}$ are defined by $\boxed{-}((u, v)) = u - v$; $\boxed{\times}((u, v)) = u \cdot v$; and $\boxed{\div}((u, v)) = u \div v$, where the domains for $\boxed{-}$ and $\boxed{\times}$ are the same as $\mathcal{D}(\boxed{+})$; and $\mathcal{D}(\boxed{\div}) = \{(x, y) \mid x \in \mathbf{R}, y \in \mathbf{R} \text{ and } y \neq 0\}$.

In order to evaluate a given binary operation function with a calculator, it is necessary to enter an ordered pair of numbers and then instruct it to become the binary operation machine. We accomplish this in a natural way in the RPN calculators by using the *Enter* key $\boxed{\text{ENT}}$ to separate the two numbers; it serves as the comma of the ordered pair. For example, if it is necessary to evaluate $\boxed{-}((6, 2)) = 6 - 2$, we press 6 $\boxed{\text{ENT}}$ 2, and we tell the calculator to become a *subtraction machine* by pressing the $\boxed{-}$ key; the display will then show 4. With calculators that use the algebraic system of entry, we press the $\boxed{-}$ key between the two numbers: 6 $\boxed{-}$ 2; this separates the 6 and 2 and prepares the calculator to become a $\boxed{-}$ machine. Pressing the $\boxed{=}$ key activates the pending $\boxed{-}$ function, and the display shows the result. Thus, with algebraic calculators we evaluate $\boxed{-}((6, 2)) = 6 - 2$ by pressing 6 $\boxed{-}$ 2 $\boxed{=}$.

Another calculator key that can be considered as a function defined on ordered pairs of real numbers is labeled $\boxed{y^x}$. When the ordered pair (u, v) is entered and the calculator is instructed to become a $\boxed{y^x}$ machine, it will process (u, v) by raising u to the v power; that is, the output will be u^v. Thus we write $\boxed{y^x}\,((u, v)) = u^v$. For example, for the ordered pair $(4, 3)$ the output will be 4^3, which we write in function notation as $\boxed{y^x}\,((4, 3)) = 4^3 = 64$.

The method of calculator entry for the $\boxed{y^x}$ function is similar to that for the binary operation functions.

For RPN calculator: press u $\boxed{\text{ENT}}$ $v\,\boxed{y^x}$, and the display will show u^v.

For algebraic calculator: press $u\,\boxed{y^x}\,v\,\boxed{=}$, and the display will show u^v.

On many calculators the $\boxed{y^x}$ function will process an ordered pair (u, v) *only when u is positive.** That is, the domain of the $\boxed{y^x}$ function is given by

$$\mathfrak{D}(\boxed{y^x}) = \{(u, v) \,|\, u \in \mathbf{R}, v \in \mathbf{R} \text{ and } u > 0\}.$$

If u is positive, then u^v is a positive number for any real number v. Therefore the range of $\boxed{y^x}$ is given by

$$\mathfrak{R}(\boxed{y^x}) = \{w \,|\, w > 0\}.$$

Note: We shall encounter problems in which we wish to evaluate an exponential with a negative base number. For example, if we want to find the value of $(-1.43)^3$, we note that $(-1.43)^3 = -(1.43)^3$. Now use a calculator to evaluate $(1.43)^3$ and then change the sign of the result. Thus $(-1.43)^3 = -2.924207$.

Combining calculator functions to get new functions   As we have now seen, we can combine two given functions to get a new function, such as the sum, difference, product, quotient, or composition function. We can use these ideas with the basic calculator functions, those given directly by calculator keys, to get almost any of the functions that we shall encounter. It will not serve our purpose to pursue this in detail in a general setting, but here we illustrate by a specific example.

Suppose f and g are functions given by $f(u) = u^2$ and $g(u) = \sqrt{u}$. Then in calculator notation f is given by $\boxed{x^2}\,(u) = u^2$ and g is given by $\boxed{\sqrt{x}}\,(u) = \sqrt{u}$. The sum function $f + g$ is then given by

$$(\boxed{x^2} + \boxed{\sqrt{x}})(u) = \boxed{x^2}(u) + \boxed{\sqrt{x}}\,(u) = u^2 + \sqrt{u}.$$

Thus to evaluate $f + g$ at u, we apply the $\boxed{x^2}$ function to u and the $\boxed{\sqrt{x}}$ function to u, obtaining an ordered pair $(u^2, \sqrt{u})$. Then we apply the $\boxed{+}$ function to this ordered pair to obtain $u^2 + \sqrt{u}$. This is precisely the sequence of steps used in having a calculator evaluate $u^2 + \sqrt{u}$.

---

*   Here we are disregarding the trivial case $u = 0$ and $v > 0$. On some calculators the $\boxed{y^x}$ can be used to evaluate u^v when u is a negative number if v is any integer.

Throughout most of this text we encounter numerous examples in which the calculator is used to evaluate functions, some of which are rather complicated, by combining basic calculator functions in an appropriate manner. In any given situation the combination of functions used is described in terms of the sequence of calculator keys used.

Example 15   Suppose f is a function given by $f(x) = \dfrac{1}{x^2} + \sqrt{x}$. Using a calculator, evaluate the given expressions and round off answers to four decimal places.

a) $f(4)$ **b)** $f(3.48)$

Solution   We can consider the given function in terms of a combination of functions $g(x) = 1/x$, $h(x) = x^2$, and $q(x) = \sqrt{x}$. Then $f(x) = g(h(x)) + q(x)$; that is, the function f is equal to the function $g \circ h + q$. We will not trouble ourselves with these details but rather will illustrate how they are used to evaluate the given expressions.

a) $f(4) = 1/4^2 + \sqrt{4}$. This can be evaluated as follows:

For RPN calculator: press [4] [x²] [1/x] [4] [√x] [+], and the display will show 2.0625.*

For algebraic calculator: press [4] [x²] [1/x] [+] [4] [√x] [=], and the display will show 2.0625.

b) To evaluate $f(3.48)$ we can follow the same sequence of steps as in (a) with 3.48 in place of 4. The result is $f(3.48) = 1.9480$.   ▪

Exercises 1.2

In problems 1 through 12, functions f and g are given by

$$f(x) = 2x - 3, \qquad g(x) = x^2 + 3x.$$

Evaluate the given expressions and give answers in exact form. If the expression is not defined, explain why.

1. $(f + g)(3)$ **2.** $(f - g)(3)$ **3.** $(f \cdot g)(5)$ **4.** $(g \cdot f)(0.5)$

5. $\left(\dfrac{f}{g}\right)(-4)$ **6.** $\left(\dfrac{g}{f}\right)(-1)$ **7.** $\left(\dfrac{f}{g}\right)(-3)$ **8.** $\left(\dfrac{g}{f}\right)(1.5)$

9. $(f \circ g)(-2)$ **10.** $(g \circ f)(0)$ **11.** $(f \circ g)(0)$ **12.** $(f \circ f)(4)$

In problems 13 through 21, functions f, g, and h are given by

$$f(x) = \frac{x - 1}{x}, \qquad g(x) = x^2 + 3, \qquad h(x) = \sqrt{x - 4}.$$

---

*   It is not necessary to press the [ENT] key between the [1/x] and [4] because after a function key the calculator is automatically ready to accept a new number.

Evaluate the given expressions and given answers in exact form. If an expression is undefined, explain.

13. $(g - f)(2)$ **14.** $(g + h)(5)$ **15.** $\left(\dfrac{h}{f}\right)(8)$ **16.** $\left(\dfrac{g}{h}\right)(3)$ **17.** $(g \circ f)(1)$

18. $(f \circ h)(4)$ **19.** $(g \circ h)(3)$ **20.** $(f \cdot g)(-3)$ **21.** $(h \circ h)(20)$

In problems 22 through 33, functions f, g, and h are given by
$$f(x) = \sqrt{x}, \qquad g(x) = x - 3, \qquad h(x) = x^2 + 4.$$
In each case, determine a formula for the given function and its domain.

22. $f + g$ **23.** $\dfrac{h}{g}$ **24.** $\dfrac{g}{h}$ **25.** $f \cdot g$

26. $f \cdot h$ **27.** $f \circ g$ **28.** $g \circ f$ **29.** $h \circ g$

30. $g \circ g$ **31.** $h \circ f$ **32.** $\dfrac{f}{g}$ **33.** $f \circ h$

In problems 34 through 39, solve the given equations, where f, g, and h are given by
$$f(x) = x + 3, \qquad g(x) = 1 - x^2, \qquad h(x) = \sqrt{x}.$$

34. $f(x) - 4 = 0$ **35.** $(f - g)(x) - x^2 = 0$ **36.** $h(x) - 2 = 0$
37. $(g \circ h)(x) - 3 = 0$ **38.** $(f \circ h)(x) - 4 = 0$ **39.** $(g \circ h)(x) - f(x) = 0$

40. If $f(x) = 2x - 3$ and $g(x) = \dfrac{x + 3}{2}$, is $f \circ g = g \circ f$?

41. If $f(x) = 2x + 5$ and $g(x) = \dfrac{x - 5}{2}$, is $f \circ g = g \circ f$?

In problems 42 through 47, for the given function f find a function g such that $(f \circ g)(x) = x$ for each x in $\mathbf{R}$.

42. $f(x) = x - 4$ **43.** $f(x) = 2x + 3$ **44.** $f(x) = 3 - 4x$
45. $f(x) = 4 - 2x$ **46.** $f(x) = 1.5x + 3$ **47.** $f(x) = 6 - 1.5x$

In problems 48 through 56, functions f, g, and h are given by
$$f(x) = \sqrt{x}, \qquad g(x) = x^2 + 4, \qquad h(x) = \dfrac{x}{x - 1}.$$
Evaluate the given expressions and give answers rounded off to two decimal places.

48. $f(5)$ **49.** $g(1.43)$ **50.** $h(\sqrt{3})$
51. $h(\pi)$ **52.** $(f \circ g)(2.4)$ **53.** $(h \circ f)(3)$
54. $(f + g)(\pi)$ **55.** $\left(\dfrac{f}{g}\right)(2)$ **56.** $(f + h)(1.6)$

57. If functions f and g are given by $f(x) = x^2$ and $g(x) = \sqrt{x}$, are functions $f \circ g$ and $g \circ f$ equal? Justify your answer.

58. Given that $h(x) = (2x + 1)^4$, find functions f and g so that $f \circ g = h$. Note that there may be more than one solution.

59. Given that $h(x) = x^2 + 2x + 1$, find functions f and g such that $f \circ g = h$.

60. Given that f is a function *defined on the set of positive integers* by the rule $f(x)$ is equal to the *number* of prime numbers that are less than or equal to x, and g is a function defined by the formula $g(x) = \sqrt{9 - x^2}$, find the set of ordered pairs that belong to

a) $g \circ f$ **b)** $f \circ g$

c) State the domain and range of each of the functions $g \circ f$ and $f \circ g$.

61. A function machine accepts any real number and processes it by squaring it, then multiplying the result by 5, and then subtracting 4 from that result. Find the corresponding output numbers for each of the input numbers: -3, 0, 2, 5.

62. A function machine is given to you without instructions as to what it does except that it will accept any real number. Suppose you wish to find out what kind of machine it is by entering numbers and observing the outputs. Entering -1 gives 1, 0 gives 4, 1 gives 7, and 2 gives 10.

   a) On the basis of this information give a rule that might describe what the machine is doing.

   b) If 16 is entered, what do you think should come out?

   c) If we decide to test one more number by entering 3 and we get 37, what do you conclude about the machine?

63. A function machine is designed so that it will accept any real number x, and the corresponding output will be the *largest integer that is less than or equal to x*. For each of the following numbers, give the corresponding output number:

$$x = 4; \ x = -6; \ x = 2.47; \ x = -1.32; \ x = 4 + \sqrt{2}.$$

64. A function machine is designed so that it can accept any positive integer and process it as follows: If a positive integer n is entered into it, the machine will try each prime number that is less than or equal to n to see if it divides n evenly, and it will keep count of *how many do*. The corresponding output will be the "how many" number. For example, if 12 is entered, the machine will determine that the only prime numbers that divide 12 evenly are 2 and 3, and therefore output corresponding to 12 is 2.

   a) Find the corresponding output of this machine for each of the following input numbers: 1, 3, 8, 15, 30, 52, 256, 420.

   b) What is the smallest number that can be entered into this machine so that the output will be 4?

65. Suppose p and q are two different prime numbers. Use the function machine described in Problem 64 to determine the corresponding output when each value below is entered.

a) p **b)** p^2 **c)** pq **d)** $p^2 q^4$

1.3 LINES AND LINEAR FUNCTIONS

One of the basic concepts in geometry is that of a line. We say two distinct lines are parallel if and only if they are inclined in the same direction. The inclination of a line L can be described in terms of a number called *slope of L,* which we now define.

Definition 1.7   *Slope of a line*

Given two distinct points (x_1, y_1) and (x_2, y_2), there is exactly one line L that passes through them. The *slope* of L is the number m given by

$$m = \frac{y_2 - y_1}{x_2 - x_1}.$$ (1.7)

Here we are assuming that $x_1 \neq x_2$.

In Fig. 1.22 we illustrate four possible cases, in which the slope is positive, negative, zero, or undefined. In (a) the slope of line L_1 is positive since $y_2 - y_1 > 0$ and $x_2 - x_1 > 0$. In (b) the slope of L_2 is negative since $y_2 - y_1 < 0$ and $x_2 - x_1 > 0$. In (c) line L_3 is horizontal, and since $y_1 = y_2$, the slope of L_3 is zero. In (d) the line L_4 is vertical; since $x_1 = x_2$, we see that Eq. (1.7) involves division by zero, and so we do not associate a slope with L_4 but describe its inclination by saying "L_4 is a vertical line."

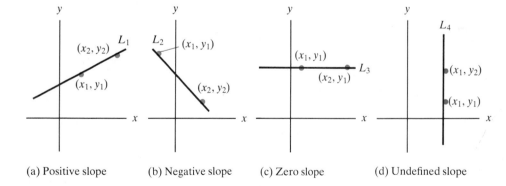

Fig. 1.22

(a) Positive slope   (b) Negative slope   (c) Zero slope   (d) Undefined slope

Equation of a Line

A line L is determined if either (1) two points through which L passes or (2) one point and the slope of L are given. Case 1 reduces to Case 2 if L is not a vertical line, since we can find its slope by using Eq. (1.7). Hence it is sufficient to assume that a point and the slope are given. Let us proceed to find an equation for L.

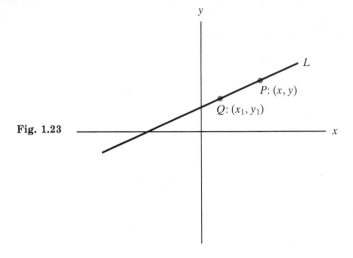

Fig. 1.23

Suppose line L passes through the given point Q: (x_1, y_1) and has slope m. Let P: (x, y) be *any point* on L, as shown in Fig. 1.23. We want to determine an equation that relates variables x and y. By using the coordinates of points P and Q in Eq. (1.7), we get $(y - y_1)/(x - x_1)$ (for P different from Q) as the slope of L; this must be equal to the given slope m. Therefore $(y - y_1)/(x - x_1) = m$. This can be written as

$$y - y_1 = m(x - x_1). \tag{1.8}$$

Equation (1.8) is called the point–slope form of the equation of L. In this form we see that Eq. (1.8) is also satisfied by the coordinates of Q; thus we have an equation satisfied by *any* point on L. Equation (1.8) can be written as $y = mx + (y_1 - mx_1)$, where $y_1 - mx_1$ is a constant, which we shall denote by b. Hence an equation for L is given by

$$y = mx + b. \tag{1.9}$$

Substituting 0 for x in Eq. (1.9) gives $y = b$, so the point $(0, b)$ is on line L. Since $(0, b)$ is on the y axis, it is called the *y-intercept* of L, and Eq. (1.9) is called the *slope–intercept form* of the equation of L.

Suppose we begin with the equation

$$Ax + By + C = 0, \tag{1.10}$$

where A, B, and C are given numbers. If $B \neq 0$, solving for y gives

$$y = \left(-\frac{A}{B}\right)x + \left(-\frac{C}{B}\right).$$

Comparing this with Eq. (1.9), we conclude that Eq. (1.10) represents a line with slope $m = -A/B$ and y intercept $-C/B$. We call Eq. (1.10) the *equation of a line in standard form*.

Vertical Lines

Suppose L is a vertical line passing through the point (c, d), as shown in Fig. 1.24.

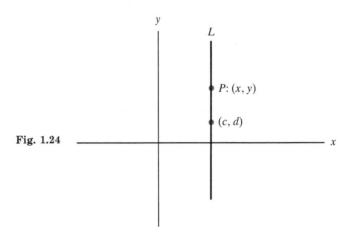

Fig. 1.24

Let P: (x, y) be *any point* on L. Since L is vertical, x must be equal to c, and y can be any number. Therefore the equation of L is simply

$$x = c. \qquad (1.11)$$

Parallel Lines; Perpendicular Lines

Suppose m_1 and m_2 are slopes of lines L_1 and L_2, respectively. The following useful relationships can be proved.

If $m_1 = m_2$, then L_1 and L_2 are *parallel* lines.

If $m_2 = -\dfrac{1}{m_1}$ (or $m_1 m_2 = -1$), then L_1 and L_2 are *perpendicular* to each other.

The converse of each of these statements is also true whenever the slopes of the lines are defined.

Linear Functions

For given values of m and b, Eq. (1.9) describes a function,

$$f(x) = mx + b. \tag{1.12}$$

Since such a function is related to a line, we call it a *linear function*. Thus a nonvertical line is associated with a linear function. Conversely, it can be argued that the graph of any linear function is a line.

Example 1   Find an equation of the line L that passes through the points P_1: $(-1, 2)$ and P_2: $(3, 4)$.

Solution   First find the slope of L by substituting into Eq. (1.7):

$$m = \frac{4 - 2}{3 - (-1)} = \frac{1}{2}.$$

Thus Eq. (1.8) gives

$$y - 2 = \tfrac{1}{2}(x + 1),$$

which can be written as

$$y = \tfrac{1}{2}x + \tfrac{5}{2} \qquad \text{or} \qquad x - 2y + 5 = 0.$$

Either of these is an acceptable equation of L.   ■

Example 2   Suppose line L is given by the equation $2x - 3y = 6$.

a) Find the slope of L and the coordinates of the x and y intercept points.

b) Sketch a graph of L.

Solution   **a)** Solving the given equation for y gives

$$y = \tfrac{2}{3}x - 2.$$

In this form the coefficient of x gives the slope $m = 2/3$. The x-intercept point can be obtained by replacing y by 0 in the given equation and solving for the corresponding value of x; this gives $(3, 0)$, the x-intercept point. Similarly, $x = 0$ gives $y = -2$, and so the y-intercept point is $(0, -2)$.

b) To sketch a graph we can plot any two points satisfying the given equation— in this case, say, the x- and y-intercept points—and then draw a line through them. The graph is shown in Fig. 1.25.   ■

Example 3   Suppose point Q is $(-3, 1)$ and line L is given by $4x - 2y = 3$. Find the equation of

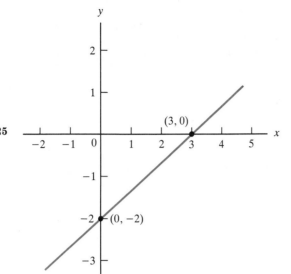

Y=(2/3)*X−2

Fig. 1.25

Graph of $2x - 3y = 6$

a) line L_1 that passes through Q and is parallel to L;

b) line L_2 that passes through Q and is perpendicular to L.

Solution   We can solve the given equation for y to get $y = 2x - \frac{3}{2}$; thus the slope of L is given by $m = 2$ (the coefficient of x).

a) Let m_1 denote the slope of L_1. Since L_1 is parallel to L, $m_1 = m$; thus $m_1 = 2$. Therefore an equation for L_1 is

$$y - 1 = 2(x + 3) \qquad \text{or} \qquad y = 2x + 7.$$

b) Since L_2 is perpendicular to L, $m_2 = -1/m = -1/2$. An equation for L_2 is

$$y - 1 = -\tfrac{1}{2}(x + 3) \qquad \text{or} \qquad x + 2y + 1 = 0. \qquad \blacksquare$$

Exercises 1.3

In problems 1 through 4, points P and Q are given. Determine the slope of the line through P and Q.

1. a) P: $(-2, 4)$; Q: $(0, 1)$

   b) P: $(3, 5)$; Q: $(-4, 5)$

3. a) P: $(4, -3)$; Q: $(1, -1)$

   b) P: $(-2, 5)$; Q: $(-2, 3)$

2. a) P: $(-4, 1)$; Q: $(-3, -2)$

   b) P: $(1, 3)$; Q: $(1, -4)$

4. a) P: $(-3, 4)$; Q: $(5, -1)$

   b) P: $(1, 3)$; Q: $(4, 3)$

In problems 5 through 8, points P and Q are given. Find an equation of the line passing through P and Q.

5. a) P: $(3, 4)$; Q: $(-1, 2)$ **6. a)** P: $(-3, 1)$; Q: $(-1, -2)$
   b) P: $(-2, 4)$; Q: $(1, 4)$ **b)** P: $(-1, 3)$; Q: $(2, 3)$
7. a) P: $(1, -2)$ Q: $(0, 0)$ **8. a)** P: $(0, 0)$; Q: $(-3, 5)$
   b) P: $(-2, 4)$; Q: $(-2, 5)$ **b)** P: $(4, 1)$; Q: $(4, 5)$

In problems 9 through 12, a point Q and the slope m of a line L passing through Q are given. Determine an equation of L.

9. Q: $(-2, 4)$; $m = 2$ **10.** Q: $(1, -3)$; $m = -4$

11. Q: $(0, 3)$; $m = -\frac{3}{5}$ **12.** Q: $(3, 0)$; $m = \frac{3}{4}$

In problems 13 through 16, an equation of a line L is given. Determine the slope of L and the *coordinates* of x- and y-intercept points of L.

13. $3x + 2y + 6 = 0$ **14.** $3x - 2y = 6$
15. $3x - 4y = 6$ **16.** $3x + 4y + 6 = 0$

In problems 17 through 24, sketch a graph of the given equations and label the x- and y-intercept points.

17. $3x - 2y = 4$ **18.** $2x + 3y = 6$ **19.** $y = 2x - 3$
20. $y = -x + 3$ **21.** $-x + 2y = 4$ **22.** $2x = 6 - 3y$
23. $3x = 6 + 2y$ **24.** $-3x - 2y + 4 = 0$

In problems 25 through 28, a point P and an equation of a line L are given. In each case, determine:

a) Equation of line L_1 passing through P and parallel to L;

b) Equation of line L_2 passing through P and perpendicular to L.

25. P: $(-2, 1)$; $2x - 3y + 4 = 0$ **26.** P: $(1, 4)$; $x + 2y = 3$
27. P: $(-1, 3)$; $x + 4 = 0$ **28.** P: $(-1, -3)$; $y - 5 = 0$

In problems 29 and 30, determine whether or not the three given points are collinear (lie on a line).

29. A: $(2, -3)$; B: $(0, -1)$; C: $(-1, 2)$
30. A: $(0, -3)$; B: $(-2, -6)$; C: $(2, 6)$

In problems 31 and 32, three points are given. Determine whether or not they are vertices of a *right triangle*.

31. A: $(0, 0)$; B: $(1, 2)$; C: $(-4, 2)$
32. A: $(2, -2)$; B: $(5, 2)$; C: $(-6, 4)$

In problems 33 through 36, determine an equation that x and y must satisfy if the point (x, y) is always equidistant from the two given points P and Q.

33. P: $(1, 3)$; Q: $(3, -1)$ **34.** P: $(0, -2)$; Q: $(-2, -4)$
35. P: $(-3, 0)$; Q: $(5, 3)$ **36.** P: $(4, -2)$; Q: $(1, 0)$

In problems 37 through 40, an equation of a line L is given. Determine the coordinates of *all* points (x, y) that are in the *first quadrant,* lie on L, and have *integral values* of x and y.

37. $2x + 3y = 11$ **38.** $3x + 4y = 27$

39. $5x + 2y = 12$ **40.** $3x + 2y = 15$

In problems 41 through 45, f and g are linear functions given by

$$f(x) = 2x - 1, \quad g(x) = -x + 3.$$

In each of the problems, find a formula for the given function and determine whether it is a linear function.

41. $f + g$ **42.** $f \cdot g$ **43.** f/g **44.** $f \circ g$ **45.** $g \circ f$

1.4   QUADRATIC FUNCTIONS; INEQUALITIES

A function f described by a formula of the form

$$f(x) = ax^2 + bx + c, \tag{1.13}$$

where a, b, and c are given real numbers and $a \neq 0$, is called a *quadratic function.* Let us consider three related problems involving quadratic functions.

1. Solving quadratic equations, that is, finding the values of x for which $f(x) = 0$.
2. Sketching graphs of $y = f(x)$.
3. Solving inequalities involving $f(x)$.

Solving Quadratic Equations; Quadratic Formula

Two techniques commonly used in solving quadratic equations involve factoring or application of the quadratic formula. Let us first develop the quadratic formula and then illustrate with examples.

We want to determine the values of x that will satisfy the equation

$$ax^2 + bx + c = 0, \tag{1.14}$$

where $a \neq 0$.

The following steps involve *completing the square* and lead to a formula giving the desired solution.

Divide by a: $x^2 + \dfrac{b}{a}x = -\dfrac{c}{a}$.

Add $(b/2a)^2$ to both sides: $x^2 + \dfrac{b}{a}x + \left(\dfrac{b}{2a}\right)^2 = -\dfrac{c}{a} + \dfrac{b^2}{4a^2}$.

Factor the left side: $\left(x + \dfrac{b}{2a}\right)^2 = \dfrac{b^2 - 4ac}{4a^2}$.

Take square roots: $x + \dfrac{b}{2a} = \pm\sqrt{\dfrac{b^2 - 4ac}{4a^2}} = \pm\dfrac{\sqrt{b^2 - 4ac}}{2a}$.

Solving for x yields the *quadratic formula*:

$$x = \frac{-b \pm \sqrt{b^2 - 4ac}}{2a}. \tag{1.15}$$

Equation (1.15) can be used to solve any quadratic equation by substituting the coefficients a, b, and c to get the solutions. The symbol $\pm$ in Eq. (1.15) is used to denote two solutions:

$$x = \frac{-b + \sqrt{b^2 - 4ac}}{2a} \quad \text{or} \quad x = \frac{-b - \sqrt{b^2 - 4ac}}{2a}.$$

Example 1   Solve the equation $2x^2 + 5x - 3 = 0$.

Solution   The given equation can be written in factored form as $(2x - 1)(x + 3) = 0$. Since a product can be zero only if one of the factors is zero, we have $2x - 1 = 0$ or $x + 3 = 0$. This gives $x = 1/2$ or $x = -3$; and so $1/2$ and -3 are solutions of the given equation.   ■

Example 2   Solve the equation $x^2 - x - 1 = 0$.

Solution   We try to factor the left side but without success. Therefore applying the quadratic formula (with $a = 1$, $b = -1$, $c = -1$) gives

$$x = \frac{-(-1) \pm \sqrt{(-1)^2 - 4(1)(-1)}}{2(1)} = \frac{1 \pm \sqrt{5}}{2}.$$

Thus the solution set S is $S = \left\{ \dfrac{1 + \sqrt{5}}{2}, \dfrac{1 - \sqrt{5}}{2} \right\}.$ ■

Example 3   Solve the equation $2x^2 - 6x + 5 = 0$.

Solution   As in Example 2, we resort to application of the quadratic formula (with $a = 2$, $b = -6$, $c = 5$) to get

$$x = \frac{6 \pm \sqrt{(-6)^2 - 4(2)(5)}}{2(2)} = \frac{6 \pm \sqrt{-4}}{4} = \frac{6 \pm 2i}{4} = \frac{3 \pm i}{2}.$$

Thus there are no real number solutions. However, if the domain of $f(x) = 2x^2 - 6x + 5$ is assumed to be the set of complex numbers, then $(3 + i)/2$ and $(3 - i)/2$ would be solutions.   ■

Graphs of Quadratic Functions
We wish to draw a graph of the equation

$$y = ax^2 + bx + c \tag{1.16}$$

for given values of the parameters a, b, and c, where $a \neq 0$. First, let us consider two examples that can be used as models for discussing the general case.

Example 4   Draw a graph of

$$y = 2x^2 - 4x - 6. \tag{1.17}$$

Solution   Rather than make an extensive table of x, y values that satisfy the given equation, we look for some *key points*. These include the coordinate axes intercepts:

    y-intercept: let $x = 0$, then $y = -6$; thus $(0, -6)$ is on the graph;

    x-intercepts: let $y = 0$, then $2x^2 - 4x - 6 = 0$; or $2(x + 1)(x - 3) = 0$.

Thus $(-1, 0)$ and $(3, 0)$ are the x-intercept points. Next we get an equivalent equation by completing the square on the x terms as follows:

$$y = 2(x^2 - 2x) - 6 = 2(x^2 - 2x + 1) - 6 - 2 = 2(x - 1)^2 - 8. \tag{1.18}$$

Equation (1.18) can be used to get information about the graph that is not directly apparent from Eq.(1.17). Since $2(x - 1)^2 \geq 0$, $y \geq -8$ for all values of x. For $x = 1$, $y = -8$, and so $(1, -8)$ is the *lowest point* in the graph. Also, if we take any two values of x that are symmetric about the line $x = 1$ (say, $x_1 = 1 - h$ and $x_2 = 1 + h$), where h is any real number, the corresponding values of y are given by

$$y_1 = 2[(1 - h) - 1]^2 - 8 = 2h^2 - 8,$$
$$y_2 = 2[(1 + h) - 1]^2 - 8 = 2h^2 - 8.$$

Note that $y_1 = y_2$, and so the graph is symmetric about the line $x = 1$, as shown in Fig. 1.26(a).

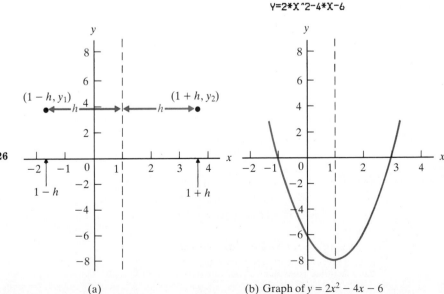

Fig. 1.26

(a)   (b) Graph of $y = 2x^2 - 4x - 6$

Using the information above one can draw a reasonably accurate graph of the given equation (Fig. 1.26b). The curve is an example of a *parabola** that *opens upward.* ■

Example 5 Draw a graph of

$$y = -2x^2 - 4x + 6. \tag{1.19}$$

Solution Following the same procedure as in the solution of Example 4, we get

y-intercept point is $(0, 6)$;

x-intercept points are $(-3, 0)$ and $(1, 0)$.

The completed square version of Eq. (1.19) is

$$y = -2(x + 1)^2 + 8. \tag{1.20}$$

Since $-2(x + 1)^2 \leq 0$, we get that $y \leq 8$ for all values of x. For $x = -1, y = 8$, and so $(-1, 8)$ is the *highest point* on the graph. Also using Eq. (1.20), one can easily show that the graph is symmetric about the line $x = -1$.

Using the information above, one can draw a reasonably accurate graph of the given equation (Fig. 1.27). As in Example 4, the curve is a parabola; in this case it *opens downward.*

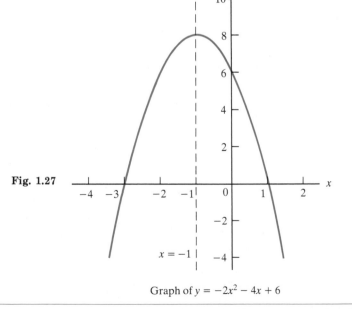

Fig. 1.27

Graph of $y = -2x^2 - 4x + 6$ ■

---

* A more detailed treatment of parabolas is given in Chapter 8.

Graphs of Quadratic Functions in General

By following a procedure similar to that used in Examples 4 and 5, we can derive the important features of the graph of any equation of the type

$$y = ax^2 + bx + c, \quad \text{where } a \neq 0. \tag{1.21}$$

The *completed-square version* of Eq. (1.21) is

$$y = a\left(x + \frac{b}{2a}\right)^2 + \frac{4ac - b^2}{4a}. \tag{1.22}$$

From Eq. (1.22) we get the following important features:

1. If $a > 0$, then $a\left(x + \dfrac{b}{2a}\right)^2 \geq 0$, and so $y \geq \dfrac{4ac - b^2}{4a}$ for all values of x.

   Thus the graph has a lowest point given by $x = \dfrac{-b}{2a}$.

2. If $a < 0$, then $a\left(x + \dfrac{b}{2a}\right)^2 \leq 0$, and so $y \leq \dfrac{4ac - b^2}{4a}$ for all values of x.

   Hence the graph has a highest point given by $x = \dfrac{-b}{2a}$.

3. Also, the curve is symmetric about the line $x = \dfrac{-b}{2a}$. In summary, we have the following.

The graph of $y = ax^2 + bx + c$, where $a \neq 0$, is a *parabola* that
1. *opens upward* if $a > 0$ and, *opens downward* if $a < 0$;

2. has a *lowest* or *highest point* given by $x = \dfrac{-b}{2a}$;

3. is *symmetric* about the vertical line $x = \dfrac{-b}{2a}$.

Example 6   Find the maximum and minimum values of the function f given by

$$f(x) = 3.1x^2 - 4.8x + 3.7 \quad \text{and} \quad \mathcal{D}(f) = \{x \,|\, 0 \leq x < 2\}.$$

Give answers rounded off to two decimal places.

Solution   We solve the problem by first drawing a graph of the given function. The graph is part of the parabola $y = 3.1x^2 - 4.8x + 3.7$ that opens upward (since $a = 3.1$, and so $a > 0$), and the lowest point is given by

$$x = -\frac{b}{2a} = -\frac{(-4.8)}{2(3.1)} = 0.774$$

(to three decimal places).

For this value of x, the corresponding value of y is given by

$$y = 3.1(.774)^2 - 4.8(.774) + 3.7 = 1.84$$

(to two decimal places).

Therefore $(0.77, 1.84)$ is the lowest point on the graph.

Now make a table of x, y values, and since 0.77 is in $\mathcal{D}(f)$, we include $x = 0.77$, $y = 1.84$:

x	0	0.5	0.77	1	1.5	1.8	1.9	1.99
y	3.7	2.08	1.84	2.00	3.48	5.10	5.77	6.42

The values of y in this table are determined with the aid of a calculator and are rounded off to two decimal places. Plotting these points and drawing a curve through them, remembering to restrict x to values in $\mathcal{D}(f)$, gives the curve shown in Fig. 1.28.

Y=3.1*X^2-4.8*X+3.7

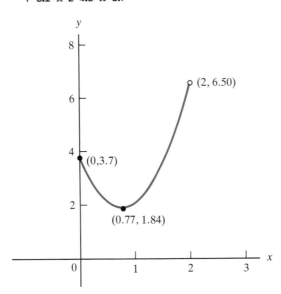

Fig. 1.28

Graph of $f(x) = 3.1x^2 - 4.8x + 3.7$, $\mathcal{D}(f) = \{x \mid 0 \le x < 2\}$

The graph in Fig. 1.28 can now be used to give an answer to the stated problem. The minimum value of the function—that is, the smallest value of y on the graph—is 1.84. From the graph we also see that as x approaches 2, y approaches 6.50, but the point $(2, 6.50)$ is not on the graph since 2 is not in $\mathcal{D}(f)$. Thus it should be clear that the function does not have a maximum value.

Example 7 Suppose right triangle ABC, shown in Fig. 1.29, has base CB of length 5 and height CA of length 3. A rectangle $CDEF$ is inscribed in the triangle, as shown. Find the dimensions of the rectangle with maximum area.

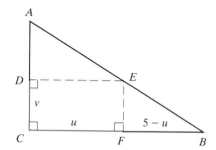

Fig. 1.29

Solution Let $u = \overline{CF}$ and $v = \overline{CD}$ denote the dimensions of the inscribed rectangle. The area K is given by

$$K = u \cdot v, \tag{1.23}$$

where u and v are variables; for example, u can be any number between 0 and 5. There is a relationship between u and v, as can be seen from the fact that triangles ACB and EFB are similar, which gives $\dfrac{\overline{EF}}{\overline{FB}} = \dfrac{\overline{AC}}{\overline{CB}}$. In terms of u and v and the given dimensions, this becomes $\dfrac{v}{5 - u} = \dfrac{3}{5}$. Therefore $v = 3 - \dfrac{3}{5}u$. Substituting this into Eq. (1.23) gives a formula for K as a function of u:

$$K = u\left(3 - \frac{3}{5}u\right).$$

Thus

$$K = -\frac{3}{5}u^2 + 3u, \tag{1.24}$$

where $0 < u < 5$.

The graph of this function, as seen in Fig. 1.30, is part of a parabola that opens downward (since the coefficient of u^2 is negative), and so it has a highest point given by

$$u = \frac{-b}{2a} = \frac{-3}{2(-3/5)} = \frac{5}{2}.$$

The corresponding value of K is given by

$$K = \frac{-3}{5}\left(\frac{5}{2}\right)^2 + 3\left(\frac{5}{2}\right) = \frac{15}{4}.$$

K=(-3/5)*U^2+3*U

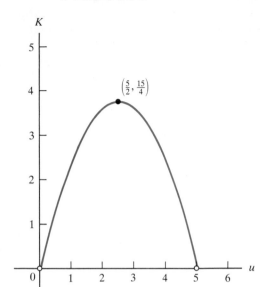

Fig. 1.30

Graph of $K = -0.6u^2 + 3u, \ \ 0 < u < 5$

This is the maximum value of K. Therefore the rectangle with dimensions

$$u = \frac{5}{2} \quad \text{and} \quad v = 3 - \frac{3}{5}\left(\frac{5}{2}\right) = \frac{3}{2}$$

will be the one that has maximum area.

Quadratic Inequalities

The following examples will illustrate techniques for solving quadratic inequalities.

Example 8   Find the solution set S for the open sentence $x^2 - 2x - 3 < 0$.

Solution   Two methods of solution are illustrated.

Method 1   Let $y = x^2 - 2x - 3$ and draw a graph. This is a parabola that opens upward, as shown in Fig. 1.31, where the x-intercept points $(-1, 0)$ and $(3, 0)$ were determined by solving the quadratic equation $x^2 - 2x - 3 = 0$.

   Solving $x^2 - 2x - 3 < 0$ is equivalent to finding the values of x in the graph of Fig. 1.31 for which y is negative. As seen from the graph, the solution includes any x between -1 and 3. Thus $S = \{x \mid -1 < x < 3\}$.

Y=X^2-2*X-3

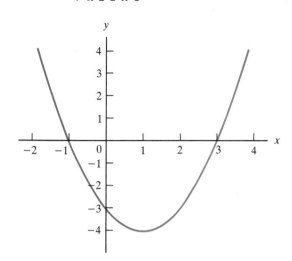

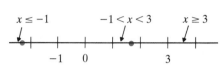

Fig. 1.32

Fig. 1.31   Graph of $y = x^2 - 2x - 3$

Method 2   The given inequality can be written as

$$(x + 1)(x - 3) < 0. \qquad (1.25)$$

We can consider all values of x by looking at three cases, as shown on the number line in Fig. 1.32.

1. If $x \leq -1$, then $(x + 1) \leq 0$ and $(x - 3) \leq 0$; thus $(x + 1)(x - 3) \geq 0$, and so x is not a solution of inequality (1.25).
2. If $-1 < x < 3$, then $(x + 1) > 0$, and $(x - 3) < 0$; hence we can write $(x + 1)(x - 3) < 0$, and so x satisfies inequality (1.25). Therefore any x for which $-1 < x < 3$ will be in S.
3. If $x \geq 3$, then $(x + 1) \geq 0$ and $(x - 3) \geq 0$; thus $(x + 1)(x - 3) \geq 0$, and so x is not a solution of inequality (1.25).

Therefore $S = \{x \mid -1 < x < 3\}$.

   Method 2 suggests the following format for solving the given inequality $x^2 - 2x - 3 < 0$.

Solve the equation:* $x^2 - 2x - 3 = 0$.

Factor: $(x + 1)(x - 3) = 0$.

Solution: $x = -1$ or $x = 3$.

---

*   In general, the cut points for $ax^2 + bx + c < 0$ can be found by using the quadratic formula to solve $ax^2 + bx + c = 0$. If the roots are not real numbers, the solution set is the empty set if $a > 0$, and **R** if $a < 0$.

Consider -1 and 3 as *cut points* on a number line, and check typical points in each region:

1. $x = -2$ **2.** $x = 0$ **3.** $x = 5$

$(-5)(-1) < 0$ $(-3)(1) < 0$ $(6)(2) < 0$

False   True   False

This suggests that the solution is given by $-1 < x < 3$.   ▪

Example 9   Find the solution set S for the open sentence $x^2 - 2x + 3 \geq 0$.

Solution   **Method 1**   Let $y = x^2 - 2x + 3$, and draw the graph shown in Fig. 1.33, where the lowest point is given by $x = -b/2a = 2/2 = 1, y = 2$. From the graph we see that for every x, the value of y is positive, and so the solution set for the given inequality is $S = \mathbf{R}$.

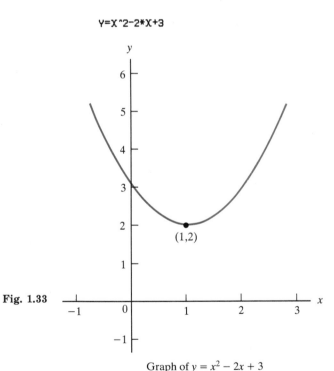

Y=X^2-2*X+3

Fig. 1.33

Graph of $y = x^2 - 2x + 3$

Method 2   By completing the square, one can write the given inequality as $(x - 1)^2 + 2 \geq 0$. Since $(x - 1)^2 \geq 0$, $(x - 1)^2 + 2 \geq 0$ for every real number x. Thus $S = \mathbf{R}$.   ▪

Exercises 1.4

In problems 1 through 12, solve the given equations. In each case,

a) express solutions in exact form;

b) state whether the roots are rational, irrational, or complex (nonreal) numbers; if the roots are irrational numbers, express them in decimal form rounded off to two decimal places.

1. $2x^2 - 3x - 2 = 0$ 2. $2x^2 + 5x - 3 = 0$ 3. $3x^2 - 4x - 2 = 0$

4. $2x^2 - 4x - 3 = 0$ 5. $x^2 - 2x - 2 = 0$ 6. $3x^2 - 4x + 1 = 0$

7. $x^2 - 2x + 2 = 0$ 8. $3x^2 + 6x + 1 = 0$ 9. $x^2 - 2\sqrt{3}x - 2 = 0$

10. $3x^2 - 2\sqrt{2}x - 1 = 0$ 11. $x^2 - 2.3x - 1.6 = 0$ 12. $2x^2 + 1.5x - 3.5 = 0$

In problems 13 through 20, sketch graphs of the given functions. In each case label the coordinates of

a) y-intercept point,   b) x-intercept points,   c) the highest or lowest point.

13. $f(x) = 2x^2 - 3x - 2$ 14. $f(x) = 2x^2 + 5x - 3$

15. $f(x) = -x^2 + 2x + 3$ 16. $f(x) = -x^2 - 2x + 8$

17. $f = \{(x, y) \mid y = x^2 + 4x + 4\}$ 18. $f = \{(x, y) \mid y = -x^2 + 2x - 1\}$

19. $f = \{(x, y) \mid y = x^2 + 4x - 2\}$ 20. $f = \{(x, y) \mid y = -x^2 - 2x + 5\}$

In problem 21 through 32, find the maximum and minimum values of the given functions. In any case where there is no such value, explain.

21. $f(x) = x^2 + 4x + 3$ 22. $f(x) = x^2 + 6x + 4$

23. $f(x) = -2x^2 + 4x - 5$ 24. $f(x) = -2x^2 - 6x + 3$

25. $f(x) = 1.5x^2 - 4.8x - 1$ 26. $f(x) = 1.2x^2 + 3.2x - 4$

27. $f(x) = -x^2 - 3x + 4;\ \mathcal{D}(f) = \{x \mid 0 \le x \le 2\}$

28. $f(x) = x^2 - 3x - 2;\ \mathcal{D}(f) = \{x \mid 0 \le x \le 2\}$

29. $f(x) = -x^2 - 3x + 3;\ \mathcal{D}(f) = \{x \mid 0 < x \le 2\}$

30. $f(x) = 2x^2 + 4x + 1;\ \mathcal{D}(f) = \{x \mid -3 \le x < 3\}$

31. $f(x) = -2x^2 - 4x + 1;\ \mathcal{D}(f) = \{x \mid 0 \le x \le 2\}$

32. $f(x) = -3x^2 - 6x + 4;\ \mathcal{D}(f) = \{x \mid 0 \le x \le 2\}$

In problems 33 through 40, find the solution set for the given inequalities. Also show the solution on a number line.

33. $x^2 - 4x + 3 > 0$ 34. $x^2 + 5x + 4 < 0$

35. $2x^2 - x - 3 \le 0$ 36. $3x^2 + 2x - 8 \ge 0$

37. $-x^2 + 2x + 4 < 0$ 38. $-2x^2 + 3x - 4 \le 0$

39. $x^2 - 4x + 4 \le 0$ 40. $x^2 - 6x + 9 \le 0.$

41. Rectangle $DEFG$ is inscribed in an isosceles triangle ABC, as shown in Fig. 1.34. Suppose $\overline{AC} = \overline{BC} = 4$ and $\overline{AB} = 2$, and denote $\overline{DE}$ by x. If K represents the area of the rectangle, find

a) a formula that gives K as a function of x, and state the domain;

b) the dimensions of the rectangle that has maximum area;

c) the maximum area.

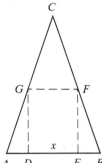

Fig. 1.34

42. A ball is thrown vertically upward from the ground with an initial speed of 36 meters per second. The position of the ball at any time t seconds after it has been thrown is given by the formula

$$s = 36t - 4.9t^2,$$

where s is the distance (in meters) of the ball from the ground.

a) Determine the location of the ball at each of the following times: $t = 1$, $t = 2$, $t = 3$, $t = 4$, $t = 5$, $t = 6$.

b) Draw a graph showing s as a function of t.

c) The graph in (b) does not give the path of the ball (since it goes straight up, then down). It merely allows us to "read off" the height of the ball at any time t. Use it to get reasonable approximations of the height of the ball at times $t = 1.5$ and $t = 4.5$.

d) How many seconds does it take for the ball to reach its highest point? How high is it at that instant?

43. A farmer purchases a rectangular plot of land 200 m by 400 m adjacent to his property line AB, along which there is an existing fence. He wants to fence in a corral in the southwest corner of the newly acquired land, as shown in Fig. 1.35, and has a total of 360 meters of fencing left over from a previous job, all of which he wants to

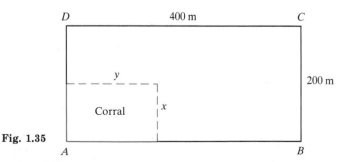

Fig. 1.35

use to fence the three sides of the corral. He is interested in making a corral of maximum area.

a) Determine the dimensions of such a corral.

b) How many square meters of the purchased land is left over after the corral has been fenced?

44. A rancher has 144 meters of fencing left over from a previous job, and he wishes to divide it into two pieces, one of which he will use to fence a square region for holding his horses, and the other will be used to fence a circular region as a training area. Suppose the fence is cut at a point x meters from one end, and the piece of length x is to be used for the training area. The rancher is interested in the amount of land that is being fenced in, that is, the total area A of the circular and square regions. Intuition tells him that there must be a value of x at which he should cut the 144 meters of fencing so that A is the smallest (thus leaving the largest possible area for grazing). Is his intuition correct? If it is, determine the place where he should make the cut, the size (radius) of the training area, and the length of the side of the holding area.

45. A travel agent is proposing a tour in which a group will travel in a plane of 150 capacity. The fare will be $1400 per person if 120 or fewer people go on the tour; but if more than 120 go, the fare per person (for the entire group) will be decreased by $10 for each person in excess of 120. For instance, if 125 go, the fare for each will be $1400 - $10(5) = $1350. Let x represent the total number of people who go on the tour and T the total revenue (in dollars) collected by the agency.

a) Find T as a function of x; be certain to indicate the domain of the function.

b) Determine the number of people that will give the largest revenue.

1.5   ABSOLUTE VALUE FUNCTION

The idea of a number line establishes a one-to-one correspondence between the set of real numbers and the set of points on a line. This gives us a basis for describing geometric ideas by relating them to numbers. For instance, if we wish to talk about the distance between two points on a line, we do so in terms of the corresponding numbers. Thus our intuitive notion of distance suggests that on the number line in Fig. 1.36 the distance between the point 3 and the origin (the point corresponding to zero) is 3, and similarly the distance between -3 and 0 is also 3.* We denote this by $|3| = 3$ and $|-3| = 3$.

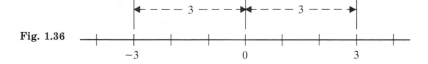

Fig. 1.36

---

*   In Section 1.0 we indicated that we would take liberties with language and say "point 3" rather than the correct but more cumbersome "the point corresponding to the number 3."

This concept suggests a rule for assigning to any real number x a corresponding nonnegative number that represents the distance between x and 0 on a number line. The rule for assigning values of $|x|$ is: if x is positive or zero, then $|x|$ is x but if x is negative, then $|x|$ is obtained by simply dropping the negative sign. This rule gives the absolute value function, which we now state as a formula.

Definition 1.8   Suppose x is any real number. The *absolute value function,* denoted by $|x|$, is defined by

$$|x| = \begin{cases} x \text{ if } x \geq 0, \\ -x \text{ if } x < 0. \end{cases} \tag{1.26}$$

The symbol $|x|$ is read *the absolute value of x.*

As indicated above, the geometric interpretation of $|x|$ is that it represents the distance between the point x and 0 on a number line. In general, the distance between any two points on a line can also be described in terms of absolute value; this we shall do after considering the following two examples.

Example 1   Given that $f(x) = |x|$, evaluate the following.

    a) $f(4)$ **b)** $f(0)$ **c)** $f(-2)$ **d)** $f(1-\sqrt{5})$

Solution   **a)** Since $4 > 0$, we use the first part of Eq. (1.26), that is, $f(4) = |4| = 4$.

    b) Similarly, $0 \geq 0$, and so $f(0) = |0| = 0$.

    c) Since $-2 < 0$, the second part of Eq. (1.26) gives $f(-2) = |-2| = -(-2) = 2$.

    d) Similarly, $1 - \sqrt{5} < 0$, and so $f(1-\sqrt{5}) = |1-\sqrt{5}| = -(1-\sqrt{5}) = \sqrt{5} - 1$.

Example 2   Draw a graph of $y = |x|$.

Solution

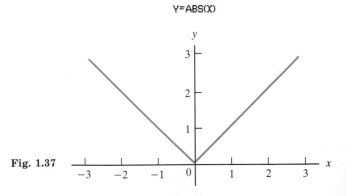

Graph of $y = |x|$

Fig. 1.37

Using Eq. (1.26), we get

$$y = |x| = \begin{cases} x & \text{if } x \geq 0, \\ -x & \text{if } x < 0. \end{cases}$$

Thus the graph consists of two half-lines: $y = x$ if $x \geq 0$ and $y = -x$ if $x < 0$. This is shown in Fig. 1.37. ■

Distance Between Two Points

To illustrate, let points A, B, and C correspond to the numbers -4, 2, and 5, respectively, on a number line, as shown in Fig. 1.38.

Fig. 1.38

The distance between A and B is 6 and is given by $|2 - (-4)| = 6$ or by $|-4 - 2| = 6$; the distance between B and C is 3 and is given by $|5 - 2| = 3$ or by $|2 - 5| = 3$. This suggests the following definition.

Definition 1.9   Suppose x and y are any two real numbers. The *distance between points x and y* on a number line is given by $|x - y|$ or by $|y - x|$.

Elementary Properties of Absolute Value

If u and v are any two real numbers, then

1. $|u| \geq 0$; that is, $|u|$ is always a nonnegative number;
2. $|-u| = |u|$;
3. $|u^2| = u^2$;
4. $\sqrt{u^2} = |u|$;   (1.27)
5. $|u \cdot v| = |u| \cdot |v|$;
6. $\left|\dfrac{u}{v}\right| = \dfrac{|u|}{|v|}$, $v \neq 0$;
7. $|u + v| \leq |u| + |v|$.

The statements given in (1.27) can be proved by considering various cases for u and v; that is, $u \geq 0$, $u < 0$, $v \geq 0$, $v < 0$. We leave it to the reader to carry out the details.

Solving Equations Involving Absolute Value

The following examples illustrate methods that can be used to find roots of an equation involving absolute-value expressions.

Example 3   Solve the equations.

     a) $|x - 2| = 3$ **b)** $|x + 3| = 4$

Solution   **a)** Geometrically, the given equality states that x is a point that is three units from 2 on the number line. Looking at Fig. 1.39, we see that $x = -1$ or $x = 5$, are such points, and so -1 and 5 are solutions.

Fig. 1.39

       Analytically, if the absolute value of a number ($x - 2$ in this case) is 3, then that number must be 3 or -3. Thus $x - 2 = -3$ or $x - 2 = 3$, and so $x = -1$ or $x = 5$.

     b) The given equation can be written as $|x - (-3)| = 4$; geometrically, this says that x is a point on the number line that is four units from -3. There are two such points given by $x = -7$ or $x = 1$. Thus -7 and 1 are solutions. Analytically, the given equation is equivalent to $x + 3 = -4$ or $x + 3 = 4$. Therefore $x = -7$ or $x = 1$.   ■

Example 4   Solve the equations.

     a) $\left| \dfrac{x - 1}{-3} \right| = 1$ **b)** $|x - 1| + 1 = 0$

Solution   **a)** Using Statement 6 of (1.27) we get $\left| \dfrac{x - 1}{-3} \right| = \dfrac{|x - 1|}{|-3|} = \dfrac{|x - 1|}{3}$. Thus the given equation is equivalent to $|x - 1| = 3$, and so $x - 1 = 3$ or $x - 1 = -3$. Therefore $x = 4$ or $x = -2$, and so 4 and -2 are solutions.

     b) Using Statement 1 of (1.27), we note that $|x - 1| \geq 0$ for every real number x. Therefore adding 1 to $|x - 1|$ cannot yield zero for any value of x. Thus the given equation has no solutions.   ■

Example 5   Find the solution set S for the following open sentences.

     a) $|x^2| - 4x + 3 = 0$ **b)** $x^2 - 2|x| - 3 = 0$

Solution   **a)** Using Statement 3 of (1.27), we can replace $|x^2|$ by x^2, and so the given equation becomes

$$x^2 - 4x + 3 = 0.$$

This is a quadratic equation that can be solved by factoring

$$(x - 1)(x - 3) = 0.$$

Thus the solution set is $S = \{1, 3\}$.

b) Using Eq. (1.26), we can replace $|x|$ by x if $x \geq 0$ and by $-x$ if $x < 0$. Consider two cases:

Case 1. For $x \geq 0$, the given equation becomes $x^2 - 2x - 3 = 0$; this can be solved by factoring, $(x - 3)(x + 1) = 0$, and so $x = 3$ or $x = -1$. Since $x = -1$ does not satisfy $x \geq 0$, we discard it as a possible solution. However, 3 is a solution.

Case 2. For $x < 0$, the given equation becomes $x^2 + 2x - 3 = 0$; this can be solved by factoring, $(x + 3)(x - 1) = 0$, and so $x = -3$ or $x = 1$. Since $x = 1$ does not satisfy $x < 0$, it is discarded. However, $x = -3$ does satisfy $x < 0$, and it yields a solution. Therefore the solution set is $\{-3, 3\}$. ■

Example 6   Find the roots of $2x - |x - 1| = 3$.

Solution   We can replace $|x - 1|$ by $x - 1$ if $x - 1 \geq 0$ (that is, $x \geq 1$), and by $-(x - 1)$ if $x - 1 < 0$ (that is, $x < 1$). Thus the given equation can be analyzed in two cases:

Case 1. $x \geq 1$ and $2x - (x - 1) = 3$; that is, $x \geq 1$ and $x = 2$. Therefore 2 is a solution.

Case 2. $x < 1$ and $2x + (x - 1) = 3$; that is, $x < 1$ and $x = \frac{4}{3}$, and so this yields no solution.

Therefore the solution set for the given equation is $\{2\}$. ■

Example 7   Find the solution set S for $|x + 3| + x + 3 = 0$.

Solution   We consider two cases: $x \geq -3$ and $x < -3$.

Case 1: $x \geq -3$ and $(x + 3) + x + 3 = 0$. This gives -3 as a solution.

Case 2: $x < -3$ and $-(x + 3) + x + 3 = 0$; since any x will satisfy this equality, we have all $x < -3$ as solutions.

Thus $S = \{x | x \leq -3\}$. ■

Solving Inequalities Involving Absolute Value

A property that is frequently useful in dealing with inequalities that involve absolute value is:

> Suppose c is a given positive number. Then
>
> **1.** $|u| < c$ is equivalent to $-c < u < c$; that is $u > -c$ *and* $u < c$;   (1.28)
>
> **2.** $|u| > c$ is equivalent to $u < -c$ *or* $u > c$.

For instance, $|u| < 3$ says that u is a point on the number line that is within three units of the origin. Thus u is any point between -3 and 3, as shown in Fig. 1.40(a), and so $-3 < u < 3$.

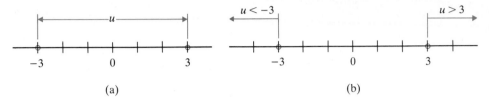

Fig. 1.40

(a)   (b)

The inequality $|u| > 3$ tells us that u is a point on the number line that is more than three units from the origin. This is shown in Fig. 1.40(b), and so $u < -3$ or $u > 3$.

Example 8   Find the solution set for each of the following inequalities.

a) $|x - 3| < 2$ **b)** $|2x + 5| \geq 3$

Solution   In each case show the result on a number line.

a) Using Statement 1 of (1.28), we can replace the given inequality by $-2 < x - 3 < 2$; that is,

$$x - 3 > -2 \quad and \quad x - 3 < 2.$$

In each of these we can add 3 to both sides to get

$$x > 1 \quad and \quad x < 5.$$

This can be written as $1 < x < 5$, and the solution set is

$$S = \{x | 1 < x < 5\}.$$

Set S is shown in Fig. 1.41, in which the open circles at 1 and 5 indicate that these values are not in S.

Fig. 1.41

b) Using Statement 2 of (1.28), we can replace the given inequality by

$$2x + 5 \leq -3 \quad or \quad 2x + 5 \geq 3.$$

This gives $x \leq -4$ or $x \geq -1$, and so the solution set is

$$S = \{x | x \leq -4 \quad or \quad x \geq -1\}.$$

The set S is shown in Fig. 1.42, in which solid circles at -4 and -1 indicate that these values are in S.

Fig. 1.42

Graphs of Functions Involving Absolute Value

Here we illustrate by examples techniques that can be applied to drawing graphs of functions involving absolute value.

Example 9   Draw a graph of the function given by $y = |x - 1| + x$.

Solution   First express the given function in terms of a formula that does not involve absolute value. This can be done by replacing $|x - 1|$ by $x - 1$ if $x \geq 1$, and by $-(x - 1)$ if $x < 1$:

$$y = \begin{cases} (x - 1) + x = 2x - 1 \text{ if } x \geq 1, \\ -(x - 1) + x = 1 \quad \text{ if } x < 1. \end{cases}$$

From this we see that the graph consists of two half-lines, as shown in Fig. 1.43.

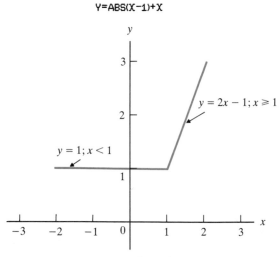

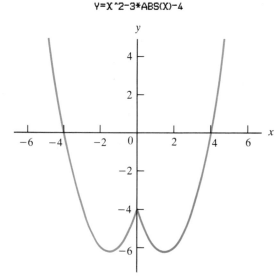

Fig. 1.43   Graph of $y = |x - 1| + x$

Fig. 1.44   Graph of $y = x^2 - 3|x| - 4$

Example 10   Draw a graph of $y = x^2 - 3|x| - 4$.

Solution   First replace $|x|$ in the given equation by x if $x \geq 0$, and by $-x$ if $x < 0$.

$$y = \begin{cases} x^2 - 3x - 4 \text{ if } x \geq 0, \\ x^2 + 3x - 4 \text{ if } x < 0. \end{cases}$$

This gives portions of two parabolas:

$$y = x^2 - 3x - 4 = (x + 1)(x - 4) \text{ for } x \geq 0,$$
$$y = x^2 + 3x - 4 = (x + 4)(x - 1) \text{ for } x < 0.$$

The graph is shown in Fig. 1.44.

Exercises 1.5

In problems 1 through 4, evaluate the given expressions. Give answers in exact form.

1. $f(x) = 1 + |x|$

    a) $f(3)$ **b)** $f(-4)$ **c)** $f(1 - \sqrt{3})$

2. $g(x) = \dfrac{|x|}{x}$

    a) $g(3)$ **b)** $g(-4)$ **c)** $g(1 - \sqrt{3})$

3. $f(x) = x^2 + 2|x - 1| - 1$

    a) $f(3)$ **b)** $f(-1)$ **c)** $f\left(\dfrac{1 - \sqrt{5}}{2}\right)$

4. $g(x) = \dfrac{1 - |x|}{1 + |x|}$

    a) $f(1)$ **b)** $f(-1)$ **c)** $f(1 - \sqrt{3})$

In problems 5 and 6, functions f and g are given by $f(x) = 3 + |x|$, $g(x) = x^2 - 2|x|$. Evaluate the given expressions in exact form.

5. a) $(f + g)(-1)$ **b)** $(g - f)(4)$ **c)** $(f \circ g)(-2)$

6. a) $(f + g)(3)$ **b)** $\left(\dfrac{f}{g}\right)(1)$ **c)** $(g \circ f)(1)$

7. Given that $f(x) = x^2 - x$ and $g(x) = |x - 1|$, find formulas for

    a) $(f \circ g)(x)$ **b)** $\left(\dfrac{f}{g}\right)(x)$ **c)** Give the domains of $f \circ g$ and $\dfrac{f}{g}$.

8. Given that $f(x) = \sqrt{x + 1}$ and $g(x) = |x + 1|$, find formulas for

    a) $(f \circ g)(x)$ **b)** $(g \circ f)(x)$ **c)** Give the domains of $f \circ g$ and $g \circ f$.

In each of the problems 9 through 24, find the solution set for the given equation.

9. $|x - 1| = 2$ **10.** $|3x - 4| = 2$ **11.** $\left|\dfrac{1 - x}{-2}\right| = 1$ **12.** $\left|\dfrac{3 - 2x}{-3}\right| = 1$

13. $2x + |x - 2| = 1$ **14.** $|x - 3| + x = 3$ **15.** $x + |x - 1| = 1$ **16.** $4x + |x - 1| + 4 = 0$

17. $|x - 2| = x - 2$ **18.** $|3 - x| = 3 - x$ **19.** $|-x| + 2 = 0$ **20.** $|-x - 1| + 1 = 0$

21. $x^2 - 3\sqrt{x^2} - 4 = 0$ **22.** $x^2 + 2\sqrt{x^2} - 3 = 0$ **23.** $x^2 - |x - 1| - 3 = 0$ **24.** $x^2 - 3|x| - 4 = 0$

In problems 25 through 30, find the solution set S for each of the given inequalities. In each case show S on a number line.

25. $|x - 1| < 1$ **26.** $|x + 2| \le 3$ **27.** $|x| + 2 \le 2x$

28. $3 - 2|1 - x| < 0$ **29.** $|x| - x \ge 1$ **30.** $|-x| - x > 0$

In problems 31 through 34, give formulas that do not use absolute value for the given functions.

31. $f(x) = |x - 2| + x$ **32.** $f(x) = 2x - |x + 1|$

33. $f(x) = |x^2| + |x + 1| + 3$ **34.** $f(x) = x^2 - |x - 1| - 4$

In problems 35 through 40, sketch graphs of the functions described by the given equations.

35. $y = |x - 2|$ 　　　　　　　　**36.** $y = |3x + 6|$ 　　　　　　　　**37.** $y = x - |x + 2|$

38. $y = |x + 3| + x$ 　　　　　　**39.** $y = x^2 - |x| - 2$ 　　　　　　**40.** $y = x^2 + 2|x| - 3$

In problems 41 and 42, determine the domain and range for each of the given functions. If necessary, draw a graph and use it to find the range.

41. a) $f(x) = |x|$ 　　　　　　　　　　　　　**b)** $g(x) = 1 - |x|$

42. a) $f(x) = 1 - |1 + x|$ 　　　　　　　　　**b)** $g(x) = |1 - x| - x$

1.6   GRAPHS OF CONIC SECTIONS

In this section graphs of relations that play a special role in geometry and certain applications are discussed. The curves considered here are *circles, parabolas, ellipses,* and *hyperbolas,* which are referred to as *conic sections* because they are the result of intersecting cones by planes. In Section 1.4 we have already noted that the graph of a quadratic function, $f(x) = ax^2 + bx + c$, where $a \neq 0$, is a parabola. Formal definitions of conic sections and detailed discussion will be included in Chapter 8. We consider only special cases here and rely on examples to illustrate the various curves and their related equations.

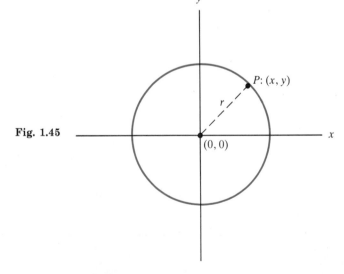

Fig. 1.45

Circles

A circle with its center at the origin and having a given radius r is the set of all points $P: (x, y)$ that are r units from the center, as shown in Fig. 1.45. Translating this into a mathematical statement gives $\sqrt{x^2 + y^2} = r$. For $r > 0$, this is equivalent to

$$\overline{x^2 + y^2 = r^2.}\tag{1.29}$$

Thus Eq. (1.29) represents a circle with center at the origin and radius r.

An equation of the type $ax^2 + by^2 = c$, where a, b, and c are nonzero numbers, all have the same sign, and $a = b$ represents a circle with center at the origin and radius $\sqrt{c/a}$ since it can be written as $x^2 + y^2 = c/a$.

Although the set of (x, y) points described by Eq. (1.29) does not define a function, we can solve for y and consider the circle as the union of two sets of points given by

$$y = f(x) = \sqrt{r^2 - x^2} \quad \text{and} \quad y = g(x) = -\sqrt{r^2 - x^2}.\tag{1.30}$$

Functions f and g have domain $\mathfrak{D} = \{x|\ -r \le x \le r\}$ and represent half-circles, as shown in Fig. 1.46.

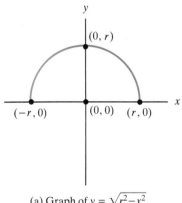

(a) Graph of $y = \sqrt{r^2-x^2}$

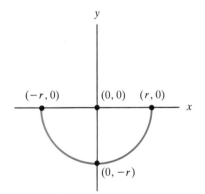

(b) Graph of $y = -\sqrt{r^2 - x^2}$

Fig. 1.46

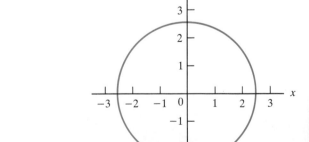

4*X^2+4*Y^2=25

Fig. 1.47   Graph of $4x^2 + 4y^2 = 25$

Example 1   Draw a graph of $4x^2 + 4y^2 = 25$.

Solution   The given equation is equivalent to

$$x^2 + y^2 = \frac{25}{4} = \left(\frac{5}{2}\right)^2.$$

This represents a circle with center at the origin and radius 5/2. The graph is shown in Fig. 1.47. ■

Parabolas

Suppose we consider graphs of equations of the form $y = ax^2$ or $x = ay^2$, where a is a given nonzero number. In Section 1.4 we noted that $y = ax^2$ represents a parabola that opens upward if $a > 0$ and downward if $a < 0$. If the roles of x and y are interchanged, the resulting equation, $x = ay^2$, will represent a parabola that opens to the right if $a > 0$ and to the left if $a < 0$. This is illustrated in the following examples.

Example 2   Draw graphs of the following: **a)** $y = 4x^2$ **b)** $y = -2x^2$

Solution   **a)** Since the coefficient of x^2 is positive, the graph is a parabola that opens upward. It is a simple matter to determine several pairs of x, y values that satisfy the given equation and then draw the graph shown in Fig. 1.48(a).

b) The coefficient of x^2 is negative, and so the parabola opens downward, as shown in Fig. 1.48(b).

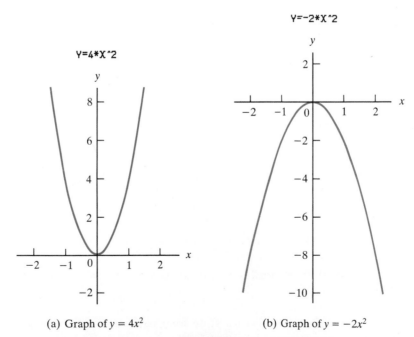

(a) Graph of $y = 4x^2$ (b) Graph of $y = -2x^2$

Fig. 1.48

Example 3   Draw graphs of the following:

a) $x = 4y^2$ **b)** $x = -y^2$

Solution   Each of the given equations represents a parabola; the one in (a) opens to the right, and the one in (b) opens to the left.

a) Solving the given equation for y gives

$$y = f(x) = \frac{1}{2}\sqrt{x} \quad \text{or} \quad y = g(x) = -\frac{1}{2}\sqrt{x},$$

where f and g are functions with domain $\mathfrak{D} = \{x \mid x \geq 0\}$; the following table gives values of each for several values of x in $\mathfrak{D}$.

x	0	0.5	1.0	1.5	2.0	3.0	4.0	5.0	9.0
$f(x)$	0	0.35	0.50	0.61	0.71	0.87	1.00	1.12	1.50
$g(x)$	0	−0.35	−0.50	−0.61	−0.71	−0.87	−1.00	−1.12	−1.50

Plotting the points given in the table and drawing the corresponding curves, we get the graphs shown in Fig. 1.49(a) and (b). The graph of the given equation is the set of all points shown in Fig. 1.49(a) along with those shown in (b); this graph is shown in Fig. 1.49(c). We see that the given equation does not define a function where x is considered the independent variable.

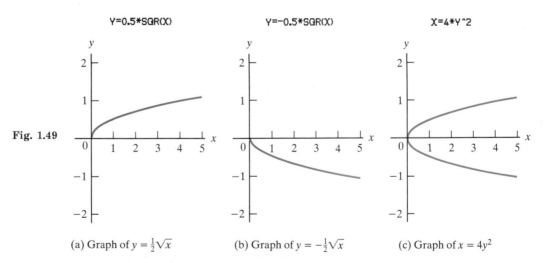

Fig. 1.49

(a) Graph of $y = \frac{1}{2}\sqrt{x}$ (b) Graph of $y = -\frac{1}{2}\sqrt{x}$ (c) Graph of $x = 4y^2$

b) This is similar to (a), in which the given equation is equivalent to

$$y = F(x) = \sqrt{-x} \quad \text{or} \quad y = G(x) = -\sqrt{-x}.$$

Here F and G are functions, each having domain $\mathfrak{D} = \{x \mid x \leq 0\}$. The graphs are shown in Fig. 1.50.

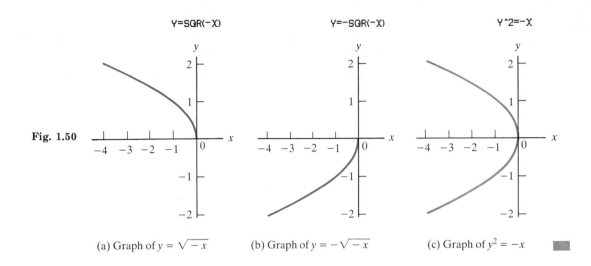

Fig. 1.50

(a) Graph of $y = \sqrt{-x}$ (b) Graph of $y = -\sqrt{-x}$ (c) Graph of $y^2 = -x$

Ellipses

Let us consider equations of the type

$$ax^2 + by^2 = c. \tag{1.31}$$

If a, b, and c are nonzero numbers all having the same sign, and if $a = b$, Eq. (1.31) represents a circle (see Example 1). Now suppose we consider the same situation, except $a \neq b$; then the graph of Eq. (1.31) is an *ellipse* with center at the origin. This is illustrated by the following example.

Example 4   Draw a graph of $4x^2 + 9y^2 = 36$.

Solution   Solving the given equation for y in terms of x gives

$$y = f(x) = \frac{2}{3}\sqrt{9 - x^2} \quad \text{or} \quad y = g(x) = -\frac{2}{3}\sqrt{9 - x^2}.$$

The domain $\mathfrak{D}$ of functions f and g is given by

$$\mathfrak{D} = \{x \,|\, 9 - x^2 \geq 0\} \quad \text{or} \quad \mathfrak{D} = \{x \,|\, -3 \leq x \leq 3\}.$$

Using several values of x from set $\mathfrak{D}$, we get the following table, which gives the corresponding values of $f(x)$ and $g(x)$; plotting the points in this table and drawing the corresponding curves gives the graphs shown in Fig. 1.51(a) and (b). It is clear that $f(-x) = f(x)$ and $g(-x) = g(x)$ for all x in $\mathfrak{D}$, and so it is sufficient to compute values for positive x only. However, we have included both in the table for emphasis.

The graph of the given equation is the ellipse shown in Fig. 1.51(c); it is the set of all points in Fig. 1.51(a) along with those in (b). We see from the graphs that f and g are functions, and that the given equation does not describe a function in which x is the independent variable.

x	-3.0	-2.5	-2.0	-1.5	-1.0	-0.5	0	0.5	1.0	1.5	2.0	2.5	3.0
$f(x)$	0	1.11	1.49	1.73	1.89	1.97	2.00	1.97	1.89	1.73	1.49	1.11	0
$g(x)$	0	-1.11	-1.49	-1.73	-1.89	-1.97	-2.00	-1.97	-1.89	-1.73	-1.49	-1.11	0

Fig. 1.51

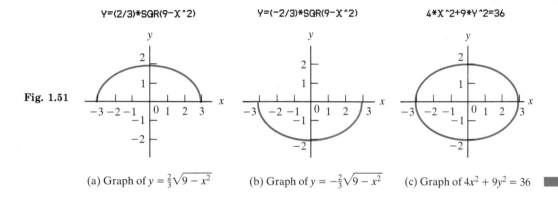

(a) Graph of $y = \frac{2}{3}\sqrt{9 - x^2}$ (b) Graph of $y = -\frac{2}{3}\sqrt{9 - x^2}$ (c) Graph of $4x^2 + 9y^2 = 36$

Hyperbolas

So far we have seen that the graph of an equation of the form

$$ax^2 + by^2 = c,$$

where a, b, and c are given nonzero numbers, is either a circle, if a, b, and c all have the same sign and $a = b$, or an ellipse, if a, b, and c all have the same sign and $a \neq b$.

Suppose *a and b have opposite signs.* Then the graph will be a *hyperbola.* The following examples illustrate such cases.

Example 5   Draw a graph of $x^2 - y^2 = 4$.

Solution   The coefficients of x^2 and y^2 have opposite signs, and so the graph is a hyperbola. Solving the given equation for y gives

$$y = f(x) = \sqrt{x^2 - 4} \quad \text{or} \quad y = g(x) = -\sqrt{x^2 - 4}. \tag{1.32}$$

All points (x, y) satisfying either of these two equations are on the graph of the given equation. Note that the domain of f and of g is

$$\mathcal{D} = \{x \mid x \leq -2 \quad \text{or} \quad x \geq 2\}.$$

The following table gives values of $f(x)$ and $g(x)$ for several values of x in set $\mathcal{D}$; $f(-x) = f(x)$ and $g(-x) = g(x)$ for all x in $\mathcal{D}$, and so each entry in the table corresponds to two points.

The graphs of $y = f(x)$ and $y = g(x)$ and the given equation are shown in Fig. 1.52. We see from the graphs that f and g are functions, and that the given equation does not describe a function in which x is the independent variable.

x	± 2.0	± 2.5	± 3.0	± 3.5	± 4.0	± 5.0	± 6.0	± 7.0	± 8.0
$f(x)$	0	1.50	2.24	2.87	3.46	4.58	5.66	6.71	7.75
$g(x)$	0	-1.50	-2.24	-2.87	-3.46	-4.58	-5.66	-6.71	-7.75

Fig. 1.52

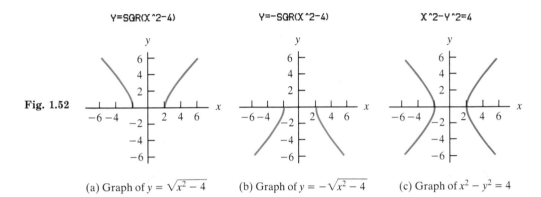

(a) Graph of $y = \sqrt{x^2 - 4}$ (b) Graph of $y = -\sqrt{x^2 - 4}$ (c) Graph of $x^2 - y^2 = 4$

Example 6   Draw a graph of $9y^2 - 4x^2 = 36$.

Solution   Solving for y gives

$$y = f(x) = \frac{2}{3}\sqrt{x^2 + 9} \quad \text{or} \quad y = g(x) = -\frac{2}{3}\sqrt{x^2 + 9}. \quad (1.33)$$

The graph of the given equation is a hyperbola consisting of all points (x, y) satisfying either of the equations in Eq. (1.33). The domain of f and of g is the set

x	0	± 0.5	± 1.0	± 1.5	± 2.0	± 2.5	± 3.0	± 4.0	± 5.0	± 6.0
$f(x)$	2.00	2.03	2.11	2.24	2.40	2.60	2.83	3.33	3.89	4.47
$g(x)$	-2.00	-2.03	-2.11	-2.24	-2.40	-2.60	-2.83	-3.33	-3.89	-4.47

Fig. 1.53

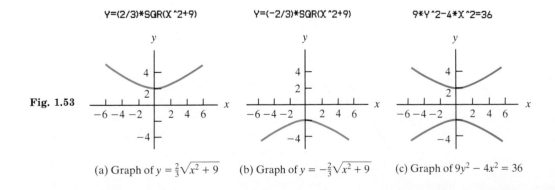

(a) Graph of $y = \frac{2}{3}\sqrt{x^2 + 9}$ (b) Graph of $y = -\frac{2}{3}\sqrt{x^2 + 9}$ (c) Graph of $9y^2 - 4x^2 = 36$

of real numbers **R**. Proceeding as in Example 5, we have the following table and the corresponding graphs shown in Fig. 1.53. Note from the graphs that f and g are functions, and that the given equation does not describe a function in which x is the independent variable.   ■

Degenerate Cases

Consider the equation $ax^2 + by^2 = c$ for the case where a and b are positive and c is negative or zero, and also the case in which a and b have opposite signs and $c = 0$. The following example illustrates these situations. Each represents what we call a degenerate case of the family of equations given $ax^2 + by^2 = c$.

Example 7   Give a graphical interpretation of each of the following equations.

    a) $x^2 + y^2 = 0$ **b)** $2x^2 + 3y^2 + 5 = 0$ **c)** $9x^2 - 4y^2 = 0$

Solution   **a)** There is only one pair of real numbers x, y that satisfy the given equation, namely, $x = 0$ and $y = 0$. Thus the graph is a single point, the origin.

    b) Since $2x^2 \geq 0$ and $3y^2 \geq 0$ for all real numbers x and y, then clearly there are no x, y pairs that satisfy the given equation. There is no graph associated with the given equation; this can be described by saying that the set $\{(x, y) | 2x^2 + 3y^2 + 5 = 0\}$ is the empty set ϕ.

    c) The given equation can be written as $(3x - 2y)(3x + 2y) = 0$. This is equivalent to

$$3x - 2y = 0 \quad \text{or} \quad 3x + 2y = 0,$$

and so the graph of the given equation consists of two lines, as shown in Fig. 1.54.

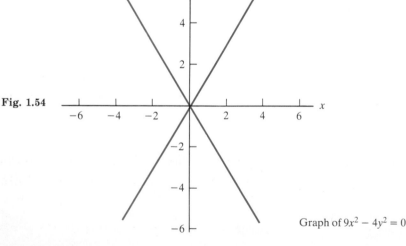

Fig. 1.54

Graph of $9x^2 - 4y^2 = 0$

■

Exercises 1.6

In each of the problems below, perform the following steps:

a) First identify the type of curve represented by the given equation.

b) Solve the given equation for y; if the result is not a function, then write it in terms of two functions, as illustrated in the examples of this section.

c) Draw graphs of the function or functions in (b), and then draw a graph of the given equation; in each case label the coordinates of the x- and y- intercept points.

If the given equation is a degenerate case, indicate that fact, and draw a graph if there is one.

1. $x^2 + y^2 = 9$	2. $9x^2 + 9y^2 = 16$	3. $x^2 + 4y^2 = 12$
4. $4x^2 + y^2 = 16$	5. $4x^2 - y = 0$	6. $3x^2 - y = 0$
7. $16x + 9y^2 = 0$	8. $9x + 4y^2 = 0$	9. $9x - y^2 = 0$
10. $16x - 9y^2 = 0$	11. $4x^2 + 4y^2 = 9$	12. $4x^2 + 9y^2 = 16$
13. $3x^2 + 3y^2 = 0$	14. $x^2 + 2y^2 + 1 = 0$	15. $x^2 - y^2 = 9$
16. $y^2 - x^2 = 4$	17. $4x^2 = 9y^2 + 36$	18. $4x^2 = 9y^2 - 36$
19. $4x^2 - 25y^2 = 0$	20. $3x^2 - 4y^2 = 0$	21. $9x^2 - 4y^2 = 36$
22. $x^2 - 4y = 0$	23. $x^2 + y^2 + 1 = 0$	24. $x^2 - y^2 + 1 = 0$

1.7 GRAPHS AND FUNCTION PROPERTIES

When you are drawing graphs of functions, it is useful to check for symmetry properties with respect to the y axis or the origin. Such properties will be described in terms of odd functions and even functions. Several examples are discussed in this section, in which graphs are used to get properties of functions that are not readily apparent from their algebraic descriptions.

Symmetry

Symmetry properties of graphs of functions are illustrated in the following examples.

Example 1   Draw a graph of the function h defined by $h(x) = x^4 - 4x^2$.

Solution   Let $y = x^4 - 4x^2$. Our first inclination is to make a table of x, y values that satisfy this equation. However, before doing that, observe the following:

$$h(-x) = (-x)^4 - 4(-x)^2 = x^4 - 4x^2 = h(x).$$

Thus $h(-x) = h(x)$ for each real number x. This tells us that if (a, b) is any point on the graph of h, then $(-a, b)$ is also on the graph. The two points (a, b) and $(-a, b)$ are symmetric about the y-axis, as illustrated in Fig. 1.55. Therefore the graph of $y = x^4 - 4x^2$ is symmetric with respect to the y-axis, and it is sufficient to draw the graph for points involving nonnegative values of x and then reflect this about the y-axis to get the remaining portion of the graph.

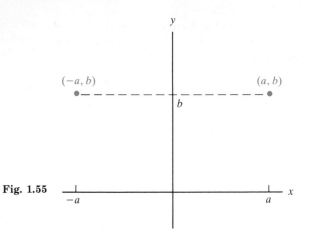

Fig. 1.55

Let us now proceed with the task of making a table of x, y values in which only nonnegative numbers for x are included.

x	0	0.25	0.50	0.75	1.00	1.25	1.50	1.75	2.00	2.25	2.50
y	0	-0.25	-0.94	-1.93	-3.00	-3.81	-3.94	-2.87	0	5.38	14.06

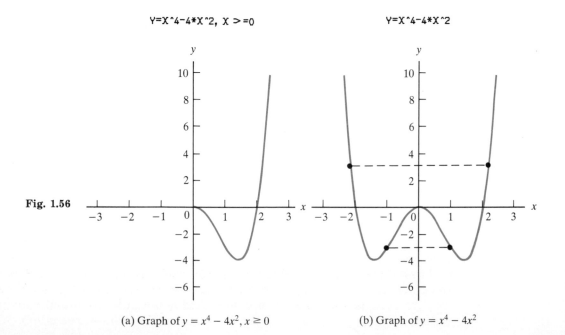

Fig. 1.56

(a) Graph of $y = x^4 - 4x^2, x \geq 0$ (b) Graph of $y = x^4 - 4x^2$

Plotting these points and drawing a curve through them gives the portion of the graph shown in Fig. 1.56(a). Reflecting this about the y-axis gives the remainder of the graph, as shown in Fig. 1.56(b); this reflection is illustrated by two pairs of points with dotted lines between them. ▄

Example 2   Draw a graph of the function g given by $g(x) = 4x - x^3$.

Solution   As in Example 1, let us first look at $g(-x)$:

$$g(-x) = 4(-x) - (-x)^3 = -4x + x^3 = -(4x - x^3) = -g(x).$$

Thus $g(-x) = -g(x)$ for each real number x; this tells us that if (a, b) is any point on the graph, then $(-a, -b)$ is also on the graph. Since the two points (a, b) and $(-a, -b)$ are symmetric about the origin (as illustrated in Fig. 1.57), the graph of g is symmetric with respect to the origin. Therefore it is sufficient to draw the portion for which the points have nonnegative values of x and then reflect this about the origin to get the remaining portion. Let us now make a table of x, y values including only nonnegative values of x.

x	0	0.25	0.50	0.75	1.00	1.25	1.50	1.75	2.00	2.25	2.50	2.75	3.00
y	0	0.98	1.88	2.58	3.00	3.05	2.63	1.64	0	-2.39	-5.63	-9.80	-15

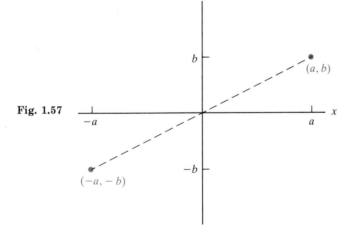

Fig. 1.57

These points are plotted and a graph is drawn, as shown in Fig. 1.58(a). Then the graph is completed by reflecting about the origin to get the curve shown in Fig. 1.58(b); this reflection is illustrated by two pairs of points with dotted lines between them.

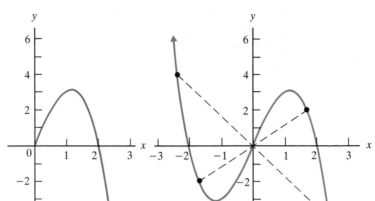

(a) Graph of $y = 4x - x^3$, $x \geq 0$ 　　　 (b) Graph of $y = 4x - x^3$

Fig. 1.58

Even and Odd Functions

In Example 1 we observed that the function $h(x) = x^4 - 4x^2$ has the property that $h(-x) = h(x)$ for every value of x. In Example 2 the function $g(x) = 4x - x^3$ has the property that $g(-x) = -g(x)$ for every value of x. The formula for h involves terms with even-number exponents only, and that for g has terms with odd-number exponents only. This suggests the following definition and terminology.

Definition 1.10　　Suppose f is a function with domain $\mathcal{D}(f)$ such that $x \in \mathcal{D}(f)$ implies $-x \in \mathcal{D}(f)$. Then

1. f is called an *even function* if $f(-x) = f(x)$ for each x in $\mathcal{D}(f)$;

2. f is called an *odd function* if $f(-x) = -f(x)$ for each x in $\mathcal{D}(f)$.

If f is an even function, the graph of $y = f(x)$ is symmetric with respect to the y-axis. If f is an odd function, the graph of $y = f(x)$ is symmetric with respect to the origin.

Example 3　　In each of the following determine whether the given function is even, odd, or neither.

a) $f(x) = \sqrt{1 - x^2}$ 　　　　　　　　　　**b)** $g(x) = x^2 + x$

c) $h(x) = \sqrt[3]{x} - 2x^3$ 　　　　　　　　**d)** $F(x) = |x| + x^2.$

Solution   **a)** $f(-x) = \sqrt{1 - (-x)^2} = \sqrt{1 - x^2} = f(x)$

Therefore $f(-x) = f(x)$ for each x in $\mathfrak{D}(f)$, and so f is an even function.

b) $g(-x) = (-x)^2 + (-x) = x^2 - x$. Since $x^2 - x$ is not equal to $g(x) = x^2 + x$ nor to $-g(x) = -x^2 - x$ for all x, then g is neither an even nor an odd function.

c) $h(-x) = \sqrt[3]{-x} - 2(-x)^3 = -\sqrt[3]{x} + 2x^3 - (\sqrt[3]{x} - 2x^3) = -h(x)$.

Therefore $h(-x) = -h(x)$ and h is an odd function.

d) $F(-x) = |-x| + (-x)^2 = |x| + x^2 = F(x)$. Thus $F(-x) = F(x)$, and so F is an even function.   ■

Example 4   Draw a graph of the function f given by $f(x) = x^2 - 2|x| - 8$.

Solution   First check to see if f is an even or odd function.

$$f(-x) = (-x)^2 - 2|-x| - 8 = x^2 - 2|x| - 8 = f(x).$$

Therefore f is an even function, and it is sufficient to draw the graph for $x \geq 0$ and then reflect that portion of the graph about the y-axis.

For $x \geq 0$, $|x| = x$ and so we want to draw the graph of $y = x^2 - 2x - 8$ for $x \geq 0$. This is part of a parabola that opens upward, and the lowest point is given by $x = -b/2a = 2/2 = 1$, $y = 1^2 - 2(1) - 8 = -9$. Thus the graph of $f(x) = x^2 - 2|x| - 8$ and $x \geq 0$ is shown in Fig. 1.59(a). This is reflected about the y-axis to get the graph of the given function, as shown in Fig. 1.59(b).

Fig. 1.59

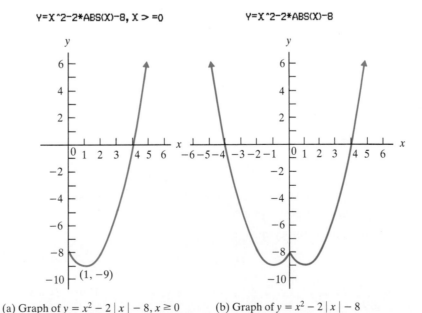

(a) Graph of $y = x^2 - 2|x| - 8$, $x \geq 0$ (b) Graph of $y = x^2 - 2|x| - 8$

Suggestions for Drawing Graphs

Throughout the remaining portion of this book graphs of functions will be used in problem solving. Although plotting points is an essential part of drawing graphs, we should be aware that there may be important features of a graph that are not easily seen from a set of isolated plotted points. Here we list some steps that frequently yield important information and in general should be followed in drawing a graph of a function given by $y = f(x)$.

1. Determine $\mathfrak{D}(f)$; this helps decide values of x to be included in a table.

2. Whenever possible find other equivalent formulas for $f(x)$ that may be more suitable for use in answering certain questions. Whenever the given formula is simplified, remember the result is valid only for x in $\mathfrak{D}(f)$.

3. Find intercept points when possible. The y-intercept is given by $[0, f(0)]$; the x-intercepts are determined by solving the equation $f(x) = 0$.

4. Determine symmetry properties if there are any. If $f(-x) = f(x)$ for all $x \in \mathfrak{D}(f)$, then the graph is symmetric with respect to the y-axis; if $f(-x) = -f(x)$, it is symmetric with respect to the origin.

5. Consider large values of x (when such are in $\mathfrak{D}(f)$) to determine features of a graph at extreme places. A convenient notation is "$x \to \infty$," which indicates "x becomes large" and is read "x approaches infinity."

The following example illustrates the use of these steps.

Example 5   Draw a graph of $y = f(x)$ given by

$$f(x) = \frac{x^2}{x^2 + 1}. \tag{1.34}$$

Solution   **a)** From Eq. (1.34) note that $\mathfrak{D}(f) = \mathbf{R}$. Also observe that $x^2 + 1 > x^2$ for any x, and so $x^2/(x^2 + 1) < 1$. Hence $0 \leq y < 1$ for any x.

b) Although it appears that Eq. (1.34) cannot be simplified, we can write $f(x)$ in equivalent form as follows (divide $x^2 + 1$ into x^2):

$$f(x) = 1 - \frac{1}{x^2 + 1}. \tag{1.35}$$

This equation is used in (e) below.

c) *y-intercept:* for $x = 0$, $y = 0/(0^2 + 1) = 0$, and so $(0, 0)$ is the y-intercept point;

x-intercept: if $y = 0$, then $x^2/(x^2 + 1) = 0$, and so $x = 0$. Thus $(0, 0)$ is the only x-intercept point.

d) Check for symmetry:

$$f(-x) = \frac{(-x)^2}{(-x)^2 + 1} = \frac{x^2}{x^2 + 1} = f(x).$$

Thus $f(-x) = f(x)$ for all $x \in \mathfrak{D}(f)$, and so f is an even function and its graph is symmetric with respect to the y-axis. Hence the table below includes only nonnegative values of x.

e) Using Eq. (1.35), we see that as $x \to \infty, f(x) \to 1$; also as $x \to -\infty, f(x) \to 1$. This tells us that as x becomes large positively or negatively, y becomes close to 1.

The considerations above essentially give the outstanding features of the desired graph. However, it is a simple matter to make a table of x, y values that will allow us to draw a more accurate graph.

x	0	0.5	1.0	1.5	2.0	2.5	3.0	3.5	4.0	5	8
y	0	0.20	0.50	0.69	0.80	0.86	0.90	0.92	0.94	0.96	0.98

Using all of the information above, we can now plot a reasonably accurate graph of the given function. This is shown in Fig. 1.60.

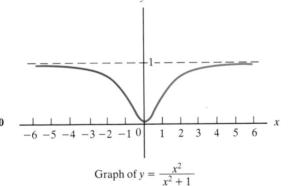

Y=X^2/(X^2+1)

Fig. 1.60

Graph of $y = \dfrac{x^2}{x^2 + 1}$

Example 6   Draw a graph of the relation given by the equation

$$2x^2 - \sqrt{3}|x|y + y^2 = 20. \tag{1.36}$$

Solution   First note that if (x_1, y_1) is any pair satisfying Eq. (1.36), then $(-x_1, y_1)$ also satisfies the equation. Thus the curve is symmetric about the y-axis, and so we can draw the curve for $x \geq 0$ and reflect that portion about the y-axis.

The given equation can be considered as a quadratic equation in y, and using the quadratic formula, we can solve for y as follows:

$$y^2 - (\sqrt{3}x)y + (2x^2 - 20) = 0, \quad x \geq 0,$$

$$y = \frac{\sqrt{3}x + \sqrt{80 - 5x^2}}{2}, \tag{1.37}$$

or

$$y = \frac{\sqrt{3}x - \sqrt{80 - 5x^2}}{2}. \tag{1.38}$$

Using a calculator and several values of x in $0 \le x \le 4$, we can determine points (x, y) that satisfy Eq. (1.37). Plotting these points gives the portion of the curve from A to B shown in Fig. 1.61. Similarly, using Eq. (1.38), we get the points of the curve from B to C. Reflecting about the y-axis gives the February 14 curve shown in Fig. 1.61.

2*X^2−SQR(3)*ABS(X)*Y+Y^2=20

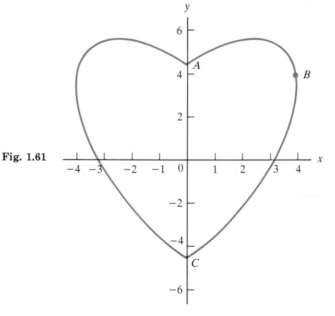

Fig. 1.61

Graph of $2x^2 - \sqrt{3}\,|x|\,y + y^2 = 20$

Properties of Functions from Graphs

Here we are interested in developing those ideas that are necessary for the introduction of inverse functions, which will be discussed in the next section. Graphs will be used to help us determine the following.

1. The range of a function
2. Increasing functions, decreasing functions
3. One-to-one functions

Range of a function   The range of a function f, denoted by $\Re(f)$, is the set of all second components of the ordered pairs that describe f. That is, if $f = \{(x, y) | y = f(x)\}$, then $\Re(f) = \{y | (x, y) \in f\}$. In geometrical terms,

$\mathcal{R}(f)$ is the set of all second coordinates (the y values) of the points that are on the graph of f.

In many cases, drawing a graph of a given function will aid in determining its range, as illustrated in the following examples.

Example 7   Find the range of the function f defined by $f(x) = x^2 + 1$.

Solution   First draw a graph of $y = x^2 + 1$, as shown in Fig. 1.62. The graph should make it clear that the y values that occur for points on the curve are those for which $y \geq 1$. Thus

$$\mathcal{R}(f) = \{y | y \geq 1\}.$$

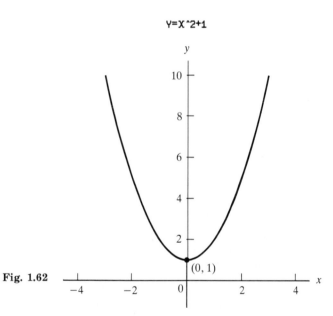

Y=X^2+1

Fig. 1.62

Range of $y = x^2 + 1$

Example 8   Determine the range of the function g given by $g(x) = -2x^2 + 4x - 3$.

Solution   This is an example of a quadratic function (see Section 1.4). The graph of $y = g(x)$ is a parabola that opens downward (since the coefficient of x^2 is negative), and the highest point is given by

$$x = -\frac{b}{2a} = -\frac{4}{2(-2)} = 1.$$

The corresponding value of y is $y = -2(1)^2 + 4(1) - 3 = -1$.

The graph of $y = -2x^2 + 4x - 3$ is shown in Fig. 1.63. From this graph we conclude that $\Re(g) = \{y | y \leq -1\}$.

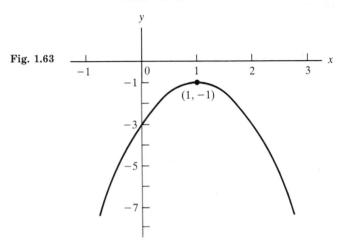

Fig. 1.63

Y=-2*X^2+4*X-3

(1, −1)

Range of $y = -2x^2 + 4x - 3$

Example 9 Find the range of the function $f(x) = \dfrac{x^2}{1 + x^2}$.

Solution The graph of this function is shown in Fig. 1.60 (Example 5). Looking at the graph, we conclude that the set of y values for points on the curve is $\{y | 0 \leq y < 1\}$. Therefore $\Re(f) = \{y | 0 \leq y < 1\}$.

Increasing, decreasing functions A function f given by $y = f(x)$ is said to be an *increasing function* if the values of y increase as x increases; similarly, f is a *decreasing* function if the values of y decrease as x increases. We state this in the following definition.

Definition 1.11 Suppose f is a function and b and c are any two numbers in $\mathfrak{D}(f)$ with $b < c$. Then f is an *increasing function* if $f(b) < f(c)$, and f is a *decreasing function* if $f(b) > f(c)$.

In geometrical terms, Definition 1.11 tells us that as we move from left to right on the graph of the function, it is increasing if the graph always rises and it is decreasing if the graph always falls. These ideas are illustrated in Fig. 1.64; in (a) the function f is increasing, and in (b) the function g is decreasing.

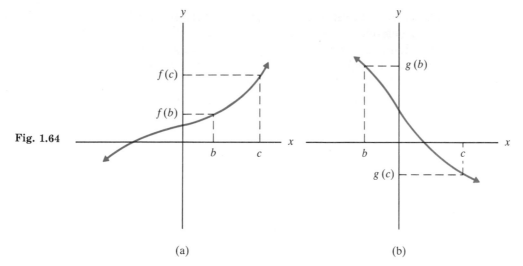

Fig. 1.64

(a)   (b)

   In general, to determine whether a function is increasing, decreasing, or neither, we shall draw a graph and rely on intuition rather than give an analytical proof to justify conclusions, as illustrated in the following example.

Example 10   In each of the following determine whether the function is increasing, decreasing, or neither.

   a) $f(x) = 2x - 1$ **b)** $g(x) = -x^3$ **c)** $h(x) = x^2 - 2x - 3$

Solution   Graphs of the given functions are shown in Fig. 1.65. Using them, we arrive at the following conclusions:

   a) f is an increasing function;   **b)** g is a decreasing function;
   c) h is neither an increasing nor a decreasing function.

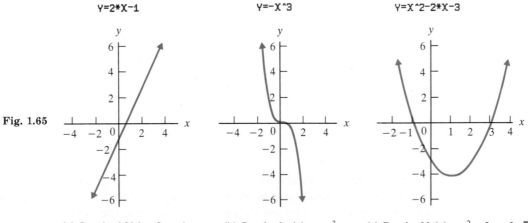

Fig. 1.65

(a) Graph of $f(x) = 2x - 1$ (b) Graph of $g(x) = -x^3$ (c) Graph of $h(x) = x^2 - 2x - 3$

One-to-one functions As indicated throughout this chapter, a function f can be considered as a mapping between two sets, the domain and range, such that each member of $\mathfrak{D}(f)$ is mapped into *exactly one* member of $\mathfrak{R}(f)$. If this mapping (or correspondence) has the additional property that no two distinct elements of $\mathfrak{D}(f)$ map into the same element of $\mathfrak{R}(f)$, then f is said to be a one-to-one function. This is stated formally in the following definition.

Definition 1.12   A function f is said to be a *one-to-one function* if every element of $\mathfrak{R}(f)$ is the image of *exactly one* element of $\mathfrak{D}(f)$; that is, no two ordered pairs of f have the same second component.

In general, we shall use a graph to determine whether or not a given function is one-to-one. Thus it will be helpful to have the following geometrical characterization of when a relation is a one-to-one function.

Suppose g is a relation (a set or ordered pairs of real numbers). If each vertical line intersects the graph of g in at most one point (so there are no ordered pairs with the same first elements), then g is a function. If, in addition, each horizontal line intersects the graph of g in at most one point (so there are no ordered pairs with the same second elements), then g is a one-to-one function.

Applying this criterion to the graphs shown in Fig. 1.66, we can conclude that the functions corresponding to the graphs of (a), (b), and (d) are one-to-one, but that of (c) is not.

Fig. 1.66

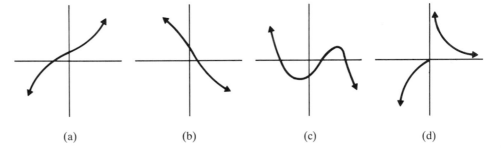

(a)   (b)   (c)   (d)

In Fig. 1.66 we observe that the function corresponding to the graph in (a) is increasing and one-to-one; similarly, in (b) the function is decreasing and one-to-one. Geometric intuition suggests the following theorem.

Theorem 1.1   **a)** If f is an increasing function, then it is one-to-one.

                 b) If f is a decreasing function, then it is one-to-one.

      Note that Theorem 1.1 gives sufficient conditions for a function to be one-to-one; it does not state that if a function is one-to-one, then it is an increasing or a decreasing function. For instance, the function corresponding to Fig. 1.66(d) is one-to-one, but it is neither increasing nor decreasing.

Example 11   In each of the following, determine whether or not the given function is one-to-one.

    a) $f(x) = x^2$ **b)** $g(x) = \sqrt{x}$ **c)** $h(x) = -\sqrt{x}$ **d)** $F(x) = \sqrt{-x}$

Solution   The graphs of the given functions are shown in Fig. 1.67 (see pages 89 and 90).

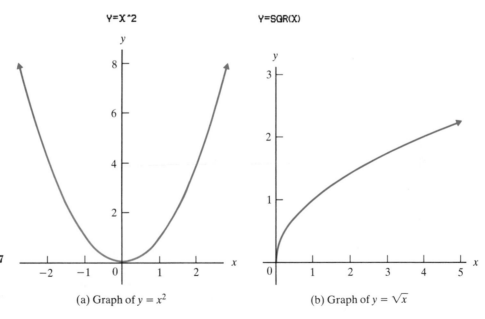

Fig. 1.67

(a) Graph of $y = x^2$ (b) Graph of $y = \sqrt{x}$

    a) From the graph of $y = x^2$ it is clear that f is not one-to-one.

    b) Looking at the graph of $y = \sqrt{x}$, we see that g is a one-to-one function.

    c) The graph of $y = -\sqrt{x}$ shows that h is a one-to-one function.

    d) The domain of F is given by $\mathcal{D}(F) = \{x | x \leq 0\}$, and the graph is shown in Fig. 1.67(d). We conclude that F is a one-to-one function.

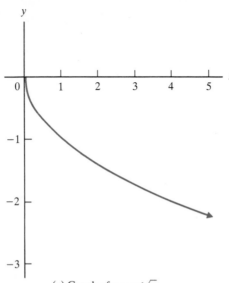

Y=-SQR(X)

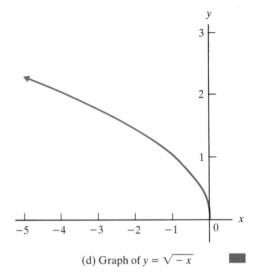

Y=SQR(-X)

Fig. 1.67

(c) Graph of $y = -\sqrt{x}$ (d) Graph of $y = \sqrt{-x}$

Exercises 1.7

In problems 1 through 14, determine whether the given function is an even function, an odd function, or neither.

1. $f(x) = 1 - 3x^2$ **2.** $g(x) = x^3 - x$ **3.** $g(x) = x^3 + 4x$

4. $f(x) = \sqrt{1 - x^2}$ **5.** $f(x) = x^2 - 3x + 5$ **6.** $g(x) = x - 3x^2$

7. $h(x) = \sqrt{x^2}$ **8.** $h(x) = \sqrt[3]{x^3 - 2x}$ **9.** $f(x) = \dfrac{1}{x}$

10. $g(x)\,\dfrac{x}{1 + x^2}$ **11.** $g(x) = x^2 + |x|$ **12.** $f(x) = x + |x|$

13. $f(x) = (x - 1)^2 + 2x$ **14.** $f(x) = (2x + 1)^2 - 4x$

In problem 15 and 26, draw a graph of the given functions. In each case determine if the graph is symmetric with respect to the y-axis or the origin, and use this information in drawing the curve.

15. $f(x) = 4 - 2x$ **16.** $f(x) = 3x + 6$ **17.** $f(x) = x^3 - x$

18. $f(x) = 3 - x^2$ **19.** $g(x) = \dfrac{x}{1 + x^2}$ **20.** $g(x) = x \cdot |x|$

21. $f(x) = (x^2 - 4) \cdot |x|$ **22.** $f(x) = 4x - x^3$ **23.** $f(x) = |x + 3| - x$

24. $g(x) = \dfrac{x^2 + x}{x + 1}$ **25.** $f(x) = \dfrac{x^2 - 2x}{x - 2}$ **26.** $f(x) = x^2 + 2|x| + 1$

In problems 27 through 38, determine the domain and the range of given functions. If necessary, draw a graph of the function and use it to get your answer.

27. $f(x) = 1 + 3x$ **28.** $g(x) = 1 - 2x$ **29.** $g(x) = 1.53 - 2.47x$

30. $f(x) = 1.27 + 3.56x$ **31.** $g(x) = x^2 - 4$ **32.** $f(x) = x^2 + 3x - 4$

33. $f(x) = -\sqrt{-x}$ **34.** $f(x) = \dfrac{1}{1 + x^2}$ **35.** $f(x) = \sqrt{4 - |x|}$

36. $f(x) = \dfrac{x^2 - x}{x - 1}$ **37.** $f(x) = \dfrac{x^2 + 2x}{x + 2}$ **38.** $f(x) = \sqrt{1 - x}$

In each of the problems 39 through 50, determine whether the given function is

a) increasing, decreasing, or neither,   **b)** one-to-one.

39. $f = \{(-2, 0), (0, 1), (2, 3), (4, 4)\}$ **40.** $f = \left\{(0, 1), \left(1, \dfrac{\sqrt{6} - \sqrt{2}}{4}\right), \left(2, \dfrac{\sqrt{2} - \sqrt{3}}{2}\right)\right\}$

41. $g(x) = 5 - 3x$ **42.** $g(x) = 2x - 5$ **43.** $f(x) = \sqrt{x + 3}$

44. $f(x) = -\sqrt{x - 1}$ **45.** $f(x) = 4 - x - 3x^2$ **46.** $g(x) = 3 + x - 2x^2$

47. $g = \{(x, y) \mid y = x^2 + 4x - 1 \text{ and } x \geq -2\}$ **48.** $f = \{(x, y) \mid y = -x^2 - 2x \text{ and } x < 2\}$

49. $f(x) = \begin{cases} -x + 1 & \text{if } 0 \leq x \leq 1 \\ -x - 1 & \text{if } -1 \leq x < 0 \end{cases}$ **50.** $f(x) = \begin{cases} x^2 + 2 & \text{if } x \geq 0 \\ x & \text{if } x < 0 \end{cases}$

51. Draw a graph of the relation given by the equation $2x^2 - 2|x|y + y^2 = 16$. (See Example 6.)

1.8   **INVERSE FUNCTIONS**

In the preceding section we introduced the idea of a function f being one-to-one; that is, each element of $\mathcal{D}(f)$ is mapped into exactly one element of $\mathcal{R}(f)$, and no two elements of $\mathcal{D}(f)$ are mapped into the same element. This suggests that if the first and second components of each pair of f are interchanged, the resulting set of ordered pairs will also be a function. Such a function will be called the inverse of f and denoted by f^{-1}. Thus we have the following definition.

Definition 1.13   If f is a one-to-one function, then the set of ordered pairs obtained by interchanging the first and second components of each pair in f is called the *inverse function of f* and is denoted by f^{-1}.*

The following examples illustrate Definition 1.13.

Example 1   Suppose function f is given by $f = \{(1, 3), (2, 5), (3, 7)\}$. Find f^{-1} and state the domain and range of f^{-1}.

---

*   The symbol f^{-1} is used here to denote a function that is related to f; it does not imply that -1 is a negative exponent.

Solution   First note that f is a one-to-one function, and so it has an inverse function. Interchanging the first and second components of each ordered pair in f gives

$$f^{-1} = \{(3, 1), (5, 2), (7, 3)\}.$$
$$\mathfrak{D}(f^{-1}) = \{3, 5, 7\}, \qquad \mathfrak{R}(f^{-1}) = \{1, 2, 3\}.$$
■

Example 2   Suppose f is a function given by the formula $f(x) = 2x - 1$.

a) Show that f is a one-to-one function.

b) Determine f^{-1}.

Solution   **a)** The graph of $y = 2x - 1$ is a line, shown in Fig. 1.68. Geometric intuition tells us that each vertical line intersects the graph at one point, and each horizontal line intersects the graph at one point. Therefore f is a one-to-one function.

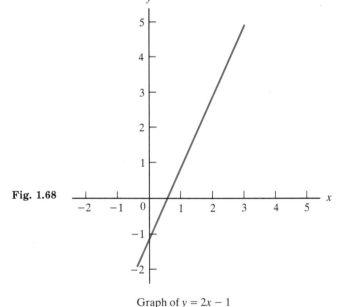

Fig. 1.68

Graph of $y = 2x - 1$

b) The rule of correspondence that gives ordered pairs (x, y) for f is given by

$$f: y = 2x - 1 \quad \text{or} \quad f = \{(x, y)|y = 2x - 1\}. \tag{1.39}$$

In order to get the rule of correspondence that gives the (x, y) ordered pairs for f^{-1}, we can interchange the roles of x and y in Eq. (1.39) and then solve for y in terms of x. Hence,

$$f^{-1}: x = 2y - 1 \quad \text{or} \quad f^{-1}: y = \frac{x + 1}{2}.$$

Therefore

$$f^{-1}(x) = \frac{x+1}{2} \quad \text{or} \quad f^{-1} = \left\{(x,y)\,|\,y = \frac{x+1}{2}\right\} \qquad (1.40)$$

To illustrate the interchange of components, note that the ordered pair $(3, 5)$ is in f (as seen from Eq. (1.39)), and $(5, 3)$ is in f^{-1} (as seen from Eq. (1.40)).

Example 3   The area A of a circle of radius r is given by the formula

$$A = \pi r^2. \qquad (1.41)$$

This defines a function g as the set of ordered pairs

$$g = \{(r, A)\,|\,A = \pi r^2\}.$$

a) Show that g is a one-to-one function.

b) Find a formula for g^{-1}.

Solution   **a)** The fact that r represents the radius of a circle implies that $r > 0$. Thus $\mathcal{D}(g) = \{r\,|\,r > 0\}$. The graph of g is shown in Fig. 1.69; from it we conclude that g is a one-to-one function.

A=(PI)*R^2

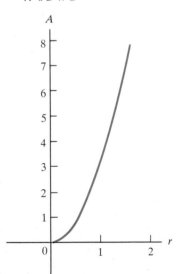

Fig. 1.69

Graph of $A = \pi r^2$

b) The function given by Eq. (1.41) is used in situations in which the radius of a circle is given and we want to find the corresponding area. However, there are situations in which the area of a circle is known and we wish to find the corresponding radius. This will be given by the inverse of g. Thus, in considering g^{-1}, we shall think of the set of ordered pairs (A, r), where A and r are related by

$$g^{-1} = \{(A, r) | A = \pi r^2\}. \tag{1.42}$$

This can be written as

$$g^{-1} = \left\{ (A, r) | r = \sqrt{\frac{A}{\pi}} \right\}.$$

Hence a formula for g^{-1} is $g^{-1}(A) = \sqrt{\dfrac{A}{\pi}}$. ■

It is worth taking a closer look at the two techniques used in finding the inverse functions in Examples 2 and 3. In Example 2, f^{-1} was determined as a set of (x, y) ordered pairs by interchanging the roles of x and y in the equation defining f (Eq. (1.39)) and then solving for y in terms of x. In Example 3, g was given as a set of (r, A) ordered pairs and we found g^{-1} as a set of (A, r) ordered pairs by simply solving the equation defining g (Eq. (1.41)) for r in terms of A. Thus, in Example 2 the interchange of the components of the ordered pairs in f was achieved by leaving the names of the ordered pairs as is, that is (x, y), and interchanging the roles of the variables in the defining equation of f. In Example 3 the interchange of components was accomplished by interchanging the ordered pair variables.

In most cases in which variables x and y are used to describe a one-to-one function, we shall consider x the independent variable (the first component of the ordered pairs) in both f and f^{-1}; thus we shall follow the procedure illustrated in Example 2 to determine f^{-1}. However, when a function occurs in applications, it is preferable to follow the pattern of Example 3.

Properties of Inverse Functions
The following properties follow immediately from Definition 1.13. Suppose f is a one-to-one function; then

a) $\mathcal{D}(f^{-1}) = \mathcal{R}(f)$ and $\mathcal{R}(f^{-1}) = \mathcal{D}(f)$
b) $(f^{-1})^{-1} = f$
c) $(f^{-1} \circ f)(x) = x$ for any $x \in \mathcal{D}(f)$ and $(f \circ f^{-1})(x) = x$ for any $x \in \mathcal{D}(f^{-1})$

Example 4   Suppose function f is given by

$$f(x) = -\sqrt{x}. \tag{1.43}$$

a) Show that f is a one-to-one function.
b) Find f^{-1} and state its domain and range.

Solution   **a)** The graph of

$$y = -\sqrt{x} \tag{1.44}$$

is shown in Fig. 1.70; from it we can see that f is a one-to-one function. Also $\mathcal{D}(f) = \{x | x \geq 0\}$ and $\mathcal{R}(f) = \{y | y \leq 0\}$.

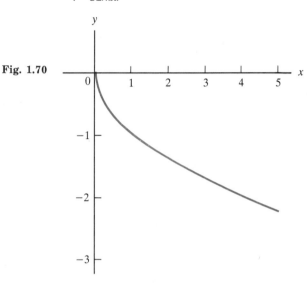

Y=-SQR(X)

Fig. 1.70

Graph of $y = -\sqrt{x}$

b) To find f^{-1} we can interchange the roles of x and y in Eq. (1.44) and then solve for y as follows:

$$x = -\sqrt{y}, \qquad x^2 = y.$$

Therefore $y = f^{-1}(x) = x^2$ where

$$\mathcal{D}(f^{-1}) = \mathcal{R}(f) = \{x | x \leq 0\} \qquad \text{and} \qquad \mathcal{R}(f^{-1}) = \mathcal{D}(f) = \{y | y \geq 0\}. \qquad \blacksquare$$

Example 5   Suppose function f is given by $f(x) = 2x - 1$. Show that $(f^{-1} \circ f)(x) = x$ and $(f \circ f^{-1})(x) = x$.

Solution   In Example 2 we found that $f^{-1}(x) = \dfrac{x+1}{2}$. Hence

$$(f^{-1} \circ f)(x) = f^{-1}(f(x)) = f^{-1}(2x - 1) = \frac{(2x - 1) + 1}{2} = x,$$

$$(f \circ f^{-1})(x) = f(f^{-1}(x)) = f\left(\frac{x+1}{2}\right) = 2\left(\frac{x+1}{2}\right) - 1 = x. \qquad \blacksquare$$

Graphs of Inverse Functions

Suppose f is a one-to-one function. Here we illustrate how the graph of $y = f^{-1}(x)$ is related to the graph of $y = f(x)$. First note the following: Points (a, b) and (b, a) are symmetric about the line $y = x$, as illustrated in Fig. 1.71. This suggests that the graph of $y = f^{-1}(x)$ can be obtained by reflecting the graph of $y = f(x)$ about the line $y = x$; this is indeed so.

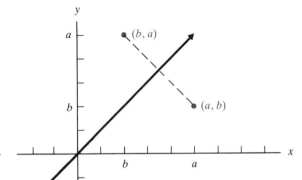

Fig. 1.71

Example 6   On the same system of coordinates draw graphs of f and f^{-1}, where f is given by $f = \{(-1, -2), (2, -1), (3, 4)\}$.

Solution   Note that f^{-1} is given by $f^{-1} = \{(-2, -1), (-1, 2), (4, 3)\}$. The graph of f consists of three points, shown as solid circular dots in Fig. 1.72. The graph of f^{-1} consists of the three points shown as open circular dots. Each point of f^{-1} is a reflection about the line $y = x$ of a corresponding point of f.

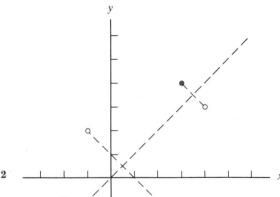

Fig. 1.72

Example 7   Suppose function g is given by $g(x) = x^2$, where $\mathfrak{D}(g) = \{x|x \le 0\}$. Draw graphs of $y = g(x)$ and $y = g^{-1}(x)$ in the same system of coordinates.

Solution   The graph of $y = g(x) = x^2$ is a half-parabola, as shown by the solid black curve in Fig. 1.73. The inverse function is given by solving $x = y^2$ for y and remembering that we want $y \le 0$ (since $x \le 0$ in $g(x)$). Thus $y = -\sqrt{x}$, and so $g^{-1}(x) = -\sqrt{x}$. Reflecting the graph of $y = g(x)$ about the line $y = x$ gives the graph of $y = g^{-1}(x)$, as shown by the colored curve in Fig. 1.73. This is the graph of $y = -\sqrt{x}$, $x \ge 0$.

Y=INV(G(X))

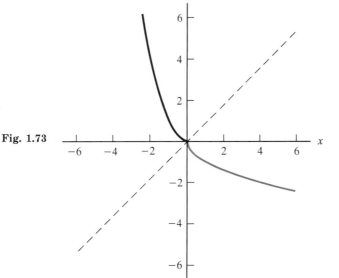

Fig. 1.73

Exercises 1.8

In problems 1 through 4, a function f is given as a set of ordered pairs. First convince yourself that f is a one-to-one function; then give f^{-1} as a set of ordered pairs.

1. $f = \{(-2, 0), (0, 1), (2, 3), (4, 4)\}$ **2.** $f = \{(1, 2), (3, 4), (4, 5), (5, 3)\}$

3. $f = \{(1, 3), (2, 4), (3, \sqrt{5}), (4, \sqrt{6})\}$ **4.** $f = \{(0, 1), (-1, \sqrt{2}), (-2, \sqrt{3}), (-3, 2)\}$

In problems 5 through 24,

a)   determine whether or not the given function f is one-to-one; if it is, then

b)   find a formula giving the inverse function as $y = f^{-1}(x)$, and state the domain of f^{-1}.

5. $f(x) = 3 - 2x$ **6.** $f(x) = 1 - 2x$ **7.** $f(x) = x^2 + 1$ **8.** $f(x) = \sqrt{x}$

9. $f(x) = \sqrt{-x}$ **10.** $f(x) = 4 - x^2$ **11.** $f(x) = \sqrt{x - 1}$ **12.** $f(x) = \sqrt{x + 1}$

13. $f(x) = x^3$ **14.** $f(x) = -x^3$ **15.** $f(x) = |x| - 2x$ **16.** $f(x) = |x| - x + 1$

17. $f(x) = \dfrac{1}{x^2 + 1}$ and $x \ge 0$ **18.** $f(x) = x^2 - 1$ and $x \ge 0$

19. $f(x) = \dfrac{1}{x}$ and $x \geq 1$ **20.** $f(x) = \dfrac{1}{x^2}$ and $x > 0$

21. $f(x) = x^2 - 2x$ and $x \geq 2$ **22.** $f(x) = x^2 - 2x$ and $x \leq 0$

23. $f(x) = x^2 - 2x + 1$ and $x \geq 1$ **24.** $f(x) = x^2 - 2x + 1$ and $x \leq 1$

In problems 25 through 30, a one-to-one function f is given. Find f^{-1}; then show that $(f^{-1} \circ f)(x) = x$ for $x \in \mathfrak{D}(f)$ and $(f \circ f^{-1})(x) = x$ for $x \in \mathfrak{D}(f^{-1})$.

25. $f(x) = 4 - 2x$ **26.** $f(x) = 2x + 3$ **27.** $f(x) = x^2 - 1$ and $x \geq 0$

28. $f(x) = \sqrt{x + 2}$ **29.** $f(x) = \sqrt{1 - x}$ **30.** $f(x) = \dfrac{1}{x}$ and $x > 0$

In problems 31 through 36, a one-to-one function is given. On the same system of coordinates draw graphs of the function and the corresponding inverse function.

31. $f(x) = 2x$ **32.** $f(x) = 2x + 3$ **33.** $f(x) = x^2$, $x \geq 1$

34. $f(x) = x^2$, $x \leq -1$ **35.** $f(x) = \sqrt{x - 1}$ **36.** $f(x) = \sqrt{1 - x}$, $x \leq 0$

37. The volume V of a sphere of radius r is given by the formula

$$V = \frac{4}{3}\pi r^3.$$

This defines V as a function f of r. Find a formula for f^{-1} that expresses r as a function of V; then use the result to complete the following table. Give results rounded off to two decimal places.

V	1	2	3.72	5.64
r				

38. A ball is dropped from the top of a building 32 meters high. The position of the ball at any time t seconds after it has been dropped is given by the formula $s = 4.9t^2$, where s is the distance in meters from the top of the building. Thus s is a function f of t. Find

a) $\mathfrak{D}(f)$,   **b)** a formula for f^{-1} in terms of $t = f^{-1}(s)$;

   c) use the result to determine the time it takes for the ball to fall 4 m, 12 m, 23.4 m. Give results rounded off to two decimal places.

39. A ball is thrown vertically upward from the ground. Its position t seconds after it has been thrown is given by $s = 39.2t - 4.9t^2$, where s is the distance in meters from the ground. Suppose we restrict our discussion to the time interval during which the ball is going up; then the formula defines a one-to-one function for $s = f(t)$.

   a) Find $D(f)$. (*Hint:* Recall the discussion related to "highest point" in Section 1.4.)

   b) Determine a formula that gives $t = f^{-1}(s)$, and use it to find how many seconds it takes for the ball to reach a distance 10 m, 48 m, 60 m, 75 m. Give results rounded off to two decimal places.

40. Suppose rectangle $DEFG$ is inscribed in an isosceles triangle ABC, as shown in Fig. 1.74, where $\overline{AC} = \overline{BC} = 3$ and $AB = 2$. Let $\overline{DE} = x$.

a) Find a formula that gives the area K of the rectangle as a function of x.

b) Give the domain and range of f.

c) Is f a one-to-one function?

d) Find the value(s) of x, rounded off to two decimal places, that will give a rectangle of area 0.5.

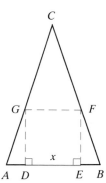

Fig. 1.74

Review Exercises

In the following problems give numerical answers in exact form unless otherwise specified.

1. Does the set $g = \{(-1, 2),\ (0, 4),\ (1, 6),\ (2, 8)\}$ determine a function? State the domain and range of g.

2. Suppose set $\mathfrak{D}$ is given by $\mathfrak{D} = \{-3,\ -1,\ 1,\ 3\}$ and the set of ordered pairs f is
$$f = \{(x, y)\,|\,x \in \mathfrak{D} \text{ and } y \text{ is 2 less than } x\}.$$
Write out the set of ordered pairs for f. Is f a function?

In problems 3 through 16, functions f, g, and h are given by $f(x) = 2x - 1$, $g(x) = \sqrt{1 - x^2}$, and $h(x) = |x + 1| - x$.

3. Give the domains of f, g, and h.

4. Determine a formula for $(f + g)(x)$ and give the domain of $f + g$.

5. Determine a formula for $(g \circ f)(x)$ and give the domain of $g \circ f$.

6. Give a formula for $(h \circ f)(x)$ and state the domain of $h \circ f$.

7. Give a formula for $(f/g)(x)$ and state the domain of f/g.

8. Express $h(x)$ in terms of formulas that do not involve absolute value.

In problems 9 through 12, evaluate the given expressions.

9. a) $(f + g)(0)$ **b)** $(f/h)(-1)$

10. a) $(h - f)(2)$ **b)** $(f \cdot h)(-3)$

11. a) $(f \circ h)(-2)$ **b)** $(g \circ f)(-1)$

12. a) $(h \circ f)(0)$ **b)** $(g \circ h)(-1)$

In problems 13 through 16, evaluate the given expressions and state results rounded off to three decimal places.

13. $g(0.483)$ **14.** $(f/g)(0.287)$ **15.** $(g \circ f)(0.372)$ **16.** $(h/g)(0.631)$

In problems 17 through 20, solve the given equations, where functions f and g are given by $f(x) = 2x - 3$, $g(x) = x^2 - 2x - 4$.

17. $(f + g)(x) + 6 = 0$ **18.** $(g/f)(x) + 4 = 0$

19. $(g \circ f)(x) + 1 = 0$ **20.** $(f \circ g)(x) + 11 = 0$

21. Find an equation of the line that passes through points $(-1, 2)$ and $(1, -3)$.

22. Determine the slope of the line given by $3x - 6y + 5 = 0$.

23. Find an equation of the line passing through $(1, 4)$ and perpendicular to the line $2x + y = 3$.

24. Find an equation of the line passing through $(-2, 1)$ and parallel to the line $2x - 3y = 5$.

25. Is there a line that passes through the three points $(-1, 3)$, $(3, 2)$, and $(7, 1)$? If there is, find its equation.

26. Is the point $(2, -3)$ on the curve given by $x^2 - 3x - 2y + 1 = 0$?

27. Suppose function f is given by $f(x) = \sqrt{3 - x}$. Is f an increasing function? Is f a one-to-one function?

28. Does the curve represented by the equation $y = x^3 - 2\sqrt{x^2} + x$ pass through the point $(-1, -4)$?

In problems 29 through 32, draw graphs of the given equations. In each case, label the coordinates of the x- and y-intercept points.

29. $2x - y = 4$ **30.** $3x + 2y = 6$

31. $y = x^2 - 4x + 3$ **32.** $y = x^2 - 4x + 4$

In problems 33 through 36, determine the coordinates of (a) the x-intercept, (b) the y-intercept, and (c) the highest or lowest points.

33. $y = 3x^2 - 6x + 3$ **34.** $y = 2x^2 - 3x + 1$

35. $y = -2x^2 + 5x - 3$ **36.** $y = -x^2 + 3x + 10$

In problems 37 through 44, find the solution set for the given equations or inequalities.

37. $x^2 + 3x - 1 = 0$ **38.** $x^2 + 4x - 5 = 0$

39. $|x + 3| + 1 = 5$ **40.** $|2x + 3| + x = 0$

41. $x^2 - 3x - 4 > 0$ **42.** $2x^2 + x - 3 < 0$

43. $|x - 3| + x \leq 0$ **44.** $x + |x - 1| \leq 3$

In problems 45 through 52, sketch the curves represented by the given equations.

45. $4x^2 + 9y^2 = 36$ **46.** $x^2 - y^2 = 4$

47. $x^2 - 3x - y - 4 = 0$ **48.** $y = \sqrt{4 - x^2}$

49. $y = -\sqrt{9 - x^2}$ **50.** $y = \dfrac{x^2 - 1}{x - 1}$

51. $y = \dfrac{4 - x^2}{x + 2}$

52. $y = |x - 2| - x$

In problems 53 through 60, determine whether the given function is one-to-one. If it is, find the corresponding inverse. In each case specify the domain and range of the inverse function.

53. $f(x) = 5 - x$

54. $f(x) = \sqrt{x + 2}$

55. $f(x) = -\sqrt{-x}$

56. $f(x) = x^2 - 4$

57. $f(x) = x^2 - 2x + 1$ and $x \geq 2$

58. $f(x) = 4 - x$ and $x < 0$

59. $f(x) = 6 - 3x$ and $x \leq 0$

60. $f(x) = x^2 - x - 2$ and $x > 1$

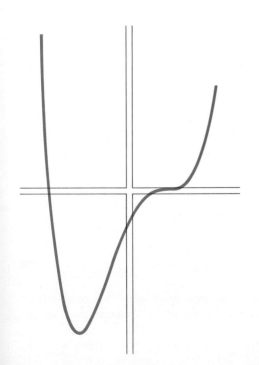

Polynomial
Functions

Polynomial functions occur frequently in both theoretical and applied mathematics. They assume an even greater significance in the modern world of high-speed computing since their values are calculated using only the arithmetic operations of addition, subtraction, and multiplication. In Section 2.1 we shall discuss a useful technique for evaluating polynomial functions using the calculator. Many complicated functions that occur in applications can be approximated by polynomials, as the student will see in calculus when studying the topic of power series.

We assume that the student has already developed some skills in working with polynomials. However, in Section 2.0 we provide a review of some techniques that are needed in working through this chapter.

**2.0   REVIEW OF ELEMENTARY ALGEBRA
       OF POLYNOMIALS AND FRACTIONS**

A formal definition of polynomial functions will be given in the next section, but we assume that the student is already somewhat familiar with such functions. In this section we review some skills in working with algebraic expressions involving polynomials.

Factoring

Section 1.0 contains a brief review of factoring certain types of simple polynomials. Here we consider additional types of expressions that can be factored.

Factoring formulas for the sum or difference of two cubes

$$a^3 + b^3 = (a + b)(a^2 - ab + b^2), \tag{2.1}$$

$$a^3 - b^3 = (a - b)(a^2 + ab + b^2). \tag{2.2}$$

Example 1   Factor $8x^6 + 27$.

Solution   The given expression can be written as the sum of two cubes, as follows:

$$8x^6 + 27 = (2x^2)^3 + 3^3.$$

Applying Eq. (2.1) gives

$$8x^6 + 27 = (2x^2 + 3)(4x^4 - 6x^2 + 9). \qquad ■$$

Factoring by grouping terms   The next two examples illustrate how certain expressions can be factored by first grouping terms in an appropriate manner.

Example 2   Factor $3x^3 - 4x^2 - 3x + 4$.

Solution   Grouping the first two terms and the last two terms gives a common factor, as follows:

$$3x^3 - 4x^2 - 3x + 4 = x^2(3x - 4) - (3x - 4) = (3x - 4)(x^2 - 1)$$
$$= (3x - 4)(x + 1)(x - 1). \qquad\blacksquare$$

Example 3   Factor $x^4 - x^2 + 6x - 9$.

Solution   Grouping the last three terms gives a difference of two squares that can be factored.

$$x^4 - x^2 + 6x - 9 = x^4 - (x^2 - 6x + 9) = (x^2)^2 - (x - 3)^2$$
$$= [x^2 + (x - 3)][x^2 - (x - 3)]$$
$$= (x^2 + x - 3)(x^2 - x + 3). \qquad\blacksquare$$

Elementary Operations with Polynomials

In the exercise set at the end of this section, several problems involving addition, subtraction, multiplication, and division of polynomials are included. Here we give an example in which one polynomial is divided by another.

Example 4   Divide $3x^4 - x^3 + 2x^2 - 5$ by $x^2 - 2x - 1$.

Solution   We can divide polynomials by using a long-division algorithm analogous to that for dividing numbers:

$$
\begin{array}{r}
3x^2 + 5x + 15 \\
x^2 - 2x - 1\,\overline{\smash{)}\,3x^4 - x^3 + 2x^2 - 5} \\
\underline{3x^4 - 6x^3 - 3x^2} \qquad 3x^2(x^2 - 2x - 1)\\
5x^3 + 5x^2 - 5 \qquad \text{Subtract}\\
\underline{5x^3 - 10x^2 - 5x} \qquad 5x(x^2 - 2x - 1)\\
15x^2 + 5x - 5 \qquad \text{Subtract}\\
\underline{15x^2 - 30x - 15} \qquad 15(x^2 - 2x - 1)\\
35x + 10 \qquad \text{Subtract}
\end{array}
$$

Thus we get a quotient $q(x) = 3x^2 + 5x + 15$ and a remainder $r(x) = 35x + 10$. This can be expressed in equation form as follows:

$$\frac{3x^4 - x^3 + 2x^2 - 5}{x^2 - 2x - 1} = 3x^2 + 5x + 15 + \frac{35x + 10}{x^2 - 2x - 1}. \qquad\blacksquare$$

Algebraic Fractions

Elementary operations with algebraic expressions of the type

$$\frac{f(x)}{g(x)},$$

where $f(x)$ and $g(x)$ are polynomials, occur frequently. A fundamental property in working with fractions is

$$\frac{f(x)h(x)}{g(x)h(x)} = \frac{f(x)}{g(x)}. \tag{2.3}$$

The property stated in Eq. (2.3) is useful in two situations:

1. Replacing $\dfrac{f(x)h(x)}{g(x)h(x)}$ by $\dfrac{f(x)}{g(x)}$ is involved in *simplifying* or *reducing* a fraction to lowest terms.

2. Replacing $\dfrac{f(x)}{f(x)}$ by $\dfrac{f(x)h(x)}{g(x)h(x)}$ is used in *addition* and *subtraction* of fractions, in which the first step involves getting equivalent fractions with common denominators.

The following examples illustrate techniques for simplifying fractions and for adding, subtracting, multiplying, and dividing fractions.

Example 5   Simplify $\dfrac{8x^3 - 64}{4x^2 - 16}$.

Solution   First factor the numerator and the denominator, and then apply the property given by Eq. (2.3), as follows:

$$\frac{8x^3 - 64}{4x^2 - 16} = \frac{8(x^3 - 8)}{4(x^2 - 4)} = \frac{8(x^3 - 2^3)}{4(x^2 - 2^2)} = \frac{2 \cdot 4(x - 2)(x^2 + 2x + 4)}{4(x - 2)(x + 2)}$$

$$= \frac{2(x^2 + 2x + 4)}{x + 2}.$$

Thus we have $\dfrac{8x^3 - 64}{4x^2 - 16} = \dfrac{2(x^2 + 2x + 4)}{x + 2}$ for all x except 2 and -2.   ▪

Example 6   Perform the indicated subtraction: $\dfrac{2}{x - 1} - \dfrac{2x - 1}{x + 2}$.

Solution   In adding or subtracting two fractions, we first express each as an equivalent fraction with common denominators. Using the property stated in Eq. (2.3), we get

$$\frac{2}{x - 1} = \frac{2(x + 2)}{(x - 1)(x + 2)} \qquad \text{and} \qquad \frac{2x - 1}{x + 2} = \frac{(2x - 1)(x - 1)}{(x + 2)(x - 1)}.$$

When these are substituted into the given expression, we get the subtraction of two fractions with common denominators. The result will be a fraction in which the numerator is the difference of the corresponding numerators and the denominator is the common denominator, as indicated in the following.

$$\frac{2}{x-1} - \frac{2x-1}{x+2} = \frac{2(x+2)}{(x-1)(x+2)} - \frac{(2x-1)(x-1)}{(x+2)(x-1)}$$

$$= \frac{2(x+2) - (2x-1)(x-1)}{(x-1)(x+2)}$$

$$= \frac{2x+2 - (2x^2 - 3x + 1)}{(x-1)(x+2)}$$

$$= \frac{-2x^2 + 5x + 1}{(x-1)(x+2)} = \frac{-2x^2 + 5x + 1}{x^2 + x - 2}.$$

Formulas for the product and quotient of two algebraic fractions follow.

$$\frac{f(x)}{g(x)} \cdot \frac{p(x)}{q(x)} = \frac{f(x)p(x)}{g(x)q(x)}. \tag{2.4}$$

$$\frac{f(x)}{g(x)} \div \frac{p(x)}{q(x)} = \frac{f(x)}{g(x)} \cdot \frac{q(x)}{p(x)} = \frac{f(x)q(x)}{g(x)p(x)}. \tag{2.5}$$

Example 7   Perform the indicated division and simplify: $\dfrac{x+2}{2x^2 + 5x - 3} \div \dfrac{x^2 + x - 2}{x^2 + x - 6}.$

Solution   First apply Eq. (2.5), then factor the resulting numerator and denominator, and finally simplify, as follows:

$$\frac{x+2}{2x^2 + 5x - 3} \div \frac{x^2 + x - 2}{x^2 + x - 6} = \frac{x+2}{2x^2 + 5x - 3} \cdot \frac{x^2 + x - 6}{x^2 + x - 2} \qquad \text{(by Eq. (2.5))}$$

$$= \frac{(x+2)(x^2 + x - 6)}{(2x^2 + 5x - 3)(x^2 + x - 2)} \qquad \text{(by Eq. (2.4))}$$

$$= \frac{(x+2)(x+3)(x-2)}{(2x-1)(x+3)(x+2)(x-1)} \qquad \text{(by factoring)}$$

$$= \frac{x-2}{(2x-1)(x-1)} \qquad \text{(by Eq. (2.3))}$$

$$= \frac{x-2}{2x^2 - 3x + 1}.$$

Example 8 Express $\left(1 - \dfrac{3}{x^2 - 2x}\right) \div \left(x - 4 + \dfrac{3}{x}\right)$ as a simple fraction in lowest terms.

Solution First express the numerator and denominator as single fractions; then the problem becomes one of dividing two fractions:

$$\left(1 - \frac{3}{x^2 - 2x}\right) \div \left(x - 4 + \frac{3}{x}\right) = \frac{x^2 - 2x - 3}{x^2 - 2x} \div \frac{x^2 - 4x + 3}{x}$$

$$= \frac{x^2 - 2x - 3}{x^2 - 2x} \cdot \frac{x}{x^2 - 4x + 3}$$

$$= \frac{x(x^2 - 2x - 3)}{(x^2 - 2x)(x^2 - 4x + 3)}$$

$$= \frac{x(x - 3)(x + 1)}{x(x - 2)(x - 3)(x - 1)}$$

$$= \frac{x + 1}{(x - 2)(x - 1)} = \frac{x + 1}{x^2 - 3x + 2}.$$

Hence

$$\left(1 - \frac{3}{x^2 - 2x}\right) \div \left(x - 4 + \frac{3}{x}\right) = \frac{x + 1}{x^2 - 3x + 2}$$

for all x except 0, 1, and 2.

Exercises 2.0

Factoring
In problems 1 through 10, factor the given expressions as far as possible.

1. $x^3 + 27$

2. $27 - 8x^3$

3. $8 - (2 - x)^3$

4. $27 - (x - 3)^3$

5. $x^3 - 2x^2 + x - 2$

6. $4x^3 + 6x^2 + 2x + 3$

7. $x^4 - x^3 - x^2 + x$

8. $x^3 - 2x^2 - x + 2$

9. $x^4 - x^2 - 4x - 4$

10. $4x^2 - 4x + 1 - 9x^4$

Operations with Polynomials
In problems 11 through 18, perform the indicated operations on the given polynomials. Express answers in simplified form.

11. $(4x^2 - 3x + 1) + (x^2 + 2x - 1)$

12. $(-3x^3 - 5x + 4) + (x^2 - 4)$

13. $(2x^3 - 3x^2 + 4x + 5) - (-x^3 + 4x - 1)$

14. $(x^4 - 5x^2 - 3x + 5) - (x^4 - 2x^3 - 3x + 1)$

15. $(x^2 + 2x - 4) \cdot (x^2 - 2x + 4)$

16. $(2x^2 - x) \cdot (x^3 + 1)$

17. $(x^3 - 2x + 4) - (x - 1)(x^2 + x - 1)$

18. $(2x^3 - 3x^2 + x - 1) - (2x + 1)(x^2 - 3)$

19. Divide $2x^4 - 3x^3 + x^2 - 3x + 4$ by $x^2 - 2x - 1$, and determine the quotient and remainder.

20. Perform the indicated division, and determine the quotient and remainder:

$$(x^4 - 2x^3 - x^2 + x - 3) \div (x^2 - x + 2)$$

21. Find a number k such that when $x^3 + 3x^2 - 2x + 3k$ is divided by $x + 3$, the remainder will be zero.

22. Find a number k such that $(2x^3 + x^2 + 4x - 2k) \div (x - 3)$ will have a remainder of zero.

Operations with Fractions

In problems 23 through 30, express the given fractions in lowest terms.

23. $\dfrac{x^2 - 2x + 1}{x^2 - x}$

24. $\dfrac{x^2 - 4x + 3}{x^2 - 2x - 3}$

25. $\dfrac{1 - x^2}{2x^2 + x - 1}$

26. $\dfrac{x^3 - 27}{2x^2 - 5x - 3}$

27. $\dfrac{8x^3 + 125}{2x^2 + 5x}$

28. $\dfrac{8 - (x - 1)^3}{(x^2 + 3)(x - 3)}$

29. $\dfrac{27 - (1 - x)^3}{8 + 2x - x^2}$

30. $\dfrac{x^3 - 3x^2 - x + 3}{x^2 - 4x + 3}$

In problems 31 through 48, perform the indicated operations, and express answers as a single fraction in lowest terms.

31. $\dfrac{3}{x - 1} + \dfrac{x}{x + 2}$

32. $x + \dfrac{x - 3}{x - 1}$

33. $\dfrac{3x - 1}{x} - \dfrac{1 - 2x}{x + 1}$

34. $\dfrac{x + 2}{x - 1} - \dfrac{x - 4}{2x + 1}$

35. $\dfrac{x^2 + 2x}{x^2 - 1} \cdot \dfrac{x^2 - 2x + 1}{x + 2}$

36. $\dfrac{x^2 - 3x - 4}{x^2 - 16} \cdot \dfrac{x + 4}{x - 1}$

37. $\dfrac{x^3 - 1}{x^2 + 2x} \cdot \dfrac{2x^3 + 4x^2}{x^2 - 1}$

38. $\dfrac{x^2 + x + 1}{x} \cdot \dfrac{x^2 + x}{x^3 - 1}$

39. $\dfrac{x^2 - 4}{2x + 1} \div \dfrac{x + 2}{x - 2}$

40. $\dfrac{9x^2 + 6x + 1}{x} \div \dfrac{3x + 1}{x^2 - 2x}$

41. $\dfrac{x}{8 - x^3} \div \dfrac{3x - x^2}{x - 2}$

42. $\dfrac{x^4 - 1}{x + 2} \div \dfrac{x^3 - 2x^2 + x - 2}{2x + 4}$

43. $\left(\dfrac{x}{x - 1} - \dfrac{x + 1}{x} \right) \div \left(3 - \dfrac{2}{x^2 - x} \right)$

44. $\left(\dfrac{x^2 + 1}{x} - 3 \right) \div \left(4 - \dfrac{x - 1}{x^2 - x} \right)$

45. $\left(1 - \dfrac{2}{x} + \dfrac{3x}{x + 1} \right) \div \left(4 - \dfrac{5x + 2}{x^2 + x} \right)$

46. $\left(x - \dfrac{3x - 1}{x + 1} \right) \div (x^2 - 1)$

47. $x - \dfrac{2 + \dfrac{x}{3}}{6 + x}$

48. $3 - \dfrac{1 + \dfrac{x}{x - 2}}{x^2 - 1}$

2.1   POLYNOMIAL FUNCTIONS: INTRODUCTION AND DEFINITIONS

Perhaps the most useful functions that we study are the polynomial functions. They occur frequently in applications, and they also form a basis for the theoretical study of more complicated functions that can be approximated by polynomials. The latter are studied in calculus under infinite series.

We have already encountered examples of polynomial functions in the preceding chapter. The linear and quadratic functions discussed in Sections 1.3 and 1.4 are examples of polynomial functions. Let us now state a formal definition of polynomial functions, introduce some related terminology, and then consider several examples.

Definition 2.1

Polynomial functions

Any function P that can be expressed by a formula of the type

$$P(x) = a_n x^n + a_{n-1} x^{n-1} + \cdots + a_1 x + a_0, \tag{2.6}$$

where n is a positive integer or zero, and $a_n, a_{n-1}, \ldots, a_1, a_0$ are given numbers, is called a *polynomial function*. If $a_n \neq 0$, we say that P has *degree n*. The numbers $a_n, a_{n-1}, \ldots, a_1, a_0$ are the *coefficients of P*. The coefficient of x^n is called the *leading coefficient* of P, and a_0 is called the *constant term*.

When $P(x)$ is written as in Eq. (2.6) with the highest-degree term first and with successive terms decreasing in degree, we say the polynomial is in *standard form*.

If $P(x) = a_0$, where $a_0 \neq 0$, then we call $P(x)$ a *constant polynomial* with *degree zero*. The special case in which $P(x) = 0$ for all $x \in \mathbf{R}$ is called the *zero polynomial*; we do not associate any degree with it.

Unless otherwise stated, for all polynomials discussed in this text *the coefficients will be real numbers.** Sometimes we want to restrict the coefficients to rational numbers or to integers; such cases will be referred to as "polynomials over the rational numbers," or "polynomials over the integers."

Note that for every real number x the formula given by Eq. (2.6) yields a real number $P(x)$. This implies, according to the convention stated in Section 1.1 (p. 19), that the *domain of any polynomial function is the set of real numbers.*

Example 1 For each of the following, determine whether the given function is a polynomial function:

$$f(x) = 2x - 5, \qquad g(x) = 1 - 2x - 3x^2, \qquad h(x) = \frac{1}{x^2 - 2x}.$$

Solution Functions f and g are polynomial functions; the standard form for $g(x)$ is $g(x) = -3x^2 - 2x + 1$. Function $h(x)$ cannot be written in the form given by Eq. (2.6), and so h is not a polynomial function.

---

*   Polynomials with complex-number coefficients are studied in courses in complex variables.

Note that $\mathcal{D}(h) = \{x|x \neq 0, x \neq 2\}$ whereas the domain of any polynomial function is **R**. ■

Example 2 For each of the following polynomial functions, determine its degree, leading coefficient, and constant term. In each case write the polynomial in standard form.

a) $f(x) = 3x^4 - 2x^3 + 5$ b) $g(x) = 1 - 2x - 3x^2$

c) $h(x) = 4(2x^3 - x^2 + x - 3)$ d) $F(x) = x^5 + 2x^4 - x$

Solution a) $f(x)$ is in standard form as given. The degree of f is 4, leading coefficient is 3, and constant term is 5.

b) $g(x) = -3x^2 - 2x + 1$ is the standard form for $g(x)$. The degree of g is 2, leading coefficient is -3, and constant term is 1.

c) $h(x) = 8x^3 - 4x^2 + 4x - 12$ is the standard form for $h(x)$. Degree of h is 3, leading coefficient is 8, and constant term is -12.

d) $F(x)$ is in standard form as given. Degree of F is 5, leading coefficient is 1, and constant term is 0. ■

Example 3 Suppose f, g, and h are polynomial functions given by

$$f(x) = 2x^3 - x + 1, \quad g(x) = x^3 - \frac{4}{3}x^2 + x - \frac{1}{2}, \quad h(x) = x^4 + \sqrt{3}x^2 - \sqrt{2}.$$

In each case determine whether the polynomial is "over the rational numbers," "over the set of integers," or neither.

Solution Function f is a polynomial over the rational numbers and over the set of integers; g is a polynomial over the rational numbers; h is neither. ■

Example 4 Suppose f and g are functions given by $f(x) = 2x - 1$ and $g(x) = x^2 - 1$. For each of the following functions, determine whether it is a polynomial function; if it is, give the degree.

a) $f + g$ b) $f \cdot g$ c) f/g d) $f \circ g$

Solution a) $(f + g)(x) = f(x) + g(x) = (2x - 1) + (x^2 - 1) = x^2 + 2x - 2.$
Hence $f + g$ is a polynomial function of degree 2.

b) $(f \cdot g)(x) = f(x) \cdot g(x) = (2x - 1)(x^2 - 1) = 2x^3 - x^2 - 2x + 1.$
Function $f \cdot g$ is a polynomial function of degree 3.

c) $\left(\frac{f}{g}\right)(x) = \frac{f(x)}{g(x)} = \frac{2x - 1}{x^2 - 1}$. Function $\frac{f}{g}$ is not a polynomial function.

d) $(f \circ g)(x) = f(g(x)) = f(x^2 - 1) = 2(x^2 - 1) - 1 = 2x^2 - 3.$
Hence $f \circ g$ is a polynomial function of degree 2. ■

Using a Calculator to Evaluate Polynomial Functions

In subsequent sections of this chapter we shall be interested in drawing graphs and finding zeros of polynomials. This requires evaluating functions at several values of the independent variable. Thus there is a need to learn techniques for doing this efficiently with the aid of a calculator. Let us consider an example to illustrate the method of expressing a polynomial in nested form and using it to evaluate the function.

Example 5   Suppose f is a polynomial function given by

$$f(x) = 3x^3 + 2x^2 - 5x + 4. \tag{2.7}$$

Evaluate:

a) $f(2.4)$ **b)** $f(-1.6)$

Solution   Function $f(x)$ can be evaluated by using the formula in Eq. (2.7) and the $\boxed{y^x}$ calculator key. However, as noted in Section 1.2, the $\boxed{y^x}$ key on many calculators cannot be used directly to evaluate an exponential expression when the base number is negative, such as $(-1.6)^3$. Although we can express $(-1.6)^3$ as $-(1.6)^3$ and then use the $\boxed{y^x}$ key, we can avoid the need for using the $\boxed{y^x}$ key by first expressing $f(x)$ as a formula in nested form. Equation (2.7) can be written as

$$f(x) = [(3x + 2)x - 5]x + 4. \tag{2.8}$$

The reader should expand this equation to see that it is equivalent to Eq. (2.7) and should also observe the pattern for getting Eq. (2.8) in terms of the coefficients of f.

The formula given in Eq. (2.8) is called the *nested form* for the polynomial f. Now suppose c is any given number, and we want to evaluate

$$f(c) = [(3c + 2)c - 5]c + 4.$$

Starting with the inside parentheses, we have the following sequence of steps: multiply 3 by c and add 2, multiply the result by c and subtract 5, multiply the result by c and add 4. Thus we have an easy-to-follow pattern in which we repeatedly multiply by c and add a coefficient (subtracting 5 is equivalent to adding -5, the coefficient of x).

a) According to Eq. (2.8),

$$f(2.4) = [(3(2.4) + 2)(2.4) - 5](2.4) + 4.$$

Since the number 2.4 occurs several times, it is good practice to store it by using the $\boxed{STO}$ key and recall it as needed with the $\boxed{RCL}$ key.* The following are suggested key stroke sequences for algebraic and RPN calculators.

---

*   For calculators that have more than a single memory it is necessary to follow $\boxed{STO}$ and $\boxed{RCL}$ with a memory address; see your owner's manual if necessary.

Using an AOS calculator, press

2.4 (STO) ( ) ( ) 3 (×) (RCL) (+) 2 ( ) (×) (RCL) (−) 5 ( ) (×) (RCL) (+) 4 (=)

The result is $f(2.4) = 44.992$.
Using a RPN calculator, press

2.4 (STO) 3 (RCL) (×) 2 (+) (RCL) (×) 5 (−) (RCL) (×) 4 (+)

The display will show 44.992.

b) To evaluate $f(-1.6)$ we can follow a sequence of key strokes similar to that in (a), except that we replace the 2.4 by 1.6, (+/−) or 16, (CHS) at the start. This gives $f(-1.6) = 4.832$. ■

Example 6   Express $f(x) = -3x^4 + 2x^3 + 5x - 6$ in nested form.

Solution   The coefficient of x^2 is zero, and so we have

$$f(x) = \{[(-3x + 2)x + 0]x + 5\}x - 6.$$ ■

Exercises 2.1

In problems 1 and 2, determine whether the given functions are polynomial functions.

1. a) $f(x) = x^3 + 1$ **b)** $g(x) = \dfrac{1}{x} + 3$ **c)** $h(x) = x + 2x^2 - x^3$

2. a) $f(x) = x^4 - x$ **b)** $g(x) = 1 - x + x^2 - x^3$ **c)** $h(x) = \dfrac{x + 3}{x^2 - 1}$

In problems 3 through 10, determine the degree, leading coefficient, and constant term for the given polynomial functions. In each case write the polynomial in standard form, and tell whether or not it is "over the rational numbers."

3. $f(x) = 3x^4 - 2x^3 + x^2 - x + 3$ **4.** $g(x) = 5 + x - x^3$

5. $g(x) = x + x^2 - x^3$ **6.** $f(x) = \sqrt{2}x^3 - 3x^2 + x$

7. $f(x) = 2(x - 1)(x^2 + 1)$ **8.** $f(x) = 3(1 - x)(x^3 + 1)$

9. $g(x) = \sqrt{3}(1 - 2x + 3x^3)$ **10.** $f(x) = \sqrt{5}(x + 1)(x + 1)(x^2 - x)$

In problems 11 through 16, suppose f and g are polynomial functions given by $f(x) = 3x^2 + 2x$ and $g(x) = x^2 - 1$. In each case, determine whether the given function is a polynomial; if it is, give its degree and its leading coefficient.

11. $f + g$ **12.** $\dfrac{g}{f}$ **13.** $\dfrac{f}{g}$ **14.** $f \cdot g$ **15.** $f \circ g$ **16.** $g \circ f$

In problems 17 through 24, round off answers to two decimal places.

17. Given that $f(x) = 2x^3 - x^2 + 5x + 1$, determine:
  a) $f(1.5)$ **b)** $f(-2.3)$

18. Given that $g(x) = 4x^3 + x^2 - 6x + 3$, determine:
  a) $g(2.7)$ **b)** $g(-0.8)$

19. Given that $g(x) = 2 + 3x - 5x^2 - 2x^3$, find:

    a) $g(0.7)$ **b)** $g(-2.1)$

20. Given that $f(x) = 3 - 5x + 2x^2 - 4x^3$, find:

    a) $f(1.7)$ **b)** $f(-1.2)$

21. Given that $f(x) = 3x^4 - 2x^3 + 5x^2 + x - 3$, find $f\left(\dfrac{1 + \sqrt{5}}{2}\right)$.

22. Given that $g(x) = 2x^4 - 4x^3 - 2x^2 - x - 1$, find $g\left(\dfrac{1 - \sqrt{5}}{2}\right)$.

23. Given that $f(x) = x^3 - 2x + 1$, find $f(1.36)$.   **24.** Given that $f(x) = 2x^3 - 3x - 4$, find $f(-0.53)$.

In problems 25 through 28, complete the given table, in which $y = g(x)$. In each case round off answers to two significant digits.

25. $g(x) = x^3 - x^2 - 3x - 4$

x	-1.48	-0.43	0	0.83	1.64
y					

26. $g(x) = 2x^3 + 3x^2 - 4x - 6$

x	1.3	1.4	1.41	1.42	1.414	1.415
y						

27. $g(x) = x^3 - 5x + 1$

x	-2.3	-2.4	-2.33	-2.34	-2.331
y					

28. $g(x) = x^4 - 2x^3 - x^2 - 3x - 2$

x	-0.5	-0.6	-0.59	-0.595	-0.599
y					

In problems 29 through 31, find a polynomial that satisfies the given conditions.

29. P has degree 1, leading coefficient 4, and $P(-1) = -3$.

30. P has degree 2, leading coefficient -3, and $P(3) = -4$.

31. P has degree 3, leading coefficient 2, and $P(2) = -1$.

32. Suppose f and g are polynomial functions of degree m and n, respectively, with $m > n$. For each of the following what can be said about the given function in reference to whether it is always, sometimes, or never a polynomial function. In each case, what meaningful statement can be made about the degree if it is a polynomial?

    a) $f + g$ **b)** $f - g$ **c)** $f \cdot g$ **d)** f/g **e)** $f \circ g$

2.2   **GRAPHS OF POLYNOMIAL FUNCTIONS**

Drawing an accurate graph of a polynomial function of degree greater than 2 can become an involved process. About all we can do, without the tools of calculus, is to determine several points on the graph and connect them with a smooth curve. Whenever possible, we look for "key points," such as the x- and y-intercept points, and we also take advantage of any symmetry properties that may occur. In any particular case the degree of difficulty in drawing a graph depends on the formula that describes the polynomial function. This fact is illustrated in examples below.

Polynomial Functions of Degree 0, 1 or 2

Polynomial functions of degree 0, 1, or 2 have already been discussed in Sections 1.3 and 1.4. There we called polynomials of degree 0 or 1 linear functions because their graphs are straight lines. Polynomials of degree 2 are called quadratic functions, and we saw that their graphs are parabolas that open upward or downward. We shall not repeat the discussion in Sections 1.3 and 1.4, but the reader may wish to review that material at this point.

Polynomial Functions of Degree Greater than 2

In the next three examples we illustrate graphs of polynomial functions in which the formula is given in factored form.

Example 1   Draw a graph of the polynomial function given by

$$f(x) = (x - 1)(x - 3)(x + 2). \tag{2.9}$$

Solution   The polynomial function given by Eq. (2.9) has degree 3, as one can see by multiplying the factors on the right to get

$$f(x) = x^3 - 2x^2 - 5x + 6. \tag{2.10}$$

We are interested in getting a table of x, y values that satisfy the equation $y = f(x)$. First, let us get some key points.

y-intercept point: Substituting $x = 0$ into Eq. (2.10) gives $y = 6$. Thus the y-intercept point is $(0, 6)$.

x-intercepts: Let $y = 0$ and use Eq. (2.9) to obtain $(x - 1)(x - 3)(x + 2) = 0$. Thus 1, 3, and -2 are the values of x that give $y = 0$. That is, the x-intercept points are $(1, 0)$, $(3, 0)$, and $(-2, 0)$. These are included in the table below.

Symmetry: $f(-x) = (-x)^3 - 2(-x)^2 - 5(-x) + 6 = -x^3 - 2x^2 + 5x + 6$. Thus $f(-x) \neq f(x)$ and $f(-x) \neq -f(x)$, and so the function is neither even nor odd. Hence the graph is not symmetric with respect to the y-axis or the origin.

To get a table of x, y values, first get a formula for f in nested form. From Eq. (2.10) we have

$$f(x) = [(x - 2)x - 5]x + 6. \tag{2.11}$$

Using a calculator when necessary, along with formula (2.11), we construct the following table.

x	-3	-2.5	-2	-1.5	-1	-0.5	0	0.5	1	1.5	2	2.5	3	3.5	4
y	-24	-9.63	0	5.63	8	7.88	6	3.13	0	-2.63	-4	-3.38	0	6.88	18

Now plot (x, y) points from this table and draw a smooth curve through them to get the graph shown in Fig. 2.1.

Y=(X−1)*(X−3)*(X+2)

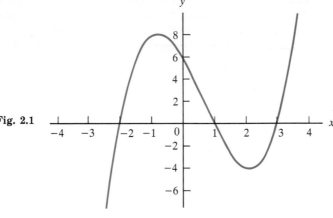

Fig. 2.1

Graph of $f(x) = (x-1)(x-3)(x+2)$

Example 2   Draw a graph of $y = (x-1)^2(x+2)$.

Solution   This example is similar to Example 1, except that the factor $(x-1)$ occurs twice. How does this influence the graph? In Example 1 the graph crosses the x-axis at each of the x-intercept points, as Fig. 2.1 shows.

    Drawing a graph of the given equation by proceeding in a manner similar to that of Example 1 gives the curve shown in Fig. 2.2. The reader is encouraged to make a table of x, y values and verify the curve shown.

Y=(X−1)^2*(X+2)

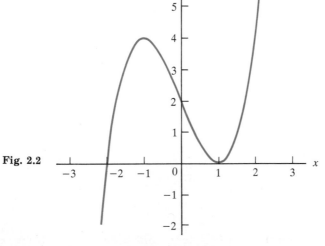

Fig. 2.2

Graph of $y = (x-1)^2(x+2)$

Note that the curve in Fig. 2.2 passes through the point $(1, 0)$ but does not cross the x-axis at that point. This always happens when the given equation has a linear factor to an even power, such as $(x - 1)^2$ in this case. ■

Example 3   Draw a graph of $y = (x - 1)^3(x + 2)$.

Solution   This example is similar to Example 2, except that $(x - 1)$ occurs as a factor three times in the given equation. Following a procedure similar to that in Example 1, we get the graph shown in Fig. 2.3. The curve crosses the x-axis at the point $(1, 0)$. This will always happen when a linear factor occurs to an odd power, such as $(x - 1)^3$ in this case. The reader should make a table of x, y values, including several values of x near 1, to verify the graph shown. Near the point $(1, 0)$ the curve becomes almost horizontal.

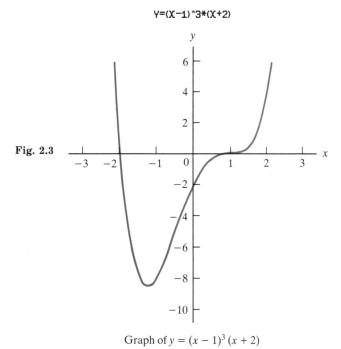

Fig. 2.3

Graph of $y = (x - 1)^3 (x + 2)$ ■

Example 4   **a)** Draw a graph of the function given by $f(x) = x^4 + x^3 - 9x^2 - 8x + 14$.

   b) From the graph, what can we conclude about the x-intercept points and the range of f?

Solution   **a)** This example is similar to the preceding three examples, except that the formula describing f is not in factored form, and therefore it is not a simple matter to determine the x-intercept points of $y = f(x)$. However, we can at least get the y-intercept point easily. If $x = 0$, then $y = 14$; hence $(0, 14)$ is the y-intercept point.

In order to evaluate f at given values of x, first get the formula for f in nested form:

$$f(x) = \{[(x + 1)x - 9]x - 8\}x + 14.$$

The next question is: What values of x should we include in the table of x, y values? It is probably wise to plot some integer values first and then see where it might be helpful to include in-between values of x. This is how the following table was constructed.

x	-4	-3	-2.7	-2.5	-2	-1.5	-1	-0.7	-0.5	0
y	94	11	3.5	1.2	2	7.4	13	15.1	15.7	14

0.5	1	1.5	1.7	2	2.5	3	4
7.9	-1	-9.8	-12.3	-14	-7.6	17	158

Plotting the (x, y) points from this table and drawing a curve through them gives the graph shown in Fig. 2.4.

Y=X^4+X^3-9*X^2-8*X+14

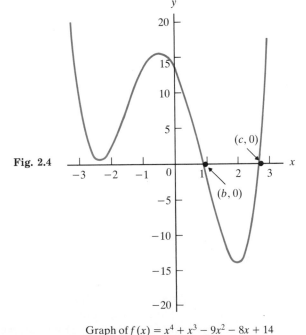

Fig. 2.4

Graph of $f(x) = x^4 + x^3 - 9x^2 - 8x + 14$

b) From the graph in Fig. 2.4 we conclude the following, although tools of calculus would be required for rigorous proofs.

1. There are exactly two x-intercept points; the first point is at $(b, 0)$, where b is slightly less than 1, and the other is $(c, 0)$, where c is approximately 2.6.

2. $\mathcal{R}(f) = \{y | y \geq -14\}$. At this point we must accept this as a conjecture. Although we could include more values of x near 2 in our table (see Problem 30) and see that the corresponding values of y are greater than -14, we would still not have a proof. Techniques of calculus are required to prove that our conjecture is indeed true. ■

Exercises 2.2

In problems 1 through 24, draw a graph of the given function. In those cases in which the given formula is not in factored form, factor it if possible; doing so will help in determining the x-intercept points.

1. $f(x) = 3x - 1$

2. $g(x) = x^2 - 3x - 4$

3. $g(x) = -x^2 + 2x + 3$

4. $f(x) = 3 - 2x$

5. $f(x) = x^3 - x$

6. $g(x) = -x^2 + 4$

7. $f(x) = x^4 - 2x^2 + 1$

8. $f(x) = x - x^3$

9. $g(x) = -x^3$

10. $g(x) = x^4 - 1$

11. $f(x) = x^4$

12. $f(x) = x(x - 1)(x + 3)$

13. $g(x) = (1 - x)(x + 3)(x - 2)$

14. $g(x) = x^3 - 2x^2 + x$

15. $f(x) = x^3 - 1$

16. $f(x) = x^3 + 2x^2 - 3x - 6$

17. $g(x) = x^3 - 2x^2 - x + 2$

18. $f(x) = (x - 1)(x^2 + 1)$

19. $g(x) = (x + 1)^2(x - 3)$

20. $f(x) = (x + 1)(x - 2)^3$

21. $f(x) = (x - 1)^4$

22. $f(x) = x^4 - 4x^2$

23. $f(x) = x^3 - 4x^2 + 4x$

24. $f(x) = x^4 - 2x^3 + x^2$

In problems 25 through 28, draw graphs of the given functions.

25. $f(x) = x^3 - x - 6$

26. $g(x) = x^3 - x^2 - 5x - 3$

27. $f(x) = x^4 + x^3 + 2x^2 + x - 2$

28. $f(x) = x^4 - 7x^2 - 2x + 4$

29. Draw a graph of $f(x) = x^3 - 2x^2 - x - 1$. Use your graph to arrive at a reasonable conjecture concerning the x-intercept points. Also determine the range of f.

30. In the solution of Example 4(b) on p. 118 we arrived at a conjecture that $\mathcal{R}(f) = \{y | y \geq -14\}$. Complete the following table to give additional support for this conclusion. Recall that $f(x) = x^4 + x^3 - 9x^2 - 8x + 14$.

x	1.9	1.99	2.1	2.01
$f(x)$				

**2.3   SYNTHETIC DIVISION;
        REMAINDER AND FACTOR THEOREMS**

Synthetic Division

One of the basic properties of integers is stated in the Euclidean algorithm:

If m and n are any integers, where $n > 0$, then there exist unique integers q and r such that $m = n \cdot q + r$, where $0 \le r < n$.

Dividing both sides of this equation by n gives

$$\frac{m}{n} = q + \frac{r}{n}.$$

This is the familiar form of expressing the result of dividing m by n and getting quotient q and remainder r. For instance, dividing 21 by 5, we get a quotient of 4 and remainder of 1. This can be expressed as $21/5 = 4 + 1/5$.

For polynomials, a result that is analogous to the Euclidean algorithm is stated in the following theorem.

Theorem 2.1   *Division algorithm*

Suppose $p(x)$ is any polynomial and c is a real number. There is a polynomial $q(x)$ and a real number r such that

$$p(x) = (x - c)q(x) + r. \tag{2.12}$$

Dividing both sides of Eq. (2.12) by $x - c$, we get

$$\frac{p(x)}{x - c} = q(x) + \frac{r}{x - c}$$

Thus, if $p(x)$ is divided by $x - c$, the *quotient* is $q(x)$ and the *remainder* is r.

Example 1   Find the quotient and remainder when $p(x) = 4x^3 - 3x^2 + x - 5$ is divided by $x - 2$.

Solution   Using the method of long division, we have

$$
\require{enclose}
\begin{array}{r}
4x^2 + 5x\ + 11 \\
x - 2 \enclose{longdiv}{4x^3 - 3x^2 +\ \ x - 5} \\
\underline{4x^3 - 8x^2} \\
5x^2 +\ \ \ x - 5 \\
\underline{5x^2 - 10x} \\
11x - 5 \\
\underline{11x - 22} \\
17
\end{array}
$$

Thus, $q(x) = 4x^2 + 5x + 11$, and $r = 17$. ■

Synthetic Division

The long-division algorithm of dividing $p(x)$ by $x - c$ can be condensed into a process called *synthetic division*. This is illustrated in the following example.

Example 2   Divide $p(x) = 4x^3 - 3x^2 - 12x + 8$ by $x - 2$, and find the quotient $q(x)$ and the remainder r.

Solution   Applying the long-division algorithm gives

$$
\require{enclose}
\begin{array}{r}
4\,x^2 + 5x\; - 2 \\
x - 2 \enclose{longdiv}{④x^3 - 3\,x^2 - 12x + 8} \\
\underline{4\,x^3 - 8\,x^2} \\
⑤x^2 - 12x + 8 \\
\underline{5\,x^2 - 10x} \\
⑦x + 8 \\
\underline{-2\,x + 4} \\
④
\end{array}
$$

Thus $q(x) = 4x^2 + 5x - 2$, and $r = 4$.

Note that the coefficients of $q(x)$ and the remainder occur in the long-division process in the order shown by the circled numbers. Inspection shows that the first coefficient of $q(x)$ is the same as the leading coefficient of $p(x)$, and then each subsequent coefficient is the result of multiplying the preceding coefficient by -2 and subtracting the product from the corresponding coefficient of $p(x)$. For example, the coefficient 5 is the result of $-3 - (-2)(4) = 5$; -2 comes from $-12 - (-2)(5)$; and the remainder 4 comes from $8 - (-2)(-2) = 4$. Thus we have a simple recursive pattern for getting the coefficients of $q(x)$ and r. The arithmetic in this pattern can be simplified if we multiply by 2 rather than -2 and add instead of subtract.

All of this suggests that the long-division process shown above can be condensed by eliminating all superfluous writing, as follows:

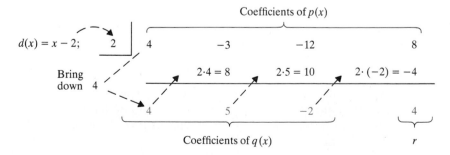

We call the method illustrated in Example 2, the *synthetic division* technique for dividing $p(x)$ by $d(x)$.

Example 3   Use synthetic division to find the quotient and remainder when

$$p(x) = 2x^4 - 16x^2 + 5x - 4$$

is divided by $x + 3$.

Solution   The coefficient of x^3 in $p(x)$ is 0, and it must be included in the listing of coefficients of $p(x)$.

$$
\begin{array}{r|rrrrr}
-3 & 2 & 0 & -16 & 5 & -4 \\
   &   & -6 & 18 & -6 & 3 \\
\hline
   & 2 & -6 & 2 & -1 & -1 \\
\end{array}
$$

$$\underbrace{\qquad\qquad\qquad}_{q(x)} \qquad\quad r$$

Thus, $q(x) = 2x^3 - 6x^2 + 2x - 1$, and $r = -1$.   ■

Synthetic Division and Nested Form

In Section 2.1 we introduced the idea of expressing a polynomial in nested form, which we used to evaluate $p(x)$ at given values of x.

     We now take a closer look at the steps involved in this evaluation process and see how it is related to synthetic division.

Example 4   Suppose $p(x) = 2x^3 - 5x^2 + 4x + 1$. Compare the synthetic division process of dividing $p(x)$ by $x - c$ with that of evaluating $p(c)$ by using nested form for $p(x)$.

Solution   In the synthetic-division computations, let us denote the coefficients of the quotient by b_0, b_1, and b_2 so that we can see where they occur in the nested-form computation.

$$
\begin{array}{r|cccc}
c & 2 & -5 & 4 & 1 \\
  &   & 2c & (2c-5)c & [(2c-5)c+4]c \\
\hline
  & 2 & 2c-5 & (2c-5)c+4 & [(2c-5)c+4]c+1 \\
  & \uparrow & \uparrow & \uparrow & \downarrow \\
  & b_0 & b_1 & b_2 & r = p(c)
\end{array}
\qquad (2.13)
$$

We now look at the sequence of steps involved in evaluating $p(c)$ by using the nested form for $p(x)$; that is,

$$p(x) = [(2x - 5)x + 4]x + 1. \qquad (2.14)$$

Comparing (2.13) and (2.14), we see that the nested-form algorithm for evaluating $p(c)$ also gives us the quotient and remainder when dividing $p(x)$ by $(x - c)$. We get the coefficients of $q(x)$ by looking at the results in the steps just preceding the multiplication by c.

     Also note in (2.13) that the remainder is equal to $p(c)$. This illustrates the Remainder theorem, which is given later in this section.

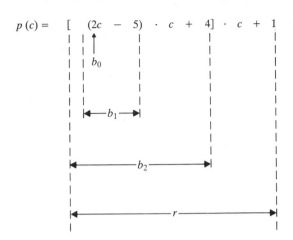

$$p(c) = [(2c - 5) \cdot c + 4] \cdot c + 1$$

Example 5 Using a calculator, find the quotient and remainder when

$$p(x) = 2.4x^3 - 3.5x^2 + 5.1x - 3.2$$

is divided by $x - 1.6$.

Solution Let the quotient be denoted by $q(x) = b_0 x^2 + b_1 x + b_2$. Coefficients b_0, b_1, b_2 and also the remainder can be found by using the formula for p in nested form and evaluating $p(1.6)$ as follows:

$$p(x) = [(2.4x - 3.5)x + 5.1]x - 3.2.$$

$$p(1.6) = [(2.4(1.6) - 3.5) (1.6) + 5.1] (1.6) - 3.2$$

With an algebraic calculator we can carry out the sequence of computations above as follows:

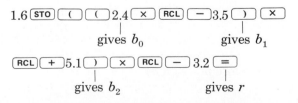

With a RPN calculator we can use the following sequence of steps:

$$1.6 \ \boxed{\text{STO}} \ 2.4 \ \boxed{\times} \ 3.5 \ \boxed{-} \ \boxed{\text{RCL}} \ \boxed{\times} \ 5.1 \ \boxed{+} \ \boxed{\text{RCL}} \ \boxed{\times} \ 3.2 \ \boxed{-}$$

$$\underset{\text{gives } b_0}{\big|} \qquad \underset{\text{gives } b_1}{\big|} \qquad\qquad \underset{\text{gives } b_2}{\big|} \qquad\qquad \underset{\text{gives } r}{\big|}$$

The result from either sequence is $q(x) = 2.4x^2 + 0.34x + 5.644$ and $r = 5.8304$. ∎

Remainder Theorem

Suppose $p(x)$ is a polynomial and c is a real number. By Theorem 2.1 there is a polynomial $q(x)$ and a number r such that

$$p(x) = (x - c)q(x) + r.$$

If this is used to evaluate the function p at c, we get

$$p(c) = (c - c)q(c) + r = 0 + r = r.$$

Hence $r = p(c)$. This result is stated as a theorem.

Theorem 2.2
　　　　　　　　　　　　　Remainder theorem

If polynomial $p(x)$ is divided by $x - c$, then the remainder is equal to $p(c)$. That is, $r = p(c)$ and $p(x) = (x - c)q(x) + p(c)$.

Example 6　If $p(x) = 3x^{16} + 2x^7 - x^4 + 1$ is divided by $x + 1$, what is the remainder?

Solution　Performing the long division to get the remainder r would be a lengthy procedure. Using Theorem 2.2, we can find r by evaluating $p(-1)$. Thus

$$r = p(-1) = 3(-1)^{16} + 2(-1)^7 - (-1)^4 + 1 = 3 - 2 - 1 + 1 = 1. \quad ∎$$

Example 7　Suppose $p(x) = 2x^4 - 4x^2 + 12x - 6$. Determine $p(-3)$ by using the Remainder theorem.

Solution　We can find $p(-3)$ by evaluating directly:

$$p(-3) = 2(-3)^4 - 4(-3)^2 + 12(-3) - 6.$$

However, in general it is simpler to use synthetic division, in which $p(x)$ is divided by $x - (-3)$; by Theorem 2.2 the remainder will be equal to $p(-3)$.

$$
\begin{array}{r|rrrrr}
-3 & 2 & 0 & -4 & 12 & -6 \\
   &   & -6 & 18 & -42 & 90 \\
\hline
   & 2 & -6 & 14 & -30 & \boxed{84}
\end{array}
$$

Therefore $p(-3) = 84$. ∎

In the synthetic-division algorithm for dividing polynomial $p(x)$ by $x - c$, the Remainder theorem tells us that the number in the lower right corner can be interpreted as either the remainder of the division problem $p(x) \div (x - c)$ or the value of $p(x)$ at c, that is, $p(c)$.

Factor Theorem

We say that 3 is a *factor* of 6 because 6 can be expressed as a product of 3 and another integer; that is, $6 = 3 \cdot 2$. In an analogous manner we say that polynomial $g(x)$ is a *factor* of polynomial $p(x)$ if there is a *polynomial* $h(x)$ such that $p(x) = g(x)h(x)$.

For instance, $x^2 - 1$ is a factor of $2x^3 + 3x^2 - 2x - 3$ because

$$2x^3 + 3x^2 - 2x - 3 = (x^2 - 1)(2x + 3).$$

Suppose $p(x)$ is a given polynomial and c is a given number. We wish to determine if $x - c$ is a factor of $p(x)$. The following theorem will be helpful.

Theorem 2.3

Factor theorem

Suppose $p(x)$ is a polynomial and c is a given number; $x - c$ is a factor of $p(x)$ if and only if $p(c) = 0$.

The proof of this theorem comes in two parts.

1. The *if* part: If $p(c) = 0$, then by Theorem 2.2 $r = 0$. Equation (2.12) gives $p(x) = (x - c) \cdot q(x)$, and so $x - c$ is a factor of $p(x)$.

2. The *only if* part: If $x - c$ is a factor of $p(x)$, then by definition of factor there is a polynomial $g(x)$ such that $p(x) = (x - c)g(x)$. Substituting c for x in this equation gives $p(c) = (c - c)g(c) = 0$.

Example 8   Use the Factor theorem to show that $x + 3$ is a factor of $p(x) = 2x^3 + 5x^2 - 2x + 3$.

Solution   If we can show that $p(-3) = 0$, then by the *if* part of the Factor theorem $x + 3$ is a factor of $p(x)$:

$$p(-3) = 2(-3)^3 + 5(-3)^2 - 2(-3) + 3 = -54 + 45 + 6 + 3 = 0. \quad \blacksquare$$

Example 9   Find a polynomial $p(x)$ of degree 3, with leading coefficient 1, that satisfies $p(-1) = 0$, $p(1) = 0$, and $p(2) = 0$.

Solution   By the *if* part of the Factor theorem we conclude that $(x + 1)$, $(x - 1)$, and $(x - 2)$ are factors. Therefore

$$p(x) = (x + 1)(x - 1)(x - 2) = x^3 - 2x^2 - x + 2. \quad \blacksquare$$

Exercises 2.3

In problems 1 through 6, use synthetic division to find the quotient $q(x)$ and remainder r when $p(x)$ is divided by $d(x)$. In each case write the result in the form given by Eq. (2.12): $p(x) = d(x) \cdot q(x) + r$.

1. $p(x) = 3x^2 - 2x + 1$; $d(x) = x - 2$

2. $p(x) = 4x^3 + 3x^2 - 5x - 4$; $d(x) = x - 3$

3. $p(x) = 3x^4 - 2x^2 + 3x - 5$; $d(x) = x + 2$

4. $p(x) = 2x^4 + 4x^3 - x + 6$; $d(x) = x + 3$

5. $p(x) = 3 - 2x + 5x^2 - 4x^3$; $d(x) = x - 3$

6. $p(x) = 5 - 4x - 3x^2 - x^3$; $d(x) = x - 2$

In problems 7 through 10, use synthetic division to find the quotient and remainder in the given division problems.

7. $(2x^3 - 3x^2 + 4x + 1) \div (x - 3)$

8. $(x^4 - 2x^3 - 2x^2 - 2x - 3) \div (x - 3)$

9. $(x^5 + 32) \div (x - 2)$

10. $(x^6 - 64) \div (x - 2)$

In problems 11 through 14, use the nested-form technique illustrated in Example 5 to find the quotient and remainder when $p(x)$ is divided by $d(x)$.

11. $p(x) = 3x^3 - 2x^2 + 3x - 4$; $d(x) = x + 2.4$

12. $p(x) = 2x^3 + 2x^2 - 6x - 5$; $d(x) = x + 1.4$

13. $p(x) = x^3 + 2x - 3$; $d(x) = x - 1.2$

14. $p(x) = -2x^3 + 3x^2 - 5$; $d(x) = x - 2.3$

In problems 15 through 18, use the Remainder theorem to help solve the given problems.

15. Find the remainder when $p(x) = 4x^{12} - 3x^8 + 5x^3 - 2x + 3$ is divided by $x + 1$

16. Find the remainder when $p(x) = x^{10} - 64x^4 + 3$ is divided by $x - 2$.

17. Given that $p(x) = 2x^3 - 3x^2 + 5x - 4$, find $p(-3)$.

18. Given that $p(x) = 5x^4 - 4x^2 + 2x - 4$, find $p(2)$.

In problems 19 through 28, use the Factor theorem to help solve the given problems.

19. Show that $x + 1$ and $x - 1$ are factors of $x^6 - 1$.

20. Show that $x - 1$ is a factor of $x^5 - 1$, and $x + 1$ is not a factor.

21. Is $x + 2$ a factor of $x^{10} - 8x^6 + 16x^5$?

22. Is $x + 3$ a factor of $x^6 + 3x^5 + x^2 + 3x$?

23. For what positive integers n is $x + 1$ a factor of $x^n + 1$?

24. For what positive integers n is $x - 1$ a factor of $x^n + 1$?

25. Find k such that $x + 2$ is a factor of $2x^3 + 4x^2 + kx - 3$.

26. Find k such that $x + 1$ is a factor of $x^{25} - 3x^{17} + kx^4 - 5x$.

27. Find k such that $x - 1$ is a factor of $x^{36} - kx^{21} + 7x^{10} - 3k$.

28. Find k such that $x - 3$ is a factor of $27x^4 - 81x^3 + kx$.

In problems 29 through 32, use the Factor theorem to help find a polynomial $p(x)$ of degree 3 with leading coefficient 1 and satisfying the given conditions. Give answers in expanded form.

29. $p(-2) = 0$, $p(-1) = 0$, and $p(1) = 0$.

30. $p(0) = 0$, $p(2) = 0$, and $p(4) = 0$.

31. $p(-1) = 0$, $p(1) = 0$, and $p(1.4) = 0$.

32. $p(-2) = 0$, $p(2) = 0$, and $p(3.5) = 0$.

2.4   ZEROS OF POLYNOMIAL FUNCTIONS

A number c is called a *zero of the polynomial function p* if the value of p at c is zero; that is, $p(c) = 0$. We also say that c is a *solution* or *root of the equation* $p(x) = 0$.

In our discussion of graphs of polynomial functions in the preceding section, we were interested in locating the x-intercept points. One does so by finding the roots of $p(x) = 0$, and that is precisely our concern in this section.

If p is a constant (nonzero) polynomial, then obviously p has no zeros. If p is the zero polynomial, then any real number is a zero of p.

Zeros of Polynomial Functions of Degree 1

Any polynomial function p of degree 1 is given by a formula of the type $p(x) = ax + b$, where $a \neq 0$. In this case p has exactly one zero, which is given by $x = -b/a$; that is,

$$p\left(-\frac{b}{a}\right) = a\left(-\frac{b}{a}\right) + b = 0.$$

Zeros of Polynomial Functions of Degree 2

Polynomial functions of degree 2 are of the form $p(x) = ax^2 + bx + c$, where $a \neq 0$. The zeros of p are the roots of the equation $ax^2 + bx + c = 0$. The quadratic formula can be applied to get two roots given by

$$x = \frac{-b + \sqrt{b^2 - 4ac}}{2a} \quad \text{and} \quad x = \frac{-b - \sqrt{b^2 - 4ac}}{2a}.$$

In general, these are two different numbers, but in special cases, where $b^2 - 4ac = 0$, they are equal.

**Zeros of Polynomial
Functions of Degree Greater Than 2**

We already have general formulas for finding the roots of polynomial equations of degree 1 or 2. Formulas can be derived for solving the general polynomial equations for degree 3 and degree 4, but they are complicated to state and cumbersome to use in any particular problem. Mathematicians have spent considerable effort attempting to find formulas for solving general polynomial equations of degree 5 or greater, but amazingly, there are no such formulas. This fact was first proved by the Norwegian mathematician Niels Henrik Abel (1802–1829).

In discussing methods for finding zeros of polynomials of a degree greater than 2, we shall therefore be satisfied with developing some techniques that are helpful in solving the problem for any given polynomial. This is the approach taken in this section and in the next section. We shall use theorems of the preceding section and state some additional theorems that will be helpful in the search for zeros of a given polynomial. Let us first consider some examples.

Example 1   Determine whether or not -1 and 2 are zeros of $p(x) = x^3 + x^2 - 2x - 2$.

Solution $p(-1) = (-1)^3 + (-1)^2 - 2(-1) - 2 = 0$, and so -1 is a zero.

$p(2) = 2^3 + 2^2 - 4 - 2 = 6$, and so 2 is not a zero.   ■

Example 2   Find the zeros of $p(x) = (x - 3)(x + 1)(x + 2)$.

Solution   Since $x - 3$ is a factor of $p(x)$, by the Factor theorem 3 is a zero of p. Similarly -1 and -2 are also zeros of p. Thus p has exactly three zeros.   ■

Example 3   Find the zeros of $p(x) = (x - 2)^2(x + 3)^3$.

Solution   We can write $p(x)$ as $p(x) = (x - 2)(x - 2)(x + 3)(x + 3)(x + 3)$. Since $x - 2$ is a factor of $p(x)$, by the Factor theorem 2 is a zero of p. Since $x - 2$ occurs twice as a factor of $p(x)$, we say that 2 is a zero of *multiplicity 2*. Similarly, -3 is a zero of *multiplicity 3*. We say that p has *five zeros,* even though there are only two distinct zeros.   ■

As noted in Example 3, a polynomial function can have *multiple zeros*. In general, if $(x - c)^k$ is a factor of $p(x)$ and $(x - c)^{k+1}$ is not a factor, then we say c is a zero of p of *multiplicity k*.

In Examples 2 and 3 it was a simple matter to find the zeros of a polynomial function because the formula for $p(x)$ was given in factored form. If $p(x)$ is given in standard form, finding the zeros of p is in general considerably more difficult. The next two theorems will be helpful in solving such problems.

Theorem 2.4   A polynomial function of degree n has at most n zeros.*

The statement in this theorem can be proved by using the Factor theorem. Suppose p has more than n zeros; call the first $n + 1$ of them $c_1, c_2, \ldots, c_{n+1}$. Applying the Factor theorem gives

$$p(x) = (x - c_1)(x - c_2) \cdots (x - c_{n+1})g(x), \qquad (2.15)$$

where $g(x)$ is a polynomial. Expanding Eq. (2.15), we see that $p(x)$ is a polynomial of degree greater than n. This contradicts the hypothesis that $p(x)$ is a degree n, and so p cannot have more than n zeros.

---

*   In advanced courses, polynomial functions are defined over the set of complex numbers. In that setting it can be shown that every polynomial of degree n has *exactly* n zeros in the set of complex numbers, where any zero of multiplicity k is counted k times.

Theorem 2.5

Rational-root theorem

Suppose p is a polynomial function given by

$$p(x) = a_n x^n + a_{n-1} x^{n-1} + \cdots + a_1 x + a_0,$$

where $a_n, a_{n-1}, \ldots, a_1, a_0$ *are integers*. If the rational number k/m (in lowest terms) is a root of $p(x) = 0$, then k divides a_0 and m divides a_n evenly.

This theorem can be proved by using the Factor theorem and divisibility properties of integers. The following examples illustrate its application.

Example 4   Suppose $p(x) = x^3 - 6x - 4$. Find the roots of $p(x) = 0$.

Solution   We first look for rational roots. If there is a rational root k/m, then by Theorem 2.5, k must divide the constant term -4 and m must divide the leading coefficient, which is 1. Therefore the only rational numbers that are possible roots are $1, -1, 2, -2, 4, -4$. We can now try these to see if any one actually is a root. It is easy to check 1 and -1 by substituting directly into the given formula to get $p(1) = -9$ and $p(-1) = 1$, and so neither is a root.

We can now try the remaining possibilities by substituting into the given formula or by synthetic division, as we now illustrate.

$$
\begin{array}{r|rrrr}
2 & 1 & 0 & -6 & -4 \\
  &   & 2 & 4 & -4 \\
\hline
  & 1 & 2 & -2 & \boxed{-8} \\
\end{array}
\qquad
\begin{array}{r|rrrr}
-2 & 1 & 0 & -6 & -4 \\
   &   & -2 & 4 & 4 \\
\hline
   & 1 & -2 & -2 & \boxed{0} \\
\end{array}
$$

$$p(2) \qquad\qquad \text{Factor of } p(x) \quad p(-2)$$

Thus $p(-2) = 0$, and so -2 is a root. Synthetic division also gives factors of $p(x)$:

$$p(x) = (x + 2)(x^2 - 2x - 2).$$

Hence the roots of $p(x) = 0$ are given by solving $(x + 2)(x^2 - 2x - 2) = 0$. Since a product can be zero only if one of the factors is zero, we see that other roots are given by solving the quadratic equation $x^2 - 2x - 2 = 0$. Applying the quadratic formula gives

$$x = \frac{2 \pm \sqrt{(-2)^2 - 4(1)(-2)}}{2} = 1 \pm \sqrt{3}.$$

Therefore the solutions to $x^3 - 6x - 4 = 0$ are -2, $1 + \sqrt{3}$, and $1 - \sqrt{3}$. The given equation has one rational root and two irrational roots.   ■

Example 5   Find the rational roots of the equation

$$x^4 + \tfrac{3}{2}x^2 + \tfrac{1}{2}x + \tfrac{3}{2} = 0.$$

Solution   Multiplying both sides of the given equation by 2, we get

$$2x^4 + 3x^2 + x + 3 = 0. \tag{2.16}$$

Applying the Rational-root theorem to Eq. (2.16) gives us the following *possible rational roots:* $1, -1, 3, -3, \tfrac{1}{2}, -\tfrac{1}{2}, \tfrac{3}{2}, -\tfrac{3}{2}$. Since all coefficients in Eq. (2.16) are positive, it is obvious that there are no positive roots (this eliminates four of the eight possible rational roots). We now try $-\tfrac{1}{2}$ by synthetic division.

$$
\begin{array}{r|rrrrr}
-\tfrac{1}{2} & 2 & 0 & 3 & 1 & 3 \\
              &   & -1 & \tfrac{1}{2} & -\tfrac{7}{4} & \tfrac{3}{8} \\
\hline
              & 2 & -1 & \tfrac{7}{2} & -\tfrac{3}{4} & \boxed{\tfrac{27}{8}}
\end{array}
$$

Note that the numbers in the bottom row alternate in sign. Convince yourself that there is no need to try numbers smaller than $-\tfrac{1}{2}$ since such a number will give a result larger than $\tfrac{27}{8}$ in the lower right corner in synthetic division. Hence $-1, -\tfrac{3}{2}, -3$ cannot be roots of the given equation.

We conclude that the given equation has no rational roots; if there are any real-number roots, they are irrational. In the next section we shall consider techniques for finding irrational roots.   ■

Example 6   Find the roots of $p(x) = 0$, where $p(x) = 3x^3 - x^2 + 2x - 8$.

Solution   First look for rational roots. According to Theorem 2.6, if there are any rational roots, they will come from the set

$$\pm 1, \pm 2, \pm 4, \pm 8, \pm\tfrac{1}{3}, \pm\tfrac{2}{3}, \pm\tfrac{4}{3}, \pm\tfrac{8}{3}.$$

Note that the coefficients of $p(x)$ alternate in sign. Hence, if x is a negative number, the corresponding value of $p(x)$ will be negative, and so there will be no negative roots. Thus we can eliminate eight of the possible rational roots. Trying the positive numbers, we eventually get

$$
\begin{array}{r|rrrr}
\tfrac{4}{3} & 3 & -1 & 2 & -8 \\
             &   & 4 & 4 & 8 \\
\hline
             & \underbrace{3 \quad 3 \quad 6} & & & \boxed{0}
\end{array}
$$

$$\text{Factor of } p(x) \qquad p(\tfrac{4}{3})$$

Thus $\tfrac{4}{3}$ is a root, and the other roots will come from solution of $3x^2 + 3x + 6 = 0$. Dividing by 3 and solving $x^2 + x + 2 = 0$ by using the quadratic formula gives two complex numbers:

$$x = \frac{-1 \pm \sqrt{-7}}{2}.$$

Therefore the given equation has only one real-number solution, which is $\frac{4}{3}$.

■

Algebraic Numbers

In this section we have been discussing zeros of polynomial functions over the set of integers. Actually we are talking about an important subset of the real numbers called the algebraic numbers, defined as follows:

Definition 2.2   An *algebraic number* is a real number that is a root of a polynomial equation with *integer coefficients*.

Clearly, any rational number a/b is an algebraic number since it is a root of $bx - a = 0$, where a and b are integers. In Example 4 we saw that $1 + \sqrt{3}$ and $1 - \sqrt{3}$ are roots of $x^3 - 6x - 4 = 0$, and so they are algebraic numbers. A natural question is: Are there any real numbers that are not algebraic numbers? The answer is: Yes, there are many, although at this point you may be acquainted with only a few such numbers, the most famous of which is the number π. As your knowledge of the real-number system increases, you will learn that there are infinitely many real numbers that are not algebraic. The set of nonalgebraic numbers, which is an important subset of $\Re$, is called the set of *transcendental numbers.**

Example 7   Prove that $2 + \sqrt{5}$ is an algebraic number.

Solution   We need to find a polynomial function p with integer coefficients such that $2 + \sqrt{5}$ is a root of $p(x) = 0$. The Factor theorem tells us that $[x - (2 + \sqrt{5})]$ must be a factor of $p(x)$. Suppose we try $[x - (2 - \sqrt{5})]$ as another factor of $p(x)$. Will the product give a polynomial with *integral coefficients*?

$$\begin{aligned} p(x) &= [x - (2 + \sqrt{5})][x - (2 - \sqrt{5})] \\ &= [x - 2 - \sqrt{5}][x - 2 + \sqrt{5}] \\ &= [(x - 2) - \sqrt{5}][(x - 2) + \sqrt{5}] \\ &= (x - 2)^2 - (\sqrt{5})^2 \\ &= x^2 - 4x + 4 - 5 \\ &= x^2 - 4x - 1. \end{aligned}$$

Therefore the equation $x^2 - 4x - 1 = 0$ has integral coefficients with $2 + \sqrt{5}$ as a root, and so $2 + \sqrt{5}$ is an algebraic number. The reader should note that in the above steps we regrouped terms of the two factors to get a product of the form $(a - b)(a + b)$, which is equal to $a^2 - b^2$. This simplifies the process of multiplying the two factors.

■

---

*   In general, it is difficult to prove that a number is transcendental. In 1882 Lindemann proved that π is transcendental.

Exercises 2.4

In problems 1 through 9, find a polynomial function p with leading coefficient 1 and the given numbers as zeros of p. Express answers in expanded form.

1. $-2, 1, 3$ **2.** $0, 2, 4$ **3.** $-4, -1, 4$

4. $-1, 2, 2$ **5.** $-3, 1, 1, 1$ **6.** $-\sqrt{2}, -1, 0, \sqrt{2}$

7. $1, 1 + \sqrt{5}, 1 - \sqrt{5}$ **8.** $-1, 1, 1 + \sqrt{2}, 1 - \sqrt{2}$ **9.** $0, -1, \sqrt{3}$

In problems 10 through 13, determine the zeros of the given polynomial; give the multiplicity of each zero.

10. $p(x) = (x - 1)(x + 2)(x + 5)$ **11.** $p(x) = x(x + 2)(x - \sqrt{5})$

12. $p(x) = (x + 2)^2(x - 3)^4$ **13.** $p(x) = x^2(x - 1)^2(x + 2)$

In problems 14 through 17, a polynomial $p(x)$ and a number c are given. Determine whether or not c is a zero of p.

14. $p(x) = x^3 + 4x^2 - x - 4; \ c = -4$ **15.** $p(x) = x^3 - \sqrt{2}x^2 + 3x - 3\sqrt{2}; \ c = \sqrt{2}$

16. $p(x) = x^{10} - 32; \ c = -\sqrt{2}$ **17.** $p(x) = x^4 + 3x^3 + x^2 - 4x - 4; \ c = -2$

In problems 18 through 20, determine whether or not the Rational-root theorem can be applied to yield any information concerning roots of $p(x) = 0$ for each of the given polynomials.

18. a) $p(x) = 2x^4 - 3x^2 + 5x - 1$ **b)** $p(x) = 4x^3 - 2x^2 + \sqrt{3}x + 3$

19. a) $p(x) = 5x^3 + \sqrt{3}x - 4$ **b)** $p(x) = 2x^4 - 3x^3 + 1.4x - 3.2$

20. a) $p(x) = x^3 - 3.5x^2 + 7.3x - 3$ **b)** $p(x) = 2x^3 - \pi x^2 + 4x - 3$

In problems 21 through 32, find all rational roots of the given polynomial equations.

a) Use Theorem 2.5 to list all *possible* rational roots.

b) Remove any that can obviously be eliminated if the polynomial has special features, such as in Examples 5 and 6.

c) Check the remaining ones.

21. $x^3 - 4x^2 + 2x - 8 = 0$ **22.** $4x^3 - 4x^2 - 19x + 10 = 0$

23. $x^3 - 2.5x^2 - 7x - 1.5 = 0$ **24.** $3x^3 + 11x^2 + 12x + 4 = 0$

25. $x^3 - \frac{5}{3}x^2 - \frac{11}{3}x - 1 = 0$ **26.** $x^3 - 3.5x^2 + 0.5x + 5 = 0$

27. $5x^3 + x^2 - 15x - 3 = 0$ **28.** $x^3 - 3x + 2 = 0$

29. $18x^3 + 27x^2 + 13x + 2 = 0$ **30.** $x^3 + \frac{14}{3}x^2 + \frac{17}{3}x + 2 = 0$

31. $x^4 + 4x^3 - 5x^2 - 36x - 36 = 0$ **32.** $4x^4 + 24x^3 + 35x^2 - 6x - 9 = 0$

In problems 33 through 36, use the Rational-root theorem to factor the polynomials.

33. $p(x) = 2x^3 + 3x^2 + 3x + 1$ **34.** $p(x) = 3x^3 - 4x^2 + 7x - 2$

35. $p(x) = 2x^4 + x^3 + x^2 + x - 1$ **36.** $p(x) = 3x^4 + 5x^3 - 4x^2 - 10x - 4$

In problems 37 through 40, show that the given number is algebraic.

37. $-1 - \sqrt{2}$ **38.** $2 + \sqrt{5}$ **39.** $\dfrac{1 + \sqrt{5}}{2}$ **40.** $\dfrac{1 - \sqrt{3}}{3}$

2.5 IRRATIONAL ZEROS OF POLYNOMIAL FUNCTIONS

In the preceding section we saw that if the coefficients of a polynomial $p(x)$ are rational numbers, then by applying the Rational-root theorem and using synthetic division we can determine all the rational roots of the equation $p(x) = 0$. Irrational roots can also be found in exact form if the problem can be reduced to that of solving quadratic equations (see Example 4 of the preceding section). If the degree of $p(x)$ is greater than 2 and there are no rational roots, then other techniques for finding irrational roots are needed. Although there are many iterative methods studied in numerical analysis courses, it is not in our interest to present them here. We shall see how to find decimal approximations for irrational roots by considering properties of graphs. Before stating an important property of graphs of polynomial functions, we first state a theorem that will give some information concerning how many roots a polynomial equation may have.

In advanced courses in the theory of complex variables, polynomial functions are studied in a more general setting in which they are defined over the set of complex numbers. In such situations it can be shown that a polynomial function of degree n has *exactly* n zeros. Also, if the coefficients are real numbers then the nonreal complex-number zeros come in pairs. This implies the following theorem.

Theorem 2.6   Suppose $p(x)$ is a polynomial of degree n. The number of real zeros of p is equal to n or is less than n by an even number.*

Example 1   Using Theorem 2.6, what can you conclude about the number of zeros in each of the following:

   a) $p(x) = 4x^3 - 3x^2 + 2x - 1$ **b)** $f(x) = 5x^4 - 3x^3 + 2x - 2$

Solution   **a)** The degree of p is 3, and so the number of zeros is 3 or 1. That is, there is definitely at least one real-number zero.

   b) The degree of f is 4, and so the number of zeros is 4, 2, or 0.   ■

As suggested above, properties of graphs of polynomial functions will be used to aid in the search for zeros. Let p be a polynomial function, and consider the graph of $y = p(x)$. To determine the x-intercept points, let $y = 0$ and find the values of x that satisfy the equation $p(x) = 0$. These are precisely the zeros of p. Thus we shall look for points where the graph crosses the x-axis. The following property will be helpful in finding such points.

---

*   Recall that in this text our discussion is restricted to polynomial functions with real-number coefficients and the set of real numbers as domain.

Important Property of
Graphs of Polynomial Functions

The graph of a polynomial function is a smooth, continuous (unbroken) curve, such as those shown in Fig. 2.5. The graph will have no gaps, holes, or sharp corners. For instance, none of the curves shown in Fig. 2.6 can be graphs of polynomial functions.

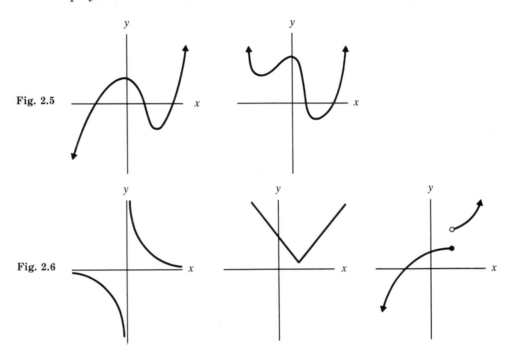

Fig. 2.5

Fig. 2.6

As a consequence of this continuity property, we get the following useful criterion for locating zeros, which is stated as a theorem.

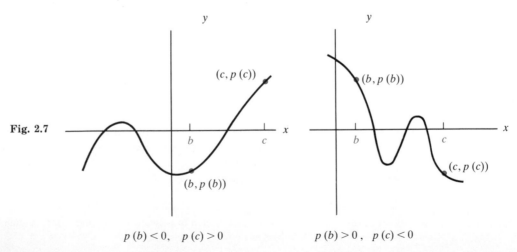

Fig. 2.7

$p(b) < 0, \quad p(c) > 0$ $p(b) > 0, \quad p(c) < 0$

Theorem 2.7   Suppose p is a polynomial function and b and c are numbers such that $p(b)$ and $p(c)$ have opposite signs. Then p has at least one zero between b and c.

Geometrically, this theorem states that the graph of $y = p(x)$ crosses the x-axis at least once between the point $(b, p(b))$ and $(c, p(c))$. This is illustrated in Fig. 2.7.

The following examples illustrate a procedure for finding irrational zeros.

Example 2   Locate the zeros of the function given by $f(x) = x^3 - 5x - 1$.

Solution   Theorem 2.6 tells us that the number of zeros is either 3 or 1. Let us first look for rational-number zeros. Applying Theorem 2.5, we find that the only *possible* rational zeros are 1 and -1. Evaluating f at these values gives $f(1) = -5$ and $f(-1) = 3$.

Hence, neither is a zero, and so there are no rational zeros. However, since $f(1) < 0$ and $f(-1) > 0$, Theorem 2.7 tells us that there is at least one zero between -1 and 1.

We now begin our search for irrational zeros by drawing a graph of $y = f(x)$. The (x, y) points given by the following table are plotted and the graph is drawn, as shown in Fig. 2.8.

x	-3	-2	-1	0	1	2	3
y	-13	1	3	-1	-5	-3	11

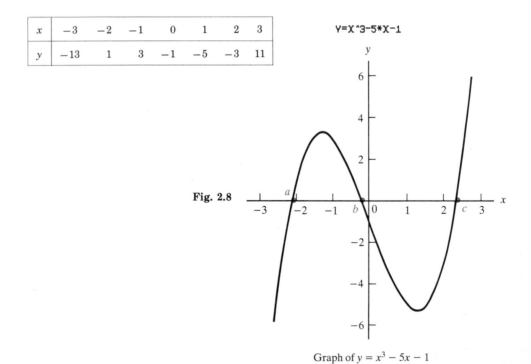

Fig. 2.8

Graph of $y = x^3 - 5x - 1$

The zeros of f correspond to the x-intercept points of the graph. From Fig. 2.8 we see that f has three zeros, labeled a, b, and c. All three are irrational numbers; a is between -3 and -2, b is between -1 and 0, and c is between 2 and 3. This problem is continued in the next example. ■

Example 3   In Example 2, three zeros of $f(x) = x^3 - 5x - 1$ were located. Each is an irrational number. Find a two-decimal-place approximation for the largest of these.

Solution   The portion of the graph for x between 2 and 3 has been enlarged in Fig. 2.9. From this diagram our first estimate of c is 2.3. Evaluating f at 2.3 gives $f(2.3) = -0.33$. Since $f(2.3) < 0$, we conclude from the diagram that $c > 2.3$, and so we try 2.4. We find that $f(2.4) = 0.82$, and again from the diagram we see that $c < 2.4$. Thus c is bracketed between 2.3 and 2.4. Figure 2.10 shows an enlarged version of the graph of $y = f(x)$ for $2.3 < x < 2.4$. From this diagram our next guess for c is 2.33. Evaluating f at 2.33 gives $f(2.33) = -0.0007$; since $f(2.33) < 0$, $c > 2.33$, as the diagram shows. We try 2.34 and get $f(2.34) = 0.1129$. Since $f(2.34) > 0$, we see from the diagram that $c < 2.34$. We now try 2.335 and get $f(2.335) = 0.0559$. Hence we conclude that $2.330 < c < 2.335$, and so 2.33 is the desired approximation for c.

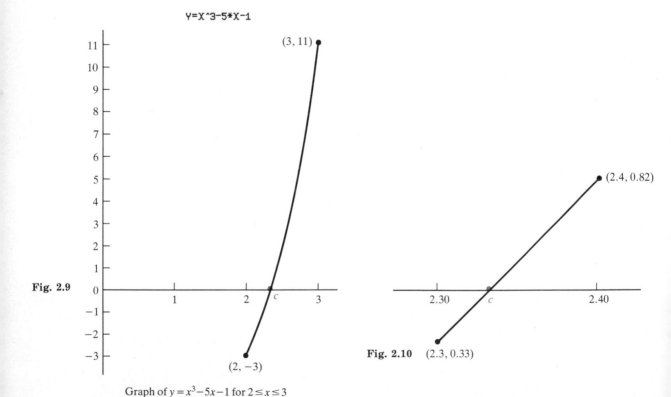

Fig. 2.9

Graph of $y = x^3 - 5x - 1$ for $2 \le x \le 3$

Fig. 2.10 $(2.3, 0.33)$

Note that the graph drawn in Fig. 2.10 appears to be a line segment; actually it is not, but a straight-line approximation provides us with a good guess for the next approximate value of c.   ■

Example 4 Locate the zeros of $f(x) = x^4 - 3x^3 - x^2 + 3x + 3$. If any of these is an irrational number, determine the smallest, correct to one decimal place.

Solution According to the Rational-root theorem, the *possible* rational roots are $-3, -1, 1, 3$. First evaluate f at each of these to see if any may be a zero, and at the same time include the results in the following table in preparation for drawing the graph of $y = f(x)$, as shown in Fig. 2.11. We express $f(x)$ in nested form and use a calculator to get the y values. Note that in the table we started with integral values of x and then included additional values of x where the curve appears to be turning.

x	-2	-1	0	1	2	3	4	-0.5	0.5	1.5	2.5
y	33	3	3	3	-3	3	63	1.69	3.94	0.19	-3.56

Y=X^4-3*X^3-X^2+3*X+3

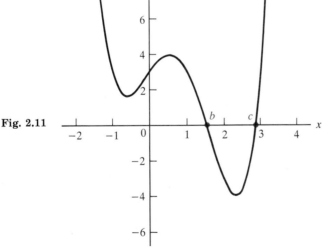

Fig. 2.11

Graph of $x^4 - 3x^3 - x^2 + 3x + 3$

From the graph in Fig. 2.11 we conclude that f has two irrational zeros, labeled b and c; b is between 1 and 2, and c is between 2 and 3. To find a decimal approximation for b, we note from the table that $f(1.5) = 0.19$ and $f(2) = -3$, and so b is bracketed between 1.5 and 2. We now try 1.55 and get $f(1.55) = -0.15$; thus b is bracketed between 1.50 and 1.55. Therefore 1.5 is the desired one-decimal-place approximation for b.   ■

Exercises 2.5

In problems 1 through 8, do the following.

a) Using Theorem 2.6, what can you conclude about the possible number of zeros of f?

b) Draw a graph of $y = f(x)$. Include in your table of x, y pairs all values of x that might possibly be rational zeros of f.

c) From the graph what can you now conclude about the number of zeros of f?

1. $f(x) = x^3 - 5x - 2$ **2.** $f(x) = x^3 - 3x - 1$

3. $f(x) = x^3 + 2x^2 - 3x - 1$ **4.** $f(x) = x^3 - x^2 - 3x + 1$

5. $f(x) = x^4 - 3x^3 - x^2 + 3x + 2$ **6.** $f(x) = x^4 - 3x^3 - x^2 + 3x - 1$

7. $f(x) = x^4 - x^3 - 2x^2 + 3$ **8.** $f(x) = x^4 - x^3 - 2x^2 - 1$

In problems 9 through 14, locate the irrational zeros of f between consecutive integers; then determine the smallest, correct to one decimal place.

9. $f(x) = x^3 - 2x - 5$ **10.** $f(x) = x^3 - 3x^2 - 3$

11. $f(x) = x^3 + 2x^2 - 3x + 2$ **12.** $f(x) = x^3 - 2x^2 - x + 3$

13. $f(x) = x^4 - 4x^3 - 2x^2 + 12x + 1$ **14.** $f(x) = x^4 - 3x^3 - x^2 + 5x - 1$

In problems 15 through 20, locate the irrational zeros of f between consecutive integers; then determine the largest, correct to two decimal places.

15. $f(x) = x^3 - 3x^2 - 3$ **16.** $f(x) = x^3 - 3x - 1$

17. $f(x) = x^3 + 2x^2 - 3x - 2$ **18.** $f(x) = x^3 + 3x^2 - x - 4$

19. $f(x) = x^4 - x^3 - 2x^2 + x - 1$ **20.** $f(x) = x^4 - 2x^3 - x^2 + 3x - 3$

2.6   **RATIONAL FUNCTIONS**

Suppose f and g are polynomial functions. In Chapter 1 we saw how new functions can be obtained by combining f and g in any of five ways. It is easy to show that $f + g, f - g, f \cdot g$, and $f \circ g$ are also polynomial functions. In general, however, f/g is not a polynomial function. Let us now investigate such functions.

Definition 2.3   Suppose f and g are polynomial functions, where $g(x)$ is a nonzero polynomial. The function f/g is called a *rational function*. That is, the function h given by $h(x) = f(x)/g(x)$ is a rational function.

The following example illustrates some properties of rational functions.

Example 1   Draw a graph of the rational function f given by $f(x) = \dfrac{2x}{x - 1}$.

Solution   First get a table of x, y values that satisfy the equation $y = \dfrac{2x}{x - 1}$. Included in the table are the x- and y-intercept points:

$$\textit{y-intercept: } x = 0, \text{ then } y = \frac{2(0)}{0-1} = \frac{0}{-1} = 0$$

$$\textit{x-intercept: } y = 0, \text{ then } \frac{2x}{x-1} = 0, \text{ and so } x = 0.$$

Therefore $(0, 0)$ is both an x- and a y-intercept point. The function f is not defined at 1, but we include several values of x near 1 in the table in order to see the behavior of the graph in that region. Also included are some large values of x (both positive and negative), even though the corresponding points will not be drawn in our diagram. These give us some idea of the graph at extreme values of x.

x	-100	-10	-5	-3	-2	-1	0	0.5	0.9	0.99	1.01	1.1	1.5	2	3	10	100
y	1.98	1.82	1.67	1.5	1.3	1	0	-2	-18	-198	202	22	6	4	3	2.22	2.02

Y=2*X/(X−1)

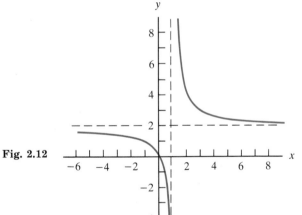

Fig. 2.12

Graph of $y = \dfrac{2x}{x-1}$

Plotting the points given in the table and drawing a curve through them gives the graph shown in Fig. 2.12.

The graph in Fig. 2.12 illustrates some features that are in general peculiar to rational functions. For values of x slightly less than 1, the corresponding values of y are negatively large; similarly, when x is slightly greater than 1, y is positively large. This is denoted by

$$\lim_{x \to 1^-} f(x) = -\infty \qquad \text{and} \qquad \lim_{x \to 1^+} f(x) = \infty.$$

These equations are read: "As x approaches 1 from below, $f(x)$ approaches negative infinity" and "As x approaches 1 from above, $f(x)$ approaches infinity," respectively. The vertical line whose equation is $x = 1$ is called a *vertical asymptote* of the graph of $y = f(x)$.

From the graph in Fig. 2.12 we note that for large values of x (positive or negative) the corresponding values of y are near 2. This feature is denoted by

$$\lim_{x \to \infty} f(x) = 2 \quad \text{and} \quad \lim_{x \to -\infty} f(x) = 2.$$

The first of these equations is read: "As x approaches infinity, $f(x)$ approaches the number 2"; the second is read in a similar way. The horizontal line $y = 2$ is called a *horizontal asymptote* of the graph of $y = f(x)$.

Graphs of Rational Functions

Here we summarize a few steps that should be included in drawing graphs of rational functions. Suppose h is a rational function given by $h(x) = f(x)/g(x)$. Also suppose polynomials $f(x)$ and $g(x)$ have no common zeros (we shall illustrate the case in which there are common zeros in Example 6); this condition is described by saying $f(x)/g(x)$ is in *lowest terms*.

1. Find the roots of $f(x) = 0$; these will give the *x-intercept points*.

2. Find the roots of $g(x) = 0$; these will give vertical asymptotes. For instance, if $g(b) = 0$, then the graph of h will approach the line $x = b$ as x gets close to b. The line $x = b$ is called a *vertical asymptote* of the graph.

3. Determine what happens to values of $h(x)$ as x becomes large positive or large negative. For instance, if $\lim_{x \to \infty} h(x) = c$, then the graph gets close to the

   horizontal line $y = c$. Such a line is called a *horizontal asymptote* of the graph.

The next example illustrates a technique for determining the behavior of rational functions for large values of $|x|$.

Example 2   Discuss the behavior of the given rational functions for large values of $|x|$.

     a) $f(x) = \dfrac{4x^3 - 3x + 1}{2x^3 - 3}$ **b)** $g(x) = \dfrac{x^2 - 4}{2x^4 - x}$ **c)** $h(x) = \dfrac{x^2 + 1}{x}$

Solution   In each case, first divide the numerator and denominator by x^k, where k is the degree of the denominator. Using the resulting formula, we can determine the behavior of the function when $|x|$ is large.

     a) Dividing numerator and denominator by x^3 gives

$$f(x) = \frac{4x^3 - 3x + 1}{2x^3 - 3} = \frac{4 - \dfrac{3}{x^2} + \dfrac{1}{x^3}}{2 - \dfrac{3}{x^3}}$$

We see that $-3/x^2$, $1/x^3$, $-3/x^3$ become small as x becomes large (positive or negative). Hence $f(x)$ is very nearly equal to $4/2 = 2$ for large values of x. This is denoted by $\lim_{x\to\infty} f(x) = 2$ and $\lim_{x\to-\infty} f(x) = 2$.

b) Dividing numerator and denominator by x^4 gives

$$g(x) = \frac{x^2 - 4}{2x^4 - x} = \frac{\dfrac{1}{x^2} - \dfrac{4}{x^4}}{2 - \dfrac{1}{x^3}}.$$

Since $1/x^2$, $-4/x^4$, $-1/x^3$ become small as x becomes large, $g(x)$ approaches $0/2$. Hence $\lim_{x\to\infty} g(x) = 0$ and $\lim_{x\to-\infty} g(x) = 0$.

c) Dividing numerator and denominator by x gives

$$h(x) = \frac{x^2 + 1}{x} = \frac{x + \dfrac{1}{x}}{1} = x + \frac{1}{x}.$$

The term $1/x$ approaches zero when x becomes large, and so $h(x)$ approaches x. In terms of a graph, we say that the graph of $y = h(x)$ approaches the line $y = x$ as x becomes large. We call the line $y = x$ an *oblique asymptote* for the graph of $y = h(x)$. ■

In general, *the graph of a rational function will have an oblique asymptote whenever the degree of the numerator is one greater than that of the denomina·tor.*

Example 3   Draw a graph of $f(x) = \dfrac{1}{(x - 1)^2}$. Give equations of any asymptotes.

Solution   *x-intercept:* There is no value of x that will give $y = 0$, and so there is no x-intercept point.

y-intercept: For $x = 0$, $y = \dfrac{1}{(0 - 1)^2} = \dfrac{1}{1} = 1$; thus $(0, 1)$ is the y-intercept point.

Vertical asymptotes: The denominator is zero for $x = 1$, and so $x = 1$ is a vertical asymptote. Also note that $\lim_{x\to1^-} f(x) = \infty$ and $\lim_{x\to1^+} f(x) = \infty$. Thus y becomes positively large when x approaches 1 from below or above.

Horizontal asymptotes: $\lim_{x\to\infty} \dfrac{1}{(x - 1)^2} = 0$ and $\lim_{x\to-\infty} \dfrac{1}{(x - 1)^2} = 0$. Hence, the line $y = 0$ (the x-axis) is a horizontal asymptote.

Incorporating all of the above information along with a few plotted points gives the graph shown in Fig. 2.13.

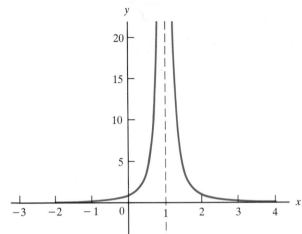

Fig. 2.13

Graph of $y = \dfrac{1}{(x-1)^2}$

Example 4   Draw a graph of $y = g(x)$, where

$$g(x) = \frac{x^2 - x}{x + 1}.$$

Give equations of any asymptotes.

Solution   *x-intercepts:* These are given by solving $x^2 - x = 0$. Thus $(0, 0)$ and $(1, 0)$ are x-intercept points.

y-intercept: If $x = 0$, then $y = 0$, and so $(0, 0)$ is the y-intercept point.

Asymptotes: The denominator is zero for $x = -1$. Therefore the line $x = -1$ is a vertical asymptote. Horizontal or oblique asymptotes are determined by considering the behavior of the function g when x becomes large.

Behavior for large x: Instead of dividing numerator and denominator by x to a power, let us divide $x + 1$ into $x^2 - x$. This gives $x - 2$ as the quotient and 2 as the remainder. Therefore $g(x)$ can be written as

$$g(x) = x - 2 + \frac{2}{x + 1}$$

From this result note that the difference between $g(x)$ and $(x - 2)$ approaches zero as x becomes large, positively or negatively. That is,

$$\lim_{x \to \infty} [g(x) - (x - 2)] = \lim_{x \to \infty} \frac{2}{x + 1} = 0.$$

This tells us that when x becomes large, the graph of $y = g(x)$ comes very near the line $y = x - 2$.

In the following table we include large values of x to substantiate this conclusion. The line $y = x - 2$ is an oblique asymptote for the graph of $y = g(x)$. By including the features discussed above and plotting the points given in the table, we get the graph shown in Fig. 2.14.

x	-4	-3	-2.5	-2	-1.1	-1.01	-0.99	-0.9	-0.5	0	0.5	1	2	3	4
$g(x)$	-7	-6	-5.8	-6	-23	-203	197	17	1.5	0	-0.17	0	0.67	1.5	2.4

x	10	100	1000	. . .	-10	-100	-1000
$y = g(x)$	8.18	98.02	998.002	. . .	-12.22	-102.02	-1002.002
$y = x - 2$	8	98	998	. . .	-12	-102	-1002

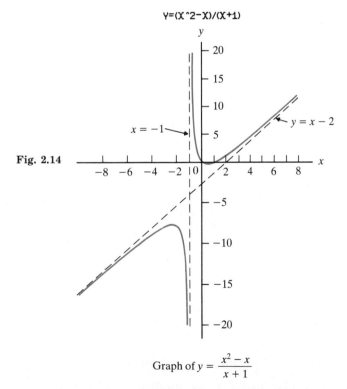

Fig. 2.14

Graph of $y = \dfrac{x^2 - x}{x + 1}$

Example 5   Draw a graph of $y = f(x)$, where $f(x) = \dfrac{x}{x^2 + 1}$. Give equations of any asymptotes.

Solution   *x-intercepts:* The numerator is zero only when $x = 0$. Thus $(0, 0)$ is the only x-intercept point.

y-intercept: If $x = 0$, $y = 0/(0^2 + 1) = 0$, and so $(0, 0)$ is the y-intercept point.

Asymptotes: The denominator is not equal to zero for any value of x; hence there are no vertical asymptotes. Dividing numerator and denominator of $f(x)$ by x^2 gives

$$f(x) = \frac{1/x}{1 + 1/x^2}.$$

From this we see that $f(x)$ approaches zero when x becomes large, positively or negatively. Thus $y = 0$ (the x-axis) is a horizontal asymptote.

Symmetry: $f(-x) = \dfrac{-x}{(-x)^2 + 1} = \dfrac{-x}{x^2 + 1} = -f(x).$

Therefore $f(-x) = -f(x)$ for all values of x, and so the graph is symmetric with respect to the origin. This suggests that it is sufficient to include only nonnegative values of x in the following table. Plotting points given in this table and incorporating the features discussed above, we get the graph shown in Fig. 2.15.

x	0	0.5	1.0	1.5	2.0	3	4	10	100
y	0	0.40	0.50	0.46	0.40	0.30	0.24	0.10	0.01

Fig. 2.15

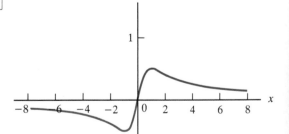

Graph of $y = \dfrac{x}{x^2 + 1}$

Example 6   Draw a graph of the rational function $y = f(x)$ given by $f(x) = (x^2 - 1)/(x + 1)$.

Solution   We can get a simpler formula for $f(x)$ as follows:

$$f(x) = \frac{x^2 - 1}{x + 1} = \frac{(x + 1)(x - 1)}{x + 1} = x - 1 \quad \text{for} \quad x \neq 1.$$

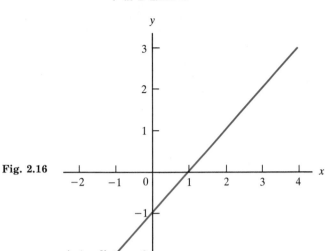

Y=(X^2−1)/(X+1)

Fig. 2.16

Graph of $y = \dfrac{x^2 - 1}{x + 1}$

Therefore the graph of $y = f(x)$ consists of the same points as the graph of $y = x - 1$ without the point $(-1, -2)$. That is, the graph is a line with a "hole" in it, as shown in Fig. 2.16. The open circle about the point $(-1, -2)$ indicates it is not included in the graph. ▪

Exercises 2.6

In problems 1 through 6, determine the behavior of the given rational function for positively large and negatively large values of x.

1. $f(x) = \dfrac{4x^2 + 2x - 1}{x^2 - 4}$ **2.** $f(x) = \dfrac{x^2 + 2x + 3}{1 - x^2}$ **3.** $g(x) = \dfrac{2x - 3}{x^2 + 3x - 4}$

4. $g(x) = \dfrac{2x^2 - 3}{x^3 + 2x - 1}$ **5.** $f(x) = \dfrac{3x^2 - 2x + 1}{x + 2}$ **6.** $f(x) = \dfrac{2x^2 - 3x - 4}{x - 1}$

In problems 7 through 16, for the given functions determine the equations of

a) the vertical asymptotes   **b)** the horizontal asymptotes.

7. $f(x) = \dfrac{3x}{2x + 4}$ **8.** $f(x) = \dfrac{2x - 3}{x + 3}$ **9.** $g(x) = \dfrac{3x^2 + 2x + 1}{x^2 - 1}$

10. $g(x) = \dfrac{3x^2 + 2x + 1}{x^2 + 1}$ **11.** $f(x) = \dfrac{2x + 1}{x^2 + x - 2}$ **12.** $f(x) = \dfrac{2x^2 + 4}{x^2 - 2x - 3}$

13. $f(x) = \dfrac{x^2 - 1}{x^2 + x - 2}$

14. $f(x) = \dfrac{x^2 - 2x}{x^2 - 4}$

15. $f(x) = \dfrac{1}{x^3 + x}$

16. $f(x) = \dfrac{x^2}{x^3 + x}$

In each of the problems 17 through 28, draw a graph of the given rational function. Indicate any asymptotes with dotted lines in your diagram, including oblique asymptotes. Also label the x- and y-intercept points on your graph.

17. $f(x) = \dfrac{2x}{x - 2}$

18. $f(x) = \dfrac{2x + 3}{x - 1}$

19. $f(x) = \dfrac{x^2}{x^2 + 3x - 4}$

20. $f(x) = \dfrac{x^2}{2x^2 + x - 3}$

21. $g(x) = \dfrac{3}{x^2 + 1}$

22. $g(x) = \dfrac{3}{x^2 + 2}$

23. $f(x) = \dfrac{x^2 - 4}{x + 2}$

24. $f(x) = \dfrac{x^2 - 2x - 3}{x^2 - 1}$

25. $f(x) = \dfrac{x^3}{x^2 + 1}$

26. $f(x) = \dfrac{2x^2 + x}{x - 1}$

27. $f(x) = \dfrac{x^3 - 4x}{x^2 - 4}$

28. $f(x) = \dfrac{x^2}{x + 1}$

In problems 29 through 32, draw a graph of the rational function f/g.

29. $f(x) = x - 1;\ g(x) = x + 1$

30. $f(x) = 1 - x^2;\ g(x) = x^2$

31. $f(x) = x^2 + 2x + 1;\ g(x) = x^2 + x$

32. $f(x) = x^2;\ g(x) = x^3 - 1$

In problems 33 through 36, a function f is defined by the given set of ordered pairs. Draw a graph of f.

33. $f = \{(x, y) \mid x + xy - y = 0\}$

34. $f = \{(x, y) \mid x^2 + x^2y - y = 0\}$

35. $f = \{(x, y) \mid x^2 - x + xy - y = 0\}$

36. $f = \{(x, y) \mid x^2 + x + x^2y - 2xy - 3y = 0\}$

2.7   *Looking Ahead to Calculus*

The single most important concept in the study of calculus involves the notion of limit and limiting processes. Limits occur in a variety of different settings, and they are basic to the fundamental ideas of differentiation and integration. It is not our intention to present a formal study of limits here. However, through numerical examples we can help the reader to develop an intuitive feeling for limits in preparation for the abstract ideas that will be presented in calculus.

In Section 1.3 we introduced the idea of *slope of a line*. Here we look at the more general notion of *slope of a curve*.

A fundamental idea in calculus is that the slope of a tangent line to a curve can be approximated to any desired degree of accuracy by the slope of a secant line. A *secant line* through point P on the curve is a line passing through P and another nearby point Q on the curve, as illustrated in Fig. 2.17 (a). Now suppose Q moves along the curve toward P. As it continues to approach P, the corresponding secant lines approach a fixed line called the *tangent line to the curve at*

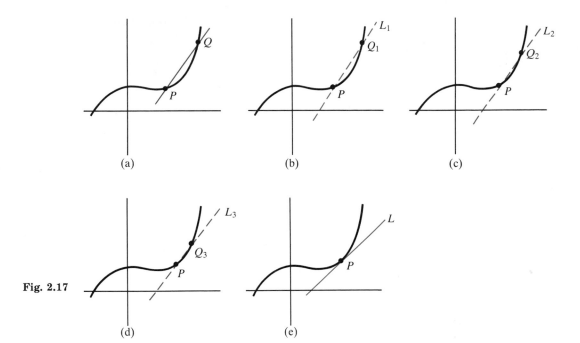

Fig. 2.17

P. This is illustrated in the sequence of diagrams shown in Fig. 2.17 (b), (c), and (d), in which the secant lines L_1, L_2, L_3 approach the tangent line L shown in (e).

Note that in each of the diagrams in Fig. 2.17, point Q is drawn to the right of *P.* We want also to consider the situation in which Q approaches P from the left; we leave it to the reader to draw corresponding diagrams.

The following definition will serve to illustrate the notion of slope of a curve.

Definition 2.4   The *slope of the tangent line* to a curve at a point P on the curve is the limiting value of the slopes of the secant lines as point Q approaches $P.$

The *slope of the curve* at point P is the slope of the tangent line at $P.$

Notation: Suppose a curve is given by the equation $y = f(x)$ and $P{:}(c, f(c))$ is a point on the curve. Let nearby points Q be given by $(c + h, f(c + h))$, where h assumes small values (positive and negative). The slope m_h of the secant line through P and Q is given by

$$m_h = \frac{f(c + h) - f(c)}{(c + h) - c} = \frac{f(c + h) - f(c)}{h}.$$

We are interested in determining the limiting value of this *difference quotient* as h approaches zero.* This limiting value is denoted by

$$\lim_{h \to 0} \frac{f(c + h) - f(c)}{h}.$$

Thus the slope m of the tangent line to the curve at point $(c, f(c))$ is given by

$$m = \lim_{h \to 0} \frac{f(c + h) - f(c)}{h}. \qquad (2.17)$$

Example 1   Suppose $f(x) = x^2 - 4x + 5$.

a) Draw a graph of $y = f(x)$ and locate the point P on the curve given by $x = 3$.

b) Make a table of values of the difference quotient for several values of h approaching 0; include both positive and negative values of h.

c) From the table in (b) guess the limiting value of the difference quotients as h approaches 0, thus getting the slope of the tangent line to the curve at P.

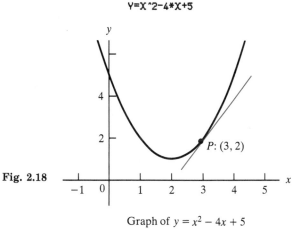

Y=X^2-4*X+5

Fig. 2.18

Graph of $y = x^2 - 4x + 5$

Solution   **a)** $f(3) = 3^2 - 4 \cdot 3 + 5 = 2$. Hence P is given by $(3, 2)$. The graph is shown in Fig. 2.18.

b) $\dfrac{f(3 + h) - f(3)}{h} = \dfrac{[(3 + h)^2 - 4(3 + h) + 5] - [3^2 - 4 \cdot 3 + 5]}{h} = h + 2$

---

*   In calculus this limiting value is called the *derivative* of $f(x)$ at c.

h	0.5	0.1	0.01	0.001	-0.5	-0.1	-0.01	-0.001
$\dfrac{f(3+h)-f(3)}{h}$	2.5	2.1	2.01	2.001	1.5	1.9	1.99	1.999

c) We see that the limiting value is 2. Hence

$$m = \lim_{h \to 0} \frac{f(3+h)-f(3)}{h} = 2.$$ ■

Example 2   Find an equation of the line that is tangent to the curve $y = \sqrt{x+1}$ at the point P: $(1, \sqrt{2})$. Give answer with numbers rounded off to two decimal places.

Solution   First find the slope of the tangent line by evaluating the limit given in Eq. (2.17).

$$\frac{f(1+h)-f(1)}{h} = \frac{\sqrt{(1+h)+1}-\sqrt{1+1}}{h} = \frac{\sqrt{2+h}-\sqrt{2}}{h}.$$

h	0.5	0.1	0.01	0.001	-0.5	-0.1	-0.01	-0.001
$\dfrac{\sqrt{2+h}-\sqrt{2}}{h}$	0.334	0.349	0.353	0.3535	0.379	0.358	0.354	0.3536

From the values in the table* we conclude that

$$\lim_{h \to 0} \frac{f(1+h)-f(1)}{h} = 0.354 \quad \text{(to three decimal places)}.$$

Hence the slope of the tangent line is 0.354, and its equation is given by

$$y - \sqrt{2} = 0.354(x-1).$$

This can be written as

$$y = 0.35x + 1.06,$$

where the numbers are rounded off to two decimal places. ■

Example 3   Find an equation of the tangent line to the curve $y = \sqrt{4-x^2}$ at the point $(-1, \sqrt{3})$. Draw a graph and show the tangent line.

Solution   First determine the slope of the tangent line by evaluating the limit given in Eq. (2.17):

$$\frac{f(-1+h)-f(-1)}{h} = \frac{\sqrt{4-(-1+h)^2}-\sqrt{4-(-1)^2}}{h} = \frac{\sqrt{3+2h-h^2}-\sqrt{3}}{h}.$$

---

*   Note that we have not included extremely small values of x in the table. Calculators cannot handle such numbers without introducing substantial round-off errors.

h	0.5	0.1	0.01	0.001	-0.5	-0.1	-0.01	-0.001
$\dfrac{\sqrt{3 + 2h - h^2} - \sqrt{3}}{h}$	0.41	0.54	0.574	0.5770	0.82	0.62	0.581	0.5777

Y=SQR(4−X^2)

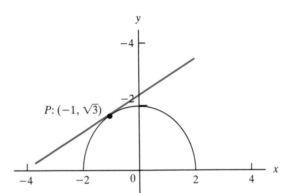

Fig. 2.19

Graph of $y = \sqrt{4 - x^2}$

From the values in the table we conclude that

$$\lim_{h \to 0} \frac{f(-1 + h) - f(-1)}{h} = 0.577 \quad \text{(to three decimal places)}.$$

Thus the slope of the tangent line is 0.577 and its equation is given by $y - \sqrt{3} = 0.577(x + 1)$. This can be written as $y = 0.577x + 2.309$. The graph is shown in Fig. 2.19.

Example 4 Find the slope of the curve given by $y = x/(x - 1)$ at the point P: $(3, 3/2)$. Draw a graph and show the tangent line at P.

Solution Let $f(x) = x/(x - 1)$, and first get the difference quotient,

$$\frac{f(3 + h) - f(3)}{h} = \frac{1}{h}\left[\frac{3 + h}{(3 + h) - 1} - \frac{3}{2}\right] = \frac{1}{h}\left[\frac{3 + h}{2 + h} - \frac{3}{2}\right].$$

The following table gives values of the difference quotient corresponding to values of h approaching 0.

From the values given in the table we conclude that

$$\lim_{h \to 0} \frac{f(3 + h) - f(3)}{h} = -0.25.$$

Thus the slope of the curve at point P is -0.25. The graph is shown in Fig. 2.20.

h	0.2	0.05	0.001	-0.1	-0.02	-0.001
$\dfrac{f(3 + h) - f(3)}{h}$	-0.227	-0.244	-0.250	-0.263	-0.253	-0.250

Y=X/(X−1)

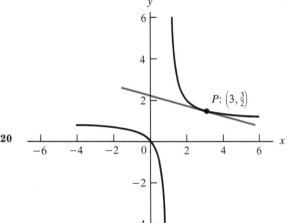

Fig. 2.20

$$\text{Graph of } y = \frac{x}{x - 1}$$

Exercises 2.7

In each of the problems 1 through 12, a function f and a number c are given.

a) Draw a graph of $y = f(x)$.

b) Locate the point P: $(c, f(c))$ on the graph and draw a tangent line to the curve at P.

c) Determine the slope of the tangent line.

1. $f(x) = 2x^2 + x - 3$; $c = -1$ **2.** $f(x) = x^3 + 3x^2 - x - 3$; $c = 2$

3. $f(x) = 4x - x^3$; $c = 1.5$ **4.** $f(x) = (x + 1)(x + 2)(x - 3)$; $c = -2$

5. $f(x) = \sqrt{x - 1}$; $c = 5$ **6.** $f(x) = -\sqrt{x - 1}$; $c = 5$

7. $f(x) = \sqrt{9 - x^2}$; $c = -2$ **8.** $f(x) = \sqrt{x^2 - 9}$; $c = 5$

9. $f(x) = \dfrac{2}{x + 2}$; $c = 0$ **10.** $f(x) = \dfrac{1}{\sqrt{x}}$; $c = 4$

11. $f(x) = 1 + \sqrt{x}$; $c = 4$ **12.** $f(x) = 1 + \sqrt{-x}$; $c = -1$

In problems 13 through 24, find an equation of the line tangent to the curve at the point P. In each case draw a graph and show the tangent line.

13. $y = x^2 - 1$; P:$(1, 0)$ **14.** $y = 3 + 2x - x^2$; P:$(1, 4)$

15. $y = \sqrt{x} + 1$; $P:(4, 3)$ **16.** $y = \dfrac{2}{\sqrt{x}}$; $P:(4, 1)$

17. $y = \dfrac{4}{\sqrt{1 - x}}$; $P:(-3, 2)$ **18.** $y = \sqrt{9 - x^2}$; $P:(-1, \sqrt{8})$

19. $y = -\sqrt{9 - x^2}$; $P:(-1, -\sqrt{8})$ **20.** $y = \dfrac{3x}{x + 2}$; $P:(1, 1)$

21. $y = \dfrac{x^2}{x + 2}$; $P:(2, 1)$ **22.** $y = \dfrac{x^2}{x - 2}$; $P:(1, -1)$

23. $y = \sqrt{1 - x}$; $P:(-3, 2)$ **24.** $y = 1 + \dfrac{2}{\sqrt{x}}$; $P:(4, 2)$

Review Exercises

In all problems involving numerical answers, give results in exact form unless otherwise specified.

1. Given that $f(x) = x^3 - 3x^2 - 2x + 4$, evaluate

    a) $f(-2)$ **b)** $f(1.6)$

2. Given that $g(x) = -4x^3 + 5x^2 + 6x - 3$, evaluate

    a) $g(3)$ **b)** $g(-2.5)$

In problems 3 and 4, give answers rounded off to two decimal places.

3. Suppose $f(x) = -2x^4 + x^3 + 2x^2 - 3x + 5$; write $f(x)$ in nested form and evaluate

    a) $f(1 + \sqrt{3})$ **b)** $f(2 - 3\sqrt{2})$

4. Suppose $f(x) = x^4 - 4x^3 - 3x^2 + 5x - 4$; write $f(x)$ in nested form and evaluate

    a) $f(\pi + 1)$ **b)** $f(\sqrt{3} - \pi)$

5. Complete the following table for $f(x) = -2x^3 + 4x^2 + 5x - 1$. Then sketch a graph of $y = f(x)$.

x	-3	-2	-1	0	0.5	1	2	3	4
$f(x)$									

6. Sketch a graph of $y = (x + 1)^2(x - 2)(x - 4)$. Label the x- and the y-intercept points.

7. Sketch a graph of $y = x^5 - 4x^3$. Label the x- and the y-intercept points.

8. Suppose $f(x) = x^3 + 2x^2 - 4x - 8$. Factor and then sketch a graph of $y = f(x)$; label the x- and the y-intercept points.

9. Find a polynomial of degree 3, having leading coefficient 1 and zeros -2, -1, and 3. Give answer in expanded form.

10. Find a polynomial of degree 4 with leading coefficient 1, having -2 as a double zero and -1 as a double zero. Give answer in expanded form.

11. Determine all zeros of the polynomial $f(x) = x^3 - 5x^2 + 4x$.

12. Determine all zeros of the polynomial $f(x) = x^3 - 4x^2 - 4x + 16$.

13. Sketch a graph of $y = f(x)$, where $f(x) = x^3 - 2x^2 + 3x + 1$. From your graph determine the *number* of roots of $f(x) = 0$ and their location between consecutive integers.

14. **a)** List all *possible* rational roots of $2x^3 - 3x^2 - 5x + 3 = 0$.

   b) Are any of these roots?

15. Find the quotient and remainder when $3x^3 - 4x^2 - 5x - 2$ is divided by $x + 1$.

16. If $x^4 - 3x^2 + 2x - 5$ is divided by $x - 3$, what are the quotient and remainder?

17. Determine the remainder when $3x^{16} + 2x^{10} - 5x^3 + 3x^2 - 1$ is divided by $x + 1$.

18. Determine the value of k so that when $x^3 - 2x^2 - kx + 5$ is divided by $x + 1$, the remainder will be -3.

19. Find all rational-number zeros of $f(x) = 2x^3 - 3x^2 - 12x - 5$.

20. Find all rational-number zeros of $f(x) = 2x^3 + 9x^2 + 7x - 6$.

21. Locate the roots of $x^3 - 5x + 3 = 0$ between consecutive integers. Determine the largest, correct to one decimal place.

22. Locate the zeros of $f(x) = 3x^3 - 2x^2 - x + 1$ between consecutive integers. Determine the smallest, correct to one decimal place.

23. Determine the domain of the function f, where

$$f(x) = \frac{x}{x^2 - x - 2}.$$

24. Determine the equations of the horizontal and vertical asymptotes for

$$y = \frac{2x^2}{x^2 - 4}.$$

25. Sketch a graph of $y = 4/(x + 1)$.

26. Sketch a graph of $y = x/(x - 1)$.

In problems 27 through 30, do the following.

a) Give the coordinates of the x- and the y-intercept points.

b) Determine equations of any horizontal and vertical asymptotes.

c) Sketch a graph of $y = f(x)$.

27. $f(x) = \dfrac{x + 3}{2 - x}$ 28. $f(x) = \dfrac{2x^2 - 8}{x^2 - x - 2}$ 29. $f(x) = \dfrac{x^2 - 2x - 3}{x - 1}$ 30. $f(x) = \dfrac{x + 1}{x^2 - x - 2}$

Exponential and Logarithmic Functions

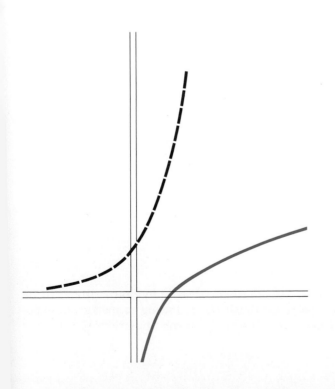

In Chapter 2 we studied an important class of functions called the polynomial functions. An example of such a function is $f(x) = x^3$. Suppose g is a function given by $g(x) = 3^x$. Note the difference between f and g: In the formula for g the independent variable x occurs as an exponent and the base is constant, whereas for f the exponent is constant. Functions such as g are called *exponential functions*. In this chapter we shall study such functions. After exploring properties of exponential functions, we shall see that each such function has an inverse function associated with it. These inverse functions are called *logarithmic functions*.

In order to understand exponential functions, we first consider definitions and properties of quantities of the type b^u, where b and u are real numbers. This is done in three stages: (1) u is an integer; (2) u is a rational number; (3) u is an irrational number. The student should already be somewhat familiar with case 1, but a brief review is included in Section 3.0 along with a review of square roots; cases 2 and 3 are discussed in the following two sections. This will give us the basics of exponential functions, and then we can proceed with the introduction of logarithmic functions in Section 3.3.

3.0   REVIEW OF INTEGRAL EXPONENTS AND SQUARE ROOTS

Integral Exponents

The expression b^m, where m is any integer, is defined in three stages ($m > 0$, $m < 0$, $m = 0$), as follows:

Definition 3.1

Suppose b is a real number.

a) *Positive integer exponents:* If $m > 0$, then

$$b^m = b \cdot b \cdot b \cdot \,\cdots\, b \quad (m \text{ factors of } b) \tag{3.1}$$

b) *Negative integer exponents:* If $n > 0$ and $b \neq 0$, then

$$b^{-n} = \frac{1}{b^n} \tag{3.2}$$

c) *Zero exponent:* If $b \neq 0$, then

$$b^0 = 1 \tag{3.3}$$

Note that 0^0 is not defined.

The definitions above, along with mathematical induction (a topic to be studied in Chapter 7), permit us to establish the following properties of exponents. These form the rules that govern the algebraic manipulations of exponential expressions.

Theorem 3.1

Rules of exponents

Suppose m and n are integers, and b and c are nonzero real numbers. Then

(E1) $b^m \cdot b^n = b^{m+n}$ **(E2)** $\dfrac{b^m}{b^n} = b^{m-n}$ **(E3)** $(b^m)^n = b^{mn}$

(E4) $(b \cdot c)^m = b^m \cdot c^m$ **(E5)** $\left(\dfrac{b}{c}\right)^m = \dfrac{b^m}{c^m}$

Simplify

Example 1   **a)** $3^6 \cdot 3^{-2}$ **b)** $\dfrac{4^2}{2^3}$ **c)** $(81)^2(3^{-2})^4$

Solution   **a)** $3^6 \cdot 3^{-2} = 3^{6+(-2)} = 3^4 = 3 \cdot 3 \cdot 3 \cdot 3 = 81$

b) $\dfrac{4^2}{2^3} = \dfrac{(2^2)^2}{2^3} = \dfrac{2^4}{2^3} = 2^{4-3} = 2^1 = 2$

c) $(81)^2 \cdot (3^{-2})^4 = (3^4)^2 \cdot (3^{-2})^4 = 3^8 \cdot 3^{-8} = 3^{8+(-8)} = 3^0 = 1$

Example 2   Express without negative exponents and simplify.

a) $(x^{-2} \cdot y^3)^{-2}$ **b)** $\dfrac{x^{-3} + 1}{x + 1}$ **c)** $\dfrac{x^{-2} - 4x^{-1} - 5}{x^{-2}(5x - 1)}$

Solution   **a)** $(x^{-2} \cdot y^3)^{-2} = (x^{-2})^{-2} \cdot (y^3)^{-2} = x^4 \cdot y^{-6} = x^4/y^6$

b) $\dfrac{x^{-3} + 1}{x + 1} = \dfrac{\dfrac{1}{x^3} + 1}{x + 1} = \dfrac{\dfrac{1 + x^3}{x^3}}{x + 1} = \dfrac{1 + x^3}{x^3} \cdot \dfrac{1}{x + 1}$

$$= \dfrac{(1 + x)(1 - x + x^2)}{x^3(x + 1)} = \dfrac{1 - x + x^2}{x^3}$$

c) $\dfrac{x^{-2} - 4x^{-1} - 5}{x^{-2}(5x - 1)} = \dfrac{\dfrac{1}{x^2} - \dfrac{4}{x} - 5}{\dfrac{5x - 1}{x^2}} = \dfrac{1 - 4x - 5x^2}{x^2} \cdot \dfrac{x^2}{5x - 1}$

$$= \dfrac{(-1)(5x - 1)(1 + x)}{5x - 1} = -1 - x$$

Example 3   Evaluate $(0.0000384)(13600000)$ and express result rounded off to three significant digits.

Solution   First write each of the given numbers in scientific notation (see Appendix C), then multiply, and round off, as follows:

$$(0.0000384)(13600000) = (3.84 \cdot 10^{-5})(1.36 \cdot 10^{7}) = (3.84)(1.36) \cdot 10^{2} = 522.$$

Therefore the product is 522 rounded off to three significant digits.   ■

Square Roots

Definition 3.2   Suppose c is a nonnegative real number. Then $\sqrt{c}$ is the *nonnegative real number* that, when squared, yields c. That is, $(\sqrt{c})^{2} = c$.

Note that for $c > 0$, Definition 3.2 states that $\sqrt{c}$ is a *positive* number. For example, $\sqrt{4} = 2$ and not ± 2; this is part of the definition. However, when we ask for solutions of $x^{2} = 4$, there are two answers, 2 and -2, whereas 2 is the only solution for the equation $x = \sqrt{4}$. That is, the two equations $x^{2} = 4$ and $x = \sqrt{4}$ are not equivalent; $x^{2} = 4$ is equivalent to $x = \sqrt{4}$ or $x = -\sqrt{4}$.

If c is any real number, then c^{2} is nonnegative. It is a temptation to write $\sqrt{c^{2}} = c$. If c is replaced by, say, 4, we get $\sqrt{4^{2}} = 4$, which is a true statement; but if c is replaced by, say, -4, then we get $\sqrt{(-4)^{2}} = -4$, which is a false statement since the left side is $\sqrt{(-4)^{2}} = \sqrt{16} = 4$. This suggests the following.

For any real number c,

$$\sqrt{c^{2}} = |c|. \tag{3.4}$$

Analogous to rules **(E4)** and **(E5)** for exponents, the following two rules are useful in working with square roots:

(S1) If $a \geq 0$ and $b \geq 0$, then $\sqrt{a \cdot b} = \sqrt{a} \cdot \sqrt{b}$.

(S2) If $a \geq 0$ and $b > 0$, then $\sqrt{a/b} = \sqrt{a}/\sqrt{b}$.

Simplify

Example 4   **a)** $\sqrt{2} \cdot \sqrt{8}$ **b)** $\dfrac{\sqrt{18} + \sqrt{32}}{\sqrt{2}}$ **c)** $\sqrt{5^{2} + 12^{2}}$

Solution   **a)** $\sqrt{2} \cdot \sqrt{8} = \sqrt{2 \cdot 8} = \sqrt{16} = 4$

b) $\dfrac{\sqrt{18} + \sqrt{32}}{\sqrt{2}} = \dfrac{\sqrt{9 \cdot 2} + \sqrt{16 \cdot 2}}{\sqrt{2}} = \dfrac{\sqrt{9} \cdot \sqrt{2} + \sqrt{16} \cdot \sqrt{2}}{\sqrt{2}}$

$$= \dfrac{3\sqrt{2} + 4\sqrt{2}}{\sqrt{2}} = \dfrac{7\sqrt{2}}{\sqrt{2}} = 7$$

c) $\sqrt{5^2 + 12^2} = \sqrt{25 + 144} = \sqrt{169} = 13$

Note that $\sqrt{5^2} = \sqrt{25} = 5$ and $\sqrt{12^2} = \sqrt{144} = 12$, and so

$$\sqrt{5^2 + 12^2} \neq \sqrt{5^2} + \sqrt{12^2}.$$

Example 5 **a)** Rationalize the denominator of $\dfrac{\sqrt{2} - 3}{1 - \sqrt{2}}$.

b) Rationalize the numerator of $\dfrac{\sqrt{2} - 3}{1 - \sqrt{2}}$.

Solution **a)** To rationalize the denominator means to express the given number in terms of a fraction without square roots in the denominator. This can be done by multiplying the numerator and denominator of the given fraction by $1 + \sqrt{2}$, as follows:

$$\frac{\sqrt{2} - 3}{1 - \sqrt{2}} = \frac{(\sqrt{2} - 3)(1 + \sqrt{2})}{(1 - \sqrt{2})(1 + \sqrt{2})} = \frac{\sqrt{2} + 2 - 3 - 3\sqrt{2}}{1^2 - (\sqrt{2})^2} = \frac{-1 - 2\sqrt{2}}{-1} = 1 + 2\sqrt{2}.$$

b) $\dfrac{\sqrt{2} - 3}{1 - \sqrt{2}} = \dfrac{(\sqrt{2} - 3)(\sqrt{2} + 3)}{(1 - \sqrt{2})(\sqrt{2} + 3)} = \dfrac{(\sqrt{2})^2 - (3)^2}{1 - 2\sqrt{2}} = \dfrac{-7}{1 - 2\sqrt{2}} = \dfrac{7}{2\sqrt{2} - 1}.$

Example 6 Find the solution set for the equation $\sqrt{1 - 3x} - x = 9$.

Solution First isolate the radical by adding x to both sides of the given equation: $\sqrt{1 - 3x} = x + 9$. Squaring both sides will get rid of the square root, but it may introduce extraneous solutions, so it will be necessary to check the resulting answers:

$$(\sqrt{1 - 3x})^2 = (x + 9)^2.$$

Expanding and simplifying, we get

$$1 - 3x = x^2 + 18x + 81,$$
$$x^2 + 21x + 80 = 0,$$
$$(x + 5)(x + 16) = 0,$$
$$x = -5 \quad \text{or} \quad x = -16.$$

Therefore -5 and -16 are possible solutions. Substituting these into the given equation, we get:

For $x = -5$: LHS $= \sqrt{1 - 3(-5)} - (-5) = 4 + 5 = 9$; so -5 is a solution.

For $x = -16$ we get LHS $= \sqrt{1 - 3(-16)} - (-16) = \sqrt{49} + 16 = 7 + 16 = 23$, so -16 is not a solution.

Thus the solution set is $S = \{-5\}$. ■

Example 7   Find the solution set for the equation $x^2 - 2\sqrt{x^2} - 3 = 0$.

Solution   Replacing $\sqrt{x^2}$ by $|x|$ gives $x^2 - 2|x| - 3 = 0$. Consider two cases:

1. For $x \geq 0$, we have $x^2 - 2x - 3 = 0$, which gives $x = -1$ or $x = 3$. Thus 3 is a solution.

2. For $x < 0$, we get $x^2 + 2x - 3 = 0$, which gives $x = 1$ or $x = -3$. Thus -3 is a solution.

The solution set is $S = \{-3, 3\}$. ■

Example 8   Solve the equation $\sqrt{3x + 4} + \sqrt{x} + 3 = 0$.

Solution   Here we get the answer by observing that $\sqrt{3x + 4} \geq 0$ and $\sqrt{x} \geq 0$. If such numbers are added to 3, the result cannot be equal to zero. Therefore the solution set is the empty set $S = \emptyset$ ■

Exercises 3.0

In problems 1 through 20, simplify the given expressions. In each case state answers in exact form, and tell whether the given number is rational or irrational.

1. $\dfrac{4 \cdot 3^{-1} \cdot 2^{-2}}{6^{-1}}$ **2.** $\dfrac{9 \cdot 3^{-3}}{27^{-1}}$ **3.** $(\sqrt{5})^3 (\sqrt{125})^{-1}$

4. $\left(\dfrac{4 \cdot 3^{-1} \cdot 2^{-2}}{8^{-1}}\right)^{-1}$ **5.** $(2^{-3} + 4^{-1})^{-3}$ **6.** $\left(\dfrac{4^0 \cdot 3^{-4} \cdot 5}{9^{-2}}\right)^2$

7. $\dfrac{2^{-3} - 5}{2^{-3}}$ **8.** $\sqrt{3} \cdot \sqrt{12}$ **9.** $\dfrac{\sqrt{1001} \cdot \sqrt{7}}{\sqrt{143}}$

10. $\sqrt{3} \cdot \sqrt{27} \cdot \sqrt{64}$ **11.** $\dfrac{\sqrt{105} \, (\sqrt{35})^{-1}}{\sqrt{3}}$ **12.** $(1 + \sqrt{2})(2 - \sqrt{8})$

13. $(3 + \sqrt{5} + \sqrt{11})^0$ **14.** $(\sqrt{2} - \sqrt{3})(\sqrt{2} + \sqrt{3})$ **15.** $\left(\dfrac{1 - \sqrt{5}}{2}\right)^2 - \left(\dfrac{1 - \sqrt{5}}{2}\right) - 1$

16. $\left(\dfrac{1 + \sqrt{5}}{2}\right)^2 - \left(\dfrac{1 + \sqrt{5}}{2}\right)$ **17.** $\sqrt{(\sqrt{5} + 1)(\sqrt{5} - 1)}$ **18.** $(1 + \sqrt{3})^2 - 2(1 + \sqrt{3}) - 2$

19. $(\sqrt{3} - 1)^2 + 2(\sqrt{3} - 1) - 2$ **20.** $(\sqrt{3})^4 - 2(\sqrt{3})^3 - 4(\sqrt{3})^2 + 3$

In problems 21 through 30, evaluate the given expressions and round off answers to four significant digits.* Appendix A includes instructions for using a calculator.

21. $(0.0000004385) \cdot (6534200000)$

22. $(3.74 \cdot 10^{-8})(5.43 \cdot 10^7)$

23. $\dfrac{(2.47 \cdot 10^{-4})(3.42 \cdot 10^2)}{4.36 \cdot 10^{-3}}$

24. $(6.54 \cdot 10^4)(3.57 \cdot 10^{-3} + 8.56 \cdot 10^{-2})$

25. $\sqrt{348.76}$

26. $\sqrt{(3.57 \cdot 10^2)(8.36 \cdot 10^{-4})}$

27. $\sqrt{(2.4)^2 + (3.6)^3}$

28. $\dfrac{(2.43)^3 - (1.58)^4}{(2.56)^3}$

29. $\dfrac{\sqrt{2} + \sqrt{3}}{\sqrt{5} - \sqrt{2}}$

30. $\left(\dfrac{3 + \sqrt{2}}{\sqrt{7} + \sqrt{19}}\right)^3$

In problems 31 through 42, simplify the given expressions. In each case determine the set of values of x for which the given expression is defined.

31. $x^2 \cdot x^4$

32. $(4x^2)(2x)^{-4}$

33. $\dfrac{x + x^{-1}}{x^2 + 1}$

34. $\sqrt{12x^3} \cdot \sqrt{3x}$

35. $(x^2 - 1)^0$

36. $(x^2 - 5x + 6)^0$

37. $(1 + \sqrt{x})(2 - \sqrt{x})$

38. $\sqrt{1 - 2x + x^2}$

39. $\sqrt{1 - (1 + x)(1 - x)}$

40. $2\sqrt{x} - (3\sqrt{x} - 4)(\sqrt{x} + 2)$

41. $(x^3 + 5x^2 + 6x)^0$

42. $\dfrac{x^2(x^{-2} - 2x^{-1} + 1)}{(x - 1)^2}$

In problems 43 through 46, rationalize the denominator.

43. $\dfrac{1}{\sqrt{2} + 1}$

44. $\dfrac{\sqrt{3} - 1}{\sqrt{3} + 1}$

45. $\dfrac{27}{1 - 2\sqrt{7}}$

46. $\dfrac{\sqrt{2} + \sqrt{5}}{\sqrt{3} - \sqrt{2}}$

In problems 47 through 50, rationalize the numerator.

47. $\dfrac{\sqrt{3}}{6}$

48. $\dfrac{\sqrt{5} - 1}{4}$

49. $\dfrac{\sqrt{3} - \sqrt{2}}{\sqrt{2} + \sqrt{3}}$

50. $\dfrac{\sqrt{18} - 3}{3}$

In problems 51 through 60, find the solution set for the given equations. Check for extraneous solutions when necessary.

51. $3x^{-1} + 4 = 4x^{-1} + 2$

52. $3x^{-2} + x^{-1} - 4 = 0$

53. $\dfrac{x^{-2} + x^{-1} + 1}{x^{-1}} = 3$

54. $\dfrac{x^{-4} - 3x^{-3}}{x^{-2}} = 4$

55. $\sqrt{3x + 1} = 2$

56. $\sqrt{x^2 - 2x} = 3$

57. $\sqrt{2x + 1} + \sqrt{x + 1} = 0$

58. $\sqrt{1 - x} = x + 5$

59. $3x^2 + \sqrt{x^2} - 4 = 0$

60. $x^2 - 5\sqrt{x^2} + 6 = 0$

In problems 61 through 70, determine which of the given statements are true. In each case give a reason for your answer.

---

*   See Appendix C for discussion of significant digits.

61. The solution set for $\sqrt{2x+3} + \sqrt{x+1} = 0$ is the empty set.

62. The only real numbers for which $\sqrt{-x}$ is defined is $x = 0$.

63. There are no real numbers x for which the expression $\sqrt{2-x} + \sqrt{x-4}$ is defined.

64. The equation $\sqrt{2-x} + \sqrt{x-4} = 1$ has no solutions.

65. The equation $x^2 - 4\sqrt{x^2} = 0$ has three solutions.

66. The equation $x^2 + \sqrt{x} + 1 = 0$ can have no solutions.

67. The equation $\sqrt{x^2} = \sqrt{-x^2}$ has no solutions.

68. The expression $\sqrt{x^3} + \sqrt{x}$ yields an irrational number for every positive real number x.

69. The equation $\sqrt{x^2} = -x$ has only 0 as a solution.

70. The expression $(\sqrt{x} - 1)^0$ is defined and equal to 1 for every x greater than or equal to zero.

3.1   **RATIONAL-NUMBER EXPONENTS**

In Section 3.0 properties of integral exponents were reviewed. Now we focus our attention on exponents that are rational numbers.

How are expressions of the type $2^{1/3}$, $4^{-2/3}$, $5^{1/4}$, and so on, defined? The definitions we make are guided by a desire to have the rules of exponents **(E1)** through **(E5)** stated in Section 3.0, valid when the exponents are any rational numbers. After all, integers are rational numbers, and we want our definitions and properties to be consistent with those discussed in Section 3.0. This is illustrated in the following example.

Example 1   How should each of the following be defined so that property **(E3)** is valid for rational-number exponents?

a) $3^{1/2}$ **b)** $4^{1/3}$

Solution   **a)** Since **(E3)** is to hold for rational numbers,

$$(3^{1/2})^2 = 3^{(1/2)\cdot(2)} = 3^1 = 3.$$

Therefore $3^{1/2}$ is a number whose square is 3. Suppose we denote this number by c; that is, $c = 3^{1/2}$. Then $c^2 = 3$, and c is a root of the equation $x^2 - 3 = 0$. In Chapter 2 we learned that the only possible rational roots of this equation are 1, -1, 3, and -3. It is easy to show that none of these actually is a root, and so $3^{1/2}$ must be an irrational number. We define $3^{1/2}$ as the *positive* solution of the equation $x^2 - 3 = 0$ and techniques studied in Section 2.5 can be applied to find a decimal approximation to $3^{1/2}$.

b) Since **(E3)** is to hold for rational numbers,

$$(4^{1/3})^3 = 4^{(1/3)(3)} = 4^1 = 4.$$

Thus $4^{1/3}$ is a number whose cube is 4. By a procedure similar to that in (a) it can be argued that $4^{1/3}$ is an irrational number; that is, it is a root of

$x^3 - 4 = 0$. We can show that this equation has only one solution, and we define $4^{1/3}$ as that number. Decimal approximations to $4^{1/3}$ can be found by using techniques discussed in Section 2.5.   ■

Example 2   How should the following be defined so that property **(E3)** is valid for rational-number exponents?

a) $(-4)^{1/2}$ b) $(-8)^{1/3}$

Solution   a) Let $c = (-4)^{1/2}$. Assuming property **(E3)** to be valid for rational numbers, we get

$$c^2 = [(-4)^{1/2}]^2 = (-4)^{(1/2)(2)} = (-4)^1 = -4.$$

Thus $c^2 + 4 = 0$, and so c is a root of the equation $x^2 + 4 = 0$. Since $x^2 + 4 > 0$ for every real number x, we conclude that c is not a real number. Thus, within the context of real numbers, $(-4)^{1/2}$ is undefined.

b) Let $d = (-8)^{1/3}$. In a manner similar to that in (a), we conclude that d must be a root of the polynomial equation $x^3 + 8 = 0$. By methods of Chapter 2 it can show that this equation has *exactly one* real root, which is -2. We define $(-8)^{1/3}$ to be equal to -2 and write $(-8)^{1/3} = -2$.   ■

The discussion in Examples 1 and 2 will help guide us in defining $u^{1/n}$ in general, where n is any positive integer and u is a real number. The expression $u^{1/n}$ is to be a root of the polynomial equation

$$x^n - u = 0. \tag{3.5}$$

Using techniques discussed in Chapter 2, we can show the following.

If $u \geq 0$, then Eq. (3.5) has *exactly one* nonnegative root.

If $u < 0$, then Eq. (3.5) has *exactly one real root for n odd,* and has *no solutions for n even.*

This leads us to the following definition, in which we first define $u^{1/n}$ and then $u^{m/n}$.

Definition 3.3   Let u be a real number and n be a positive integer. Then $u^{1/n}$ is given by the following.

1. If $u \geq 0$, $u^{1/n}$ is that *nonnegative number* satisfying $x^n - u = 0$.
2. If $u < 0$ and n is odd, $u^{1/n}$ is that unique real number satisfying $x^n - u = 0$.

If $u < 0$ and n is even, then $u^{1/n}$ is not defined (as a real number).

Now suppose r is any rational number; that is, $r = m/n$, where m and n are integers, $n > 0$, and m/n is in lowest terms. Then u^r is given by

$$u^r = u^{m/n} = (u^{1/n})^m, \tag{3.6}$$

provided that $u^{1/n}$ is a real number as defined above.

Definition 3.3 states that $u^{m/n}$ is defined for all cases except when u is a negative number and n is an even integer.

Note that in Eq. (3.6), $(u^{1/n})^m$ involves a number $u^{1/n}$ raised to an integer power m. We can show that if we raise u to the m power first to get u^m and then apply Definition 3.3, the resulting number is the same as that given by Eq. (3.6). Thus u^r is also given by

$$u^r = u^{m/n} = (u^m)^{1/n}. \tag{3.7}$$

We can illustrate the need for requiring m/n to be in *lowest terms* in Definition 3.3 by considering $(-8)^{1/3}$ and $(-8)^{2/6}$. As shown in Example 2, $(-8)^{1/3} = -2$. According to Eq. (3.7), $(-8)^{2/6}$ would be equal to $((-8)^2)^{1/6} = (64)^{1/6} = 2$, and so $(-8)^{2/6}$ would not be equal to $(-8)^{1/3}$. Therefore, when expressions of the type $(-8)^{2/6}$ are encountered, it is necessary to first write the exponent in lowest terms.

By Definition 3.3 and Theorem 3.1, it can be proved that the rules of exponents are valid for rational-number exponents.

Theorem 3.2   If r and s are any rational numbers, and u and v are any nonzero real numbers, for which u^r, u^s, v^r, and v^s are real numbers (as defined above), then

(E1) $u^r \cdot u^s = u^{r+s}$ **(E2)** $u^r/u^s = u^{r-s}$ **(E3)** $(u^r)^s = u^{rs}$

(E4) $(u \cdot v)^r = u^r \cdot v^r$ **(E5)** $(u/v)^r = u^r/v^r$

Example 3   Evaluate each of the following.

a) $3^{1/2} \cdot 3^{3/2}$ **b)** $(2^{1/3}) \div (2^{-5/3})$ **c)** $\dfrac{4^{5/3} - 4^{2/3}}{2^{1/3}}$

Solution   Here we use the rules of exponents as stated in Theorem 3.2.

a) $3^{1/2} \cdot 3^{3/2} = 3^{(1/2+3/2)} = 3^2 = 9$

b) $(2^{1/3}) \div (2^{-5/3}) = 2^{1/3-(-5/3)} = 2^{(1/3+5/3)} = 2^2 = 4$

c) $\dfrac{4^{5/3} - 4^{2/3}}{2^{1/3}} = \dfrac{(2^2)^{5/3} - (2^2)^{2/3}}{2^{1/3}} = \dfrac{2^{10/3} - 2^{4/3}}{2^{1/3}} = \dfrac{2^{1/3}(2^{9/3} - 2^{3/3})}{2^{1/3}}$

$$= 2^3 - 2^1 = 8 - 2 = 6$$

Radical Notation

Expressions involving rational-number exponents are frequently written in radical notation as stated in the following definition.

Definition 3.4   Suppose n is an integer greater than 1,

$$\sqrt[n]{u} = u^{1/n}, \tag{3.8}$$

where u is any real number for which $u^{1/n}$ is defined.

We read $\sqrt[n]{u}$ as "the nth root of u." When $n = 2$, it is conventional to write $\sqrt{u}$ rather than $\sqrt[2]{u}$.

Using Eq. (3.7) and Eq. (3.8), we get

$$u^{m/n} = \sqrt[n]{u^m}. \tag{3.9}$$

In working with radicals, the following properties are frequently useful.

$$\sqrt[n]{u \cdot v} = \sqrt[n]{u} \cdot \sqrt[n]{v}, \tag{3.10}$$

$$\sqrt[n]{u/v} = \sqrt[n]{u}\big/\sqrt[n]{v}. \tag{3.11}$$

These are generalizations of corresponding properties **(S1)** and **(S2)** for square roots given on p. 158.

In Eqs. (3.10) and (3.11) we are assuming that u, v, and n are numbers such that $\sqrt[n]{u}$ and $\sqrt[n]{v}$ are defined (as real numbers); that is $u^{1/n}$ and $v^{1/n}$ are defined as stated in Definition 3.3.

Example 4   Write each of the following in radical form and simplify.

a) $3^{4/3}$ **b)** $(-4)^{2/3}$

Solution   **a)** $3^{4/3} = \sqrt[3]{3^4} = \sqrt[3]{81} = \sqrt[3]{27 \cdot 3} = \sqrt[3]{27} \cdot \sqrt[3]{3} = 3\sqrt[3]{3}$

b) $(-4)^{2/3} = \sqrt[3]{(-4)^2} = \sqrt[3]{16} = \sqrt[3]{8 \cdot 2} = \sqrt[3]{8}\sqrt[3]{2} = 2\sqrt[3]{2}$

Example 5   Rationalize the denominator of $\dfrac{x-1}{\sqrt{x}-1}$.

Solution   Multiplying the numerator and denominator by $\sqrt{x}+1$ gives

$$\frac{x-1}{\sqrt{x}-1} = \frac{(x-1)(\sqrt{x}+1)}{(\sqrt{x}-1)(\sqrt{x}+1)} = \frac{(x-1)(\sqrt{x}+1)}{x-1} = \sqrt{x}+1.$$

Thus

$$\frac{x-1}{\sqrt{x}-1} = \sqrt{x}+1,$$

where the equality is valid for all $x \geq 0$ and $x \neq 1$.   ∎

Example 6   Use a calculator to find decimal approximations for the following. Give answers rounded off to three decimal places.

   a) $\sqrt{5}$ **b)** $(-4)^{3/5}$ **c)** $(2.43)^{2.56}$

Solution   **a)** Pressing the 5 and $\boxed{\sqrt{x}}$ keys gives $\sqrt{5} = 2.236$.

   b) Here we can use the $\boxed{y^x}$ key, but the base number must be positive. Therefore we first get

$$(-4)^{3/5} = \sqrt[5]{(-4)^3} = \sqrt[5]{-(4)^3} = -\sqrt[5]{4^3} = -(4^{3/5})$$

   and then evaluate $4^{3/5}$ by using the $\boxed{y^x}$ key. This gives $(-4)^{3/5} = -2.297$.

   c) Using the $\boxed{y^x}$ key gives $(2.43)^{2.56} = 9.708$.   ∎

Example 7   Prove that $1 + \sqrt[3]{5}$ is an irrational number.

Solution   First find a polynomial equation with *integer* coefficients having $1 + \sqrt[3]{5}$ as a root; then use the Rational-root theorem given in Section 2.4 to show that it has no rational roots. Then we can conclude that $1 + \sqrt[3]{5}$ must be an irrational number.

   Let $x = 1 + \sqrt[3]{5}$. Isolating the radical, $x - 1 = \sqrt[3]{5}$, and raising both sides to the third power gives

$$(x-1)^3 = (\sqrt[3]{5})^3 = 5.$$

Expanding the left side* and collecting like terms gives

$$x^3 - 3x^2 + 3x - 6 = 0. \tag{3.12}$$

From the way Eq. (3.12) was developed, it should be clear that $1 + \sqrt[3]{5}$ is a root.

   We can now check to see if Eq. (3.12) has any rational roots. Using the Rational-root theorem and the fact that the coefficients alternate in sign, we

---

*   Use the formula $(a-b)^3 = a^3 - 3a^2b + 3ab^2 - b^3$.

need to check only 1, 2, 3, and 6 as possibilities. It is easy to verify that none of these is a root. Therefore $1 + \sqrt[3]{5}$ is an irrational number. ■

Example 8   A water tank has the shape of an inverted right circular cone of radius 1.6 m and height 4.8 m. Suppose water flows into the tank at the constant rate of 0.25 m³ per minute. As the water level rises, the surface has a circular shape of increasing radius r, as shown in Fig. 3.1. Suppose the tank is empty at the start.

a) Find a formula for r (in meters) as a function of time t (in minutes) after the start.

b) Use the result in (a) to find the radius of the water surface at time $t = 8$ minutes.

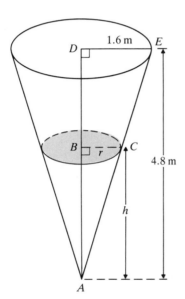

Fig. 3.1

Solution   a) Water flows into the tank at a constant rate of 0.25 m³ per minute, and so the volume V in the tank at time t is given by

$$V = 0.25\, t. \tag{3.13}$$

Volume V is also given by the formula for volume of a cone:

$$V = \tfrac{1}{3}\pi r^2 h. \tag{3.14}$$

But r and h are related, as we can see from ratios of corresponding sides of similar triangles, $\triangle ABC$ and $\triangle ADE$. That is, $h/r = 4.8/1.6$, or $h = 3r$. Substituting into Eq. (3.14) gives

$$V = \pi r^3. \tag{3.15}$$

Equating the two expressions for V from (3.13) and (3.15), we get $\pi r^3 = 0.25\, t$, and solving this for r gives the desired function

$$r = \frac{\sqrt[3]{t}}{\sqrt[3]{4\pi}}.$$

(3.16)

Since $0 \le r \le 1.6$, the domain of the function given by Eq. (3.16) is restricted to $0 \le t \le 4\pi(1.6)^3$; that is, $0 \le t \le 51.47$ (to two decimal places).

b) Substituting $t = 8$ into Eq. (3.16) gives the corresponding radius,

$$r = \frac{\sqrt[3]{8}}{\sqrt[3]{4\pi}} = 0.86 \text{ m.}$$

Exercises 3.1

In problems 1 through 15, perform the given arithmetic operations and simplify your results. Give answers in exact form.

1. $3^{5/2} \cdot 3^{-3/2}$
2. $4^{-1/2} \cdot 8^{3/2}$
3. $3^{5/2} \div 3^{-1/2}$

4. $5^{4/3} \div 5^{-2/3}$
5. $(3^6)^{2/3}$
6. $(5^{-1/3})^6$

7. $5^{1/3} \cdot 135^{2/3}$
8. $(21^{3/2} \cdot 3^{1/2}) \div 7^{1/2}$
9. $8 \cdot 16^{-3/4}$

10. $(\sqrt{8} - \sqrt{2})^2$
11. $(\sqrt[3]{16} + \sqrt[3]{2})^3$
12. $(\sqrt[3]{2} \cdot \sqrt[3]{12}) \div \sqrt[3]{3}$

13. $\dfrac{7^{5/2} - 63^{3/2}}{\sqrt{7}}$
14. $\left(\dfrac{\sqrt[3]{4} \cdot \sqrt[4]{2}}{\sqrt[6]{8}}\right)^{12}$
15. $\sqrt{1007^2 + 1224^2}$

In problems 16 through 27, evaluate the given expressions and give answers rounded off to three decimal places.

16. $5^{3/4}$
17. $4^{2/3} - 3^{5/4}$
18. $(1.6)^{-2.4}$

19. $\sqrt[3]{5} - \sqrt[4]{5}$
20. $(-16)^{3/5}$
21. $(-1.47)^{2/3}$

22. $(1 + \pi)^{-4/5}$
23. $\sqrt[4]{1 + \sqrt{5}}$
24. $(1 - \sqrt{3})^{2/5}$

25. $(\sqrt{2} + \sqrt{5})^{-1/2}$
26. $\sqrt{2.4^2 + 3.4^2}$
27. $(4 - \sqrt{32})^{1/5}$

In problems 28 through 33, perform the indicated algebraic operations and simplify your answers. Give results without negative exponents.

28. $x^{5/2} \cdot x^{-3/2}$
29. $\left(\sqrt{x} + \dfrac{1}{\sqrt{x}}\right)^2$
30. $x^{-2}\left(\sqrt[3]{x^6} + \sqrt[4]{x^8}\right)$

31. $x + x^{-1} - \left(\sqrt{x} - \dfrac{1}{\sqrt{x}}\right)^2$
32. $\sqrt{x} \div \sqrt{x^3}$
33. $(\sqrt{x} + \sqrt{2})(\sqrt{x} - \sqrt{2})$

34. Rationalize the denominators of

a) $\dfrac{8}{\sqrt{5} + 1}$
b) $\dfrac{2}{(1 + \sqrt{5})^2}$
c) $\dfrac{x - 4}{\sqrt{x} - 2}$

35. Rationalize the numerators of

a) $\dfrac{1 - \sqrt{3}}{2}$
b) $\dfrac{(1 + \sqrt{3})^2}{4}$
c) $\dfrac{\sqrt{x} - 3}{x - 9}$

In problems 36 through 41, evaluate the given expressions, where functions f and g are given by

$$f(x) = x^2 - 2, \qquad g(x) = x + \frac{1}{x}.$$

In each case in which the result is a rational number, give the answer in exact form; otherwise give the answer in decimal form rounded off to two places.

36. $f(\sqrt[3]{5})$ 　　　　　　　　　　 **37.** $(f+g)(\sqrt{3})$ 　　　　　　　　 **38.** $(f \circ g)(\sqrt{2})$

39. $(g \circ f)(\sqrt{5})$ 　　　　　　　　 **40.** $(f \cdot g)(\sqrt{2})$ 　　　　　　　 **41.** $(g \circ f)\left(\sqrt{3} + \dfrac{1}{\sqrt{3}}\right)$

In problems 42 through 45, prove that the given numbers are irrational.

42. $\sqrt{5} - 2$ 　　　　　 **43.** $\sqrt{3} + 2$ 　　　　　 **44.** $\sqrt[3]{4} + 1$ 　　　　　 **45.** $\sqrt[3]{2} - 1$

46. The volume V of a sphere of radius r is given by the formula $V = \frac{4}{3}\pi r^3$.

    a) Solve for r and get a formula that gives r as a function of V.

    b) Use the result in (a) to find the radius of a sphere whose volume is 148.4 cm³. Give answer rounded off to one decimal place.

47. A spherical balloon is being inflated in such a manner that its radius r is given by $r = 1 + 2\sqrt{t}$, where t represents the time in seconds after inflation begins and r is measured in centimeters. The balloon will burst if the radius exceeds 15 cm.

    a) Find a formula that gives the volume of the balloon as a function of t, and state the domain of this function.

    b) What is the volume at the end of 4 seconds? What is it at the end of one minute?

48. Suppose an inverted circular cone of height 20 cm and base of radius 10 cm is partially filled with water. Let r represent the radius of the water surface and h the height, as shown in Fig. 3.2.

    a) Find a formula for the volume V of water as a function of r.

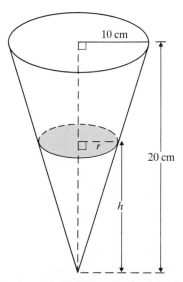

Fig. 3.2

b) Use (a) to find r as a function of V. Then find r when the volume of water is 1200 cm³.

49. Suppose $x = 1 + \sqrt{3}$, $y = \sqrt{3 + \sqrt{13 + 4\sqrt{3}}}$, and $z = 110771/40545$.

    a) Is $x = y$?   **b)** Is $x = z$?   **c)** Is $y = z$?

3.2   EXPONENTIAL FUNCTIONS

In Section 3.1 we discussed rational number exponents. Before we introduce exponential functions, it is necessary to consider the matter of irrational exponents.

Irrational Number Exponents

In drawing a graph of the function defined by $f(x) = 3^x$, we first find pairs of numbers x, y that satisfy $y = 3^x$, and then use these to draw the graph. In Section 3.1 we defined 3^x for x-any rational number. However, for $x = \sqrt{2}$ we are confronted with the question: What does $3^{\sqrt{2}}$ mean? The answer to this question is not simple, but we can get some idea of what is involved by looking at the successive decimal approximations to $\sqrt{2}$; these are $1, 1.4, 1.41, 1.414, 1.4142, \ldots$. Since each of these is a rational number, we can apply Definition 3.3, and so each number of the following sequence has been defined:

$$3^1, \ 3^{1.4}, \ 3^{1.414}, \ 3^{1.4142}, \ldots \tag{3.17}$$

If the sequence of numbers in (3.17) gets close to (we say *converges* to) a fixed real number, then we define $3^{\sqrt{2}}$ as that number. Actually, the sequence in (3.17) does converge, but we are not prepared to argue that here. However, we can get some idea that this is so by using a calculator to get the following table, where the results are rounded off to four decimal places.

x	1	1.4	1.41	1.414	1.4142	1.41421	1.414213
3^x	3	4.6555	4.7070	4.7277	4.7287	4.7288	4.7288

Concepts from calculus are required to give a rigorous definition of 3^x for irrational numbers x. After that, it is possible to prove that the rules of exponents **(E1)** through **(E5)** are valid when the exponents are *any real numbers*.

A number such as $3^{\sqrt{2}}$ is an irrational number, and so it cannot be represented as a finite decimal. To get a decimal approximation for $3^{\sqrt{2}}$, we could evaluate several terms of the sequence given in (3.17), as we have done in the table, but fortunately the calculator is prepared to do something of this type automatically for us. The $\boxed{y^x}$ key instructs the calculator to execute such a program and thus gives an approximation correct to several decimal places. The procedure is as follows:

With an Algebraic calculator, press 3, $\boxed{y^x}$, 2, $\boxed{\sqrt{x}}$, $\boxed{=}$.

With a RPN calculator, press 3, ENT , 2, √x , yˣ . The result is $3^{\sqrt{2}} = 4.72880$ (to five decimal places).

Exponential Functions

The above discussion of $3^{\sqrt{2}}$ gives us some idea of what is involved in formulating a definition of b^x for any real number x. In Section 3.1 we allowed b to be a negative number for some rational-number exponents—for example, $(-4)^{1/3}$—but when x is an irrational number, b must be a positive number.

We are now ready to introduce the idea of an exponential function.

Definition 3.5   If b is any given *positive real number* and $b \neq 1$, then the function f given by the formula $f(x) = b^x$ is called the *exponential function* with *base b*.

The reason for the condition $b \neq 1$ is that 1^x is equal to 1 for all x; and thus 1^x is a constant function, which we prefer not to include as an exponential function.

Example 1   Draw a graph of $y = 3^x$.

Solution   First complete the following table, then plot the corresponding (x, y) points and draw a graph, as shown in Fig. 3.3.

x	-3	-2	-1	-0.5	0	0.5	1	1.5	2	3
y	0.04	0.11	0.33	0.58	1	1.73	3	5.20	9	27

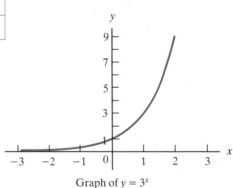

Fig. 3.3

Graph of $y = 3^x$

From the curve in Fig. 3.3, note that $f(x) = 3^x$ is an increasing function; the domain and range of f are given by

$$\mathcal{D}(f) = \{x | x \text{ is a real number}\}, \qquad \mathcal{R}(f) = \{y | y > 0\}. \qquad ■$$

Example 2   Draw a graph of $y = (1/3)^x = 3^{-x}$.

Solution   Following the pattern of Example 1, we make a table of x,y values, plot the corresponding points, and draw the curve, as shown in Fig. 3.4. It is instructive to compare the x,y values in this table with those in the table of Example 1.

x	-3	-2	-1.5	-1	-0.5	0	0.5	1	2	3
y	27	9	5.20	3	1.73	1	0.58	0.33	0.11	0.04

Y=(1/3)^X

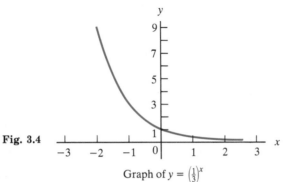

Fig. 3.4

Graph of $y = \left(\frac{1}{3}\right)^x$

From the graph of Fig. 3.4 we see that the function g given by $g(x) = (1/3)^x$ is a decreasing function with domain and range

$$\mathcal{D}(g) = \{x \mid x \text{ is a real number}\}, \qquad \mathcal{R}(g) = \{y \mid y > 0\}. \qquad ■$$

Example 3   Suppose f is a function given by $f(x) = (1 + x)^{1/x}$, where $x > -1$ and $x \neq 0$. Make a table of $f(x)$ values for several values of x near zero. From the table draw a conclusion regarding the behavior of $f(x)$ as x approaches zero.
Note: The function f given here is not an exponential function since the base, $1 + x$, is not a constant.

Solution   Use a calculator to complete the following table of $f(x)$ values for the given values of x.

x	$f(x)$
1	2
0.5	2.25
0.2	2.48832
0.1	2.59374
0.01	2.70481
0.001	2.71692
0.0001	2.71815

x	$f(x)$
-0.8	7.47674
-0.5	4
-0.2	3.05176
-0.1	2.86797
-0.01	2.73200
-0.001	2.71964
-0.0001	2.71842

From the values of $f(x)$ in the above table we conclude that $(1 + x)^{1/x}$ appears to be approaching a number (as x approaches 0) that is between 2.71815 and 2.71842.

The Number e

In Example 3 we observed that as x approaches zero, $(1 + x)^{1/x}$ appears to approach 2.718. . . as a limit. This is actually true (as is shown in calculus), and the limiting value is a transcendental number denoted by e:

$$e = 2.718281828459045235360287. . .$$

The number e is an important number that occurs frequently in applied as well as theoretical problems in mathematics.*

Example 4   Draw a graph of $y = e^x$.

Solution   First make a table of x,y values, then plot the corresponding points and draw a curve through these points, as shown in Fig. 3.5. Some calculators have an $\boxed{e^x}$ key, so the value of y can be determined directly by pressing the $\boxed{e^x}$ key after x is entered in the display. For calculators that do not have an $\boxed{e^x}$ key, we suggest using the $\boxed{y^x}$ key, where 2.718281828 (rounded off to calculator capacity) is first stored with the $\boxed{STO}$ key and recalled with the $\boxed{RCL}$ key when needed. In Section 3.4 methods for evaluating e^x without first storing e will be given. In the table the values of y have been rounded off to two decimal places.

x	-3	-2.5	-2	-1.5	-1	-0.5	0	0.5	1	1.5	2	2.5	3
y	0.05	0.08	0.14	0.22	0.37	0.61	1	1.65	2.72	4.48	7.39	12.18	20.09

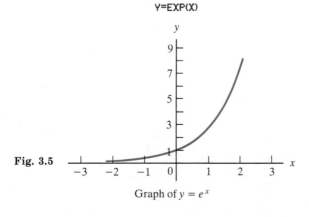

Y=EXP(X)

Fig. 3.5

Graph of $y = e^x$

---

*   The letter e is used in honor of the Swiss mathematician Leonhard Euler (1707–1783), one of the greatest mathematicians of all time.

From the graph in Fig. 3.5 we conclude that $F(x) = e^x$ is an increasing function with domain and range given by

$$\mathfrak{D}(F) = \{x | x \text{ is a real number}\}, \qquad \mathfrak{R}(F) = \{y | y > 0\}. \qquad \blacksquare$$

The above examples suggest the following general conclusions concerning exponential functions:

> The domain and range of $G(x) = b^x$ are given by
> $$\mathfrak{D}(G) = \{x | x \text{ is a real number}\},$$
> $$\mathfrak{R}(G) = \{y | y > 0\}.$$
> If $0 < b < 1$, then G is a decreasing function; if $b > 1$, G is an increasing function.

Exercises 3.2

In problems 1 through 10, use a calculator to evaluate the given expression. If your calculator indicates *Error* for any problem, explain why. Give answers rounded off to three decimal places.

1. $3^{\sqrt{2}}$ **2.** $(\sqrt{2})^2$ **3.** $5^{-\sqrt{3}}$ **4.** $\pi^{2.48}$ **5.** $e^{-\sqrt{3}}$

6. $\sqrt{4 - e^2}$ **7.** $(1 - e)^2$ **8.** $e^{-2/3}$ **9.** $(1 - e)^{2/3}$ **10.** $\sqrt[4]{e^2 - 1}$

11. Which is greater (a) $3^{\sqrt{3}}$ or $(\sqrt{3})^3$, (b) e^3 or 3^e?

12. Which is greater (a) $5^{\sqrt{5}}$ or $(\sqrt{5})^5$, (b) π^e or e^π?

In problems 13 through 27, draw a graph of the given functions.
Note: -3^x means $-(3^x)$ and not $(-3)^x$; similarly, -3^{-x} means $-(3^{-x})$.

13. $y = 2^x$ **14.** $y = (1/2)^x$ **15.** $y = (1.53)^x$

16. $y = (1.53)^{-x}$ **17.** $y = e^{-x}$ **18.** $y = 2e^x$

19. $y = (e - 1)^x$ **20.** $y = \left(\dfrac{e - 1}{2}\right)^x$ **21.** $y = \left(\dfrac{1 + \sqrt{5}}{2}\right)^x$

22. $y = \frac{1}{2}(e^x + e^{-x})$ **23.** $y = 1 + e^x$ **24.** $y = -3^x$

25. $y = -3^{-x}$ **26.** $y = 3^{-x/2}$ **27.** $y = \frac{1}{2}(e^x - e^{-x})$

28. Given that $f(x) = 1 - 5^{-x}$, evaluate each of the following correct to two decimal places.

     a) $f(0)$ **b)** $f(1)$ **c)** $f(1/2)$ **d)** $f(-2)$ **e)** $f(-0.24)$

29. Given that $g(x) = 1/(1 + e^x)$ evaluate each of the following correct to two decimal places.

     a) $g(0)$ **b)** $g(1)$ **c)** $g(2)$ **d)** $g(-3)$ **e)** $g(-0.64)$

30. The predicted population P of a certain city is given by the formula

$$P = 450000(1.08)^{n/12}$$

where n is the number of years after 1980. Find the predicted population for each of the following years. Round off answers to the nearest thousand.

a) 1985   **b)** 1990   **c)** 1995   **d)** 2000

31. A function that occurs frequently in the study of probability and statistics is given by

$$f(x) = \frac{1}{\sqrt{2\pi}} e^{-x^2/2},$$

where x is any real number. Compute the corresponding values of $f(x)$ to two decimal places for x equal to 0, 0.2, 0.4, 0.6, 0.8, 1.0, 1.2, 1.4, 1.6, 1.8, 2.0. Plot a graph of $y = f(x)$. Note that $f(-x) = f(x)$ and use this to draw the graph for negative values of x.

32. In Example 3 we discussed the function $f(x) = (1 + x)^{1/x}$, where $x > -1$ and $x \neq 0$. Explain these restrictions on values of x.

In problems 33 through 38, functions f, g, and h are given by

$$f(x) = x^2 - 3, \qquad g(x) = e^x - 1, \qquad h(x) = 3^x + 3^{-x}.$$

Evaluate each of the given expressions and round off answers to three decimal places.

33. $(f + g)(1)$ **34.** $(f \cdot g)(-1)$ **35.** $(f \circ h)(2)$

36. $(h \circ g)(-3)$ **37.** $(h \circ f)(\sqrt{2})$ **38.** $(f/g)(2)$

39. Determine the values of x that satisfy the equation $x^2 = 2^x$ as follows:

a) Complete the table.

x	-1	-0.8	-0.75	-0.5	0	1	2	3	4	5
x^2										
2^x										

b) Draw graphs of $y = x^2$ and $y = 2^x$ on the same system of coordinates, using values from the table.

c) Use the graphs to help determine the roots of $x^2 = 2^x$; give answers correct to one decimal place.

40. Determine the roots of $x^3 = 3^x$. Follow instructions similar to those of Problem 39, using values from the following table.

x	0	1	2	2.45	2.50	2.75	3	4
x^3								
3^x								

3.3   LOGARITHMIC FUNCTIONS

In the preceding section we defined exponential functions given by $f(x) = b^x$, where b is a given positive number and $b \neq 1$. We noted that f is an increasing function for $b > 1$ and a decreasing function for $0 < b < 1$. In Section 1.8 inverse functions were introduced, and it was noted that an increasing or decreasing function is one-to-one, and consequently its inverse relation is a function. Thus the *inverse of any exponential function is also a function,* which we shall call a *logarithmic function.*

In Section 1.8 we illustrated a useful technique for determining a formula for some inverse functions. For instance, suppose g is a function defined by

$$g\colon y = 3x - 2. \tag{3.18}$$

To get a formula for g^{-1} we can interchange x and y in Eq. (3.18) obtaining

$$x = 3y - 2.$$

Then solving for y gives

$$g^{-1}\colon y = \frac{x + 2}{3}.$$

Now suppose we attempt a similar approach to determine a formula for the inverse function of f, given by

$$f\colon y = 3^x. \tag{3.19}$$

Since f is increasing, we can be certain that its inverse is a function. Interchanging x and y in Eq. (3.19) gives $x = 3^y$. In this case, we cannot solve for y in terms of x in a simple manner, as we did in the above example. We could express the inverse function as a set of ordered pairs and denote it by the symbol f^{-1}, as used in general situations:

$$f^{-1} = \{(x, y) | x = 3^y\}. \tag{3.20}$$

However, inverses of exponential functions play an important role in applied as well as theoretical mathematics, and so they deserve special names. These functions are called *logarithmic functions.* The function given in Eq. (3.20) is denoted by $\log_3$, which is read "log base 3 function." For example

$$\log_3 = \{(x, y) | x = 3^y\}$$

means that $\log_3$ is a function defined by $y = \log_3 x$ if and only if $x = 3^y$.

Now let us return to inverses of exponential functions in general and state the following definition.

Definition 3.6   Suppose b is a given positive number and $b \neq 1$. The *logarithmic function base b,* denoted by $\log_b$ is defined by

$$y = \log_b x \text{ if and only if } x = b^y.$$

That is, $\log_b$ is the inverse of the exponential function with base b.

The domain and range of the $\log_b$ function are given by

$$\mathfrak{D}(\log_b) = \{x | x > 0\}, \qquad \mathfrak{R}(\log_b) = \mathbf{R}.$$

Graphs of Logarithmic Functions
The following example illustrates a procedure for drawing a graph of a logarithmic function.

Example 1   Draw a graph of the function given by

$$y = \log_3 x. \tag{3.21}$$

Solution   Since $\log_3$ is the inverse of the exponential function given by $f(x) = 3^x$, we can draw its graph by simply reflecting the graph shown in Fig. 3.3 about the line $y = x$. Or we can get a table of x, y values satisfying Eq. (3.21) by interchanging the x and y values in the table shown on p. 171. This gives the following table, which can be used to draw the graph of $y = \log_3 x$.

x	0.04	0.11	0.33	0.58	1	1.73	3	5.20	9	27
y	-3	-2	-1	-0.5	0	0.5	1	1.5	2	3

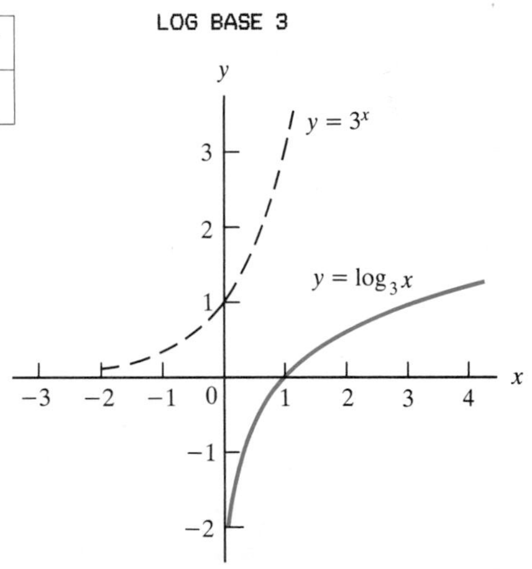

LOG BASE 3

Fig. 3.6

Graph of $y = \log_3 x$

In Fig. 3.6, the broken curve is the graph of $y = 3^x$, and the solid curve is the graph of $y = \log_3 x$.

Properties of Logarithmic Functions

We now state three useful properties that form a basis for algebraic manipulation of logarithmic functions. They are analogous to corresponding properties **(E1)**, **(E2)**, and **(E3)** for exponential functions discussed in the preceding sections.

Suppose u and v are positive numbers and t is any real number. Then

$$\textbf{(L1)}\ \log_b(u \cdot v) = \log_b u + \log_b v \qquad \textbf{(L2)}\ \log_b\!\left(\frac{u}{v}\right) = \log_b u - \log_b v$$

$$\textbf{(L3)}\ \log_b(u^t) = t(\log_b u)$$

We shall prove the statement given by property **(L1)**; proofs for the other two are similar.

Let $\log_b u = h$ and $\log_b v = k$. Using Definition 3.6, we get

$$u = b^h \qquad \text{and} \qquad v = b^k.$$

Since **(L1)** involves the product $u \cdot v$, we multiply and use property **(E1)** to get

$$u \cdot v = b^h \cdot b^k = b^{h+k}.$$

Applying Definition 3.6 to $u \cdot v = b^{h+k}$ gives

$$\log_b(u \cdot v) = h + k.$$

Replacing h by $\log_b u$ and k by $\log_b v$, we have

$$log_b(u \cdot v) = log_b u + \log_b v.$$

Note: Properties **(L1)**, **(L2)**, and **(L3)** involve logarithms of products, quotients, and powers. We do not give similar formulas for sums and differences because there are no simple results for $\log_b(u + v)$ and $\log_b(u - v)$.

There are a few additional properties of logarithmic functions that are worth noting. Let us evaluate $\log_b x$ for $x = 1$ and $x = b$.

Let $\log_b 1 = c$; then $b^c = 1$, and so $c = 0$. Thus $\log_b 1 = 0$.

Let $\log_b b = d$; then $b^d = b$, and so $d = 1$. Thus $\log_b b = 1$.

These two special cases occur frequently, and so we label them as property **(L4)** for easy reference.

$$\textbf{(L4)}\ \log_b 1 = 0 \qquad \text{and} \qquad lob_b b = 1$$

Since the $\log_b$ function and the exponential function with base b are inverses of each other, we have the following identities.

> **(L5)** $b^{\log_b x} = x$ for $x > 0$
>
> **(L6)** $\log_b(b^x) = x$ for $x \in \mathbf{R}.$

Let us consider several examples illustrating the use of properties **(L1)** through **(L6)** and Definition 3.6.

Example 2   Evaluate each of the following and give answers in exact form.

     a) $\log_2 8$ **b)** $\log_{10}(0.0001)$ **c)** $\log_{0.5}(4\sqrt{2})$

Solution   **a)** Let $\log_2 8 = r$. By Definition 3.6, $2^r = 8 = 2^3$. Thus $r = 3$, and so $\log_2 8 = 3$.

     b) Let $\log_{10}(0.0001) = q$. By Definition 3.6, $10^q = 0.0001 = 10^{-4}$, and so $q = -4$. Thus $\log_{10}(0.0001) = -4$.

     c) Let $\log_{0.5}(4\sqrt{2}) = m$. By Definition 3.6, $(0.5)^m = 4\sqrt{2} = 2^2 \cdot 2^{1/2} = 2^{5/2}$. Hence $(0.5)^m = 2^{5/2}$. But $(0.5)^m = \left(\dfrac{1}{2}\right)^m = \dfrac{1}{2^m} = 2^{-m}$. Thus $2^{-m} = 2^{5/2}$, and so $m = -5/2$. Therefore $\log_{0.5} 4\sqrt{2} = -2.5$.   ■

Example 3   Given that $\log_5 3 = 0.6826$ and $\log_5 6 = 1.1133$ (correct to four decimal places), evaluate the given expressions. Give answers rounded off to three decimal places.

     a) $\log_5 2$ **b)** $\log_5(\log_2 8)$ **c)** $(\log_5 12) \div (\log_5 3)$

Solution   **a)** $\log_5 2 = \log_5(6/3) = \log_5 6 - \log_5 3 = 1.1133 - 0.6826 = 0.4307$. Here we used property **(L2)**. Thus we have $\log_5 2 = 0.431$.

     b) From Example 1(a) we have $\log_2 8 = 3$. Therefore $\log_5(\log_2 8) = \log_5 3 = 0.6826$. Rounding off to three decimal places, we have $\log_5(\log_2 8) = 0.683$.

     c) First we evaluate $\log_5 12$. Using **(L1)** and **(L3)**, we get

$$\log_5 12 = \log_5(2^2 \cdot 3) = \log_5(2^2) + \log_5 3 = 2\log_5 2 + \log_5 3.$$

Using $\log_5 2 = 0.4307$ (from part (a)) and $\log_5 3 = 0.6826$, we get

$$\log_5 12 = 2(0.4307) + 0.6826 = 1.5440.$$

Thus $(\log_5 12) \div (\log_5 3) = (1.5440) \div (0.6826) = 2.2619$. Rounded off to three decimal places, $(\log_5 12) \div (\log_5 3) = 2.262$.   ■

Example 4 Combine $3\log_5 2 + (3/2)\log_5 8 - (1/2)\log_5 32$ and express the result as $\log_5$ of a number.

Solution
$$3\log_5 2 + (3/2)\log_5 8 - (1/2)\log_5 32 = \log_5 2^3 + \log_5 8^{3/2} - \log_5 32^{1/2}$$
by **(L3)**

$$= \log_5\left(\frac{2^3 \cdot 8^{3/2}}{32^{1/2}}\right) = \log_5\left(\frac{2^3 \cdot 2^{9/2}}{2^{5/2}}\right) = \log_5(2^5) = \log_5 32.$$
by **(L1), (L2)**   by **(E3)**   by **(E1), (E2)**

Therefore the given expression is equal to $\log_5 32$.   ■

Example 5 Suppose p and q are positive numbers. Write the following as linear combinations of $\log_b p$ and $\log_b q$.

a) $\log_b(pq^3)$ **b)** $\log_b\left(\dfrac{p\sqrt{q}}{q}\right)$

Solution   **a)** $\log_b(pq^3) = \log_b p + \log_b q^3 = \log_b p + 3\log_b q$
by **(L1)**   by **(L3)**

b) $\log_b\left(\dfrac{p\sqrt{q}}{q}\right) = \log_b\left(\dfrac{pq^{1/2}}{q}\right) = \log_b\left(\dfrac{p}{q^{1/2}}\right)$

$$= \log_b p - \log_b q^{1/2} = \log_b p - \frac{1}{2}\log_b q$$
by **(L2)**   by **(L3)**   ■

Example 6 Solve the equations:

a) $\log_3(2x + 5) - \log_3(x) = 1$ `b)` $\log_3(2x - 5) - \log_3(x) = 1$

Solution   **a)** Applying **(L2)** to the given equation, we get $\log_3\left(\dfrac{2x+5}{x}\right) = 1$. Using Definition 3.6, we get $(2x + 5)/x = 3$. Thus $2x + 5 = 3x$, and so $x = 5$.

As a check we wish to see if 5 actually satisfies the given equation. Replacing x by 5 in the left-hand side gives

$$\text{LHS} = \log_3(2 \cdot 5 + 5) - \log_3(5) = \log_3 15 - \log_3 5 = \log_3\frac{15}{5} = \log_3 3 = 1.$$

Hence 5 is a solution of the given equation.

b) Following a pattern similar to that in (a), we get $x = -5$. Now substituting -5 for x in the left-hand side of the given equation gives

$$\text{LHS} = \log_3(-10 - 5) - \log_3(-5) = \log_3(-15) - \log_3(-5).$$

Since -15 and -5 are not in the domain of the $\log_3$ function, that is, $\log_3(-15)$ and $\log_3(-5)$ are not defined, we see that -5 is not a solution of the given equation. Thus, there is no real number x that satisfies the given equation.   ■

Example 7   Solve for x: $\log_{10}x + \log_{10}(x + 48) = 2$.

Solution   Applying **(L1)**, we can write the given equation as $\log_{10}[x(x + 48)] = 2$. Using Definition 3.6 gives $x(x + 48) = 10^2$. The quadratic equation

$$x^2 + 48x - 100 = 0$$

can now be solved by factoring: $(x + 50)(x - 2) = 0$. Hence -50 and 2 are solutions to the quadratic equation.

We now check to see if these are solutions to the given equation. Replacing x by -50 yields

$$\text{LHS} = \log_{10}(-50) + \log_{10}(-50 + 48) = \log_{10}(-50) + \log_{10}(-2).$$

This gives an undefined result since -50 and -2 are not in the domain of $\log_{10}$, and so -50 is not a solution.

If we replace x by 2 in the original equation, it is easy to verify that 2 is a solution. Therefore the solution set for the given equation is $\{2\}$. ▬

Need to Check Answers

In Examples 6 and 7 we indicated a need to check answers resulting from intermediate steps to see if they actually are solutions to the given equations. To see the reason for this let us take a closer look at the steps involved in the solution of Example 7.

Let f represent the function given by the left-hand side of the given equation:

$$f(x) = \log_{10}x + \log_{10}(x + 48).$$

The domain of f is given by

$$\mathcal{D}(f) = \{x|x > 0 \quad \text{and} \quad x + 48 > 0\} = \{x|x > 0\}.$$

The first step in our solution involves the function g given by

$$g(x) = \log x(x + 48).$$

The domain of g is given by

$$\mathcal{D}(g) = \{x|x(x + 48) > 0\} = \{x|x > 0 \quad \text{or} \quad x < -48\}.$$

Since $\mathcal{D}(g) \neq \mathcal{D}(f)$, functions f and g are not equal. In fact, $\mathcal{D}(f) \subset \mathcal{D}(g)$, and so there may be values of x that are solutions to an equation involving $g(x)$ but are not solutions to the corresponding equation involving $f(x)$. This is so in Example 7.

Example 8   Determine the domains of functions f and g given by

a) $f(x) = \log_3(x^2 - 5x + 6)$ b) $g(x) = \log_3(x - 2) + \log_3(x - 3)$

Solution   Here we use the fact that $\mathfrak{D}(\log_3) = \{u \mid u > 0\}$.

a) $\mathfrak{D}(f) = \{x \mid x^2 - 5x + 6 > 0\} = \{x \mid (x - 2)(x - 3) > 0\} = \{x \mid x < 2 \text{ or } x > 3\}$.

$$\mathfrak{D}(f): \quad \begin{array}{ccc} & & \\ \hline 0 & 2 & 3 \end{array}$$

b) $\mathfrak{D}(g) = \{x \mid x - 2 > 0 \quad \text{and} \quad x - 3 > 0\} = \{x \mid x > 3\}$.

$$\mathfrak{D}(g): \quad \begin{array}{cc} & \\ \hline 0 & 3 \end{array}$$

We conclude that $\mathfrak{D}(f) \neq \mathfrak{D}(g)$, and hence functions f and g are not equal. ∎

Exercises 3.3

In problems 1 through 15, evaluate the given expressions and give answers in exact form. If the given expression is not defined, tell why.

1. $\log_2(32)$ **2.** $\log_3(1/27)$ **3.** $\log_5(125/\sqrt{5})$

4. $\log_7(49/\sqrt{7})$ **5.** $\log_{10}100$ **6.** $\log_{10}1000$

7. $\log_{10}(0.0001/\sqrt{0.0001})$ **8.** $\log_e(1/e)$ **9.** $\log_{0.3}(10/3)$

10. $\log_5(\log_5 5)$ **11.** $\log_7(\log_7 1)$ **12.** $\log_8(\log_3 3)$

13. $\log_3(\log_5(1/5))$ **14.** $\log_{10}(\log_{10}0.1)$ **15.** $\log_2(4\sqrt{2})$

In problems 16 through 24, p and q are positive numbers. Write the given expressions as linear combinations of $\log_b p$ and $\log_b q$ (see Example 5).

16. $\log_b(p^4 q^5)$ **17.** $\log_b(p^{1.5}q^2)$ **18.** $\log_b((\sqrt{p}\,q)/(p\,\sqrt{q}))$

19. $\log_b(p^{3/2}\,q^{4/3})$ **20.** $\log_b\left(\dfrac{p+q}{p^{-1}+q^{-1}}\right)$ **21.** $\log_b\left(\dfrac{p-q}{q^{-1}-p^{-1}}\right), p > q$

22. $\log_b(b\sqrt{pq})$ **23.** $\log_b(pq/b)$ **24.** $\log_b(b^2/(pq))$

In problems 25 through 40, use the following to evaluate the given expression and give answers rounded off to four decimal places:

$$\log_5 2 = 0.43068, \quad \log_5 3 = 0.68261, \quad \log_5 7 = 1.20906,$$
$$\log_3 11 = 2.18266, \quad \log_3 22 = 2.81359$$

25. $\log_5 6$ **26.** $\log_5 63$ **27.** $\log_5 75$ **28.** $\log_3 2$

29. $\log_3 66$ **30.** $\log_3 \sqrt{44}$ **31.** $\log_3 \sqrt{54}$ **32.** $\log_5(\log_3 9)$

33. $\log_5(\sqrt[4]{21})$ **34.** $\log_3 99$ **35.** $\log_5 10.5$ **36.** $\log_5(\sqrt{14}/5)$

37. $\log_5(\log_5 25)$ **38.** $(\log_3 9)(\log_5 42)$ **39.** $(\log_3 33) \div (\log_5 81)$ **40.** $\log_5 70 - \log_3 4$

In problems 41 through 45, write each of the given expressions as $\log_b$ of a number for the given b (see Example 4).

41. $\log_3 5 + \log_3 20$ **42.** $2 \log_3 5 - \log_3 4$ **43.** $\frac{1}{2}\log_7 4 + \frac{2}{3}\log_7 27 - \frac{1}{6}\log_7 64$

44. $3 \log_2 3 - 2 \log_2 9 + 2 \log_2 5$ **45.** $\frac{1}{2}\log_2 5 - \frac{1}{2}\log_2 20 + \frac{1}{4}\log_2 81$

In problems 46 through 55, solve for the indicated letter. When necessary, be certain to check to see that your solution satisfies the given equation.

46. If $\log_3 x = 4$, then $x =$ _____ .

47. If $\log_b 16 = 2$, then $b =$ _____ .

48. If $\log_5(\frac{1}{25}) = y$, then $y =$ _____ .

49. If $\log_5(3x - 1) = 1$, then $x =$ _____ .

50. If $\log_5(4x) - \log_5(2x - 1) = 2$, then $x =$ _____ .

51. If $\log_3(2x) + \log_3(5x) = \log_3 10$, then $x =$ _____ .

52. If $\log_b(\frac{1}{27}) = -3$, then $b =$ _____ .

53. If $\log_5 25 + \log_3 27 = 2x + 1$, then $x =$ _____ .

54. a) If $\log_7 x^2 - \log_7(x + 6) = 0$, then $x =$ _____ .

55. If $\log_{10} x + \log_{10}(x + 3) = 1$, then $x =$ _____ .

b) If $2\log_7 x - \log_7(x + 6) = 0$, then $x =$ _____ .

In problems 56 through 67, determine whether the given statement is true, false, or meaningless. A statement is meaningless if any part of it is undefined. Give reasons for your answers.

56. $\log_3 9 - \log_3 2 = \log_3(4.5)$

57. $\log_5(\frac{3}{2}) + \log_5 2 = \log_5 3$

58. $\log_7(3^2 + 4^2) = 2\log_7 3 + 2\log_7 4$

59. $\log_3\left(\dfrac{1 - \sqrt{3}}{2}\right) = \log_3(1 - \sqrt{3}) - \log_3 2$

60. $\log_{10} 100 - \log_{10} 0.01 = 4$

61. $\log_5(\frac{3}{2}) = \log_5 3/\log_5 2$

62. $\log_5\left(\dfrac{2}{\sqrt{5} + 1}\right) = \log_5(\sqrt{5} - 1) - \log_5 2$

63. $\log_2(\log_2 \frac{1}{2}) = -1$

64. $\log_3(1 + \log_2 4) = 1$

65. If $f(x) = \log_8(x^2 - 4)$ and $g(x) = \log_8(x - 2) + \log_8(x + 2)$, then $f = g$.

66. If $f(x) = \log_8[(1 - x)(1 + x)]$, then $f(x) \le 0$, for all x in $\mathfrak{D}(f)$.

67. If $f(x) = \log_5\left(\dfrac{x + 2}{x - 3}\right)$ and $g(x) = \log_5(x + 2) - \log_5(x - 3)$, then $f = g$.

In problems 68 through 76, state the domain of the given functions.

68. $f(x) = \log_{10}(1 + x)$

69. $f(x) = \log_5 \sqrt{25 - x^2}$

70. $g(x) = \log_3(-x)$

71. $h(x) = \log_3(-x^2 + 8x - 15)$

72. $f(x) = \log_3(x - 4) + \log_3 x$

73. $g(x) = \log_3[(x - 4)x]$

74. $f(x) = \log_{10}(1 - |x|)$

75. $h(x) = \log_2(x + 1) - \log_2 x$

76. $f(x) = \log_e(e^{-x})$

3.4   **USING A CALCULATOR TO**
         EVALUATE LOGARITHMIC FUNCTIONS

In the examples of the preceding section, we were able to evaluate logarithms by converting to exponential form. For example, to evaluate $\log_3 \sqrt{27}$ we let $\log_3 \sqrt{27} = y$. This is equivalent to $3^y = \sqrt{27} = 3^{3/2}$. Thus $y = 3/2$, and so $\log_3 \sqrt{27} = 3/2$. However, attempting a similar procedure to evaluate $\log_3 6.4 = x$, we have $3^x = 6.4$. Since 6.4 cannot be expressed as a simple power of 3, we are unable to complete the solution as we did in the first example. In this section we introduce techniques by which a calculator can be used to solve such problems.

Common and Natural Logarithms

For computational purposes, the base of logarithms that is frequently used is $b = 10$. Since it is cumbersome to write the subscript 10 in $\log_{10}$ each time, we shall write log and it is understood that the base is 10. For theoretical as well as computational purposes, it is an interesting fact that the transcendental number $e = 2.718281828\ldots$ (see Example 3 of Section 3.2) occurs naturally as a base of logarithms in the study of calculus. To avoid writing $\log_e$ each time, we replace it by ln. Thus we have the following notation.

$$\log_{10} x \text{ is written as } \log x;$$

$$\log_e x \text{ is written as } \ln x.$$

The notation adopted here is consistent with that appearing on scientific calculators.

Logarithms with base 10 are called *common logarithms,* whereas those with base e are called *natural logarithms.*

Logarithms With Calculators

Most scientific calculators have both ⌊log⌋ and ⌊ln⌋ keys. We shall consider several examples that will illustrate the use of these keys. Some calculators have the ⌊ln⌋ key but not the ⌊log⌋ key; we shall see that this is sufficient for our purposes.

The ⌊log⌋ and ⌊ln⌋ keys represent functions of one variable. If a positive number x is entered into the display of the calculator and then the ⌊ln⌋ key is pressed, the result $\ln x$ will appear almost immediately in the display; it is not necessary to press the ⌊=⌋ key on algebraic calculators. Similarly the ⌊log⌋ key gives $\log x$ for any positive number x in the display.

Example 1   Evaluate each of the following, correct to four decimal places.

a) $\ln 2$ **b)** $\log 0.0037$ **c)** $\ln\left(\dfrac{1 + \sqrt{5}}{2}\right)$ **d)** $\ln\left(\dfrac{2 - \sqrt{17}}{3}\right)$

Solution   **a)** Pressing the keys ⌊2⌋ and ⌊ln⌋ gives $\ln 2 = 0.6931$.

b) If the calculator has a ⌊log⌋ key, then entering 0.0037 into the display and pressing ⌊log⌋ gives $\log 0.0037 = -2.4318$. If there is no ⌊log⌋ key, $\log 0.0037$ can be evaluated by using Eq. (3.22), given below, with $b = 10$ and $u = 0.0037$.

c) To evaluate $\ln\left(\dfrac{1 + \sqrt{5}}{2}\right)$ we first compute $\dfrac{1 + \sqrt{5}}{2}$ and, with the result in the calculator display, press the ⌊ln⌋ key. This gives $\ln\left(\dfrac{1 + \sqrt{5}}{2}\right) = 0.4812$.

d) To evaluate $\ln\left(\dfrac{2 - \sqrt{17}}{3}\right)$ we follow a procedure similar to that in (c). In this case the calculator indicates *Error*; the reason is that $(2 - \sqrt{17})/3$ is a negative number and so is not in the domain of the ln function. That is, $\ln\left(\dfrac{2 - \sqrt{17}}{3}\right)$ is undefined. ■

Change of Base

In using a calculator to evaluate $\log_b u$, where b is a positive number and $b \neq 1$, it is necessary to convert to logarithms with base e or base 10. This can be done as follows:

Let $\log_b u = t$, which is equivalent to $b^t = u$. Taking ln of both sides of this equation gives $\ln b^t = \ln u$, which is equivalent to saying $t(\ln b) = \ln u$. Thus $t = \ln u / \ln b$. Therefore we have the following formula, which expresses $\log_b u$ in terms of $\ln u$ and $\ln b$:

$$\log_b u = \frac{\ln u}{\ln b}. \tag{3.22}$$

Similarly, using log in place of ln in the above discussion we get

$$\log_b u = \frac{\log u}{\log b}. \tag{3.23}$$

Example 2   Evaluate each of the following, and give answers rounded off to four decimal places.

a) $\log_3 7.5$ **b)** $\log_5(0.0348)$

Solution   The formula given in Eq. (3.22) or (3.23) can be used in each of these problems. We choose Eq. (3.22) since some calculators have a ⬭ ln ⬭ key but not a ⬭ log ⬭ key.

a) $\log_3 7.5 = \dfrac{\ln 7.5}{\ln 3} = 1.8340$

b) $\log_5(0.0348) = \dfrac{\ln 0.0348}{\ln 5} = -2.0865$ ■

Inverse Logarithms

In the above examples all the problems were of the following type: Given a positive number u, find log u or ln u. We are now interested in the inverse problem: Given the value of log u or ln u, determine u. For example, given that

$\log u = 0.4735$, we wish to find u. The notation that has been traditionally used is $u = \text{Antilog } 0.4735$. However, since this actually involves the inverse of the log function, we shall denote it by $u = \log^{-1} 0.4735$. This is read, "u is the inverse log of 0.4735."

As another example of notation, if $\ln v = 1.2654$, then we can write that $v = \ln^{-1} 1.2654$, and say, "v is the inverse ln of 1.2654."*

So far, in the two examples being considered here, we merely introduced some notation. Let us proceed to actually determine u and v. Since the log function is defined as the inverse of the function given by $f(x) = 10^x$, the inverse of the log function must be this exponential function. Therefore, if $\log u = 0.4735$, then

$$u = \log^{-1} 0.4735 = 10^{0.4735}.$$

This is precisely what Definition 3.6 tells us; $\log u = 0.4735$ implies $u = 10^{0.4735}$. We can now determine u by using a calculator, as follows:†

1. If your calculator has a $\boxed{10^x}$ key, then evaluate $10^{0.4735}$ by pressing $\boxed{10^x}$ after entering 0.4735 into the display. This gives $u = 2.9751$ (to four places).

2. If your calculator does not have a $\boxed{10^x}$ but has an $\boxed{\text{INV}}$ key, then with 0.4735 in the display, pressing the $\boxed{\text{INV}}$ and $\boxed{\log}$ keys gives $u = 2.9751$.

Similarly, the ln function and the function given by $f(x) = e^x$ are inverses of each other, so the solution of $\ln v = 1.2645$ is $v = \ln^{-1} 1.2654 = e^{1.2654}$. Thus v can be found pressing the $\boxed{\text{INV}}$ and $\boxed{\ln}$ keys or by using the $\boxed{e^x}$ key after entering 1.2654. Therefore $v = 3.5445$ to four decimal places.

The discussion above illustrates the following:

$\boxed{10^x}$ and $\boxed{\log}$ keys are inverses of each other;

$\boxed{e^x}$ and $\boxed{\ln}$ keys are inverses of each other.

Thus we have the following special cases of properties **(L5)** and **(L6)**.

(L7) $10^{\log x} = x$ for all $x > 0$ and $\log(10^x) = x$ for $x \in \mathbf{R}.$
(L8) $e^{\ln x} = x$ for all $x > 0$ and $\ln(e^x) = x$ for $x \in \mathbf{R}.$

---

* The notation adopted here is consistent with that used for inverse functions in general (see Section 1.8).

† If your calculator does not have $\boxed{\log}$ and $\boxed{10^x}$ keys but has $\boxed{\ln}$ and $\boxed{e^x}$ keys, proceed as follows: Express the original problem, $\log u = 0.4735$, in equivalent ln form by using the change-of-base formula given in Eq. (3.22). That is, with $b = 10$, $\ln u = (\ln 10) \log u$. Therefore $\ln u = (\ln 10) \log u = 2.30259 \log u = (2.30259)(0.4735) = 1.0903$. Thus $u = e^{1.0903}$, which can be evaluated by using the $\boxed{e^x}$ key, or the $\boxed{\text{INV}}$ and $\boxed{\ln}$ keys.

To evaluate 10^u or e^u, first enter u into the display. Pressing

$\boxed{10^{\times}}$ or $\boxed{\text{INV}}$ and $\boxed{\log}$ gives 10^u in the display;   (3.24)

$\boxed{e^{\times}}$ or $\boxed{\text{INV}}$ and $\boxed{\text{ln}}$ gives e^u in the display.   (3.25)

Example 3   In each of the following, solve for v, correct to four decimal places.

a) $v = \log^{-1}0.243$ **b)** $\ln v = 1.345$ **c)** $\log v = -1.4382$

d) $e^v = 0.456$ **e)** $10^v = 1.4837$ **f)** $\ln((2v + 1) - \ln 3 = 1.48$

Solution   **a)** Since $v = \log^{-1}0.243 = 10^{0.243}$, following instructions stated in (3.24) gives $v = 1.7498$.

b) $\ln v = 1.345$ is equivalent to $v = \ln^{-1}1.345$, or $v = e^{1.345}$. Following (3.25) gives $v = 3.8382$.

c) $\log v = -1.4382$ is equivalent to $v = \log^{-1}(-1.4382)$, or $v = 10^{-1.4382}$. Using (3.24) gives $v = 0.0365$.

d) $e^v = 0.456$ is equivalent to $v = \ln 0.456$. Enter 0.456 and press the $\boxed{\text{ln}}$ key to get $v = -0.7853$.

e) $10^v = 1.4837$ is equivalent to $v = \log 1.4837$. This can be evaluated by using the $\boxed{\log}$ key to get $v = 0.1713$.

   If the calculator does not have a $\boxed{\log}$ key, then take ln of both sides of the given equation to get $v \ln 10 = \ln 1.4837$. Thus $v = \ln 1.4837/\ln 10$, which can be evaluated by using the $\boxed{\text{ln}}$ and $\boxed{\div}$ keys.

f) The given equation is equivalent to $\ln[(2v + 1)/3] = 1.48$. Thus $(2v + 1)/3 = e^{1.48}$, and so $v = (3e^{1.48} - 1)/2$. Now use (3.25) to find $e^{1.48}$, and then continue with the remaining arithmetic operations. This gives $v = 6.0894$. ■

Example 4   Evaluate each of the following. Give answers in exact form.

a) $e^{\ln 5}$ **b)** $10^{-\log 5}$ **c)** $\log(10^{-4.5})$

Solution   **a)** By property **(L8)**, $e^{\ln 5} = 5$.

b) By properties **(L3)** and **(L7)**, $10^{-\log 5} = 10^{\log 5^{-1}} = 5^{-1} = \frac{1}{5}$.

c) By property **(L7)**, $\log(10^{-4.5}) = -4.5$. ■

Example 5   Solve the equation

$$2 \ln(2v - 1) + 2 \ln v = 1.   (3.26)$$

Solution   Dividing both sides of the given equation by 2 and then using property **(L1)**, we get $\ln[v(2v - 1)] = 1/2$. Using Definition 3.6 gives $v(2v - 1) = e^{1/2} = \sqrt{e}$. Thus we have a quadratic equation to solve,

$$2v^2 - v - \sqrt{e} = 0. \tag{3.27}$$

Applying the quadratic formula gives

$$v = \frac{1 \pm \sqrt{1 + 8\sqrt{e}}}{4}.$$

Evaluating by calculator, we get 1.1917 and -0.6917 as solutions to Eq. (3.27). It is necessary to check these to see if they are solutions to the given equation. Replacing v in Eq. (3.26) by each of these values, we can easily see that 1.1917 is the only solution. ■

Example 6   Find the domain of $f(x) = \ln\left(\dfrac{x+1}{x}\right)$. Then draw a graph of $y = f(x)$.

Solution $$\mathcal{D}(f) = \left\{ x \left| \frac{x+1}{x} > 0 \right\} = \{x | x > 0 \quad \text{or} \quad x < -1\}.$$

To draw a graph of $y = f(x)$, first make the following table. Then draw the graph shown in Fig. 3.7.

x	-10	-3	-2	-1.5	-1.1	-1.01	-1.001	0.001	0.01	0.1	0.5	1	2	3	10
y	-0.1	-0.4	-0.7	-1.1	-2.4	-4.6	-6.9	6.9	4.6	2.4	1.1	0.7	0.4	0.3	0.1

Y=LN((X+1)/X)

Fig. 3.7

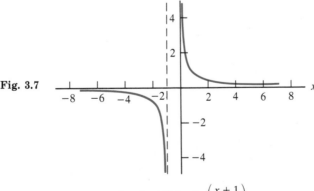

Graph of $f(x) = \ln\left(\dfrac{x+1}{x}\right)$

We see that $x = -1$ and $x = 0$ are vertical asymptotes, and $y = 0$ is a horizontal asymptote for the curve in Fig. 3.7. ■

Example 7   Solve the equation

$$5^x = 3 \cdot 4^{1-x}. \tag{3.28}$$

Give answers correct to three decimal places.

Solution   Taking ln of both sides of Eq. (3.28) gives $\ln 5^x = \ln (3 \cdot 4^{1-x})$. Applying properties **(L1)** and **(L3),** we get

$$x \ln 5 = \ln 3 + (1 - x) \ln 4,$$
$$x \ln 5 = \ln 3 + \ln 4 - x \ln 4,$$
$$x \ln 5 + x \ln 4 = \ln 3 + \ln 4,$$
$$x(\ln 5 + \ln 4) = \ln 3 + \ln 4,$$
$$x = \frac{\ln 3 + \ln 4}{\ln 5 + \ln 4}.$$

We can now evaluate this expression by using a calculator. However, we can simplify slightly by using **(L1)** to get $x = \ln 12 / \ln 20$. Using a calculator gives $x = 0.829$. Substituting 0.829 for x in Eq. (3.28), we can check to see that 0.829 is a solution. ■

Example 8   Find the roots of the equation $e^{-x} - x = 0$ correct to two decimal places.

Solution   In this example x appears in a linear term as well as in the exponent. Such equations are more difficult to solve than others considered in this section. If we write the problem as $e^{-x} = x$ and take ln of both sides (as we did in Example 7), the resulting equation is $-x = \ln x$. However, this does not help in solving for x. Therefore we use a different approach and solve by a process of estimation.

We can get information about how many roots there are and their approximate values by drawing graphs. Suppose the graphs of $y = e^{-x}$ and $y = x$ are drawn on the same set of coordinates, as shown in Fig. 3.8. Let (c, d) denote the point of intersection of these two curves; then $d = e^{-c}$ and $d = c$, so $e^{-c} = c$. Hence c is a solution to the equation $e^{-x} = x$.

Figure 3.8 shows that there is only one point of intersection. From the graph a reasonable estimate of c is 0.6. Evaluating e^{-x} for $x = 0.6$ gives $e^{-0.6} = 0.55$, and so it is clear from the graph that c is to the left of 0.6. We now try $x = 0.5$, and so $e^{-0.5} = .61$. Thus we see that c is to the right of 0.5. Trying 0.57 gives $e^{-0.57} = 0.57$. This tell us that $x = 0.57$ is the desired solution to two decimal places. (See Problem 70 of this section for an interesting way to solve this problem.)

Y=EXP(-X), Y=X

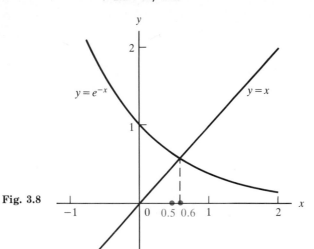

Fig. 3.8

Graphs of $y = e^{-x}$ and $y = x$

Exercises 3.4

If your calculator should indicate *Error* while you are solving any problem in this set, determine the reason.

In problems 1 through 12, evaluate the given expression and give answers rounded off to four decimal places.

1. $\ln 5$ **2.** $\ln 0.47$ **3.** $\log 1.87$ **4.** $\log 0.0435$

5. $\ln (1.56^2 + 2.73^2)$ **6.** $\log (2.43 \sqrt{5.75})$ **7.** $\ln (2 - \sqrt{5.43})$ **8.** $\log [(2 - \sqrt{6})/5]$

9. $\log [(1 + \sqrt{3})/8]$ **10.** $\log_3 6$ **11.** $\log_5 3.47$ **12.** $\log_7 (\sqrt{3} - 1)$

In problems 13 through 18, evaluate and give answers in exact form.

13. $e^{\ln (1.43)}$ **14.** $10^{\log (2.54)}$ **15.** $\log (10^{-0.42})$

16. $\ln (e^{3.2})$ **17.** $e^{-\ln 2}$ **18.** $e^{-3\ln 2}$

In each of the problems 19 through 42, determine the value of v correct to four decimal places.

19. $v = \log^{-1}(0.478)$ **20.** $v = \log^{-1}(-0.587)$ **21.** $\ln v = 1.532$ **22.** $v = \ln^{-1}(1.378)$

23. $\log v = -0.372$ **24.** $\ln v = 1 - \sqrt{3}$ **25.** $10^v = -0.473$ **26.** $e^v = 0.875$

27. $e^{-v} = 1.238$ **28.** $e^{-v} = -0.471$ **29.** $10^{-v} = 1.378$ **30.** $e^{2v} = 0.431$

31. $e^{(3v+1)} = 0.475$ **32.** $v = 10^{-0.47}$ **33.** $v = e^{-0.71}$ **34.** $e^{v+1} = 3e^{2v-1}$

35. $e^{2v-1} = 1.362$ **36.** $10^{-(4v+1)} = 3.473$

37. $\log (3v + 4) = \log 2 + \log (v^2 + 1)$

38. $\ln (2v - 5) - \ln 7 = 2.43$

39. $\ln (v - 5) + \ln 2.43 = 1.56$

40. $\ln (e^{v-1}) = e^{-1.6}$

41. $\ln (e^{1-3v}) = 4$

42. $\ln (2v + 1) + \ln v = 1$

In each of the problems 43 through 50, determine whether the given statement is true, false, or meaningless. A statement is meaningless if any part of it is undefined. Give reasons for your answers.

43. $10^{\log 8} = 8$

44. $e^{-\ln 3} = 1/3$

45. $e^{\ln (-3)} = 1/3$

46. $\log (\log^{-1} -4) = -4$

47. $\ln^{-1}(\ln 3) = 3$

48. $e^{(\ln 6 - \ln 2)} = 3$

49. $e^{(\ln 2)(\ln 3)} = 6$

50. $\ln (e^2 + e^3) = 5$

In problems 51 and 52, find the domains of the functions.

51. a) $f(x) = \ln (x - 2)$

b) $g(x) = \log(x^2 - x - 2) - \log(x + 1)$

52. a) $f(x) = \log(x + 1) + \log(x - 1)$

b) $g(x) = \log(x^2 - 1)$

In problems 53 through 59, draw graphs of the given functions. In each case label the coordinate intercept points.

53. $f(x) = \ln (x - 2)$

54. $f(x) = \log (x^2 - 1)$

55. $f(x) = \log (-x)$

56. $g(x) = 1 + \ln x$

57. $g(x) = \ln x - \ln (x - 1)$

58. $h(x) = \ln \left(\dfrac{x}{x - 1}\right)$

59. $f(x) = \ln e^{x-1}$

In problems 60 through 63, solve the given equations. Give solutions rounded off to two decimal places.

60. $8^x = 3 \cdot 5^x$

61. $5^x = 3 \cdot 8^{1 - x}$

62. $e^{-x} - 2x = 0$

63. $e^{-2x} - x = 0$

In problems 64 through 69, functions f, g, and h are given by

$$f(x) = \ln x, \qquad g(x) = e^x, \qquad h(x) = x^2.$$

Express answers correct to three decimal places.

64. Evaluate $(f + g)(1.25)$

65. Evaluate $(g \circ h)(0.5)$

66. Evaluate $(g \circ f)(0.21)$

67. Evaluate $(h \circ f)(0.68)$

68. For what value of x is $(f \circ g)(x) = x$?

69. For what value of x is $(g \circ f)(x) = x$?

70. In Example 8 we found the root of $e^{-x} - x = 0$ by an estimation process. Try the following with your calculator. Enter any number into the display of your calculator, and then press the keys in the given sequence.

   a) If your calculator has an $\boxed{e^x}$ key, press $\boxed{+/-}$ $\boxed{e^x}$ $\boxed{+/-}$ $\boxed{e^x}$, and so on. That is, press the change-sign key and the $\boxed{e^x}$ key repeatedly. After each $\boxed{e^x}$ look at the display. Continue until you see something interesting, and then given an intuitive explanation of what is happening by using graphs similar to the one in Example 8.

   b) If your calculator does not have an $\boxed{e^x}$ key, then carry out the instruction of (a), but replace the $\boxed{e^x}$ key by $\boxed{INV}$ and $\boxed{\ln}$ keys. This is equivalent to $\boxed{e^x}$, as we saw in (3.25).

3.5 **APPLICATIONS OF EXPONENTIAL FUNCTIONS**

There are many instances in real life in which experimental evidence indicates that the quantity of a certain substance varies exponentially with respect to time. For example, the population of a bacteria culture, such as yeast or *E. coli*, increases when the bacteria divide. When the population is small, it will tend to increase slowly, but as time goes on, the population becomes larger, and the rate of increase becomes greater. This is an example of a *growth process,* in which the bacteria population can be described as an exponential function of time.

As another example, the disintegration process of a radioactive substance is such that all atoms have an equal chance of disintegrating. As time passes, the number of atoms present becomes smaller, and the rate of disintegration is decreased. This is an example of a *decay process,* in which the amount of material present can be expressed as an exponential function of time.

When money is invested in a bank account in which interest is compounded continuously, the accumulated value of the investment t years later can be expressed as an exponential function of t. This is another example of a growth process.

An exponential growth or decay process can be formulated mathematically as follows. Suppose A represents the amount present at any time t. Then A is given by the formula

$$A = A_0 e^{kt}, \tag{3.29}$$

where A_0 is the amount present at time $t = 0$, and k is a constant that can be determined for any particular growth or decay process. For instance, in the example of bacteria population, k is a positive number determined experimentally and is dependent on the unit of time being used, the kind of bacteria, and the nutrient. For a growth process k is a positive number, but for a decay process k is a negative number.

The following examples illustrate application of Eq. (3.29) in specific growth or decay processes.

Example 1 The growth of a culture of bacteria *E. coli* in a solution containing inorganic salts and glucose is being observed. At the start, $t = 0$, it is determined that the population consists of 10^6 bacteria per milliliter. One hour and 12 minutes later the number is doubled. Let N represent the number of bacteria per milliliter at any time t hours after the start of observation.

a) Determine the formula that gives N as a function of t.

b) How many bacteria per milliliter will there be at the end of two hours?

c) How long will it take to have 10^7 bacteria per milliliter?

Solution   **a)** Applying Eq. (3.29) gives

$$N = N_0 e^{kt}. \tag{3.30}$$

The value of N_0 is determined by using $t = 0$, $N = 10^6$, and so $N_0 = 10^6$. Thus $N = 10^6 e^{kt}$. To find k we can substitute the given information, $t = 1.2$ hours, $N = 2 \cdot 10^6$ into this equation to get $2 \cdot 10^6 = 10^6 e^{1.2k}$. Thus we need to solve $2 = e^{1.2k}$ for k. We can do so by applying the ln function to both sides to get

$$\ln 2 = \ln e^{1.2k} = 1.2k \ln e = 1.2k, \qquad k = \frac{\ln 2}{1.2}.$$

Substituting into Eq. (3.30) gives*

$$N = 10^6 \, e^{[(\ln 2)/1.2]t} \tag{3.31}$$

b) Replacing t by 2 in Eq. (3.31) gives $N = 3\,174\,802$.

c) Replacing N by 10^7 in Eq. (3.31) and solving for t, we get

$$10^7 = 10^6 \, e^{[(\ln 2)/1.2]t}, \qquad 10 = e^{[(\ln 2)/1.2]t}, \qquad \ln 10 = \frac{\ln 2}{1.2}t,$$

$$t = \frac{1.2 \ln 10}{\ln 2} = 3.986 \text{ hours.}$$

Thus at the end of 3 hours and 59 minutes, the number of bacteria per milliliter will be increased tenfold.   ■

Carbon Dating

In chemistry and physics we learn that everything in nature is made up of atoms and each atom has a nucleus. Most materials are stable, which means that if they are left undisturbed they do not change with time. There are some materials (such as uranium) which change constantly by the emission of rays of energy and streams of atomic particles from the nuclei. Such materials are called *radioactive,* and we say that the nuclei *decay.*

Any microscopic sample of radioactive material contains a large number of radioactive nuclei. These do not all decay at once, since the decay is a random process that occurs over a period of time. The quantitative measure of the rate of decay of a given radioactive isotope is given in terms of its half-life. The *half-life* of an isotope is the time (in years, days, or seconds) that it takes for half of the given sample to decay. For example, the half-life of carbon 14 is 5730 years. If at some time a piece of petrified wood contains 10 grams of ^{14}C, then after 5730 years it will contain 5 grams; after 11460 years it will have 2.5 g of ^{14}C, and so on.

Some radioactive isotopes decay very rapidly while others take a long time. For example, the half-life of uranium 238 is 4.5 billion years while that of polo-

---

*   Since $e^{(\ln 2/1.2)t} = e^{(t/1.2)\ln 2} = e^{\ln(2^{t/1.2})} = 2^{t/1.2}$, we see that N is also given by $N = 10^6 \cdot 2^{t/1.2}$.

nium is only 0.00016 seconds. The following illustrates an application of radioactive decay which is sometimes referred to as *carbon dating*.

The element carbon has three isotopes ^{12}C, ^{13}C, and ^{14}C; the first two of which are stable, but ^{14}C is *radioactive*. The loss of ^{14}C through radioactive decay is compensated for by cosmic radiation so in living organisms the ratios between the three isotopes are maintained at approximately 100 to 1 to 0.01 for ^{12}C to ^{13}C to ^{14}C. These are the ratios for any living organism, but when an organism dies, the number of ^{14}C atoms decreases, and the amount A of ^{14}C present t years after death is given by

$$A = A_0 e^{kt}. \tag{3.32}$$

This equation forms the basis for the method used by archeologists to estimate the age of unearthed bones. That method is illustrated in the following example.

Example 2   **a)** What percentage of ^{14}C remains 4000 years after the death of an organism?

b) The ratio of ^{12}C to ^{14}C in the bones of a skeleton is measured and found to be 100 to 0.004. Determine the number of years since death occurred.

Solution   First determine k by substituting 5730 for t and $0.5\,A_0$ for A in Eq. (3.32):

$$0.5\,A_0 = A_0 e^{5730k}.$$

Solving for k gives $k = (\ln 0.5)/5730 = -0.000121$. When we replace k by this number, Eq. (3.32) becomes*

$$A = A_0 e^{[(\ln 0.5)/5730]t}. \tag{3.33}$$

a) Let A_1 represent the amount of ^{14}C present when $t = 4000$. Since A_0 is the amount of ^{14}C present at $t = 0$, the percent of ^{14}C at the end of 4000 years is given by $(A_1/A_0) \cdot 100$. Substituting 4000 for t into Eq. (3.33), we get

$$A_1 = A_0 e^{(4000 \ln 0.5)/5730}.$$

Dividing both sides of this equation by A_0 and multiplying by 100 gives

$$\frac{A_1}{A_0} \cdot 100 = 100 e^{(4000 \ln 0.5)/5730} = 61.6.$$

Hence about 62% of the ^{14}C isotope still remains in the bones of the skeleton after 4000 years.

b) Here we assume that death occurred at time $t = 0$. Suppose t_1 represents the number of years that elapsed until the skeleton was discovered. At time $t = 0$ the ratio of ^{12}C to ^{14}C is approximately 100 to 0.01, and at time t_1 the ratio is found to be 100 to 0.004. Thus the proportion of ^{14}C still remaining after t_1

---

*   Since $e^{(\ln 0.5/5730)t} = e^{(t/5730)\ln 0.5} = e^{\ln(0.5^{t/5730})} = 0.5^{t/5730} = (\frac{1}{2})^{t/5730} = 2^{-t/5730}$, A is also given by $A = A_0 \cdot 2^{-t/5730}$.

years is $0.004/0.01 = 0.4$, or 40 percent. That is, when $t = t_1$, $A = 0.4\,A_0$. Substituting this into Eq. (3.33) gives

$$0.4\,A_0 = A_0 e^{[(\ln 0.5)/5730]t_1}.$$

Solving this equation for t_1, we have

$$t_1 = \frac{5730 \ln 0.4}{\ln 0.5} = 7575.$$

Thus the skeleton is approximately 7600 years old.   ■

Another type of problem in which exponential functions occur is computation of compound interest.

Example 3   Suppose \$1000 is invested at a bank that pays interest at the rate of 8% per year.* Find the value of the investment at the end of one year if interest is compounded

a) annually   **b)** semiannually   **c)** quarterly   **d)** daily   **e)** continuously

Solution   Let A represent the value of the \$1000 investment at the end of one year. Use the formula where interest equals principle times rate times time.

a) $A = 1000 + 1000(0.08)(1) = 1000(1 + 0.08) = 1080$

b) At the end of the first six months, the investment is worth

$$B_1 = 1000 + (1000)(0.08)\frac{1}{2} = 1000\left(1 + \frac{0.08}{2}\right).$$

We now consider the amount of B_1 as being invested for the next six months to get

$$A = B_1 + B_1(0.08)\frac{1}{2} = B_1\left(1 + \frac{0.08}{2}\right) = 1000\left(1 + \frac{0.08}{2}\right)^2 = 1081.60.$$

c) We can follow the procedure used in (b) and compute the successive values of the investment at the end of 3, 6, 9, and finally 12 months. This would give

$$A = 1000\left(1 + \frac{0.08}{4}\right)^4 = 1082.43.$$

d) Following a procedure similar to that in (b) and (c), we get

$$A = 1000\left(1 + \frac{0.08}{365}\right)^{365} = 1000(1.08328) = 1083.28.$$

---

*   The interest rate is usually stated as a percent per year. For example, for an 8 percent interest, $r = 0.08$.

e) Suppose we continue the procedure illustrated above and determine the value A_m of the investment when interest is compounded m times per year. We get

$$A_m = 1000\left(1 + \frac{0.08}{m}\right)^m.$$

We now ask: "What happens to A_m when m becomes large?" The problem here reminds us of the problem discussed in Example 3 of Section 3.2, in which we saw that $f(x) = (1 + x)^{1/x}$ approaches the number e as x approaches zero. In order to express our problem in this form, let $x = 0.08/m$; then $m = 0.08/x$, and we get

$$A_m = 1000(1 + x)^{0.08/x} = 1000[(1 + x)^{1/x}]^{0.08}.$$

Therefore, when $m \to \infty$, $x \to 0$, and so $A_m \to 1000e^{0.08} = 1083.29$. Therefore, when interest is compounded continuously, the value at the end of one year of the $1000 investment is $A = 1000e^{0.08} = \$1083.29$. ▪

The preceding example suggests the following generalizations.

Suppose a sum of P dollars is invested at an interest rate of r. The value A of such an investment at the end of t years is given by the following.

1. If interest is compounded m times per year, then

$$A = P\left(1 + \frac{r}{m}\right)^{mt}. \tag{3.34}$$

2. If interest is compounded continuously, then

$$A = Pe^{rt}. \tag{3.35}$$

Example 4   Suppose $2400 is invested and the rate of interest is 8.75 percent. Find the value of this investment at the end of 10 years if interest is compounded

a) semiannually   **b)** quarterly   **c)** continuously

Solution   Here $t = 10$ and interest at 8.75% gives $r = 0.0875$.

a) We can use Eq. (3.34) with $m = 2$ to get

$$A = 2400\left(1 + \frac{0.0875}{2}\right)^{2 \cdot 10} = 5651.20.$$

b) Using Eq. (3.34) with $m = 4$, we get

$$A = 2400\left(1 + \frac{0.0875}{4}\right)^{4 \cdot 10} = 5703.25.$$

c) Here we use Eq. (3.35) to get

$$A = 2400e^{0.0875 \cdot 10} = 5757.30.$$

Exercises 3.5

1. In Example 1, use the formula $N = 10^6 \cdot 2^{t/1.2}$ to get the answers to parts (b) and (c).

2. Assume that bacteria reproduce according to the law described in Example 1, that is, they double in number every 1.2 hours. Suppose a solution contains 10 000 bacteria at the start of an experiment. How many bacteria will there be at the following times?

   a) 12 hours later   **b)** one day later   **c)** two days later

3. Using the method in Example 2, determine the following.

   a) What percentage of ^{14}C remains 10 000 years after death?

   b) Find the "age" of a skeleton in which the ratio of ^{12}C to ^{14}C is found to be 100 to 0.001.

4. How many years will it take for a given amount of ^{14}C to decay to one-fourth of the given amount?

5. A chemist finds that in three days a sample of iodine-131 decays to 77 percent of the original amount. Find the half-life of ^{131}I.

6. Radium 226 is a radioactive isotope of radium with a half-life of 1620 years. A sample of ^{226}Ra contained 10 grams in 1900. How many grams will there be in the following years?

   a) 2000   **b)** 3000

7. Strontium-90 is a radioactive isotope of strontium that occurs as a component in the fallout of thermonuclear explosions and that contaminates the soil. Thus it becomes a radiation hazard through progressive concentration in the bones of people and animals. The half-life of ^{90}Sr is 29 years. What percentage of the ^{90}Sr produced by a thermonuclear test in 1965 will still be present in 1985?

8. The population of a certain city is increasing at an exponential rate, given by Eq. (3.29). In 1950 the population was 120 000, and in 1970 it was 164 000. What is the expected population in each of the following years?

   a) 1990   **b)** 2050

9. A biologist finds that in a certain nutrient solution the number of bacteria tripled in two hours. How many times the original number will there be at the end of each of the given periods?

   a) four hours   **b)** five and a half hours

10. Given that $300 is invested at $8\frac{1}{4}$ percent interest for 20 years, find the value of this investment if interest is compounded

    a) semiannually   **b)** quarterly   **c)** continuously

11. How much money should be invested at $8\frac{1}{2}\%$ so that it will be worth $6000 twelve years from now if interest is compounded continuously?

12. What rate of interest is required so that $4500 will be worth $6400 in five years if interest is compounded

   a) continuously?   b) quarterly?

13. A father wants to purchase a savings certificate to be used for his son's college expenses 12 years from now. He has $4000 to invest and has a choice between two certificates, one that pays $8\frac{3}{4}\%$ interest compounded semiannually and one that pays $8\frac{1}{2}\%$ compounded continuously. Which plan would earn him more and by how much?

14. What rate of interest is necessary so that an investment will double itself in eight years when interest is compounded continuously?

15. Janet has a $3000 savings certificate that her mother purchased for her 10 years ago. It pays 6.5% interest compounded continuously. She needs $6000 to buy a car and wants to use the money from the savings certificate to pay cash for it. Does she have enough? If not, how much longer will she have to wait until the certificate is worth $6000? If it is enough, how much money will be left over after she pays for the car?

16. How many years does it take for a bank savings account to triple if interest is paid at the rate of 7.5% and is compounded continuously?

17. A sum of $2500 was deposited 10 years ago in a savings account that paid 6.5% interest compounded continuously. The bank has just decided to increase the rate of interest to 6.75%. How much will there be in the account five years from now?

18. An inflation rate of r percent per year means that the cost of an item is r percent more than a year ago; thus the cost at the end of t years can be calculated by using formula (3.34), where $m = 1$. That is, $A = P(1 + r)^t$. Assume that the rate of inflation is 8% and that you had to pay $50 000 for your home in 1980. What would be the cost of a corresponding home in the year 2100? Before you perform any calculations, make a guess of what you think the answer is.

19. A sum of money is invested with interest of r percent compounded annually.

   a) Complete the following table giving the number N of years it takes an investment to double in value. Give answers rounded off to the nearest whole number.

   b) Multiply r and N, as indicated in the third row of the table, and discover the *Rule of 72*.

r	4	6	8	9	12	18	24
N							
$r \cdot N$							

20. Complete a table similar to that of problem 19, but assume that interest is compounded continuously.

3.6 Looking Ahead to Calculus

In Example 3 of Section 3.2, we introduced the number e as the limiting value of the function f given by $f(x) = (1 + x)^{1/x}$ as x approaches 0. This is denoted by

$$\lim_{x \to 0} (1 + x)^{1/x} = e = 2.71828 \ldots$$

In this section we consider other problems of this type and continue the discussion of Section 2.7, in which we introduced the notion of slope of a curve. The following examples illustrate our intuitive numerical approach to limit concepts.

Example 1   Determine $\lim\limits_{x \to 0} \dfrac{2^x - 1}{x}$. Give answer rounded off to two decimal places.

Solution   Let $f(x) = (2^x - 1)/x$, and make a table giving values of $f(x)$ corresponding to values of x (both positive and negative) approaching 0.* The values of $f(x)$ are rounded off to three decimal places.

x	0.5	0.1	0.01	0.001	$\cdots$	-0.5	-0.1	-0.01	-0.001
$f(x)$	0.828	0.718	0.696	0.693	$\cdots$	0.586	0.670	0.691	0.693

From the values seen in the table, we conclude that $\lim\limits_{x \to 0} \dfrac{2^x - 1}{x} = 0.69$.   ■

Frequently we are interested in determining the behavior of a function at extreme values of the independent variable. For instance, if x assumes large positive values, do the corresponding values of $f(x)$ approach a fixed number? As an example, consider the function given by $f(x) = 2x^2/(x^2 - x)$. For large values of x, both the numerator and denominator are large numbers, and the corresponding values of $f(x)$ are not immediately obvious. However, in this case we can divide the numerator and denominator by x^2, and so $f(x)$ can be written as

$$f(x) = \frac{2}{1 - (1/x)}.$$

As x becomes large, $1/x$ approaches zero, and we conclude that $f(x)$ approaches 2. This is denoted by

$$\lim_{x \to \infty} f(x) = \lim_{x \to \infty} \frac{2x^2}{x^2 - x} = 2.$$

---

*   Note that we do not include extremely small values of x in the table. Calculators cannot handle such numbers without introducing round-off errors.

In general, we cannot get a formula for the given function in the form that will allow us to evaluate $\lim_{x \to \infty} f(x)$, as we did in the above illustration. However, in many instances by considering numerical values of $f(x)$ for large values of x, we can get a good idea of the limiting value of $f(x)$.

Example 2   Suppose f, g, and h are functions given by

$$f(x) = \left(1 + \frac{1}{x}\right)^4, \qquad g(x) = \left(1 + \frac{1}{x}\right)^x, \qquad h(x) = \left(1 + \frac{1}{x}\right)^{2x}.$$

Evaluate

a) $\lim_{x \to \infty} f(x)$ b) $\lim_{x \to \infty} g(x)$ c) $\lim_{h \to \infty} h(x)$

Solution   Let us first make a table giving values of $f(x)$, $g(x)$, and $h(x)$ for large values of x. The entries in the last two columns are given to six decimal places.

x	10	100	1000	10 000	100 000	1 000 000
$f(x)$	1.464	1.041	1.004	1.004	1.000040	1.000004
$g(x)$	2.594	2.705	2.717	2.718	2.718268	2.718280
$h(x)$	6.728	7.316	7.382	7.388	7.388982	7.389049

From the values of $f(x)$, $g(x)$, and $h(x)$ given in the table, we conclude:

a) $\lim_{x \to \infty} \left(1 + \frac{1}{x}\right)^4 = 1$ b) $\lim_{x \to \infty} \left(1 + \frac{1}{x}\right)^x = 2.71828$ c) $\lim_{x \to \infty} \left(1 + \frac{1}{x}\right)^{2x} = 7.38905$

The limiting value in (b) appears to be the number e; in fact, it is e, as will be seen in calculus. Similarly, the limiting value of $h(x)$ appears to be e^2 (check by evaluating e^2). ◼

In Example 2 one is tempted to use the following intuitive argument: $1 + (1/x)$ approaches 1 as $x \to \infty$, and so in all three cases we have a number very near 1, and 1 to a power should approach 1 as a limiting value. However, there is an important difference: the exponent of $f(x)$ is a fixed number, whereas in both $g(x)$ and $h(x)$ it involves a variable. Our intuitive argument leads us to the correct result for $f(x)$ but not for $g(x)$ or $h(x)$.

Example 3   Find the slope of the line that is tangent to the curve $y = 3^x$ at the point $P{:}(0, 1)$. Draw a graph and show the tangent line L.

Solution   Let $f(x) = 3^x$. The slope m of the tangent line at P is given by

$$m = \lim_{h \to 0} \frac{f(0 + h) - f(0)}{h} = \lim_{h \to 0} \frac{3^h - 1}{h}$$

(see Eq. (2.17), page 148). To evaluate this limit we make a table giving values of $(3^h - 1)/h$ corresponding to values of h (both positive and negative) near zero. Note that the table does not include extremely small values of h since such numbers would introduce calculator round-off errors.

From the values of $(3^h - 1)/h$ given in the table we conclude that $m = 1.099$ (to three decimal places). The graph is shown in Fig. 3.9.

h	0.5	0.1	0.01	0.001	0.0001	$\cdots$	-0.1	-0.01	-0.001	-0.0001
$\dfrac{3^h - 1}{h}$	1.46	1.16	1.105	1.0992	1.0987	$\cdots$	1.04	1.093	1.0980	1.0986

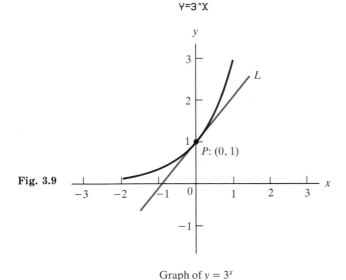

Y=3^X

Fig. 3.9

Graph of $y = 3^x$

Example 4   Find an equation of a line that is tangent to the curve $y = 4e^{-x}$ at the point $P:(1, 4/e)$. Give answer with numbers rounded off to two decimal places.

Solution   The slope m of the tangent line is given by

$$m = \lim_{h \to 0} \frac{f(1 + h) - f(1)}{h} = \lim_{h \to 0} \frac{4e^{-(1+h)} - 4e^{-1}}{h}.$$

Let us make a table giving values of the difference quotient for values of x near zero.

h	0.1	0.01	0.001	0.0001	$\cdots$	-0.1	-0.01	-0.001	-0.0001
$\dfrac{f(1+h)-f(1)}{h}$	-1.40	-1.46	-1.4708	-1.4715	$\cdots$	-1.55	-1.48	-1.4723	-1.4716

From the values appearing in the table we conclude that $m = -1.472$ (to three decimal places). Hence an equation for the tangent line is $y - (4/e) = -1.472(x - 1)$. This can be written as $y = -1.47x + 2.94$, where the numbers are rounded off to two decimal places.

Exercises 3.6

In problems 1 through 10, determine the given limits. Give answers correct to two decimal places (see Examples 1 and 2).

1. $\lim\limits_{x \to 0} \dfrac{4^x - 1}{x}$

2. $\lim\limits_{x \to 0} \dfrac{3^{-x} - 1}{x}$

3. $\lim\limits_{x \to 0} \dfrac{e^x - 1}{x}$

4. $\lim\limits_{x \to 1} \dfrac{e^x - e}{x - 1}$

5. $\lim\limits_{x \to 0} \dfrac{5^x - 2^x}{x}$

6. $\lim\limits_{x \to 0} (1 + 2x)^{1/x}$

7. $\lim\limits_{x \to 3} \dfrac{\sqrt{4 - x} - 1}{x - 3}$

8. $\lim\limits_{x \to \infty} \left(1 - \dfrac{1}{x}\right)^x$

9. $\lim\limits_{x \to \infty} (xe^{-x})$

10. $\lim\limits_{x \to 1} \dfrac{\sqrt{2 - x} - 1}{x - 1}$

In problems 11 through 15, a function f and a point P are given. Find the slope of a line that is tangent to the curve $y = f(x)$ at P. Draw a graph and show the tangent line. Give answers rounded off to two decimal places (see Example 3).

11. $f(x) = e^x$; $P{:}(1, e)$

12. $f(x) = e^{-x}$; $P{:}(-1, e)$

13. $f(x) = \ln x$; $P{:}(1, 0)$

14. $f(x) = \ln x$; $P{:}(2, \ln 2)$

15. $f(x) = \dfrac{\ln x}{x}$; $P{:}(1, 0)$

In problems 16 through 20, a function f and a number c are given. Find an equation of the line that is tangent to the curve $y = f(x)$ at the point $P{:}(c, f(c))$. Give answers with numbers rounded off to two decimal places (see Example 4).

16. $f(x) = 2e^x$; $c = 0$

17. $f(x) = \ln x$; $c = e$

18. $f(x) = x \ln x$; $c = 1$

19. $f(x) = xe^{-x}$; $c = -1$

20. $f(x) = \ln(-x)$; $c = -2$

Review Exercises

In each of the problems give answers in exact form whenever it is reasonable to do so. Otherwise express results in decimal form rounded off to three decimal places. In problems involving undefined quantities, give reasons for an "undefined" answer.

In problems 1 through 15, evaluate the given expression.

1. $\log 8$

2. $\log \sqrt{43}$

3. $\ln 23$

4. $\log(\sqrt{2} + \sqrt{3})$

5. $\ln(36^3)$

6. $\log(\ln 48)$

7. $\ln(\log 48)$

8. $\ln\left(\dfrac{\sqrt{2} + \sqrt{6}}{3}\right)$

9. $\log_5 8$

10. $\log_3(\sqrt{5} + \sqrt{12})$

11. $\log_7(\log 24)$

12. $\log_8(e^3)$

13. $\log(\ln 0.6)$

14. $\log_5(1 - \sqrt{2})$

15. $\log_3(27\sqrt{3})$

In problems 16 through 24, the functions f and g are defined by $f(x) = e^x + e^{-x}$; $g(x) = 3\ln(2x - 1)$. Evaluate the given expressions.

16. $f(0)$

17. $(f \circ g)(0.2)$

18. $(g \circ f)(-2)$

19. $f(-5/2)$

20. $g(4)$

21. $(f \cdot g)(2)$

22. $(f/g)(3)$

23. $(f + g)(\sqrt{2})$

24. $(f \circ g)(\sqrt{3})$

In problems 25 through 36, solve the given equations.

25. $\ln e^x = 3$

26. $\log e^x = 3$

27. $1 - \ln(2x + 1) = 3$

28. $\log(\ln x) = 1$

29. $\ln(\log x) = 1$

30. $e^{2x-1} = 4$

31. $e^{3x} = 10^{1-x}$

32. $\log 10^{4-3x} = 1$

33. $3^{x-1} = 4$

34. $5^x = 3(7^x)$

35. $2e^x + 1 = 0$

36. $3e^x - 1 = 0$

37. Plot a graph of $y = e^{-x}$.

38. Plot a graph of $y = 4^x$.

39. Plot a graph of $y = 1 - 3^x$.

40. Find the domain of each of the following functions.

    a) $f(x) = \ln(x - 3) + \ln x$

    b) $g(x) = \ln[x(x - 3)]$

41. Find the roots of

    a) $\ln(x - 1) + \ln x = 1$

    b) $\ln[x(x - 1)] = 1$

42. Find the roots of

    a) $\ln(x + 1) - \ln x = 1$

    b) $\ln\left(\dfrac{x + 1}{x}\right) = -1$

43. Draw a graph of $y = 1 + \ln(x - 1)$.

44. Draw a graph of $y = \ln\left(\dfrac{x + 1}{x - 1}\right)$.

45. Given that $y = x(2^{-x})$ and $x \geq 0$, make a table of x, y values that satisfy the equation, using values of x beginning with $x = 0$ and then at 0.5 units apart until you reach 3.0. Plot these points and then make a reasonable estimate of the value of x that makes y a maximum. Refine your estimate by more computations, and then find the maximum value of y.

46. Strontium-90 has a half-life of 29 years. What percent will remain after 60 years?

47. Carbon-14 has a half-life of 5730 years. How many years will it take for 75% of it to disappear?

48. A sum of $1640 was invested 16 years ago at 8% interest compounded quarterly. What is the current value of the investment?

49. Determine the current value of an investment of $3250 that was made 25 years ago at the rate of 8.25% compounded continuously.

50. In how many years will an investment be doubled if interest is 6% compounded quarterly?

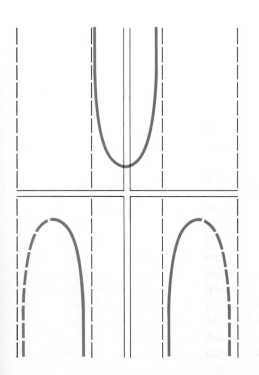

Trigonometric (Circular) Functions

In the preceding two chapters we studied some important types of functions, namely polynomial, exponential, and logarithmic functions. Equally important in applications and in theoretical mathematics are the trigonometric functions, which we shall introduce in this chapter. There are six such functions, and each is defined either on the set of real numbers **R** or a subset of **R**.

As the name indicates, trigonometry pertains to the study of measurements related to triangles. As far back as 3000 years ago, the Egyptians and Babylonians used properties of triangles to establish land boundaries and explore astronomy. In modern times the ideas related to the solution of triangles are still important in several areas of application, but trigonometric functions have become an integral part of the study of calculus, as well as many advanced courses in mathematics. Trigonometric functions play a key role in the solutions of a wide range of applied problems in physics, engineering, and several other fields.

Historical development and applications involving triangles lead us to the introduction of trigonometric functions defined on measures of angles. The unit of angular measure that has traditionally been used in such areas as surveying and navigation is the *degree*. However, for theoretical purposes in calculus it becomes necessary to define trigonometric functions on real numbers. One can conveniently do so by choosing another unit of angular measure called the *radian*. Before stating definitions of the trigonometric functions, we first discuss angular measure.

4.1   ANGLES AND UNITS OF ANGULAR MEASURE

The study of *plane trigonometry* suggests that we begin with a given plane. All the geometric figures discussed, such as lines, rays, angles, and triangles, are subsets of this plane. In geometry, a *ray* is defined as a half line together with its endpoint, and an *angle* is the union of two rays with a common endpoint. The idea of measure of angle is also introduced but usually limited to angles with measures less than or equal to 180° (or sometimes 360°).

It now becomes necessary to extend the notion of angular measure beyond that studied in geometry. Eventually we shall want to have angle measure expressed as a real number (radian measure), and it will be useful to have a correspondence between the angles in the plane and the set of real numbers. In order to do this, it is convenient to think of an angle as being generated by a ray that is rotated about its endpoint from its initial position to a final position. The ray corresponding to the initial position is called the *initial side* of the angle, and that in the final position is called the *terminal side* of the angle. The point about which rotation takes place is called the *vertex* of the angle. The definition of an angle is now extended to be the union of two rays, together with the rotation. Measure of an angle is then described in terms of "amount of rotation." This allows us to have angles with measures greater than 180° (indeed greater than 360°); we can also have angles with negative measures by using direction of rotation. A *directed angle* will have positive measure if the rotation is counterclockwise and negative measure if the rotation is clockwise. For purposes of brevity we shall frequently say "the angle is positive" to mean "the measure of the angle is positive."

In Fig. 4.1(a) angle A is shown with initial and terminal sides labeled, as well as with an arrow indicating direction of rotation. The arrow is used to indicate both the direction and the extent of rotation. Figure 4.1(b) shows angle B, in which the rotation is more than a complete revolution. Angles A and B are positive angles, and angle C (Fig. 4.1c) is negative.

Fig. 4.1

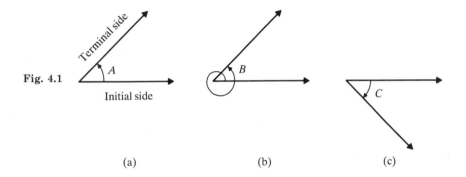

    (a)   (b)   (c)

Two widely used units of angular measure are degrees-minutes-seconds and radians. Scientific calculators frequently include a third unit of angular measure, the *grad*.* This unit is rarely encountered, however, and it will not be used in this text.

Degrees, Minutes, Seconds

If the initial side of an angle is rotated counterclockwise one complete revolution, the measure of the corresponding angle is defined to be 360 degrees, denoted by 360°. Thus an angle of 1° is one in which the initial side is rotated counterclockwise 1/360 of a revolution. For more refined measurements, the units of minutes and seconds are used; they are defined as follows:

    60 minutes equals one degree, denoted by $60' = 1°$.

    60 seconds equals one minute, denoted by $60'' = 1'$.

On a calculator, minutes and seconds must be entered as a decimal part of a degree.

For example, $30°15' = 30.25°$, and $42°12'45'' = 42.2125°$. Figure 4.2 illustrates degree measure of several angles. For brevity we write $A = 90°$ to denote that the measure of angle A is 90°, and similarly for other angles.

---

*   A grad is one hundredth of a right angle; that is, 400 grads is equivalent to a complete revolution.

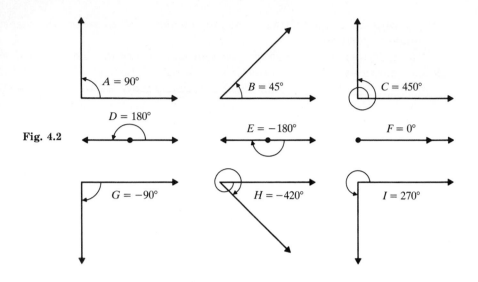

Fig. 4.2

Radian Measure

Although measure of angles in degrees is useful in some fields of application, it is more convenient to use another unit of measure for theoretical work in mathematics, as well as in many applied areas. This unit, the *radian*, is defined as follows:

> *An angle with vertex at the center of a circle and subtending an arc of length equal to the radius of the circle has measure one radian.*

In Fig. 4.3(a) an angle of measure 1 radian is shown. In this case we write $\theta = 1$ radian.* In general, the radian measure of any angle is defined as follows (see Fig. 4.3b).

Definition 4.1   Suppose α is an angle with vertex at the center of a circle of radius r and subtending an arc of length s, where r and s are measured in the same units. The radian measure of α is defined as

$$\alpha = \frac{s}{r}\text{radians.}^\dagger$$

---

*   In trigonometry angles are frequently indicated by Greek letters α (alpha), β (beta), γ (gamma), θ (theta), ϕ (phi), and so on.

†   Note that this definition is independent of the size of circle. That is, in Fig. 4.3(b) the two ratios s/r and s'/r' are equal; this is a fact from geometry.

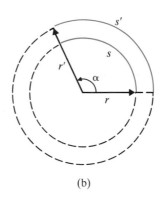

Fig. 4.3

(a)   (b)

Example 1   If $r = 4$ cm and $s = 3$ cm, then $\alpha = 3$ cm$/4$ cm $= 3/4$. Since the cm units "cancel," the result is a real number, and it is not necessary to write "radians" after $3/4$. In this text we shall write $\alpha = 3/4$ ($\alpha = 0.75$ in calculator display form) or $\alpha = 3/4$ rad to mean α is an angle having radian measure $3/4$.   ▬

> *When the measure of an angle is given as a real number (with no unit designation) it will be understood that the unit of measure is the radian.*

For example, $\theta = 15$ means that θ is an angle whose measure is 15 radians.

Example 2   Express $36°16'23''$ in decimal form correct to four decimal places.

Solution   Since $60' = 1°$, then $16' = 16/60$ degree. Also $3600'' = 1°$, so $23'' = 23/3600$ degree. Therefore

$$36°16'23'' = \left(36 + \frac{16}{60} + \frac{23}{3600}\right) \text{ degrees} = 36.2731°.$$

The computation of the result is easily done with a calculator.*   ▬

Example 3   Express $64.276°$ in degrees, minutes and seconds (to the nearest second).

Solution
$$64.276° = 64° + (0.276)(60') = 64° + 16.56'$$
$$= 64° + 16' + (0.56)(60'') = 64°16'34''.$$

Note: In order to get maximum accuracy, we suggest the following steps. Record the $64°$, enter 0.276 into the calculator, and multiply by 60. Then record the

---

*   Throughout the entire text we assume that a calculator is used to do most of the arithmetic computations. Appendix A includes calculator instructions.

whole number part of the result (16). Then subtract 16 from the display and multiply the result by 60. This gives the number of seconds.   ■

Degree-Radian Relationships

If the initial side of an angle is rotated counterclockwise one complete revolution, the measure in degrees of the corresponding angle is 360°. The same angle in radians has measure s/r, where, in this special case, s is the circumference of the circle of radius r; that is, $s = 2\pi r$, and so $s/r = 2\pi r/r = 2\pi$. Thus we have 360° and 2π radians as measures of the same angle, and we write $360° = 2\pi$ radians. Dividing both sides of this equality by 2 gives

$$180° = \pi \text{ radians.} \tag{4.1}$$

From Eq. (4.1) we get the following:

$$1° = \frac{\pi}{180} \text{ radians} = 0.017453 \text{ radians;}$$

$$1 \text{ radian} = \frac{180°}{\pi} = 57.296° = 57°17'45''. \tag{4.2}$$

Equations (4.2) can be used to convert the measure of an angle from one unit to the other. However, the decimal numbers involved are difficult to remember, and we suggest that the student remember the equality stated in (4.1) and use it as a starting point for conversions.

Example 4   Change 30° to an equivalent measure in radians.

Solution   Since $1° = \pi/180$ radians, 30° must be 30 times $\pi/180$ radians; that is,

$$30° = (30)\left(\frac{\pi}{180}\right)\text{rad} = \frac{\pi}{6}\text{rad}$$

$$= 0.5236 \text{ (to four decimal places).}   ■$$

Example 5   Express 147°32′ in radian measure correct to four decimal places.

Solution   We first convert 147°32′ to a decimal number of degrees, and then multiply the result by $\pi/180$. We have

$$147°32' = \left(147 + \frac{32}{60}\right)^° = \left(147 + \frac{32}{60}\right)\cdot\left(\frac{\pi}{180}\right)\text{radians}$$

$$= 2.5749 \text{ rad.}   ■$$

Example 6   Express 2.5 radians in terms of degrees (to three decimal places).

Solution   Since 1 radian $= \left(\dfrac{180}{\pi}\right)^{\circ}$, we have 2.5 radians $= 2.5\left(\dfrac{180^{\circ}}{\pi}\right) = 143.239^{\circ}$.   ■

Example 7   Convert $13\pi/4$ radians to degree measure.

Solution   This is similar to Example 6: $\dfrac{13\pi}{4} = \left(\dfrac{13\pi}{4}\right)\left(\dfrac{180}{\pi}\right)^{\circ} = 585^{\circ}$.   ■

It should be clear from the above examples that we have the following two rules:

To convert from degrees to radians, multiply by $\dfrac{\pi}{180}$.

To convert from radians to degrees, multiply by $\dfrac{180}{\pi}$.

Exercises 4.1

1. Illustrate by a sketch each of the following angles. A protractor may be useful, but if one is not available, then a reasonably approximate drawing will be sufficient.

   a) $A = 135^{\circ}$ **b)** $B = 720^{\circ}$ **c)** $C = -60^{\circ}$ **d)** $D = -540^{\circ}$

   e) $E = 210^{\circ}$ **f)** $F = 10^{\circ}$ **g)** $G = -300^{\circ}$ **h)** $H = 22^{\circ}30'$

2. In each part of Fig. 4.4, determine the measure (in degrees) of the angle. Use a protractor or make a reasonable estimate.

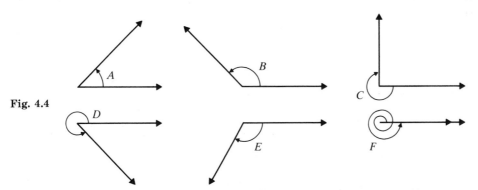

Fig. 4.4

3. Illustrate by a sketch each of the following angles given in radian measure.

   a) $A = 2\pi$ **b)** $B = \dfrac{17\pi}{6}$ **c)** $C = \dfrac{\pi}{2}$ **d)** $D = \dfrac{\pi}{4}$

   e) $E = -\dfrac{7\pi}{2}$ **f)** $F = -\dfrac{3\pi}{2}$ **g)** $G = \dfrac{9\pi}{4}$ **h)** $H = \dfrac{\pi}{3}$

4. In each part of Fig. 4.5 determine the measure in radians of the angle shown. Express answers in exact form, using π. Estimate if necessary.

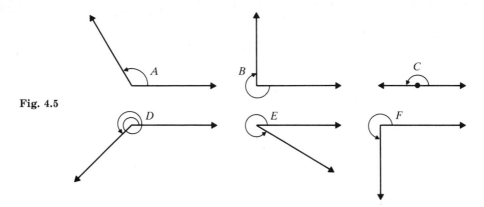

Fig. 4.5

5. For each of the following, sketch one angle that satisfies the given conditions.

a) $0 < \theta < \dfrac{\pi}{2}$ **b)** $\pi < \theta < \dfrac{3\pi}{2}$ **c)** $-\pi < \theta < -\dfrac{\pi}{2}$

d) $\dfrac{3\pi}{4} < \theta < \pi$ **e)** $\dfrac{9\pi}{4} < \theta < \dfrac{11\pi}{4}$ **f)** $\theta > 2\pi$

6. Express the given angle as a decimal number of degrees correct to three decimal places.

a) $156°37'$ **b)** $215°18'36''$

7. Express the given angle as a decimal number of degrees correct to four decimal places.

a) $48°39'42''$ **b)** $-(75°12'41'')$

8. Express the given angle in degrees and minutes correct to the nearest minute.

a) $24.36°$ **b)** $149.375°$

9. Express the given angle in degrees, minutes, and seconds correct to the nearest second.

a) $37.583°$ **b)** $321.5764°$

10. Express the given angle in radian measure. Write your answer in two forms: exact (using π), and as a decimal correct to three places.

a) $60°$ **b)** $-135°$ **c)** $225°$ **d)** $720°$

11. Follow instructions of problem 10 for

a) $120°$ **b)** $315°$ **c)** $22.5°$ **d)** $-330°$

12. Express the given angle in radian measure, and round off answers to three decimal places.

a) $23.53°$ **b)** $-48.635°$ **c)** $237°48'$ **d)** $121°40'31''$ **e)** $437°23'$

13. Convert to radian measure, and round off results to two decimal places.

a) $64.431°$ **b)** $229°47'30''$ **c)** $-(36°23'08'')$ **d)** $148.012°$ **e)** $472.37°$

14. Each of the following numbers represents the measure of an angle in radians. Convert to the corresponding measure in degrees, and express the result in exact form.

a) $\dfrac{\pi}{6}$ b) $-\dfrac{2\pi}{3}$ c) $\dfrac{3\pi}{2}$ d) $\dfrac{23\pi}{45}$ e) $\dfrac{7\pi}{18}$

15. Follow instructions of problem 14 for

a) $\dfrac{3\pi}{4}$ b) $-\dfrac{7\pi}{2}$ c) $\dfrac{11\pi}{18}$ d) -17π e) $\dfrac{15\pi}{4}$

16. In each of the following, the given number represents an angle in radian measure. Convert to degrees and express the result in two forms: decimal number correct to three decimal places, and degrees, minutes, and seconds correct to the nearest second.

a) 1.15   b) 2.48   c) 0.0493   d) -5.76 e) 64

17. Follow instructions of problem 16 for

a) 1.37   b) 0.0034   c) $\dfrac{1 + \sqrt{5}}{2}$ d) -3.45 e) 30

4.2 APPLICATIONS INVOLVING RADIAN MEASURE

In this section we consider examples that illustrate applications of radian measure.

Arc Length

In Section 4.1 radian measure of an angle was defined as follows:

$$\theta = \frac{s}{r}, \tag{4.3}$$

where the angle has its vertex at the center of a circle of radius r, and s is the length of the intercepted arc, as shown in Fig. 4.6. Equation (4.3) can be written in equivalent form as

$$s = r\theta. \tag{4.4}$$

Fig. 4.6

Example 1   Find the length of arc of a circle of radius 64.87 meters that is intercepted by a central angle of 23°37′.

Solution   We must first express the given angle in radians:

$$\theta = 23°37' = (23 + 37/60) \cdot (\pi/180) \text{ radians.}$$

Substituting into Eq. (4.4) gives

$$s = 64.87(23 + 37/60) \cdot (\pi/180) \text{ meters.}$$

The computation can be done by calculator, then rounded to two decimal places to get $s = 26.74$ m.   ■

Example 2   The distance from the earth to the moon is approximately 384 000 km. If the angle subtended by the moon from a point on the earth is measured as 30′50″, then we can approximate the diameter of the moon by assuming it to be the arc of a circle, as shown in Fig. 4.7. That is, the diameter of the moon is approximately equal to s, where

$$s = r\theta = 384\,000\left(\frac{30}{60} + \frac{50}{3600}\right) \cdot \frac{\pi}{180} \text{ km} = 3444 \text{ km.}$$

Fig. 4.7

■

Velocity of Rotation*

Suppose we have a circular wheel of radius $r = 10$ cm that is rotating about its center O, and P is a point on the circumference, as shown in Fig. 4.8. Suppose also that point P travels a distance of $s = 20$ cm each second. We say the *linear velocity* of P is 20 cm per second and write $v = 20$ cm/sec. During each second the radial line OP rotates through an angle $\theta = s/r = 20$ cm/10 cm $= 2$ radians. We say that the *angular velocity* of rotation is 2 radians per second and denote this by $\omega = 2$ rad/sec (ω is the Greek letter omega).

Fig. 4.8

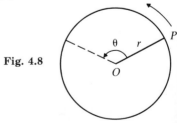

---

*   Velocity is a vector quantity. The direction of motion is assumed to be clear from the problem description.

The above example illustrates the problem of a point P moving in a circular path. We associate two types of velocity: v (linear velocity) tells us "how fast" P is moving, and ω (angular velocity) tells us how fast the central angle θ is changing (that is, how fast the radial line OP is rotating). Both v and ω represent measures of how fast P is moving at any given instant. In general, v and ω are functions of time. In the special case in which P is moving at a constant speed, we call such a motion *uniform circular motion*. We shall limit our discussion to this case and leave the general case in which v varies with time for calculus.

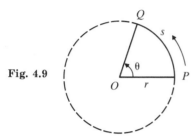

Fig. 4.9

We wish to determine the equation that gives the relationship between v and ω. Suppose point P moves to point Q, a distance of s, in time t (see Fig. 4.9). Then $v = s/t$. During the same time the radial line OP rotates through a central angle of θ, and $\omega = \theta/t$. Since $s = r\theta$, we get

$$v = \frac{s}{t} = \frac{r\theta}{t} = r\left(\frac{\theta}{t}\right) = r\omega.$$

Thus we have

$$v = r\omega, \tag{4.5}$$

where ω *is in radians per unit of time.*

Example 3　The wheel of a turbine rotates at the rate of 648 revolutions per minute, and the distance from the center to a point P on the outer edge is 96.3 cm. What is the linear velocity of point P?

Solution　Since 1 rev $= 2\pi$ rad, $\omega = 648$ rev/min $= 648(2\pi)$ rad/min. Substituting into Eq. (4.5) gives

$$v = (648)(2\pi)(96.3)\frac{\text{cm}}{\text{min}} = \frac{(648)(2\pi)(96.3)}{100}\frac{\text{m}}{\text{min}} = 3921\frac{\text{m}}{\text{min}}. \qquad \blacksquare$$

Example 4　The diameter of each wheel of a bicycle is 70 cm. Suppose a person riding the bicycle travels at a constant speed and is timed at 3 minutes over a distance of two city blocks, where the length of a block is 200 meters. Find the angular velocity of a spoke of a wheel.

Solution   Each time the wheel (or a spoke) makes one revolution, the bicycle moves forward a distance equal to the circumference of the wheel, 70π cm. Therefore, when the bicycle travels two blocks (400 m, or 40,000 cm), the number of revolutions of a wheel is $40\,000/(70\pi)$. It takes 3 minutes to make this number of revolutions, and so

$$\omega = \frac{40\,000/(70\pi)}{3}\ \frac{\text{rev}}{\text{min}} = 60.63\ \frac{\text{rev}}{\text{min}}.$$

Expressing ω in radians per second, we have

$$\omega = \frac{(60.63)(2\pi)}{60}\ \frac{\text{rad}}{\text{sec}} = 6.35\ \frac{\text{rad}}{\text{sec}}. \qquad \blacksquare$$

Area of a Sector of a Circle

A sector of a circle is defined to be the region bounded by two radial lines and the intercepted arc of the circle. Figure 4.10 shows two regions associated with the same radial lines. In order to distinguish between these two, we always indicate the central angle of the sector. Figure 4.10(a) shows the sector with central angle α, Fig. 4.10(b) the sector with central angle β.

Fig. 4.10

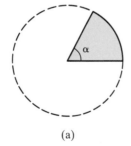 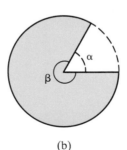

(a)   (b)

From the study of geometry we know that in any given circle the areas of two sectors are proportional to the corresponding central angles. That is, in Fig. 4.11

$$\frac{\text{Area of sector } AOB}{\theta} = \frac{\text{Area of sector } COD}{\alpha}.$$

In particular, if we let sector COD be the entire circle so that $\alpha = 2\pi$ and the area is πr^2, we get

$$\frac{\text{Area of sector } AOB}{\theta} = \frac{\pi r^2}{2\pi} = \frac{r^2}{2}.$$

That is, the area of sector AOB equals $\theta r^2/2$.

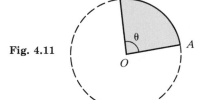
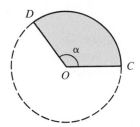

Fig. 4.11

Therefore the area of the sector of a circle of radius r and central angle θ *in radians* is given by

$$\text{Area} = \theta r^2/2. \tag{4.6}$$

Example 5   Find the area of the sector of a circle of radius 2.54 cm and central angle 73°24′.

Solution   We first convert 73°24′ to radians and then substitute into Eq. (4.6):

$$73°24' = (73 + 24/60) \cdot (\pi/180) \text{ radians.}$$

Therefore

$$\text{Area} = (1/2)(73 + 24/60) \cdot (\pi/180) \cdot (2.54)^2 = 4.13 \text{ cm}^2. \quad \blacksquare$$

Exercise 4.2

1. Suppose the radius of a circle is 37.43 cm. Find the length of arc intercepted by the given central angle. Give answers in cm rounded off to two decimal places.

   a) 36°   **b)** 73°23′   **c)** 3.58

2. The radius of a circle is 75.23 cm. Find the length of arc intercepted by the given central angle. Give answers in cm rounded off to two decimal places.

   a) 187°15′   **b)** $17\pi/12$ **c)** 18°15′35″

3. Given that the radius of a circle is 25.32 cm, find the central angle that subtends the given arc. Give answers in radians correct to two decimal places.

   a) $s = 12.47$ cm   **b)** $s = 60.53$ cm   **c)** $s = 29.45$ cm

4. Given that a central angle of 68°35′ subtends an arc of a circle of length 47.53 cm, find the radius of the circle. Give your answer in centimeters correct to two decimal places.

5. Suppose point P moves along a circular path with a radius of 3.57 m and center at O. Find the total distance traveled by P if the radial line OP sweeps out the given angle. Give answers in meters rounded off to two decimal places.

   a) 257°   **b)** 1440°   **c)** $9\pi/2$ **d)** 35π

6. In problem 5 the point P travels a distance of 47.55 m. Through what angle does OP sweep? Give your answer (a) in radians correct to four decimal places and (b) in degrees correct to two decimal places.

7. Find the velocity v of a point on the rim of a wheel of radius 24.37 cm if it is rotating at the given angular velocity.

  a) $\omega = 5.4$ rad/sec   **b)** $\omega = 1247$ rad/min   **c)** $\omega = 63.5$ rev/min   **d)** $\omega = 124$ deg/sec

8. A wheel of diameter 127.48 cm is rotating at a constant rate. Find the angular velocity if a point on the rim is moving at the given speed. Give answers correct to two decimal places in rad/sec and in rev/sec.

  a) $v = 348$ cm/sec   **b)** $v = 2.75$ m/sec

9. Find the angular velocity of the minute hand of a clock in each of the following units.

  a) rev/hr   **b)** rev/min   **c)** deg/min   **d)** rad/min

10. Find the angular velocity of the second hand of a clock in each of the following units.

  a) rev/min   **b)** deg/hr   **c)** rad/sec

11. The length of the minute hand of a clock from the pivot point to the tip is 6.5 cm. Find the linear velocity of its tip in each of the following units.

  a) cm/hr   **b)** cm/min   **c)** cm/sec

12. The length of the hour hand of a clock from the pivot point to the tip is 5.2 cm. Find how far its tip will travel in the given time.

  a) 2 hours   **b)** 3 hours and 40 minutes   **c)** 16 hours and 32 minutes

13. Find the linear velocity of the tip of a propeller blade that is 2.48 m from the pivot point to its tip and is rotating at 640 rev/min. Express your answer in m/min, rounded off to two decimal places.

14. The length of the minute hand of a clock is 8.5 cm and the length of the hour hand is 6.1 cm. Give answers in meters, rounded off to two decimal places, and find the ratio of the distance in (a) to that in (b).

  a) How far will the tip of the minute hand travel in a year? Assume 365 days in a year.

  b) How far will the tip of the hour hand travel in a year?

15. Assume that the earth is spherical with radius 6400 km and that it rotates about an axis through the north and south poles once every 24 hours. How fast is a point on the equator moving in km/hr because of rotation?

16. A trundle wheel is an instrument used to measure distance (Fig. 4.12). It consists of a wheel pivoted at one end of a handle so that it can turn freely. The operator holds the other end of the handle and rolls the wheel (without slipping) along the path whose distance is to be measured. A "meter trundle wheel" is one whose circumference is one meter. Suppose Diane wants to measure the length of a Logan city block. She rolls her meter trundle wheel the length of the block and counts 196 clicks (indicating 196 revolutions). She moves at a constant speed, and it takes her 3 minutes and 36 seconds. Give answers rounded off to two decimal places.

  a) What is the length of the block in meters?

  b) What is her linear velocity?

  c) What is the angular velocity of the wheel in rev/sec? In rad/sec?

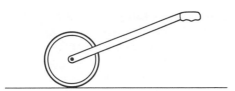

Fig. 4.12

17. A satellite travels around the earth and makes one revolution every 4.5 hours. Assuming that the orbit is a circle of radius 7240 km, find how fast it is traveling in km/hr. Give answer correct to the nearest whole number.

18. A circle has radius 17.3 cm. Find the area correct to two decimal places of the sector of the circle having the given central angle.

    a) 24°   **b)** 37°53′   **c)** $\pi/3$ **d)** 3.56

19. The radius of a circle is 1.26 m, and the area of a sector is 0.8764 square meters. Find the central angle in each of the following.

    a) radians (four decimal places)   **b)** degrees (two decimal places)

20. What is the measure in radians of the smaller angle between the hour and minute hands of a clock at the times given?

    a) 1:15 a.m.   **b)** 1:45 p.m.

21. A pulley of diameter 31.64 cm is driven by a belt (Fig. 4.13). If 32 meters of belt pass around the pulley without slipping, through what angle does a radial line OP on the pulley turn? Express the answer in each of the following.

    a) radian measure (four decimal places)

    b) degree measure (two decimal places)

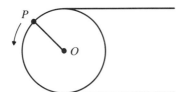

Fig. 4.13

22. In problem 21 suppose it takes 24 seconds for the 32 meters of belt to pass around the pulley. Find the angular velocity of the pulley in each of the following.

    a) rad/sec   **b)** deg/sec

23. Assume that the earth travels about the sun in a circular orbit (actually it is an ellipse that is nearly circular), and the distance between the earth and the sun is 149 million kilometers.

    a) A radial line from the sun through the earth sweeps out an angle of how many radians in a day? (Assume that it takes 365.25 days to travel once around the sun.)

    b) What is the angular velocity of the radial line in radians per hour?

    c) What is the linear velocity of the earth in kilometers per hour?

24. A treadle sewing machine is driven by two wheels with a belt passing around them, as shown in Fig. 4.14. The sewing machine used by Mot'l the tailor has the following measurements: The diameter of the larger wheel is 31 cm, and that of the smaller wheel is 7 cm. If Mot'l treadles his machine at a fixed rate so that in 45 seconds the larger wheel turns through 63 revolutions, find the angular velocity of each wheel (assume the belt does not slip). Express each answer in the following units

a) rev/sec   **b)** rad/sec

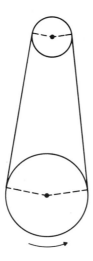

Fig. 4.14

25. Using the information of problem 24, suppose P is a point on the belt. Find the linear velocity of P in centimeters per second. Also determine how far point P travels when the sewing machine is operated at the given rate for 8 seconds.

26. The area of a given sector of a circle is 265.78 cm², and the length of the arc is 36.3 cm. Find the following.

a) The radius of the circle in centimeters to one decimal place

b) The central angle of the sector in radians to two decimal places

27. Given that the area of a circular sector is 24.32 square meters and the radius is 6.47 m, find the length of arc of the sector. Give answer in meters correct to two decimal places.

28. The front wheel of a tricycle is 51.4 cm in diameter, and each rear wheel has a diameter of 23.5 cm.

a) Through how many revolutions will each wheel turn if the tricycle travels along a straight path for a distance of 48 meters.

b) Express each answer in number of radians the wheel will turn.

29. It is between one and two o'clock, and the angle measured clockwise from the hour hand to the minute hand is 64°15′. What time is it? Give the answer correct to the nearest minute.

30. a) A certain pickup truck comes factory-equipped with standard-size tires. The diameter of such a tire is 29 inches. The speedometer is calibrated with this size tire. If the truck travels for one hour at a constant speed with the speedometer reading 55 mi/hr, how many revolutions will a wheel make?

b) The owner of the truck prefers larger tires and replaces the originals with tires of 30.75 inches diameter. Now he travels for one hour at a constant speed with the speedometer reading 55 mi/hr; thus each wheel will make the same number of revolutions as in (a). How far does he go during that hour? By how many miles per hour is he violating the 55 mi/hr speed limit?

4.3   TRIGONOMETRIC FUNCTIONS

Six important functions are introduced in this section. They occur frequently in applications and in subsequent study of mathematics, and so we give them special names: sine, cosine, tangent, cotangent, secant, and cosecant. These functions are abbreviated to sin, cos, tan, cot, sec, and csc. Before giving their definitions it is first necessary to talk about an angle in standard position.

Angles in Standard Position

Each of the trigonometric functions will be defined on a set of measures of angles. We can think of any given angle as being placed in a standard position in reference to a system of rectangular coordinates. *An angle is in standard position* when the vertex of the angle is located at the origin of a rectangular coordinate system and the initial side lies on the positive x-axis. Figure 4.15 illustrates angles in standard position. Angles α, β, and γ are positive angles, and δ is negative.

Fig. 4.15

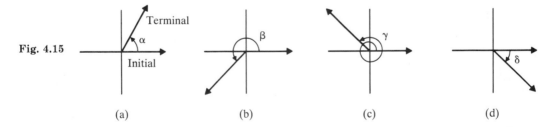

(a)   (b)   (c)   (d)

When the terminal side of an angle in standard position is located in a given quadrant, the angle is said to be in that quadrant. For example, in Fig. 4.15, α is in quadrant I, β is in quadrant III, γ is in quadrant II, and δ is in quadrant IV. If the terminal side of an angle θ coincides with one of the coordinate axes, then θ is called a *quadrantal angle* and is not said to be in any quadrant.

When two angles are placed in standard position and their terminal sides coincide, we say that the *two angles are coterminal*. For example, $\alpha = 45°$ and $\beta = 405°$ are coterminal since $405° = 360° + 45°$. Similarly $210°$ and $-150°$ are coterminal since $210° = 360° + (-150°)$. Note that θ and $\theta + k \cdot 360°$ (where k is any integer) are coterminal angles.

Example 1   For each of the following, draw a figure with the given angle shown in standard position. An approximate freehand sketch is sufficient.

a) 64°   **b)** −155°   **c)** 450°

Solution

Fig. 4.16

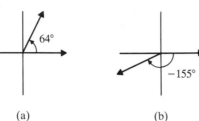

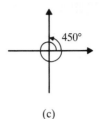

   (a)   (b)   (c)

Example 2   Determine the quadrant in which each of the given angles is located, and draw a sketch of each in standard position.

a) $A = -(3\pi/4)$ **b)** $B = -5$ **c)** $C = 7.5$

Solution   Since $-\pi < -3\pi/4 < -\pi/2$, angle A is in quadrant III. Similarly, $-2\pi < -5 < -3\pi/2$, and so B is in quadrant I; $2\pi < 7.5 < 5\pi/2$, and so C is in quadrant I. Sketches are given in Fig. 4.17.

Fig. 4.17

Example 3   Draw an angle of measure −2.48 in standard position. Then draw the smallest positive angle θ that has the same terminal side as −2.48, and determine the measure of θ correct to two decimal places.

Solution   The angle of measure −2.48 is shown in Fig. 4.18(a), and θ is shown in (b). The measure of θ is given by $\theta = 2\pi - 2.48 = 3.80$.

Fig. 4.18

            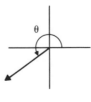

   (a)   (b)

Example 4   Determine all angles that are coterminal with 120°. Give two specific ones.

Solution   The set of all angles coterminal with 120° is given by

$$\{120° + k \cdot 360°|k \text{ is an integer}\}.$$

Examples of two specific angles from this set are:

$$k = 2 \text{ gives an angle measure } 840°,$$
$$k = -1 \text{ gives an angle of measure } -240°.$$

Trigonometric Functions

Let θ be an angle in standard position and $P{:}(x, y)$ be any point (other than the origin) on the terminal side of θ. Let r be the distance from the origin to P; that is, $r = \sqrt{x^2 + y^2}$. Note that r represents a *positive* number. Suppose a perpendicular is drawn from P *to the x-axis* and the point of intersection is called A, as shown in Fig. 4.19. Right triangle PAO is called a *reference triangle* for θ.

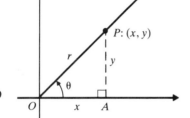

Fig. 4.19

We define the six trigonometric functions of angle θ as follows:

Definition 4.2   If θ is an angle in standard position and $P{:}(x, y)$ is a point on the terminal side of θ, then

$$\sin \theta = \frac{y}{r}, \qquad \tan \theta = \frac{y}{x}, \qquad \sec \theta = \frac{r}{x},$$

$$\cos \theta = \frac{x}{r}, \qquad \cot \theta = \frac{x}{y}, \qquad \csc \theta = \frac{r}{y},$$

(4.7)

where $r = \sqrt{x^2 + y^2}$.

Several observations can be made:

1. The above definitions are independent of the point P taken on the terminal side. That is, if $P_1{:}(x_1, y_1)$ is some other point on the terminal side and $r_1 = \sqrt{x_1{}^2 + y_1{}^2}$, then the two right triangles OAP and OA_1P_1 shown in Fig. 4.19 are similar; hence the ratios of corresponding sides are equal.

2. In Definition 4.2 each of the given rules of correspondence defines a function. That is, for any given angle θ a corresponding unique real number is determined by the ratio indicated in (4.7), whenever this ratio does not involve division by zero.

3. For quadrantal angles the reference triangle becomes a line segment. However, the above definitions are in terms of x, y, r, and so we can use them in that form. For example, for $0°$ we can take the point $(1, 0)$ on the terminal side; then $r = 1$, and we have $\sin 0° = y/r = 0/1 = 0$, $\cos 0° = x/r = 1/1 = 1$, and so on.

4. If the terminal side of θ coincides with the y-axis, then $x = 0$ and $\tan \theta = y/0$ and $\sec \theta = r/0$ are not defined. Similarly, if the terminal side of θ coincides with the x-axis, then $y = 0$ and $\cot \theta = x/0$ and $\csc \theta = r/0$ are not defined.

5. From (4.7) it is easy to see that the following reciprocal relations hold:

$$\sin \theta = \frac{1}{\csc \theta}, \qquad \cos \theta = \frac{1}{\sec \theta}, \qquad \tan \theta = \frac{1}{\cot \theta},$$

$$\cot \theta = \frac{1}{\tan \theta}, \qquad \sec \theta = \frac{1}{\cos \theta}, \qquad \csc \theta = \frac{1}{\sin \theta}. \tag{4.8}$$

Trigonometric Functions
for Special Angles: 30°, 45°, 60°

There are two right triangles in which the sides are related in a simple manner, and so the trigonometric functions for the angles of these triangles can be expressed in exact form. The reader is reminded of the following properties encountered in the study of geometry.

If one angle of a right triangle is $45°$ then the other is also $45°$, and the triangle is isosceles. Hence the lengths of the two sides are equal. Suppose both are taken to be 1 unit in length; by the Pythagorean theorem the length of the hypotenuse is given by $\sqrt{1^2 + 1^2} = \sqrt{2}$, (see Fig. 4.20a). Thus point P:$(1, 1)$ is on the terminal side of a $45°$ angle in standard position, with $r = \sqrt{2}$, as shown in Fig. 4.21(a). Values for the trigonometric functions of $45°$ can now be found. For example, $\sin 45° = 1/\sqrt{2} = \sqrt{2}/2$; using a calculator to express $\sqrt{2}/2$ in decimal form, we get $\sin 45° = 0.7071$ (to four decimal places). We say that $\sqrt{2}/2$ is an *exact form* for $\sin 45°$, and 0.7071 is the *decimal approximation* rounded off to four places.

In a right triangle with one angle equal to $30°$ and the other $60°$, the hypotenuse is twice as long as the shorter side (the side opposite the $30°$ angle). This property can be seen in Fig. 4.20(b), in which triangle ABD is an equilateral triangle and triangles ACB and DCB are congruent. Thus, if the length of side AB is taken as 2, the side opposite the $30°$ angle must be 1. By the Pythagorean theorem the length of BC is $\sqrt{2^2 - 1^2} = \sqrt{3}$.

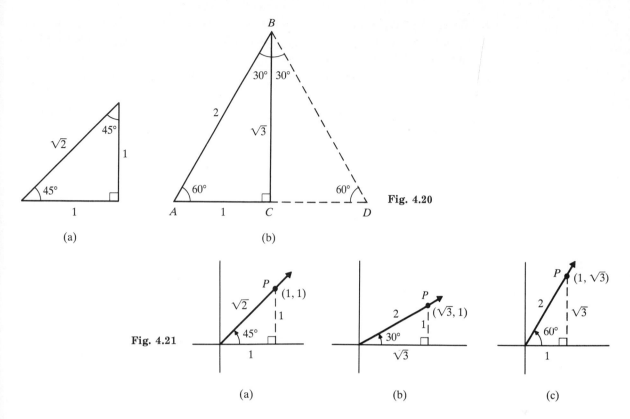

Fig. 4.20

Fig. 4.21

Now suppose an angle of 30° is placed in standard position. We can take point P on the terminal side so that $r = 2$; then P is the point $(\sqrt{3}, 1)$, as shown in Fig. 4.21(b). Trigonometric function values for 30° can now be found:

$$\sin 30° = 1/2, \quad \cos 30° = \sqrt{3}/2, \quad \tan 30° = 1/\sqrt{3},$$
$$\cot 30° = \sqrt{3}, \quad \sec 30° = 2/\sqrt{3}, \quad \csc 30° = 2.$$

In a similar manner, for an angle of 60° in standard position, point P can be taken as $(1, \sqrt{3})$ with $r = 2$, as shown in Fig. 4.21(c). Thus the trigonometric function values for 60° are given by $\sin 60° = \sqrt{3}/2$, $\cos 60° = 1/2$, $\tan 60° = \sqrt{3}$, $\cot 60° = 1/\sqrt{3}$, $\sec 60° = 2$, $\csc 60° = 2/\sqrt{3}$.

Becoming familiar with definitions of the six trigonometric functions as stated in Definition 4.2 is absolutely essential. In the remaining portion of this section we illustrate the use of these with several examples. In each case we shall proceed with the following steps.

1. Sketch the given angle in standard position, and indicate it with a curved arrow.

2. Take a convenient point $P{:}(x, y)$ on the terminal side, and draw a reference triangle to the x-axis.

   3. Use the Pythagorean theorem as needed to get x, y, or r; remember that r is always positive and the signs ($+$ or $-$) for x and y are determined by the quadrant in which P is located.

   4. Use Definition 4.2 to determine the desired trigonometric function values.

Example 5   Suppose θ is an angle in standard position and the point $(-3, 4)$ is on the terminal side of θ. Find the values of the six trigonometric functions of θ.

Solution   Figure 4.22 shows a reference triangle for θ in which point P is taken as $(-3, 4)$, and so $r = \sqrt{(-3)^2 + 4^2} = 5$. Therefore

$$\sin \theta = 4/5, \qquad \tan \theta = 4/-3, \qquad \sec \theta = 5/(-3),$$
$$\cos \theta = -3/5, \qquad \cot \theta = -3/4, \qquad \csc \theta = 5/4.$$

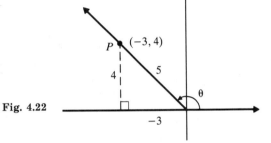

Fig. 4.22

Example 6   Evaluate the six trigonometric functions for $315°$. Express each answer first in exact form and then in decimal form correct to four places.

Solution   In Fig. 4.23 we see that the reference triangle for $315°$ is a $45°$ right triangle. It is therefore convenient to take $(1, -1)$ as the point P, and so $r = \sqrt{1^2 + (-1)^2} = \sqrt{2}$. Applying Definition 4.2 gives $\sin 315° = -1/\sqrt{2} = -\sqrt{2}/2$ (exact form).

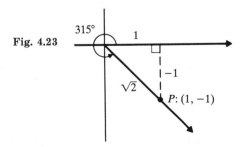

Fig. 4.23

Using the calculator to evaluate $-\sqrt{2}/2$, we get $\sin 315° = -0.7071$. Similarly,

$$\cos 315° = \frac{1}{\sqrt{2}} = \frac{\sqrt{2}}{2}; \qquad \cos 315° = 0.7071$$

$$\tan 315° = \frac{-1}{1} = -1; \qquad \tan 315° = -1.0000$$

$$\cot 315° = \frac{1}{-1} = -1; \qquad \cot 315° = -1.0000$$

$$\sec 315° = \frac{\sqrt{2}}{1} = \sqrt{2}; \qquad \sec 315° = 1.4142$$

$$\csc 315° = \frac{\sqrt{2}}{-1} = -\sqrt{2}; \qquad \csc 315° = -1.4142 \qquad ■$$

Example 7   Evaluate $\sin(-2\pi/3)$ and $\tan(-2\pi/3)$. Express answers in exact form.

Solution   Sketch $\theta = -2\pi/3$. The reference triangle for $\theta = -2\pi/3$ is a 30°, 60° right triangle, so we can take P as $(-1, -\sqrt{3})$ (see Fig. 4.24). Applying Definition 4.2 gives

$$\sin\left(-\frac{2\pi}{3}\right) = \frac{-\sqrt{3}}{2} \text{ and } \tan\left(-\frac{2\pi}{3}\right) = \frac{-\sqrt{3}}{-1} = \sqrt{3}.$$

Fig. 4.24

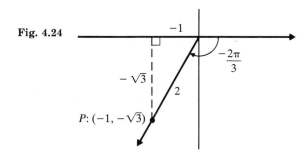

■

Example 8   Evaluate each of the following.

a) $\sin 180°$ **b)** $\cos 180°$ **c)** $\tan 90°$ **d)** $\sec(-540°)$.

Solution   We first draw the diagrams shown in Fig. 4.25.

a) Take point P as $(-1, 0)$, and so $r = 1$. Hence $\sin 180° = y/r = 0/1 = 0$.

b) Take P as in (a) and apply Definition 4.2 to get $\cos 180° = x/r = -1/1 = -1$.

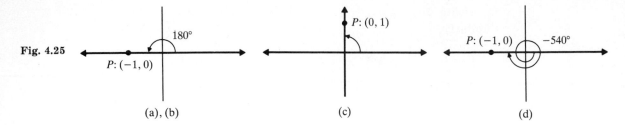

Fig. 4.25

(a), (b)   (c)   (d)

c) Let P be $(0, 1)$, and so $r = 1$. Applying Definition 4.2 gives $\tan 90° = y/x = 1/0$. Since division by zero is not defined, we say that $\tan 90°$ is undefined.

d) Figure 4.25 (d) shows $-540°$ in standard position. Note that the terminal side coincides with the negative x-axis, and so we can take the point P as $(-1, 0)$ and $r = 1$. Thus $\sec(-540°) = r/x = 1/-1 = -1$. ■

Example 9   Suppose θ is an angle in the second quadrant and $\cos\theta = -0.7$. Find $\sin\theta$ and $\tan\theta$ in (a) exact form; (b) decimal form correct to three places.

Solution   Since θ is in the second quadrant and $\cos\theta = -0.7 = -7/10$, we get the reference triangle shown in Fig. 4.26 by taking $x = -7$, $r = 10$; y is given by $y = \sqrt{10^2 - (-7)^2} = \sqrt{51}$.

a) Using Definition 4.2 gives $\sin\theta = \dfrac{\sqrt{51}}{10}$; $\tan\theta = \dfrac{\sqrt{51}}{-7}$.

b) Using a calculator to evaluate the exact-form answers, we get $\sin\theta = 0.714$; $\tan\theta = -1.020$.

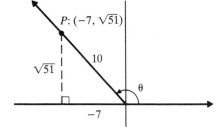

Fig. 4.26

Example 10   Suppose θ is an angle in the third quadrant and $\tan\theta = 3/4$. Find the remaining five trigonometric functions of θ.

Solution   Since $\tan\theta = 3/4 = -3/-4$ and θ is in quadrant III, we can take $(-4, -3)$ as the point to determine a reference triangle, as shown in Fig. 4.27:

$$r = \sqrt{(-3)^2 + (-4)^2} = \sqrt{25} = 5.$$

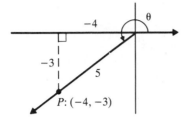

Fig. 4.27

Therefore

$$\sin \theta = \frac{-3}{5}, \qquad \cos \theta = \frac{-4}{5}, \qquad \cot \theta = \frac{-4}{-3} = \frac{4}{3},$$

$$\sec \theta = \frac{5}{-4}, \qquad \csc \theta = \frac{5}{-3}.$$

Exercises 4.3

In problems in which drawing an angle is required, a reasonable freehand sketch is sufficient.

1. In each of the following, draw a figure illustrating the given angle in standard position.

a) 40°   b) 220°   c) −220°   d) 725°   e) −460°

2. Draw a figure illustrating the given angle in standard position.

a) $\dfrac{5\pi}{4}$ b) 3.41   c) −1.80   d) 8.8   e) $-\dfrac{17\pi}{11}$

3. Determine the quadrant in which the given angle lies (that is, the quadrant in which the terminal side is located).

a) 37°   b) 335°   c) −125°   d) 580°   e) −480°

4. Determine the quadrant in which the given angle lies.

a) $-\dfrac{3\pi}{4}$ b) $\dfrac{7\pi}{3}$ c) 3.56   d) 8.47   e) −5.40

5. Draw a figure with the given angle in standard position. Then draw the smallest positive angle that has the same terminal side, and determine its measure.

a) −100°   b) 540°   c) −540°   d) $-\dfrac{3\pi}{4}$ e) −4.32

6. For each of the given pair of angles, determine whether the second one is coterminal with the first one.

a) 60°, 240°   b) −45°, 315°   c) $-\dfrac{3\pi}{4}, \dfrac{5\pi}{4}$

d) $\pi, -\pi$ e) 30°, 750°   f) $\dfrac{3\pi}{2}, -\dfrac{3\pi}{2}$

7. Find three angles that are coterminal with $\theta = 90°$.

8. Find three angles that are coterminal with $\theta = -\pi/6$.

9. Determine the set of all angles that are coterminal with $\theta = -2\pi/3$.

10. Determine the set of all angles that are coterminal with $\theta = 30°$.

11. In each of the following, find the set of all angles that are coterminal with an angle that has its terminal side passing through the given point.

    a) $(1, 1)$ **b)** $(-3, -3)$ **c)** $(-1, \sqrt{3})$

12. Determine the set of all angles that are coterminal with the angle in standard position whose terminal side passes through the given point.

    a) $(0, 3)$ **b)** $(0, -5)$ **c)** $(5, 0)$ **d)** $(-2.3, 0)$

13. Suppose θ is an angle in standard position and the point $(4, -3)$ is on the terminal side of θ. Evaluate the six trigonometric functions for θ. Express each answer in (a) exact form; (b) decimal form correct to two places.

14. The point $(2, 3)$ is on the terminal side of angle α, Find the six trigonometric functions for α, and give answers in exact form.

In problems 15 through 23, evaluate the given expression and give answers in exact form.

15. **a)** $\sin 60°$ **b)** $\cos 60°$ **c)** $\sin 210°$ **d)** $\cos 210°$

16. **a)** $\tan 30°$ **b)** $\sec 30°$ **c)** $\tan 300°$ **d)** $\sec 300°$

17. **a)** $\cot -45°$ **b)** $\csc -45°$ **c)** $\cot 405°$ **d)** $\csc 405°$

18. **a)** $\sin 225°$ **b)** $\cos 330°$ **c)** $\tan 135°$ **d)** $\cot 150°$

19. **a)** $\sin \dfrac{\pi}{6}$ **b)** $\tan \dfrac{5\pi}{6}$ **c)** $\cos\left(-\dfrac{\pi}{3}\right)$ **d)** $\cos \dfrac{2\pi}{3}$

20. **a)** $\cos\left(-\dfrac{5\pi}{4}\right)$ **b)** $\sec\left(-\dfrac{7\pi}{4}\right)$ **c)** $\tan\left(\dfrac{17\pi}{3}\right)$ **d)** $\sin\left(-\dfrac{17\pi}{6}\right)$

21. **a)** $\sin 90°$ **b)** $\cos 0°$ **c)** $\tan 270°$ **d)** $\sec 180°$

22. **a)** $\sin\left(-\dfrac{\pi}{2}\right)$ **b)** $\tan \pi$ **c)** $\cot(-\pi)$ **d)** $\sec(-4\pi)$

23. **a)** $\sec\left(-\dfrac{17\pi}{3}\right)$ **b)** $\cos(17\pi)$ **c)** $\tan\left(-\dfrac{11\pi}{6}\right)$ **d)** $\sin\left(\pi + \dfrac{5\pi}{6}\right)$

24. In the table below, write a $+$ sign or a $-$ sign to indicate the sign of the corresponding entry.

θ	$\sin \theta$	$\cos \theta$	$\tan \theta$	$\cot \theta$	$\sec \theta$	$\csc \theta$
$124°$						
$-320°$						
3.04						
-1.16						

In problems 25 through 31, give each answer in

a) exact form   **b)** in decimal form correct to three decimal places

25. Given that θ is an angle in the second quadrant and $\cos \theta = -3/5$, find the other five trigonometric functions of θ.

26. Given that $\sin \alpha = -3/4$ and the terminal side of α is in the fourth quadrant, find the remaining five trigonometric functions of α.

27. Given that $\cot \beta = 3/4$ and β is in the third quadrant, find the other five trigonometric functions of β.

28. Given that $\tan \gamma = -1.2$ and the terminal side of γ is in the second quadrant, find the remaining five trigonometric functions of γ.

29. Given that $\sin \theta = -0.25$ and $\tan \theta$ is negative, find the remaining five trigonometric functions of θ.

30. Given that $\tan \theta = -3$ and θ is a second quadrant angle, find the remaining five trigonometric functions of θ.

31. Evaluate $\dfrac{\cos \dfrac{2\pi}{3} - \sin \dfrac{4\pi}{3} + \tan \dfrac{5\pi}{4}}{\sin \dfrac{\pi}{2} - \tan \dfrac{5\pi}{3} + \sec \dfrac{2\pi}{3}}$

32. Verify that $\sin(\alpha - \beta) = \sin \alpha \cos \beta - \cos \alpha \sin \beta$ for each of the following pairs of values of α and β.

a) $\alpha = \dfrac{2\pi}{3}, \beta = \dfrac{\pi}{6}$ **b)** $\alpha = \dfrac{\pi}{2}, \beta = \pi$ **c)** $\alpha = \dfrac{3\pi}{2}, \beta = \dfrac{\pi}{2}$ **d)** $\alpha = \dfrac{5\pi}{4}, \beta = 3\pi$

Hint: In each case evaluate the left-hand side and the right-hand side of the equation for the given α and β, and then verify that the two resulting numbers are equal.

33. Verify that $(\sin \theta)^2 + (\cos \theta)^2 = 1$ for each of the given values of θ.

a) $\theta = 60°$ **b)** $\theta = 150°$ **c)** $\theta = \pi$

34. Verify that $\sin(2\theta) = 2(\sin \theta)(\cos \theta)$ for the given values of θ.

a) $\theta = 90°$ **b)** $\theta = 30°$ **c)** $\theta = \dfrac{2\pi}{3}$

35. Verify that $(\sec \theta)^2 - (\tan \theta)^2 = 1$ for the given values of θ.

a) $\theta = -\dfrac{3\pi}{4}$ **b)** $\theta = 225°$ **c)** $\theta = 495°$

36. For which of the given angles α and β is $\cos(\alpha + \beta) = \cos \alpha + \cos \beta$?

a) $\alpha = \pi, \beta = 0$ **b)** $\alpha = 0, \beta = \dfrac{\pi}{2}$ **c)** $\alpha = 45°, \beta = 45°$

d) $\alpha = 120°, \beta = 30°$

4.4 EVALUATING TRIGONOMETRIC FUNCTIONS

In the preceding section we introduced the six trigonometric functions and saw how they can be evaluated for several special angles by applying the given definitions. However, if we wanted to determine $\sin 37°$ applying the definition, we could draw an angle of measure 37° by means of a protractor and then construct a reference triangle and measure the lengths of the sides. These measurements can only give a crude approximate evaluation of sin 37°. There is, of course, an easier way, and that is by applying a calculator, which will give an accurate result correct to several decimal places.

All scientific calculators have keys labeled $\boxed{\text{sin}}$, $\boxed{\text{cos}}$, and $\boxed{\text{tan}}$. There are also keys that will allow the operator to put the calculator in degree, radian, or grad mode. The owner's manual that comes with the calculator describes this feature and should be consulted to make certain it is understood.*

Example 1   Evaluate sin 37°.

Solution   First be certain that your calculator is in degree mode. Then press keys 3, 7, $\boxed{\text{sin}}$. The display will read 0.6018 on many calculators, but if greater decimal accuracy is desired, the operator can have the calculator display more decimal digits (the owner's manual has instructions for doing this). Thus we can get, accurate to nine decimal places, sin 37° = 0.601815023.   ■

Example 2   Evaluate cot 64°.

Solution   The calculator does not have a key labeled cot. However, as we observed in (4.8), the cotangent function is the reciprocal of the tangent, and so we have cot 64° = 1/tan 64°. With the calculator in degree mode, press keys 6, 4, $\boxed{\text{tan}}$, $\boxed{\text{1/x}}$. The display will give cot 64° = 0.487732589. The student should note at this point that 1/tan 64° and tan(1/64)° are not equal. That is, the $\boxed{\text{1/x}}$ key should be pressed after the $\boxed{\text{tan}}$ key.   ■

Example 3   Evaluate cos 24°31′43″ correct to five decimal places.

Solution   We first convert 24°31′43″ into a decimal number of degrees, as follows:

$$24°31'43'' = \left(24 + \frac{31}{60} + \frac{43}{3600}\right)^\circ.$$

Be sure your calculator is in degree mode, and carry out the following sequence of steps: Evaluate 24 + 31/60 + 43/3600, and then, with the result in the display, press $\boxed{\text{cos}}$ to get cos 24°31′43″ = 0.90975.   ■

Example 4   Evaluate sin 1.2 correct to four decimal places.

Solution   Note that sin 1.2 means sine of 1.2 radians. Place the calculator in radian mode; then press 1.2, $\boxed{\text{sin}}$, and the value will appear in the display: sin 1.2 = 0.9320.   ■

---

*   See Appendix A for basic calculator instruction.

Example 5   Evaluate $\sec(-2.47)$ and round off answer to five decimal places.

Solution   Since the calculator does not have a key labeled sec, we use (4.8) to get $\sec(-2.47) = 1/\cos(-2.47)$. To evaluate this, we place the calculator in radian mode, then press 2.47, change sign, $\boxed{\cos}$, $\boxed{1/x}$. This gives $\sec(-2.47) = -1.27741$.

Example 6   Evaluate $\tan 450°$.

Solution   Place the calculator in degree mode; press 450 and $\boxed{\tan}$, and the display will indicate *Error*. Applying Definition 4.2, we see that $\tan 450°$ is undefined. Some calculators give 9.9×10^{99} as the value of $\tan 450°$; such a large number should alert us to the fact that the calculator is sending a special message.

Example 7   Given that $f(x) = \sqrt{1 - x^2}$ and $g(x) = \cos x$, evaluate each of the following. Give answers rounded off to four decimal places.

a) $(f + g)(1/2)$ **b)** $(f \circ g)(-2)$ **c)** $(f/g)(0.64)$

Solution   Note that the statement of the problem implies that x is in radians, so be certain that your calculator is in radian mode.

a) $(f + g)\left(\dfrac{1}{2}\right) = f\left(\dfrac{1}{2}\right) + g\left(\dfrac{1}{2}\right)$

$$= \sqrt{1 - \left(\dfrac{1}{2}\right)^2} + \cos\left(\dfrac{1}{2}\right) = \dfrac{\sqrt{3}}{2} + \cos(0.5) = 1.7436.$$

b) $(f \circ g)(-2) = f(g(-2)) = f(\cos(-2)) = \sqrt{1 - (\cos(-2))^2} = 0.9093.$

c) $\left(\dfrac{f}{g}\right)(0.64) = \dfrac{f(0.64)}{g(0.64)} = \dfrac{\sqrt{1 - (0.64)^2}}{\cos 0.64} = 0.9580.$

Exercises 4.4

In problems 1 through 20, use a calculator to evaluate the given expressions, and give answers rounded off to four decimal places.

1. $\sin 28°$

2. $\cos 72°$

3. $\sec 35°$

4. $\csc 17°$

5. $\sin 43°21'$

6. $\cos 12°37'41''$

7. $\tan(-31.48°)$

8. $\sec 148.16°$

9. $\cos(251°23'53'')$

10. $\sin(478°15')$

11. $\sin 0.4$

12. $\tan(\pi/3)$

13. $\sec(\pi/2)$

14. $\csc 3.23$

15. $\cos(3\pi/17)$

16. $\tan(-2\pi/47)$

17. $\csc(2.78 + 5\pi)$

18. $\cot(3\pi/8)$

19. $\tan(-8.32)$

20. $\sin(1 - \sqrt{5})$

In problems 21 through 36, evaluate the given expressions. Give answers correct to four decimal places.

21. $(2.48) \sin 73°16'$

22. $\dfrac{3.56 \sin 24°17'}{\sin 47°21'}$

23. $\dfrac{2 \tan 35°12'}{1 - (\tan 35°12')^2}$

24. $65.48 \csc 43°18'$

25. $\tan(3\pi/13)$ **26.** $\sec 1.47$ **27.** $\cos(7\pi/17)$ **28.** $\dfrac{8.54 \sin(5\pi/11)}{\sin(3\pi/7)}$

29. $(\sin 23°48')^2 + (\cos 23°48')^2$ **30.** $\sec(31°12'36'')$

31. $\cot(72°15'41'')$ **32.** $\dfrac{1}{\csc(3\pi/8)} + \dfrac{1}{\sec(3\pi/8)}$

33. $\sin\left(\dfrac{1 + \sqrt{2}}{5}\right)$ **34.** $\sin 37° \cos 56° - \sin 56° \cos 37°$

35. $\left(\dfrac{1 + \sqrt{5}}{3}\right) \sin\left(\dfrac{5\pi}{12}\right)$ **36.** $3\left(\sin \dfrac{4\pi}{5}\right)^2 - 2\left(\cos \dfrac{4\pi}{5}\right)^2$

In problems 37 through 48, evaluate the given expressions where

$$f(x) = 1 + x, \qquad g(x) = \sin x, \qquad h(x) = \sqrt{x - 1}.$$

Give answers rounded off to two decimal places. If any of the given expressions are not defined as a real number, tell why.

37. $(f + g)\left(\dfrac{\pi}{2}\right)$ **38.** $(f \circ g)\left(\dfrac{\pi}{2}\right)$ **39.** $(h \circ g)\left(\dfrac{\pi}{6}\right)$ **40.** $(g + h)(3)$

41. $\left(\dfrac{f}{g}\right)(1)$ **42.** $\left(\dfrac{g}{h}\right)(1)$ **43.** $(h \circ g)\left(\dfrac{\pi}{2}\right)$ **44.** $(g \circ f)(-4)$

45. $(g \circ f)(0)$ **46.** $(g \circ g)(4)$ **47.** $\left(\dfrac{g}{f}\right)(2.5)$ **48.** $\left(\dfrac{g}{h}\right)(1.64)$

**4.5   CIRCULAR FUNCTIONS;
PERIODIC PROPERTIES AND GRAPHS**

In Section 4.3 we defined trigonometric functions whose domains are sets of measures of angles. If angles are measured in radians, then the domains can be considered as sets of real numbers. When trigonometric functions are studied in calculus, it is assumed that their domains consist of real numbers. In this section circular functions are introduced, and we shall see that they are the same as trigonometric functions in which angular measure is in radians. In anticipation of this fact, we shall even use the same names for the circular functions as for the corresponding trigonometric functions.

An equation of a circle with center at the origin and having a radius of 1 is given by

$$x^2 + y^2 = 1. \tag{4.9}$$

Such a circle is called the *unit circle*. Consider a point P starting at point $A:(1, 0)$ and moving along the unit circle. Let S denote the *directed* distance P traverses, where distance is taken as positive if P travels counterclockwise and negative if it moves clockwise from point A. Thus with each directed distance S, we have a corresponding real number s.*

Definition 4.3

Circular sine and cosine functions

Suppose s is any *real number* and a point starts at $A:(1,0)$ and moves along the unit circle a directed distance corresponding to s and reaches point $P:(u,v)$, as shown in Fig. 4.28. Two circular functions, denoted by sin and cos, are defined by

$$\sin s = v(\text{the second component of } P),$$

$$\cos s = u(\text{the first component of } P).$$

Fig. 4.28

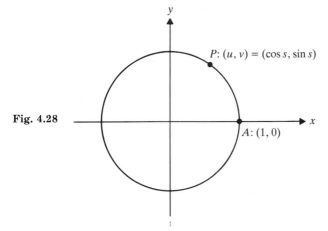

Definitions of four additional circular functions are now stated in terms of $\sin s$ and $\cos s$; these definitions are obviously motivated by corresponding properties of trigonometric functions from Section 4.3.

Definition 4.4

Let s be any real number, and $\sin s$ and $\cos s$ be as given in Definition 4.3. The four circular functions, denoted by tan, cot, sec, and csc, are defined by

$$\tan s = \frac{\sin s}{\cos s}, \qquad \cot s = \frac{\cos s}{\sin s},$$

$$\sec s = \frac{1}{\cos s}, \qquad \csc s = \frac{1}{\sin s}.$$

---

* Here we are thinking of distance S as having a unit of measure associated with it (for example, centimeters), and s is the corresponding number without the unit of measure. For instance, if $S = 3$ cm, then $s = 3$.

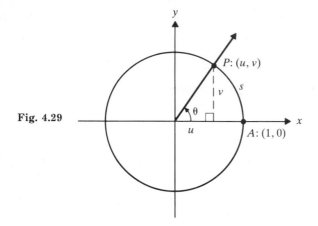

Fig. 4.29

If we take a closer look at the circular functions given in these definitions, it is easy to see that they are precisely the same as the corresponding trigonometric functions defined in Section 4.3 in which angles are measured in radians. For instance, let θ be an angle in standard position, as shown in Fig. 4.29. (The diagram shows θ as an acute angle, but the discussion holds for angles of any size.) If θ is in radians, then the corresponding arc length s is given by $s = r \cdot \theta = 1 \cdot \theta = \theta$. According to Definition 4.2, where point P on the terminal side is taken on the unit circle, the trigonometric function values for $\sin \theta$ and $\cos \theta$ are given by

$$\sin \theta = \frac{v}{r} = \frac{v}{1} = v,$$

$$\cos \theta = \frac{u}{r} = \frac{u}{1} = u.$$

But by Definition 4.3, $v = \sin s$ and $u = \cos s$. Therefore

$$\sin \theta = \sin s \qquad \text{and} \qquad \cos \theta = \cos s.$$

We are not interested here in carrying out all the details to show that each circular function is equal to the corresponding trigonometric function. The important fact is that in either case we have defined six functions that have *domains consisting of sets of real numbers*. It is in this setting that the student will encounter trigonometric (or circular) functions in calculus.

We shall refer to the six functions we are discussing here interchangeably as either trigonometric or circular functions. One might ask: Why talk about the same thing in two different contexts? The reason is that in the setting in which trigonometric functions are introduced, it is convenient to relate the functions to triangles and use them in problems related to triangles, but using circular functions provides us with simple means for deriving several important properties of these functions. This is illustrated in the following discussion.

Periodic Properties

Suppose s is any real number. Then the same point P on the unit circle is associated with s and with $s + k(2\pi)$, where k is any integer (zero, positive, or negative). According to Definition 4.3, we get

$$\sin(s + 2k\pi) = \sin s \quad \text{and} \quad \cos(s + 2k\pi) = \cos s. \qquad (4.10)$$

This tells us that the sine function (the cosine function as well) repeats itself every 2π units, and so it is called a *periodic function*. Using Eq. (4.10) and Definition 4.4, we can make similar statements for the other circular functions.

Identities for Circular Functions

An equation is an *identity* if its solution set is the set of all real numbers for which both sides of the equation are defined. Here we illustrate derivation of a few basic identities for circular functions; several others will be discussed in Chapter 5.

1. Suppose point $P:(u, v)$ on the unit circle is associated with a real number s, and M is the corresponding point associated with $s + \pi$, as shown in Fig. 4.30. Since the circumference of the unit circle is $2\pi r = 2\pi \cdot 1 = 2\pi$, point M is diametrically opposite P, and so it is given by $(-u, -v)$. Using Definition 4.3 as related to M gives $\sin(s + \pi) = -v$ and $\cos(s + \pi) = -u$. But in relation to point P, $\sin s = v$ and $\cos s = u$. Therefore, we get the identities

$$\sin(s + \pi) = -\sin s \quad \text{and} \quad \cos(s + \pi) = -\cos s. \qquad (4.11)$$

Fig. 4.30

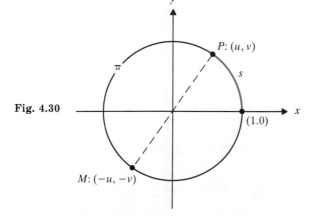

Using Eq. (4.11) and Definition 4.4, we easily derive the identities

$$\tan(s + \pi) = \tan s \quad \text{and} \quad \cot(s + \pi) = \cot s. \qquad (4.12)$$

2. In a manner similar to (1), suppose $P:(u, v)$ is associated with real number s; then the corresponding point associated with $s + \pi/2$ is $\theta:(-v, u)$, as shown in Fig. 4.31. Hence, by Definition 4.3,

$$\sin\left(s + \frac{\pi}{2}\right) = u = \cos s \quad \text{and} \quad \cos\left(s + \frac{\pi}{2}\right) = -v = -\sin s.$$

Therefore we get the identities

$$\sin\left(s + \frac{\pi}{2}\right) = \cos s \quad \text{and} \quad \cos\left(s + \frac{\pi}{2}\right) = -\sin s. \qquad (4.13)$$

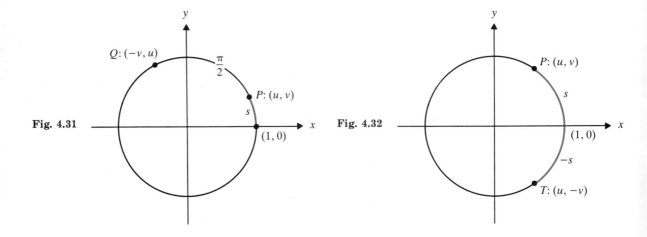

Fig. 4.31   **Fig. 4.32**

3. Figure 4.32 shows points P and T related to numbers s and $-s$, respectively. From this and Definition 4.3 we get the identities

$$\sin(-s) = -\sin s \quad \text{and} \quad \cos(-s) = \cos s. \qquad (4.14)$$

Applying Eq. (4.14) and Definition 4.4 gives

$$\tan(-s) = -\tan s, \quad \cot(-s) = -\cot s,$$
$$\sec(-s) = \sec s, \quad \csc(-s) = -\csc s. \qquad (4.15)$$

From Eqs. (4.14) and (4.15) we conclude that sin, tan, cot, and csc are *odd functions,* and that cos and sec are *even functions.*

4. Suppose s is any real number; the associated point on the unit circle is $(\cos s, \sin s)$. Since this point must satisfy Eq. (4.9), we get the identity

$$(\cos s)^2 + (\sin s)^2 = 1. \tag{4.16}$$

We shall see numerous occasions in which this identity is applied. It will also be written as

$$\cos^2 s + \sin^2 s = 1.$$

In a similar manner several other identities for circular functions can be derived. Some of them are included in Exercises 4.5. We shall delay further consideration of identities until Chapter 5.

Periodic Functions

In the above discussion we noted that if g is any of the circular functions, then $g(s + 2\pi) = g(s)$ for every real number s for which g is defined. Thus function g repeats itself over an interval of length 2π. Any function that repeats itself over consecutive intervals of fixed length is called a *periodic function.*

Many scientific investigations involve phenomena that are of a cyclic nature and that can be described in terms of periodic functions. It is an interesting and important fact that practically all periodic functions can be expressed as linear combinations of sine and cosine functions.* It is this fact that makes trigonometry extremely useful in application of mathematics to many real-life problems.

Definition 4.5   Suppose f is any function with the property that there is a positive number p such that

$$f(x + p) = f(x) \tag{4.17}$$

for all x in the domain of f. We say that f is a *periodic function.* Let p be the *smallest positive number* for which Eq. (4.17) holds; then p is called the *period* of f.

We shall now draw a graph of each of the trigonometric functions and use it to determine the period of that function.

---

*   This is the basis for a broad topic in advanced mathematics called Fourier series.

Graph of the Sine Function

We could make a table of x, y values that satisfy the equation $y = \sin x$ and then use the corresponding (x, y) points to draw the graph. However, we can gain considerable insight into the behavior of the sine function by considering it as a circular function.

When one draws a graph in the x, y rectangular system of coordinates, it is customary to call x the independent variable; that is, we draw a graph of $y = \sin x$. Thus, in applying Definition 4.3, we shall replace s by x and think of x as being associated with arc length (not as the x coordinate of P). In order not to get variables confused, we shall denote the coordinates of P by (u, y), as illustrated in Fig. 4.33. Therefore, according to Definition 4.3, we have $y = \sin x$, where x is any real number associated with the directed arc length of a point moving from A to P.

We can now proceed to draw the graph of $y = \sin x$ by letting point $P:(u, y)$ start at A and move along the unit circle (counterclockwise for $x \geq 0$). We record the corresponding points $T:(x, y)$ on the graph shown in Fig. 4.34.

When P is at A, $x = 0$, $y = 0$; so T is at $A_1:(0, 0)$. As P moves from A to B, x increases from 0 to $\pi/2$, and the corresponding values of y increase from 0 to 1;

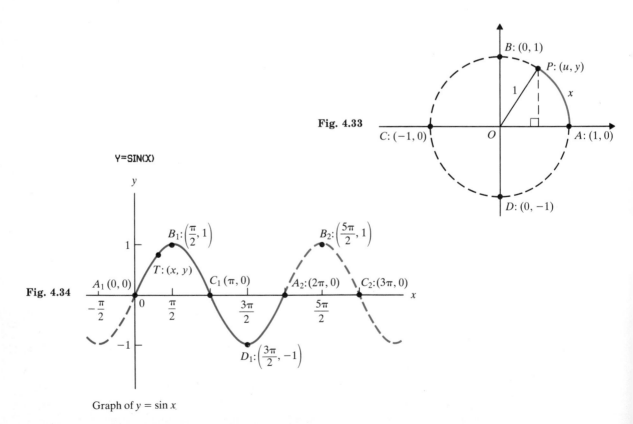

Fig. 4.33

Fig. 4.34

Graph of $y = \sin x$

then point T moves from $A_1:(0, 0)$ to $B_1:(\pi/2, 1)$. As P moves from B to C, x increases from $\pi/2$ to π, and the corresponding values of y decrease from 1 to 0; this gives the points of the graph from B_1 to C_1. As P moves from C to D, x increases from π to $3\pi/2$, and y decreases from 0 to -1; this gives the points on the graph between C_1 and D_1. As P moves from D to A, x increases from $3\pi/2$ to 2π, and y increases from -1 to 0, giving the corresponding points T between D_1 and A_2 in Fig. 4.34.

The procedure above gives us one complete cycle of the sine curve. Since we know that $\sin(x + 2\pi) = \sin x$ for each real number x, we can continue the graph as indicated by the broken portion of the curve.

From the graph in Fig. 4.34 we see that $p = 2\pi$ is the smallest positive number p such that $\sin(x + p) = \sin x$ for each real number x. Thus we can conclude the following from the graph.

> The sine function is periodic with period 2π. The domain and range of the sine function are given by
>
> $$\mathfrak{D}(\sin) = \mathbf{R}, \qquad \mathfrak{R}(\sin) = \{y \mid -1 \leq y \leq 1\}.$$

Graph of the Cosine Function

We can draw a graph of $u = \cos x$ by following a procedure similar to that used above to draw the graph of the sine function. Note that we are calling the dependent variable u. In Fig. 4.33 the first coordinate of $P:(u, y)$ gives the value of $\cos x$ for any given real number x; that is, $u = \cos x$. We omit the details and draw the curve shown in Fig. 4.35, with the solid portion corresponding to the points (x, u) that we get as point P moves counterclockwise around the unit circle from point A in Fig. 4.33.

Fig. 4.35

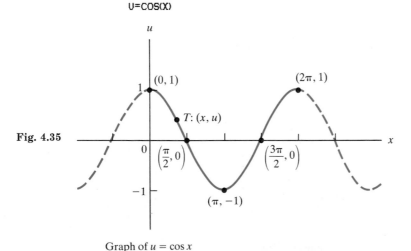

Graph of $u = \cos x$

From the curve in Fig. 4.35, we see that $p = 2\pi$ is the smallest positive number p such that $\cos(x + p) = \cos x$ for every real number x. Therefore we conclude the following.

> The cosine function is periodic with period 2π. The domain and range of the cosine function are given by
>
> $$\mathfrak{D}(\cos) = \mathbf{R}, \qquad \mathfrak{R}(\cos) = \{u \,|\, {-}1 \le u \le 1\}.$$

Graph of the Tangent Function

We shall draw a graph of the tangent function by first making a table of x, y values that satisfy $y = \tan x$. By plotting these points, we shall get the curve shown in Fig. 4.36. In selecting what values of x to use in the table, note that Eq. (4.12) gives $\tan(x + \pi) = \tan x$ for each real number x for which $\tan x$ is defined. Thus it is sufficient to make a table in which x is between $-\pi/2$ and $\pi/2$. Also, from Eq. (4.15) we have $\tan(-x) = -\tan x$ for each x in $\mathfrak{D}(\tan)$; this tells us that the graph of $y = \tan x$ is symmetric about the origin. Therefore it is sufficient to make a table for $0 \le x < \pi/2$. The tangent function is not defined at $\pi/2$; we include several values of x near $\pi/2 = 1.570796 \ldots$.

x	0	0.25	0.50	0.75	1.00	1.25	1.50	1.52	1.55	1.56	1.57
y	0	0.26	0.55	0.93	1.56	3.01	14.1	19.7	48.1	92.6	1256

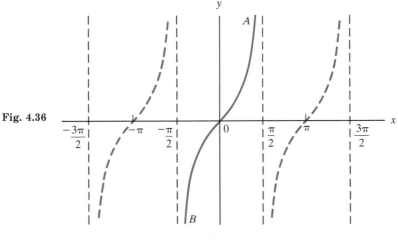

Fig. 4.36

Graph of $y = \tan x$

In Fig. 4.36 the portion of the curve from 0 to A corresponds to the points in the table. Since the curve is symmetric about the origin, the portion of the curve from 0 to B is obtained by reflecting points from 0 to A about the origin. The remaining branches (broken portions) of the curve come from the periodic property given by $\tan(x + \pi) = \tan x$.

From the graph in Fig. 4.36 we conclude the following.

The tangent function is periodic with period π. The domain and range of the tangent function are given by

$$\mathcal{D}(\tan) = \{x \mid x \neq \frac{\pi}{2} + k\pi, \quad k \text{ is an integer}\}, \qquad \mathcal{R}(\tan) = \mathbf{R}.$$

We also note from the graph that the dashed vertical lines through x equal $\pi/2, 3\pi/2, \ldots, -\pi/2, -3\pi/2, \ldots$ are *vertical asymptotes* to the graph of the tangent function.

Graph of the Cotangent Function

We can draw a graph of $y = \cot x$ by following a procedure similar to that used above in drawing a graph of the tangent function. We omit the details and give the graph (Fig. 4.37).

From the graph in Fig. 4.37, we conclude the following.

Fig. 4.37

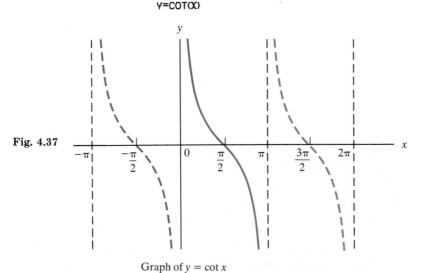

Graph of $y = \cot x$

> The cotangent function is periodic with period π. The domain and range of the tangent function are given by
>
> $$\mathfrak{D}(\cot) = \{x \mid x \neq k\pi, \quad k \text{ is an integer}\}, \qquad \mathfrak{R}(\cot) = \mathbf{R}.$$

We also note that the cotangent curve has vertical asymptotes, given by $x = k\pi$, where k is an integer.

Graph of the Secant Function

Since $\sec(x + 2\pi) = \sec x$, in making a table of x, y values that satisfy $y = \sec x$, it is sufficient to include values of x in the interval $-\pi$ to π. From Eq. (4.15) we have $\sec(-x) = \sec x$ for every x in $\mathfrak{D}(\sec)$; this tells us that the graph is symmetric about the y-axis; hence it is sufficient to include in our table values of x between 0 and π. The secant function is not defined at $\pi/2$, but we include several values of x near $\pi/2 = 1.57 \ldots$ in the following table.

x	0	0.25	0.50	0.75	1.00	1.25	1.50	1.56	1.57
y	1	1.03	1.14	1.37	1.85	3.17	14.1	92.6	1256

x	1.58	1.60	1.75	2.00	2.25	2.50	2.75	3.00	π
y	-109	-34.2	-5.61	-2.40	-1.59	-1.25	-1.08	-1.01	-1

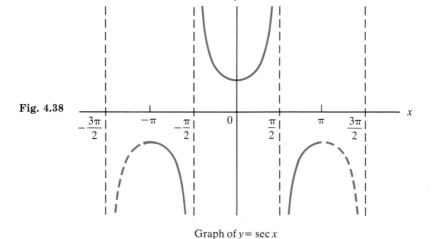

Fig. 4.38

Graph of $y = \sec x$

We now plot the points given in this table and draw the curve corresponding to x between 0 and π; then from symmetry about the y-axis we draw the curve for x between 0 and $-\pi$. This gives us the solid portion of the curve shown in Fig. 4.38. The remainder of the curve (broken portion) can now be drawn by using the identity $\sec(x + 2\pi) = \sec x$.

From Fig. 4.38 we conclude the following.

The secant function is periodic with period 2π. The domain and range are given by

$$\mathcal{D}(\sec) = \{x \mid x \neq \frac{\pi}{2} + k\pi, \quad k \text{ is an integer}\},$$

$$\mathcal{R}(\sec) = \{y \mid y \leq -1 \quad \text{or} \quad y \geq 1\}.$$

Note that the vertical lines given by $x = (2k + 1)\pi/2$, where k is any integer, are vertical asymptotes of the secant curve.

Graph of the Cosecant Function

Following a procedure similar to that used to draw a graph of the secant function, we get the graph of $y = \csc x$, shown in Fig. 4.39.

From the graph in Fig. 4.39, we conclude the following.

Fig. 4.39

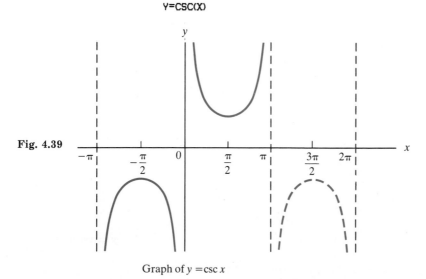

Graph of $y = \csc x$

The cosecant function is periodic with period 2π. The domain and range are given by

$$\mathcal{D}(\csc) = \{x \mid x \neq k\pi, \quad k \text{ is an integer}\},$$
$$\mathcal{R}(\csc) = \{y \mid y \leq -1 \quad \text{or} \quad y \geq 1\}.$$

Note that $y = \csc x$ has infinitely many vertical asymptotes, given by $x = k\pi$, where k is any integer.

Exercises 4.5

In problems 1 through 8, use Definitions 4.3 and 4.4 or identities derived in this section to derive the given identities.

1. a) $\sin(\pi - s) = \sin s$ 　　　　　　　　　　**b)** $\cos(\pi - s) = -\cos s$

2. a) $\sin\left(\dfrac{\pi}{2} - s\right) = \cos s$ 　　　　　　**b)** $\cos\left(\dfrac{\pi}{2} - s\right) = \sin s$

3. a) $\sin\left(\dfrac{3\pi}{2} + s\right) = -\cos s$ 　　　　**b)** $\cos\left(\dfrac{3\pi}{2} + s\right) = \sin s$

4. a) $\sin\left(\dfrac{3\pi}{2} - s\right) = -\cos s$ 　　　　**b)** $\cos\left(\dfrac{3\pi}{2} - s\right) = -\sin s$

5. a) $\tan(\pi + s) = \tan s$ 　　　　　　　　**b)** $\cot(\pi + s) = \cot s$

6. a) $\sec(\pi + s) = -\sec s$ 　　　　　　　**b)** $\csc(\pi + s) = -\csc s$

7. a) $\tan\left(\dfrac{\pi}{2} + s\right) = -\cot s$ 　　　　　**b)** $\cot\left(\dfrac{\pi}{2} + s\right) = -\tan s$

8. a) $\sec\left(\dfrac{\pi}{2} + s\right) = -\csc s$ 　　　　　**b)** $\csc\left(\dfrac{\pi}{2} + s\right) = \sec s$

9. Draw a graph of the sine function by first making a table of x, y values that satisfy the equation $y = \sin x$; plot these points, and then draw the curve. Use the identities derived in this section, such as $\sin(x + 2\pi) = \sin x$ and $\sin(-x) = -\sin x$, to convince yourself that it is sufficient to include in the table values of x in $0 \leq x \leq \pi$. For values of x, use $0, 0.25, 0.50, 0.75, \ldots$.

10. Follow instructions similar to those in problem 9 for $y = \cos x$.

11. Draw a graph of $y = \tan x$ by considering $y = \sin x / \cos x$ and using information from the graphs of $y = \sin x$ and $y = \cos x$ in Figs. 4.34 and 4.35. Determine the vertical asymptotes by noting the values of x for which $\cos x = 0$.

In problems 12 through 14, follow instructions similar to those given in problem 11 to draw a graph of the given function.

12. $y = \cot x = \dfrac{\cos x}{\sin x}$ 　　　　**13.** $y = \sec x = \dfrac{1}{\cos x}$ 　　　　**14.** $y = \csc x = \dfrac{1}{\sin x}$

15. If $f(x) = 1 - \sin x$, is $f(\pi - x) = f(x)$ for every $x \in \mathbf{R}$?

16. If $f(x) = \sin x + \cos x$, is $f(\pi + x) = -f(x)$?

17. Given that $g(x) = \sqrt{(\sin x)^2 - 1}$, find $\mathfrak{D}(g)$.

18. Given that $g(x) = \sqrt{(\cos x)^2 - 1}$, find $\mathfrak{D}(g)$.

19. If $f(x) = (\sin x)^2 + (\cos x)^2$, what is the range of f?

20. If $f(x) = x \sin x$, is f an odd function? An even function?

4.6 SOLVING RIGHT TRIANGLES

As noted earlier, the word trigonometry implies the study of measurements related to triangles. The historical development of the subject was indeed motivated by practical needs in areas such as surveying, navigation, and architecture.

Let us first describe a situation that involves triangles in its solution. Suppose we wish to determine the height of a mountain peak, and there is no convenient way to measure it directly. One approach is to locate two points A and B on the ground, as shown in Fig. 4.40, and measure the distance between them. With surveying instruments, angles α and β can be measured. With this much information we can determine the height h by using trigonometric properties of right triangles (see problem 24).

A triangle has six parts: three angles and three sides. When we say "angle of a triangle," we mean the angle formed by the two rays that contain two sides of the triangle and have the vertex as their common end point. To "solve a triangle" means that measurements of some of these parts are given (usually sufficient to describe a unique triangle), and the remaining parts are to be determined from the given information. In this section problems involving right triangles only are considered. Solving general triangles is discussed in the following section.

Standard labeling of parts of a right triangle is shown in Fig. 4.41, in which the right angle is at vertex C, side a is opposite angle α, side b is opposite angle β, and the hypotenuse is denoted by c. Note that a letter is used to refer to a part of the triangle or to its measure; for instance, b denotes the side AC or the length of side AC.

Since the sum of the angles of any triangle is $180°$, angles α and β of a right triangle are acute angles. It will be convenient to state definitions of trigonometric functions of these angles by referring to the sides of the triangle.

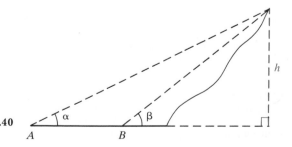

Fig. 4.40

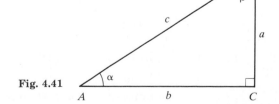

Fig. 4.41

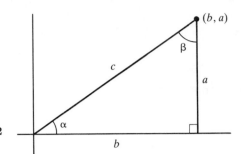

Fig. 4.42

Suppose a coordinate system is chosen such that right triangle ABC is situated as shown in Fig. 4.42, with angle α in standard position. Point $B:(b, a)$ is on the terminal side of angle α; applying Definition 4.2 gives

$$\sin \alpha = \frac{a}{c}, \qquad \cos \alpha = \frac{b}{c}, \qquad \tan \alpha = \frac{a}{b},$$

$$\cot \alpha = \frac{b}{a}, \qquad \sec \alpha = \frac{c}{b}, \qquad \csc \alpha = \frac{c}{a}. \tag{4.18}$$

In order to avoid the need to place an angle of a right triangle in standard position to determine its trigonometric function values, let us introduce the following notation: $\text{opp}(\alpha)$, $\text{adj}(\alpha)$, and hyp will denote the side opposite α, the side adjacent to α, and the hypotenuse, respectively. The results given in (4.18) can be written as

$$\sin \alpha = \frac{\text{opp}(\alpha)}{\text{hyp}}, \qquad \cos \alpha = \frac{\text{adj}(\alpha)}{\text{hyp}}, \qquad \tan \alpha = \frac{\text{opp}(\alpha)}{\text{adj}(\alpha)},$$

$$\cot \alpha = \frac{\text{adj}(\alpha)}{\text{opp}(\alpha)}, \qquad \sec \alpha = \frac{\text{hyp}}{\text{adj}(\alpha)}, \qquad \csc \alpha = \frac{\text{hyp}}{\text{opp}(\alpha)}. \tag{4.19}$$

In a similar manner, the trigonometric function values for angle β can be stated in terms of the sides of the triangle. For instance,

$$\sin \beta = \frac{\text{opp}(\beta)}{\text{hyp}} = \frac{b}{c} \qquad \text{and} \qquad \cos \beta = \frac{\text{adj}(\beta)}{\text{hyp}} = \frac{a}{c}.$$

In the following examples, a calculator will be useful for numerical computations.

Example 1   In a right triangle $a = 32.4$ cm, $\alpha = 40°$. Find b, c, and β.

Solution   Draw a right triangle and denote the given parts as shown in Fig. 4.43. To determine side b the first step is to look for an equation that involves b and the given

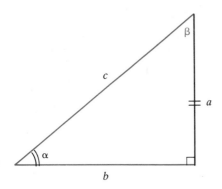

Fig. 4.43

parts. We could use either $\tan \alpha = a/b$, which gives $b = a/\tan \alpha$, or $\cot \alpha = b/a$ to get $b = a \cot \alpha$. Since the calculator does not have a key labeled cot, we choose the first of these:

$$b = a/\tan \alpha = 32.4/\tan 40° = 38.6 \text{ cm.}$$

To determine the hypotenuse c, we could use any of three equations: $\sin \alpha = a/c$; $\csc \alpha = c/a$; $c = \sqrt{a^2 + b^2}$. In general, it is a good practice to use a relationship that involves only the given parts, if possible. That is, the third option has a slight disadvantage because we might make an error in solving for b. The second has the disadvantage of involving $\csc \alpha$, and our calculator does not have a csc key. Therefore, we decide on the first:

$$c = a/\sin \alpha = 32.4/\sin 40° = 50.4 \text{ cm.}$$

To determine β, we know from geometry that the sum of the measures of the three angles of a triangle is $180°$; that is, $\alpha + \beta + 90° = 180°$. Therefore $\beta = 180° - 90° - \alpha = 90° - 40° = 50°$. ■

Example 2   Given $c = 16.25$ cm and $\beta = 68°24'$. Find the area of the triangle.

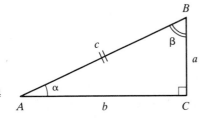

Fig. 4.44

Solution   Area $= (1/2)ab$, so let us first find sides a and b. From $\sin \beta = b/c$ we get $b = c \sin \beta$, and from $\cos \beta = a/c$ we get $a = c \cos \beta$. Hence

$$\text{Area} = (1/2)(c \cos \beta)(c \sin \beta).$$

Substituting the given information and evaluating gives

$$\text{Area} = \frac{c^2 \sin \beta \cos \beta}{2} = \frac{(16.25)^2(\sin 68°24' \cos 68°24')}{2} = 45.19 \text{ cm}^2. \quad ■$$

Example 3   Given that $a = 37.4$ cm, $b = 63.3$ cm, find c, α, and β.

Solution $c = \sqrt{a^2 + b^2} = \sqrt{(37.4)^2 + (63.3)^2} = 73.5$ cm.

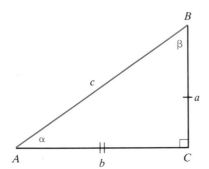

Fig. 4.45

For angle α we use $\tan \alpha = a/b = 37.4/63.3$, or $\tan \alpha = 0.59084$. We are now confronted with the problem of finding α when we know $\tan \alpha$. This is the inverse of the problem of finding $\tan \alpha$ when α is given. The subject of inverse trigonometric functions will be discussed formally in Chapter 5, and here we merely point out that scientific calculators can be used to find an angle corresponding to a given value of a trigonometric function. The calculator keys that correspond to inverse functions are usually labeled $\boxed{\sin^{-1}}$, $\boxed{\cos^{-1}}$, $\boxed{\tan^{-1}}$, or there is an $\boxed{\text{INV}}$ key, which is to be followed by the appropriate $\boxed{\sin}$, $\boxed{\cos}$, $\boxed{\tan}$ key. This is illustrated by the above problem in which $\tan \alpha = 0.59084$ and we wish to determine α.

If the calculator has an $\boxed{\text{INV}}$ key, enter the number 0.59084 into the display, and with calculator in degree mode, press the $\boxed{\text{INV}}$ and $\boxed{\tan}$ keys in that order. The display will read 30.5763° (to four decimal places).

If the calculator has a $\boxed{\tan^{-1}}$ key, then with 0.59084 in the display and with the calculator in degree mode, press $\boxed{\tan^{-1}}$. The display will read 30.5763°. Thus $\alpha = 30.5763° = 30°35'$. To find β, use $\beta = 90° - \alpha$, and so $\beta = 59°25'$.   ■

Example 4   Given that α is an acute angle and

a) $\sin \alpha = 0.4835$, find α in degrees correct to two decimal places.

b) $\cos \alpha = 0.6897$, find α in radians correct to three decimal places.

Solution   **a)** Place the calculator in degree mode, enter the number 0.4835 into the display, and then press $\boxed{\text{INV}}$ and $\boxed{\sin}$ (or $\boxed{\sin^{-1}}$). The display will show 28.91°. Thus $\alpha = 28.91°$.

b) Place the calculator in radian mode, enter the number 0.6897 into the display, and then press (INV) and (cos) (or (cos⁻¹)). The display will show 0.810. That is, $\alpha = 0.810$ radians. ■

In certain applications it is necessary to use angles measured from a horizontal line of sight. An angle formed by a horizontal ray and an observer's line of sight to an object above the horizontal is called the *angle of elevation*. If the object is below the horizontal, then the angle between the horizontal and the line of sight is called the *angle of depression*. These terms are illustrated in Fig. 4.46.

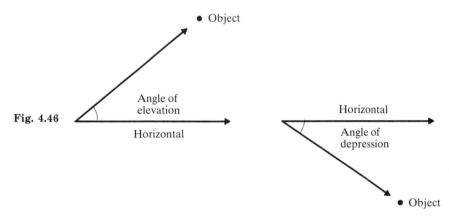

Fig. 4.46

Example 5   From a window 25 meters above the ground, the angle of elevation to the top of a nearby building is 24°20′, and the angle of depression to the bottom of the building is 14°40′. Find the height of the building.

Solution   In Fig. 4.47 we wish to find h, which is equal to $\overline{BC} + \overline{CD}$. We know $\overline{CD} = \overline{AE} = 25$ m, so $h = \overline{BC} + 25$ m. From triangle ACD we have $\overline{AC} = \overline{CD}/\tan 14°40′ = 25/\tan 14°40′$. Using triangle ABC, we have $\overline{BC} = \overline{AC} \tan 24°20′$. Thus

$$h = 25 + \frac{25 \tan 24°20′}{\tan 14°40′} = 68.20 \text{ m}.$$

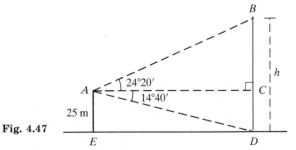

Fig. 4.47

■

Accuracy of Measurements

Note that in Example 3 angle β was determined to the nearest minute. This was done primarily to illustrate the technique for getting such accuracy. In applied work the degree of accuracy of computed values as well as measured values will depend on several factors and on the way the results will ultimately be used. It is pointless to calculate the height of a mountain peak in meters to four decimal places and use such a number on a map.

In practical applications involving measurements of angles and lengths, along with the associated computations, one of the first questions that must be resolved is: *What degree of accuracy should be used?* Naturally, the answer depends on the particular problem and the way the results will subsequently be applied. We cannot expect computed values to be reliable to more significant digits than the starting data, which in applications are usually physical measurements.

The accuracy of computation in problems that involve approximate numbers is discussed in Appendix C. It should be understood that the rules stated there form a practical guide to be used in applied problems. *In this text, as in most mathematics texts, no effort is made to be completely consistent with these rules.* Most of our problems are mathematical in nature, and our primary goal is to provide the student with examples that will lead to a better understanding of the basic mathematical concepts being discussed. Thus, in most of the problems that involve computations, the reader is asked to find a result correct to a given number of decimal places or to a given number of significant digits. Furthermore, in many problems we say, for example, that the length of a side of a triangle is 24.3, and we do not even specify the units. In practical applications, such as in physics, chemistry, or engineering, the units will be specified, and there should be no problem in following the rules given in Appendix C, which govern computations with approximate numbers.

Exercises 4.6

Unless otherwise specified, supply answers involving lengths in the given units correct to two decimal places, angles in degrees and minutes correct to the nearest minute, and areas rounded off to the nearest whole number.

In problems 1 through 10, standard notation as described in this section is used to denote sides and angles of right triangles.

1. $\alpha = 35°24'$; $a = 3.27$ cm; find b, c, β.
2. $a = 56$ cm, $b = 33$ cm; find c, α, β.
3. $a = 175$ cm, $c = 337$ cm; find b, α, β.
4. $\beta = 65.72°$, $a = 32.5$ m; find b, c, α, and the area of the triangle.
5. $\alpha = 27°17'$, $c = 56.5$ cm; find a, b, β, and the area of the triangle.
6. $b = 2730$ m, $c = 4666$ m; find a, α, β.
7. $a = 24208$ m, $b = 10575$ m; find c, α, β.

8. $\beta = 42°30'$, $b = 3.25$ cm; find a, c, α.

9. $b = 73.56$ cm, $c = 131.42$ cm; find a, α, β, and the area of the triangle.

10. $\alpha = 37.43°$, $c = 64.56$ cm; find a, b, β, and the area of the triangle.

11. A line passes through points $(5, 2)$ and $(8, 15)$. Find the acute angle at which it intersects the x-axis.

12. Find the area in cm^2 correct to two decimal places of an equilateral triangle having a side of length 12.56 cm.

13. Find the area in m^2 correct to two decimal places of an isosceles triangle that has two sides of length 2.47 m; the angle opposite one of them is $41°37'$.

14. The lengths of sides of a parallelogram are 38.4 cm and 64.8 cm, and an interior angle is 115.65°. Find the area in cm^2 correct to one decimal place of the parallelogram.

15. A regular polygon is inscribed in a circle of radius 57 cm. Find the area in cm^2 correct to one decimal place of the polygon of the following description.

a) four sides (a square)   **b)** six sides (a hexagon)

c) eight sides (an octagon)   **d)** n sides

16. You wish to fence a triangular piece of land with dimensions given by $a = 236$ m and $\alpha = 70°$ (see Fig. 4.48). Find the total amount of fencing you must purchase.

Fig. 4.48

17. You wish to mount an antenna and have purchased a tower that is 12.48 meters tall. The tower is to be anchored from the top by three guy wires, each of which is to be 7.36 meters from the base (Fig. 4.49). How much guy wire do you need?

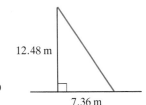

Fig. 4.49

18. If the elevation of the sun is 17.48° at 5 P.M. on December 21, the shortest day of the year, how far east of a retaining wall 5.48 meters tall should one locate plants requiring year-round full sun?

19. The distance from the base to the top of the Leaning Tower of Pisa is 54.6 m, and it makes an angle of $84°45'$ with the horizontal. How far does the top overhang the base?

20. In Fig. 4.50 line segment AB is a diameter of the circle of radius 24 cm, and C is a point on the circle with length of arc $\widehat{AC}$ equal to 27.3 cm. Find the length of chord AC.

Hint: Let θ be the central angle shown in Fig. 4.50; use definition of radian measure to find θ. Recall facts from geometry about measures of central and inscribed angles in a circle.

21. A segment of a circle of radius 4.56 cm is shown as the shaded region between chord AB and arc $\widehat{AB}$ (Fig. 4.51). If the central angle θ is 1.15 radians, what is the area of the segment?

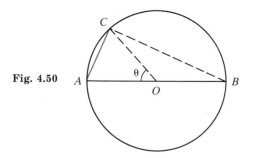
Fig. 4.50

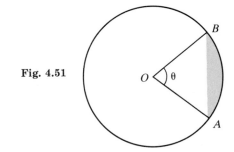
Fig. 4.51

22. A surveyor starts at point A in Fig. 4.52 and measures $\overline{AB} = 41.32$ m, $\overline{BC} = 37.53$ m, $\theta = 137.44°$. Find the distance from A to C and angle α.

23. A triangular piece of land is bounded by two farm roads that intersect at right angles and a highway that intersects one of the roads at an angle of 24.5°, as shown in Fig. 4.53. You wish to purchase the property and know that the previous owner required 843 meters of fencing to enclose it. Land sells at \$2.50 per square meter in this region. How much does the property cost?

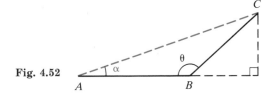
Fig. 4.52

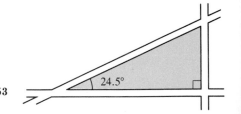
Fig. 4.53

24. A surveyor wishes to determine the height h of a mountaintop above the horizontal ground, as shown in Fig. 4.54. He observes the angles of elevation from two points A and B on the ground and in line with the mountaintop, and he measures the distance from A to B. These measurements are: $\alpha = 43°30'$, $\beta = 32°20'$, and $\overline{AB} = 256$ m. Find the height of the mountaintop above the horizontal ground level. Give answer to the nearest meter.

25. From point A, which is 8.1 meters above the horizontal level of the ground, the angle of elevation of the top of a tower CB is $\alpha = 32°30'$, and the angle of depression of the base is $\beta = 16°40'$ (see Fig. 4.55). Find the height $\overline{CB}$ of the tower. Give the answer in meters correct to one decimal place.

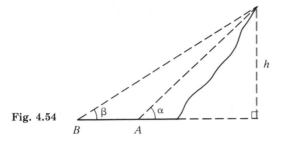

Fig. 4.54

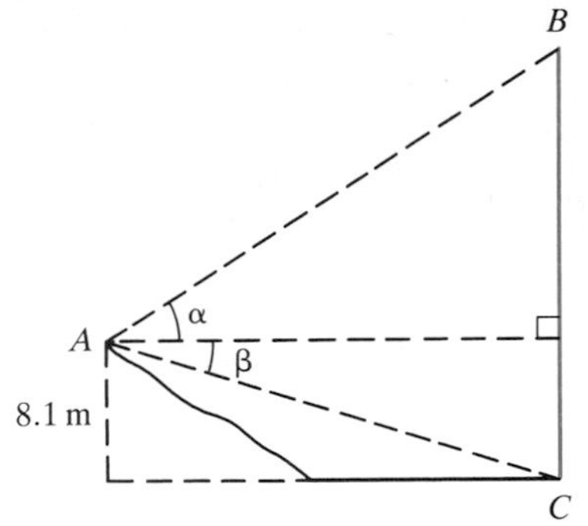

Fig. 4.55

26. A sector with central angle 72° is cut out of a circular piece of tin of radius 16.48 cm. The edges of the remaining piece are joined together to form a cone. Find the volume of the cone. See inside front cover for a formula giving volume of a cone.

27. Suppose A, B, and C are vertices of a right triangle, and α is the measure of the angle at A, as shown in Fig. 4.56. Also suppose the length of AB is 1. Extend side CA to point D such that the length of AD is also 1.

a) Show that the angle CDB is equal to $\alpha/2$.

b) Use right triangle BCD to find $\tan \alpha/2$. Specifically, show that it can be expressed in the form $\tan \alpha/2 = \sin \alpha/(1 + \cos \alpha)$. This is a useful identity, which will be seen again in Chapter 5.

28. In problem 24 of Exercises 4.2 Mot'l's treadle sewing machine was described (see Fig. 4.57). The radii of the two wheels are $r_1 = 3.5$ cm and $r_2 = 15.5$ cm. The distance between the centers is $\overline{EF} = 56$ cm. Find the length of the belt that goes around the two wheels. In the diagram, E and F are centers of the wheels, points A, B, C, and D are points at which the belt is tangent to the respective wheels, and line BG is parallel to EF. Give answer in centimeters correct to one decimal place.

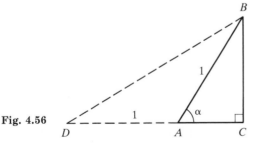

Fig. 4.56

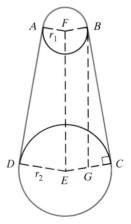

Fig. 4.57

29. A right triangle is inscribed in a circle of radius 5.6 cm. One angle of the triangle is 64°. Find the lengths of the two sides.

30. The area of a right triangle is 6.73 cm², and one of its angles is 36°. Find the length of the hypotenuse.

31. The perimeter of a right triangle is 8.56 m, and one of its angles is 23°30′. Find the lengths of the two sides.

32. One angle of a right triangle is 47°30′, and its perimeter is 15.48 cm. Determine the area of the triangle.

4.7   **LAW OF COSINES AND LAW OF SINES**

Techniques used in the preceding section apply to the solution of right triangles. We now consider the general case, in which triangles are not necessarily right triangles. Although it is true that solving a general triangle can be reduced to problems involving right triangles, it is desirable to have formulas that can be applied directly.

Suppose A, B, and C are vertices of a triangle, as shown in Fig. 4.58. Greek letters α, β, and γ are used to denote the three angles and a, b, c to represent the three sides. As indicated in Fig. 4.58, angle α has vertex at A and side a is opposite α; likewise for B, β, b and C, γ, c. Thus a triangle has six parts: three angles and three sides. In general, three given parts, at least one of which is a side, is sufficient to describe a specific triangle, and our problem is to determine the remaining three parts. First we develop two sets of formulas; these are called the *Law of cosines* and the *Law of sines*.

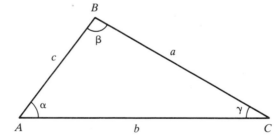

Fig. 4.58

Law of Cosines

Figure 4.59 shows triangle ABC in which D is the base of the altitude from vertex A.

Let $h = \overline{AD}$ and $x = \overline{CD}$. From right triangle ADC we get $x = b \cos \gamma$ and $h = b \sin \gamma$. Applying the Pythagorean theorem to right triangle ADB, we have

$$c^2 = h^2 + (a - x)^2 = h^2 + a^2 - 2ax + x^2.$$

Substituting $x = b \cos \gamma$ and $h = b \sin \gamma$ gives

$$c^2 = (b \sin \gamma)^2 + a^2 - 2a(b \cos \gamma) + (b \cos \gamma)^2$$
$$= a^2 + b^2[(\sin \gamma)^2 + (\cos \gamma)^2] - 2\,ab \cos \gamma$$
$$= a^2 + b^2 - 2\,ab \cos \gamma,$$

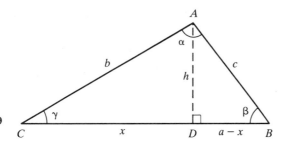

Fig. 4.59

where in the last step the identity given in Eq. (4.16) was used to replace $(\sin \gamma)^2 + (\cos \gamma)^2$ by 1. Thus we have*

$$c^2 = a^2 + b^2 - 2ab \cos \gamma.$$

In a similar manner we can develop analogous formulas for a^2 and b^2. The set of three equations, listed in (4.20), is called the *Law of cosines* for triangle *ABC*.

$$a^2 = b^2 + c^2 - 2bc \cos \alpha,$$
$$b^2 = a^2 + c^2 - 2ac \cos \beta, \qquad (4.20)$$
$$c^2 = a^2 + b^2 - 2ab \cos \gamma.$$

The technique used to solve a triangle depends upon the given information. Problems can be classified into the following four cases, in which the given three parts are:

1. Two sides and the included angle;

2. Three sides;

3. Two sides and an angle opposite one of them;

4. One side and two angles.

The Law of cosines is particularly suitable for solving triangles described by cases 1 and 2, but the Law of sines is better suited for case 4. Case 3 presents a special problem in that it is possible for the given information to describe either one triangle, two triangles, or no triangle. For this reason, case 3 is usually referred to as the "ambiguous case." We shall illustrate through example how this case can be handled by using the Law of cosines involving solution of a quadratic equation; due to calculators, this problem presents no special difficulty in computation of answers.

---

*   In the derivation of this formula, the diagram used shows γ as an acute angle. Actually the final result holds if γ is any angle between 0° and 180°. See problem 16 of Exercises 4.7.

Example 1   *Given two sides and the included angle.* Suppose $a = 33.24$, $b = 47.37$, and $\gamma = 38°15'$. Find c, α, and β.

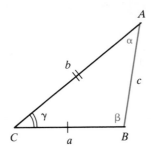

Fig. 4.60

Solution   To find c, apply the third equation of (4.20) to get

$$c^2 = (33.24)^2 + (47.37)^2 - 2(33.24)(47.37)\cos 38°15'.$$

Using a calculator to evaluate the right-hand side and then pressing the $\boxed{\sqrt{x}}$ key gives $c = 29.59$.

Suggestion: In order to get maximum calculator accuracy, store the full decimal value of c in the calculator, and then use that value in subsequent computations in determing α and β.

To determine α, use the first equation of (4.20) in the form

$$\cos \alpha = \frac{b^2 + c^2 - a^2}{2bc}.$$

Evaluating this expression gives $\alpha = 44.0589° = 44°04'$.

In a similar manner the second equation of (4.20) can be used to get $\beta = 97°41'$. We could have determined β by using $\beta = 180° - (\alpha + \gamma)$, but we prefer to use this as a check on our computations. That is, we see that

$$\alpha + \beta + \gamma = 44°04' + 97°41' + 38°15' = 180°. \qquad \blacksquare$$

Example 2   *Given three sides.* Suppose $a = 56.84$, $b = 83.45$, and $c = 51.63$. Find angles α, β, and γ.

Solution   Apply the first equation of (4.20) to get

$$\cos \alpha = \frac{b^2 + c^2 - a^2}{2bc} = \frac{(83.45)^2 + (51.63)^2 - (56.84)^2}{2(83.45)(51.63)}.$$

This gives $\alpha = 42.0491° = 42°03'$. In a similar manner the second and third equations of (4.20) give $\beta = 100°29'$ and $\gamma = 37°28'$.

As a check, adding the computed values of α, β, γ gives

$$\alpha + \beta + \gamma = 42°03' + 100°29' + 37°28' = 180°. \qquad \blacksquare$$

Example 3   *Given two sides and an angle opposite one of them.* Suppose $a = 17.48$, $b = 25.63$, and $\alpha = 37°48'$. Find c, β, and γ.

Solution   If the given values of a, b, and α are substituted into the first equation of (4.20),

$$a^2 = b^2 + c^2 - 2bc \cos \alpha,$$

the result is a quadratic equation in c,

$$c^2 - (2b \cos \alpha)c + (b^2 - a^2) = 0.*$$

Applying the quadratic formula gives

$$c = [2b \cos \alpha \pm \sqrt{(-2b \cos \alpha)^2 - 4(b^2 - a^2)}] \div 2$$
$$= b \cos \alpha \pm \sqrt{a^2 - b^2[1 - (\cos \alpha)^2]}.$$

From Eq. (4.16), $1 - (\cos \alpha)^2$ is identically equal to $(\sin \alpha)^2$, and so

$$c = b \cos \alpha \pm \sqrt{a^2 - (b \sin \alpha)^2}. \tag{4.21}$$

Substituting the given values of a, b, and α into Eq. (4.21) gives

$$c = 25.63 \cos 37°48' \pm \sqrt{(17.48)^2 - (25.63 \sin 37°48')^2}.$$

We can now evaluate this result by calculator. To avoid recording any intermediate computations, first evaluate the square root part, store it by using the (STO) key, and recall it when needed by using the (RCL) key.[†] Thus we get two answers: $c_1 = 27.91873$ and $c_2 = 12.58462$. In order to be consistent with the given data, we round off to two decimal places: $c_1 = 27.92$ and $c_2 = 12.58$.

In this example we see that there are two solutions; these are illustrated in Fig. 4.61. The second triangle in Fig. 4.61 is obtained from the first by rotating side a about the top vertex, as indicated in the diagram.

Fig. 4.61

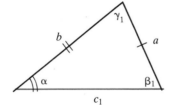 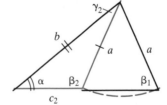

---

* Substituting the values of a, b, and α at this point gives
$$c^2 - [2(25.63)\cos 37°48']c + (25.63^2 - 17.48^2) = 0.$$
That is, $c^2 - 40.5033\,c + 351.3465 = 0$. This is the quadratic equation that determines c, but it is not necessary to record the intermediate numbers that appear as the coefficient of c and the constant term. It is simpler to solve the quadratic equation for the general case and then substitute the values of a, b, and α into the final result, shown in Eq. (4.21).

† The (STO) and (RCL) keys may be labeled differently on some calculators (see Appendix A or owner's manual).

Angles β_1 and γ_1 can now be found by using

$$\cos \beta_1 = \frac{a^2 + c_1^2 - b^2}{2ac_1} \quad \text{and} \quad \cos \gamma_1 = \frac{a^2 + b^2 - c_1^2}{2ab}.$$

These give $\beta_1 = 63°59'$ and $\gamma_1 = 78°13'$.

As a check, we note that $\alpha + \beta_1 + \gamma_1 = 37°48' + 63°59' + 78°13' = 180°$. To find β_2 and γ_2, note that

$$\beta_2 = 180° - \beta_1 = 180° - (63°59') = 116°01',$$
$$\gamma_2 = 180° - (\alpha + \beta_2) = 26°11'.$$

The Ambiguous Case

In Example 3 two sides and an angle opposite one of them were given; the third side was determined by solving a quadratic equation, and two solutions, given by Eq. (4.21), were found. A geometrical interpretation of Eq. (4.21) can be given by considering the diagram shown in Fig. 4.62. From right triangle ACD, we get $\overline{CD} = b \sin \alpha$ and $\overline{AD} = b \cos \alpha$. Applying the Pythagorean theorem to right triangle BCD gives

$$\overline{DB} = \sqrt{a^2 - (\overline{CD})^2} = \sqrt{a^2 - (b \sin \alpha)^2}.$$

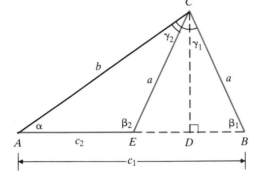

Fig. 4.62

Therefore $c_1 = \overline{AD} + \overline{DB}$ and $c_2 = \overline{AD} - \overline{ED} = \overline{AD} - \overline{DB}$, and so

$$c_1 = b \cos \alpha + \sqrt{a^2 - (b \sin \alpha)^2},$$
$$c_2 = b \cos \alpha - \sqrt{a^2 - (b \sin \alpha)^2}. \tag{4.22}$$

Evaluating c_1 and c_2 in Example 3 resulted in two positive numbers, giving two solutions. In general, however, any one of the following possibilities might occur:

1. c_1 and c_2 are positive numbers, giving two solutions;
2. $c_1 = c_2$, which implies $a^2 - (b \sin \alpha)^2 = 0$, and so $a = b \sin \alpha$, which tells us that ABC is a right triangle;
3. c_1 is positive and c_2 is negative, giving one solution since a side of a triangle cannot have negative length;
4. c_1 and c_2 are complex nonreal numbers, giving no solutions; in this case $a < b \sin \alpha$, which tells us that the given side a is not long enough to reach side c when drawn from vertex C.

Eqs. (4.22) can be applied when sides a, b, and angle α are given. If two other sides and an angle opposite one of them are given, it is a simple matter to write the equations corresponding to those given in (4.22). In any problem, we suggest that the reader draw a diagram similar to that in Fig. 4.62 and then derive corresponding formulas from geometrical considerations similar to those leading to Eqs. (4.22).

Law of Sines
Figure 4.63 shows triangle ABC, in which D is the foot of the altitude h from vertex B.

Fig. 4.63

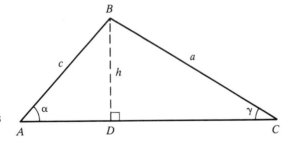

From right triangle ADB, $h = c \sin \alpha$; and from right triangle BDC, $h = a \sin \gamma$. Therefore $c \sin \alpha = a \sin \gamma$. Dividing both sides by ac, we can write this equation as

$$\frac{\sin \alpha}{a} = \frac{\sin \gamma}{c}. \tag{4.23}$$

In a similar manner (see problem 32) we can show that

$$\frac{\sin \alpha}{a} = \frac{\sin \beta}{b}. \tag{4.24}$$

Equations (4.23) and (4.24) can be written in compact form to give the *Law of sines:*

$$\frac{\sin \alpha}{a} = \frac{\sin \beta}{b} = \frac{\sin \gamma}{c}. \tag{4.25}$$

The derivation of Eq. (4.25) was based on Fig. 4.63, in which angles α and γ are acute. The Law of sines is still valid if one of the angles is obtuse (see problem 32).

Example 4   *Given one side and two angles.* Suppose $b = 5.834$, $\alpha = 64°12'$, and $\gamma = 47°47'$. Find a, c, and β.

Solution   To find β, use $\beta = 180° - (\alpha + \gamma)$ and get $\beta = 68°01'$. To determine a, apply the Law of sines in the form

$$a = \frac{b \sin \alpha}{\sin \beta} = \frac{5.834 \sin 64°12'}{\sin 68°01'}.$$

This gives $a = 5.664$. Similarly,

$$c = \frac{b \sin \gamma}{\sin \beta} = \frac{5.834 \sin 47°47'}{\sin 68°01'} = 4.659.$$

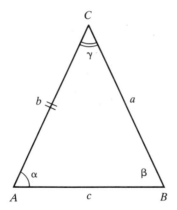

Fig. 4.64

Area of a Triangle

The area K of a triangle is given by the familiar formula from geometry,

$$K = \frac{1}{2}(\text{Base}) \cdot (\text{Altitude}). \tag{4.26}$$

Instead of deriving various formulas for K, depending on the given information, we suggest that in each case you draw a diagram showing the given parts; decide on one of the sides as the base, determine the corresponding altitude, and then use Eq. (4.26). This is illustrated by the following example, in which three sides of a triangle are given.

Example 5   Suppose $a = 34$, $b = 48$, and $c = 28$. Find the area of triangle ABC.

Solution   Triangle ABC is shown in Fig. 4.65. Suppose we take b to be the base and h the

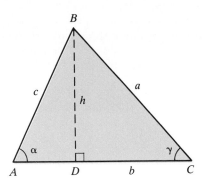

Fig. 4.65

corresponding altitude. From right triangle BCD, $h = a \sin \gamma$, where γ can be determined by using the Law of cosines. The area K is given by

$$K = \frac{1}{2}b \cdot h = \frac{1}{2}b(a \sin \gamma) = \frac{1}{2}(48)(34) \sin \gamma, \qquad (4.27)$$

where

$$\cos \gamma = \frac{a^2 + b^2 - c^2}{2ab} = \frac{34^2 + 48^2 - 28^2}{2(34)(48)}. \qquad (4.28)$$

Using a calculator, we first evaluate the right-hand side of Eq. (4.28) and then, with the result in the display, press [INV] and [cos] (or [cos⁻¹]) to get γ in the display; we then continue with Eq. (4.27) to find K. The result is $K = 467.22$ (to two decimal places). ■

In Example 5 we illustrated how the area of a triangle can be determined when the three sides are known. It is a good exercise in algebra to follow a similar pattern for the general case to get the following formula for K in terms of a, b, and c,

$$K = \sqrt{s(s-a)(s-b)(s-c)},^* \qquad (4.29)$$

where s is the semiperimeter

$$s = \frac{1}{2}(a + b + c).$$

This formula can be applied to the problem in Example 5 to check the solution there.

---

* The formula given in Eq. (4.29) is called Heron's Formula in honor of the famous Greek philosopher–mathematician Heron of Alexandria (A.D. 75).

Exercises 4.7

In problems 1 through 15, use the given data to find the remaining three parts of the triangle. Give answers involving length rounded off to the same number of significant digits* as the given data, and give angles correct to the nearest minute.

1. $a = 36$, $b = 67$, $\gamma = 43°$
2. $a = 24$, $b = 73$, $\gamma = 130°$
3. $a = 85$, $c = 42$, $\beta = 83°24'$
4. $a = 41.32$, $b = 57.56$, $\gamma = 61°12'$
5. $a = 2.48$, $b = 1.75$, $\alpha = 124°$
6. $a = 17$, $b = 45$, $c = 50$
7. $a = 31.5$, $b = 63.4$, $c = 41.6$
8. $a = 17$, $b = 25$, $\alpha = 37°$
9. $\alpha = 27°$, $\beta = 73°$, $a = 16$
10. $\beta = 67°$, $\gamma = 26°$, $a = 463$
11. $\alpha = 47°$, $\gamma = 112°$, $c = 81$
12. $\beta = 61°47'$, $\gamma = 82°15'$, $b = 63.54$
13. $a = 2730$, $c = 4666$, $\alpha = 32°$
14. $a = 47.3$, $b = 32.5$, $c = 40.5$
15. $\alpha = 73.46°$, $\beta = 23.75°$, $c = 4.875$

16. In this section the derivation of the Law of cosines was based on Fig. 4.59, in which angle γ was acute. Suppose γ is obtuse, as shown in Fig. 4.66. Derive the Law of cosines for this case. That is, show that $c^2 = a^2 + b^2 - 2ab \cos \gamma$.

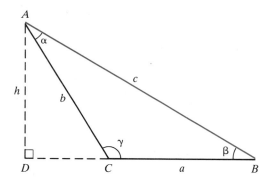

Fig. 4.66

17. If triangle ABC is a right triangle with $\gamma = 90°$, show that the third equation given in (4.20) reduces to the Pythagorean theorem.

18. A ship sails due east from point A for a distance of 48.6 km; then it changes direction by an angle of $16°40'$ toward the south, as shown in Fig. 4.67. After the ship sails 37.8 km in the new direction, how far is it from point A?

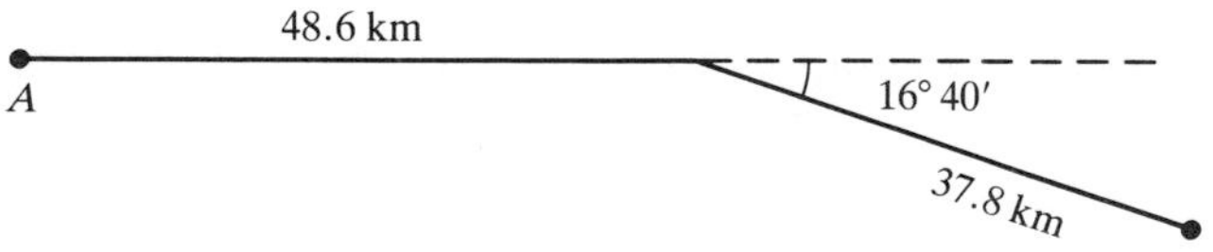

Fig. 4.67

48.6 km

A

16° 40'

37.8 km

---

*   See Appendix C for a discussion of significant digits.

19. A surveyor wants to find the distance from point A to a point C on the opposite side of the river. He locates a point B on his side of the river, measures the distance $\overline{AB}$ and the two angles α and β, as shown in Fig. 4.68. The measurements are $\overline{AB} = 132.4$ m, $\alpha = 78°$, $\beta = 53°$. Find the distance $\overline{AC}$.

20. If $a = 3.76$, $b = 5.34$, and $\gamma = 48°50'$, find the altitude to side b; then determine the area of the triangle correct to two decimal places.

21. In order to measure the height of clouds at night, two observers 136 meters apart are located at points A and B with a spotlight at point L, which is in line with A and B. A vertical beam of light from L is reflected from the bottom of the clouds at point C, and the observers measure the angles of elevation α and β at A and B. These are $\alpha = 74°$ and $\beta = 58°$, as shown in Fig. 4.69. How far above the earth is the bottom of the clouds?

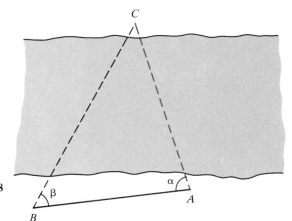

Fig. 4.68

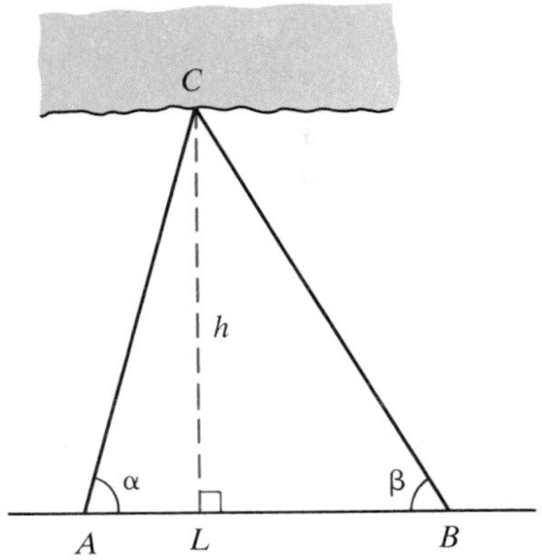

Fig. 4.69

22. From point A on top of a building, the angle of depression of a point C on the ground is observed to be $\alpha = 54°$ (see Fig. 4.70), and from a window at point B, 15 meters directly below A, the angle of depression is $\beta = 42°$. Find the height of the building.

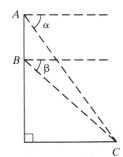

Fig. 4.70

23. An equilateral triangle is inscribed in a circle of radius 4.56. Find the perimeter of the triangle.

24. A square is inscribed in a circle of radius 4.56. Find the area of the square.

25. A surveyor wants to find the width d of a river. She notices a tree T on the opposite bank and takes two points A and B along the bank on her side of the river. She measures the distance x between A and B and the two angles α and β, as shown in Fig. 4.71, and finds $x = 19.8$ meters, $\alpha = 33°$, $\beta = 124°$. From these measurements, calculate the width of the river.

26. A technique for determining an inaccessible height is the following: A surveyor locates two points A and B and measures the distance between them. Then the angles α, β, θ are measured. This is illustrated by Fig. 4.72, in which points A, B, C are the plane of the ground, D is directly above C, angle θ is the angle of elevation of point D from B, and α and β are angles of triangle ABC. Show that

$$h = \frac{d \sin \alpha \tan \theta}{\sin[180° - (\alpha + \beta)]}.$$

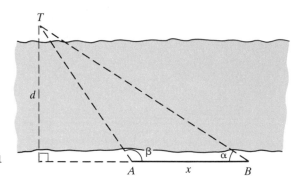

Fig. 4.71

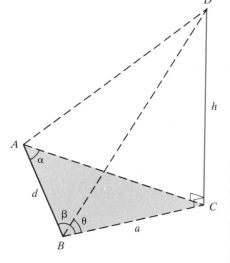

Fig. 4.72

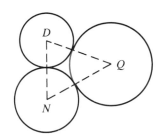

Fig. 4.73

27. In problem 26, suppose that we want to determine the height h of a mountain peak and points A and B are such that $d = 463$ meters, $\theta = 47°20'$, $\beta = 63°10'$, $\alpha = 46°40'$. Find h.

28. A dime, a nickel, and a quarter are placed on a table so that they just touch each other, as shown in Fig. 4.73. The diameters of the dime, nickel, and quarter are 1.75 cm, 2.25 cm, and 2.50 cm, respectively. Find the length of the smaller part of the circumference of the quarter between the two points where it touches the dime and the nickel. In the diagram, N, Q, and D, respectively, are the centers.

29. In problem 28 the centers of the coins form a triangle. Find the measure of the smallest angle to the nearest degree.

30. Points A and B are located on opposite sides of a lake (see Fig. 4.74). From point C, which is on a nearby hill, the angles of depression to A and B are observed to be $\alpha = 12°$ and $\beta = 17°$, respectively. If the hill is inclined at $27°$ with the horizontal and point D at the base of the hill is 48 meters from C, what is the width of the lake?

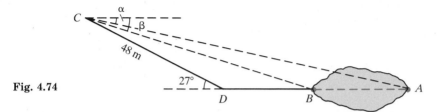

Fig. 4.74

31. On a rectangular set of coordinates, the locations of two forest ranger stations are given as A:(15, 32), B:(84, 15). A fire is spotted at point C, and angles α and β are measured, as shown in Fig. 4.75: $\alpha = 20°$, $\beta = 117°$. Locate the fire by finding the coordinates of C.

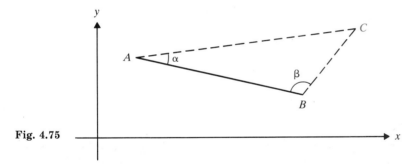

Fig. 4.75

32. In order to complete the proof of the Law of sines given in this section, it is necessary to show that $(\sin \alpha)/a = (\sin \beta)/b$. The diagram in Fig. 4.63 is adjusted as shown in Fig. 4.76. Prove that $\sin \alpha/a = \sin \beta/b$. Use $\sin(180° - \beta) = \sin \beta$ from problem 1(a), Exercises 4.5.

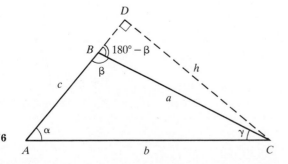

Fig. 4.76

33. Suppose a triangle ABC is inscribed in a circle, as shown in Fig. 4.77. Show that each of the ratios appearing in the Law of sines,

$$\frac{a}{\sin \alpha} = \frac{b}{\sin \beta} = \frac{c}{\sin \gamma},$$

is equal to the diameter of the circle. That is, show diameter $= a/\sin \alpha$. *Hint:* Point D is selected so that side DB passes through the center O of the circle. Recall from geometry that angle CDB is equal to angle CAB (angle α). Also BD is a diameter, and so angle BCD is a right angle.

34. To determine the distance between points A and B on opposite sides of a lake, a surveyor takes points C and D, as shown in Fig. 4.78, and gets the following measurements: $\overline{AC} = 205$ m, $\overline{CD} = 263$ m, $\overline{DB} = 185$ m, $\gamma = 126°$, and $\theta = 104°$. Using this information, find the distance across the lake correct to the nearest meter.

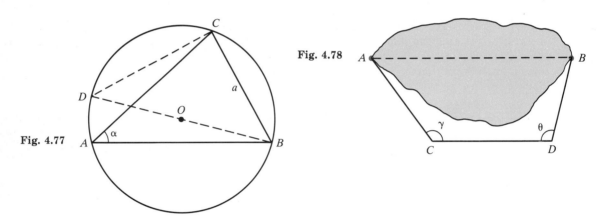

Fig. 4.78

Fig. 4.77

35. A vertical tower is located on a hill that is inclined at an angle of $12°$ with the horizontal (see Fig. 4.79). From point A, which is 43 meters down the hill from the base B of the tower, the angle of elevation of C at the top of the tower is $\alpha = 37°$. Find the height of the tower.

36. Given a circle of radius 8.435 with a central angle $\theta = 52°35'$, as shown in Fig. 4.80, find the area of the shaded region between the chord and arc.

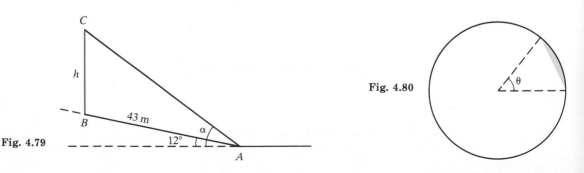

Fig. 4.80

Fig. 4.79

37. The area of triangle ABC is 246.3 m², $a = 31.4$ m, and $b = 17.5$ m. Find angle γ to the nearest minute.

38. If the area of triangle ABC is 25.46 m², $\alpha = 46°$, and $\beta = 82°$, find the lengths of the three sides. Give answers in meters correct to two decimal places.

39. Quadrilateral $OABC$ is inscribed in a quarter circle, as shown in Fig. 4.81, with length of $\overline{AB}$ equal to 2 and length of $\overline{BC}$ equal to 4. Find the area of quadrilateral $OABC$ and express the answer as $k + l\sqrt{m}$, where k, l, and m are positive integers.

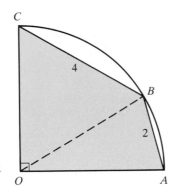

Fig. 4.81

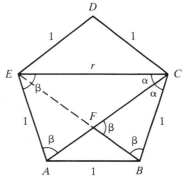

Fig. 4.82

40. Consider a regular pentagon $ABCDE$ with sides of length 1, as shown in Fig. 4.82. Let r be the length of a diagonal (such as CE).

   a) Show that each of the two angles labeled α is equal to 36°, and each of the angles labeled β is 72°. Thus triangles ACE and BCF are similar.

   b) Show that $\overline{CF} = 1$ and $\overline{BF} = r - 1$; then, using the corresponding-ratios property of similar triangles, prove that r satisfies the equation $r^2 - r - 1 = 0$. Solve this in exact form to get the well-known number called the golden ratio.

   c) Apply the Law of cosines to triangle ACE to find $\cos 72° = 1/2r = 1/(1 + \sqrt{5})$. Thus we have expressed $\cos 72°$ in exact form. Check by getting a decimal approximation of $1/(1 + \sqrt{5})$, and compare with the calculator value of $\cos 72°$.

Review Exercises

   1. In each of the following, make a sketch showing the given angle in standard position. A reasonable approximation is sufficient.

   a) 135°   **b)** $-240°$ **c)** $\dfrac{5\pi}{2}$

   d) $-137°$ **e)** -2.34 **f)** $\dfrac{17\pi}{6}$

   2. Determine the quadrants in which the given angles are located.

   a) 235°   **b)** 4.705   **c)** -2.47

   d) $-640°$ **e)** 841°   **f)** 30

In problems 3 through 10, give answers in exact form.

3. Evaluate each of the following.
   a) $\sin 90°$ **b)** $\tan 30°$ **c)** $\sec 150°$ **d)** $\cos(-240°)$
   e) $\tan(-180°)$ **f)** $\csc 450°$ **g)** $\cot(-315°)$ **h)** $\sin 270°$

4. Evaluate each of the following.
   a) $\cos 3\pi$ **b)** $\cot(-\pi)$ **c)** $\sin \dfrac{5\pi}{3}$ **d)** $\cos\left(-\dfrac{7\pi}{3}\right)$

   e) $\tan \dfrac{7\pi}{6}$ **f)** $\sec \dfrac{3\pi}{2}$ **g)** $\sec\left(\pi - \dfrac{\pi}{6}\right)$ **h)** $\csc\left(\dfrac{\pi}{3} + \dfrac{5\pi}{6}\right)$

5. If θ is an angle in the third quadrant and $\tan \theta = 4/3$, determine each of the following.
   a) $\sin \theta$ **b)** $\sec \theta$ **c)** $\cos(\theta + \pi)$ **d)** $\tan(\theta - \pi)$

   e) $\csc\left(\theta - \dfrac{\pi}{2}\right)$ **f)** $\cos\left(\theta + \dfrac{\pi}{2}\right)$

6. In each of the following, determine θ from the given information.
   a) $\sin \theta = -\dfrac{\sqrt{2}}{2}$ and $\pi < \theta < \dfrac{3\pi}{2}$ **b)** $\cos \theta = -\dfrac{1}{2}$ and $0 < \theta < \pi$

   c) $\tan \theta = -1$ and $-2\pi < \theta < -\pi$ **d)** $\sec \theta = -1$ and $0 < \theta < 2\pi$

7. In each of the following, determine α from the given information.
   a) $\sin \alpha = -1$ and $0° \le \alpha \le 360°$ **b)** $\csc \alpha = 2$ and $-90 < \alpha < 90°$

   c) $\cos \alpha = -\dfrac{1}{\sqrt{2}}$ and $0 \le \alpha \le 180°$ **d)** $\tan \alpha = -1$ and $-90° \le \alpha \le 90°$

8. Given that $\alpha = 3\pi/2$, $\beta = \pi/3$, and $\gamma = 5\pi/6$, evaluate each of the following.
   a) $\sin \alpha$ **b)** $\tan \gamma$ **c)** $\cos(\alpha - \beta)$
   d) $\sec(\beta + \gamma)$ **e)** $\sec(\gamma - \alpha)$ **f)** $\cos(\alpha + \gamma - \beta)$

9. Given that $\alpha = 30°$, $\beta = 90°$, and $\gamma = 210°$, evaluate each of the following.
   a) $\sin(\alpha + \gamma)$ **b)** $\sin \alpha + \sin \gamma$ **c)** $\cos(\alpha - \beta)$
   d) $\cos \alpha - \cos \beta$ **e)** $\tan 2\gamma$ **f)** $2 \tan \gamma$

10. Given that $\cos \theta = -0.75$ and $\tan \theta$ is negative, determine each of the following.
    a) $\sin \theta$ **b)** $\cot \theta$ **c)** $\sec\left(\theta - \dfrac{\pi}{2}\right)$ **d)** $\tan(\theta + \pi)$

In problems 11 through 16, evaluate the given expressions and give answers correct to four decimal places.

11. **a)** $\sin 43°$ **b)** $\tan 154°$ **c)** $\cos 57°16'$
    d) $\cot 48°$ **e)** $\sec 327°12'$ **f)** $\sin(-231°)$

12. **a)** $\cos 1.43$ **b)** $\sin 3.86$ **c)** $\tan\left(\dfrac{5\pi}{12}\right)$ **d)** $\cot\left(\dfrac{12}{5\pi}\right)$

13. a) $\sin(53° + 75°)$ **b)** $\sin 53° + \sin 75°$

14. a) $\tan(1.36 + 2.14)$ **b)** $\tan 1.36 + \tan 2.14$

15. a) $(\sin 153°)^2 + (\cos 153°)^2$ **b)** $(\sin 1.5)^2 + (\cos 1.5)^2$

16. a) $2\left(\sin \dfrac{\pi}{12}\right)\left(\cos \dfrac{\pi}{12}\right)$ **b)** $\left(\cos \dfrac{\pi}{3}\right)^2 - \left(\sin \dfrac{\pi}{3}\right)^2$

17. In each of the following, determine whether the given statement is true or is false.

   a) π and $-\pi$ are coterminal angles.

   b) $-\dfrac{3\pi}{2}$ and $-\dfrac{\pi}{2}$ are coterminal angles.

   c) $210°$ and $-\dfrac{5\pi}{6}$ are coterminal angles.

   d) An angle in standard position with terminal side passing through the point $(-1, 2)$ is coterminal with $150°$.

18. Draw a graph of $y = 2 \sin x$ by first making a table of several x, y pairs that satisfy the given equation. Use degree measure for the x values.

19. Follow instructions in problem 18 for $y = 2 \cos x$.

20. If $y = -\tan x$, make a table of x, y values that satisfy the equation, starting with $x = -2.0$ and then increasing by 0.2 for successive values of x up to $x = 2.0$. Plot the corresponding points, and draw a graph of $y = -\tan x$.

21. The hypotenuse of a right triangle is 37.42 cm, and one angle is $48°12'$. Find the lengths of the two sides. Give answers correct to four significant digits.

22. If ABC is an isosceles triangle with $\overline{AB} = \overline{AC} = 4.73$, and the angle opposite AB is $52°14'$, find the length of the altitude from A to BC. Then find the area of the triangle. Give answers correct to two decimal places.

23. If the hypotenuse of a right triangle is 24.3 cm, and one of the sides is 15.4 cm, find the length of the other side correct to three significant digits. Determine the angles correct to the nearest minute.

In problems 24 through 32, parts of a triangle are given (using conventional notation, as described in this chapter). First, determine if the given information is sufficient to determine a triangle. If it is, find the remaining parts. Give answers correct to the accuracy consistent with the given information.

24. $b = 32, c = 47, \alpha = 18°$

25. $a = 15, b = 20, c = 40$

26. $\alpha = 62.5°, \beta = 23.6°, c = 3.47$

27. $a = 3.4, b = 4.6, c = 4.0$

28. $\beta = 64°12', b = 32.5, c = 23.8$

29. $\alpha = 30°, \beta = 60°, \gamma = 90°$

30. $\alpha = 48°, \beta = 74°, \gamma = 58°, a = 436$

31. $\alpha = 36°, \beta = 65°, a = 36.4, b = 25.3$

32. $\beta = 32°14', \gamma = 64°18', a = 42.53$

33. In Fig. 4.83, the length of CD and angles α and β are measured and found to be: $\overline{CD} = 137\,\mathrm{m}$, $\alpha = 44°$, $\beta = 123°$. Find the distance from A to B and from A to C.

Fig. 4.83

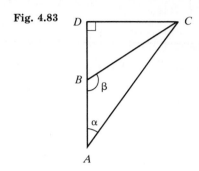

34. In Fig. 4.84, O is the center of a circle, AB is tangent to the circle at B, and C is a point on the circle and on OA, as shown. If the radius of the circle is 12 cm and the length of arc $\widehat{BC}$ is 9 cm, what is the area of the shaded region?

Fig. 4.84

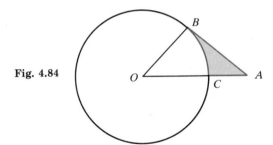

Trigonometric Identities, Inverse Functions, Equations, Graphs

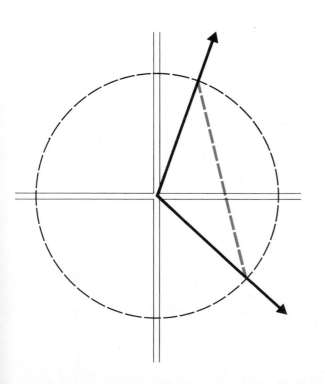

Problem solving in mathematics frequently involves a sequence of steps in which the problem is restated in a different but equivalent form until ultimately it is reduced to a form that can be solved by familiar techniques. For example, in algebra the student learns to solve the equation $x^2 - x - 6 = 0$ by replacing $x^2 - x - 6$ by $(x - 3)(x + 2)$, so that the problem then becomes one of solving $(x - 3)(x + 2) = 0$. In this form the problem can be solved by resorting to a theorem which states that if the product of two numbers is zero, then one of the two numbers must be zero. That is, $x - 3 = 0$ or $x + 2 = 0$; so 3 and -2 are the solutions.

In this example we call the equation $x^2 - x - 6 = (x - 3)(x + 2)$ an *identity,* because it is satisfied by every real number. That is, if we replace x by any given real number in the expression on the left-hand side of the equals sign and in the expression on the right-hand side, the two resulting numbers will be equal. However, the equation $x^2 - x - 6 = 0$ does not have this property, since it is satisfied by only two values of x. We call such an equation a *conditional equation.*

Definition 5.1   An equation that is satisfied by all values of the variable (or variables) for which both the left-hand side and the right-hand side are defined is called an *identity.*

For example, $(x^2 - 4)/(x - 2) = x + 2$ is an identity since it is satisfied by all real numbers x except 2, a value of x for which the left-hand side is not defined. The student has already encountered several identities, such as the factoring formulas

$$x^2 - y^2 = (x + y)(x - y), \qquad x^2 + 2xy + y^2 = (x + y)^2,$$

and so on. It is helpful to have a collection of identities involving trigonometric functions. We have already seen some of these in Section 4.5, such as $\sin^2\theta + \cos^2\theta = 1$, given by Eq. (4.16); this identity was used in deriving the law of cosines, Eq. (4.20).

This chapter includes a large number of identities with which the student should become familiar. These may be difficult to "memorize," but through frequent encounters while solving numerous problems, students eventually find that they know them. In subsequent sections of this chapter the usefulness of identities will become apparent through several examples in which a problem is solved by using identities to obtain an equivalent problem, the solution of which is possible by familiar techniques.

5.1 BASIC IDENTITIES

The following equations are satisfied by each value of θ for which both sides of the given equation are defined. That is, they are identities.

(I.1) $$\boxed{\csc \theta = \frac{1}{\sin \theta}}$$ (I.2) $$\boxed{\sec \theta = \frac{1}{\cos \theta}}$$

(I.3) $$\boxed{\cot \theta = \frac{1}{\tan \theta}}$$ (I.4) $$\boxed{\sin(-\theta) = -\sin \theta}$$

(I.5) $$\boxed{\cos(-\theta) = \cos \theta}$$ (I.6) $$\boxed{\tan(-\theta) = -\tan \theta}$$

(I.7) $$\boxed{\tan \theta = \frac{\sin \theta}{\cos \theta}}$$ (I.8) $$\boxed{\cot \theta = \frac{\cos \theta}{\sin \theta}}$$

(I.9)* $$\boxed{\sin^2\theta + \cos^2\theta = 1}$$ (I.10) $$\boxed{1 + \tan^2\theta = \sec^2\theta}$$

(I.11) $$\boxed{1 + \cot^2\theta = \csc^2\theta}$$

Note: Identities **(I.1)** through **(I.11),** as well as others developed in the next two sections, are listed inside the back cover for easy reference.

Proofs that **(I.1)** through **(I.9)** are identities have already been given in Chapter 4. Identity **(I.10)** can be derived as follows:

Dividing both sides of **(I.9)** by $\cos^2\theta$ and then using **(I.2)** and **(I.7)** gives **(I.10):**

$$\frac{\sin^2\theta}{\cos^2\theta} + \frac{\cos^2\theta}{\cos^2\theta} = \frac{1}{\cos^2\theta}, \qquad \tan^2\theta + 1 = \sec^2\theta.$$

Identity **(I.11)** can be proved in a similar manner.

Identities **(I.1)** through **(I.11)** can be used to derive or prove several other identities. The following examples illustrate techniques for proving trigonometric identities.

---

*   The notation $\sin^2\theta$ means $(\sin \theta)^2$; that is, we first get $\sin \theta$ and then square the result. However, $\sin \theta^2$ means we first square θ and then get the sine of the result. In general, $\sin^2\theta$ and $\sin \theta^2$ are not equal.

Example 1   Prove that $\cos x \tan x = \sin x$ is an identity.

Solution   Let LHS and RHS represent "left-hand side" and "right-hand side," respectively, of the given equation.

$$\underset{\substack{\uparrow \\ \text{given}}}{\text{LHS}} = \underset{\substack{\uparrow \\ \text{by (I.7)}}}{\cos x \tan x} = (\cos x)\left(\frac{\sin x}{\cos x}\right) \underset{\substack{\uparrow \\ \text{algebra}}}{=} \underset{\substack{\uparrow \\ \text{given}}}{\sin x} = \text{RHS}.$$

The transitive property of equality allows us to conclude that LHS = RHS. Hence the given equation is an identity.   ■

Example 2   Prove that $\dfrac{1 - \sec x}{1 + \sec x} = \dfrac{\cos x - 1}{\cos x + 1}$ is an identity.

Solution $$\underset{\substack{\uparrow \\ \text{given}}}{\text{LHS}} = \underset{\substack{\uparrow \\ \text{by (I.2)}}}{\frac{1 - \sec x}{1 + \sec x}} = \underset{\substack{\uparrow \\ \text{algebra}}}{\frac{1 - \dfrac{1}{\cos x}}{1 + \dfrac{1}{\cos x}}} = \underset{\substack{\uparrow \\ \text{algebra}}}{\frac{\dfrac{\cos x - 1}{\cos x}}{\dfrac{\cos x + 1}{\cos x}}} = \underset{\substack{\uparrow \\ \text{given}}}{\frac{\cos x - 1}{\cos x + 1}} = \text{RHS}.$$

Therefore LHS = RHS, and the given equation is an identity.   ■

Example 3   Prove that $(\sin x + \cos x)^2 = \dfrac{\sec x \csc x + 2}{\sec x \csc x}$ is an identity.

Solution $$\underset{\substack{\uparrow \\ \text{given}}}{\text{LHS}} = \underset{\substack{\uparrow \\ \text{algebra}}}{(\sin x + \cos x)^2} = \sin^2 x + 2 \sin x \cos x + \cos^2 x \underset{\substack{\uparrow \\ \text{by (I.9)}}}{=} 1 + 2 \sin x \cos x,$$

$$\underset{\substack{\uparrow \\ \text{given}}}{\text{RHS}} = \frac{\sec x \csc x + 2}{\sec x \csc x} \underset{\substack{\uparrow \\ \text{algebra}}}{=} \frac{\sec x \csc x}{\sec x \csc x} + \frac{2}{\sec x \csc x}$$

$$\underset{\substack{\uparrow \\ \text{algebra}}}{=} 1 + \frac{2}{\sec x \csc x} \underset{\substack{\uparrow \\ \text{by (I.1) and (I.2)}}}{=} 1 + 2 \sin x \cos x.$$

By the transitive property of the equals relation, we conclude that LHS = RHS. Thus the given equation is an identity.   ■

Example 4   Is the function $f(x) = \dfrac{x + \sin x}{|x|}$ odd, even, or neither?

Solution $$f(-x) = \frac{-x + \sin(-x)}{|-x|} = \frac{-x - \sin x}{|x|} = -\left(\frac{x + \sin x}{|x|}\right) = -f(x).$$

Here we used **(I.4)** to justify the second equals sign. Since $f(-x) = -f(x)$, f is an odd function.

■

Example 5   Determine the domain and range of the function $f(x) = \cos x \sec x$.

Solution   Note that $\cos x$ is defined for all x in **R**, and $\sec x$ is defined for all x in **R** except $\pi/2, 3\pi/2, \ldots, -\pi/2, -3\pi/2, \ldots$. Therefore

$$\mathfrak{D}(f) = \left\{ x \mid x \in \mathbf{R} \text{ and } x \neq \frac{(2k+1)\pi}{2}, k \text{ is an integer} \right\}.$$

Using **(I.2)** and algebra, we get

$$f(x) = (\cos x)(\sec x) = (\cos x)\left(\frac{1}{\cos x}\right) = 1 \text{ for all } x \in \mathfrak{D}(f).$$

Thus $\mathfrak{R}(f) = \{1\}$.

■

Technique for Proving Identities

Note that in the above examples we did not begin our proof with the given equation and manipulate it until we got an obvious equality. Here we emphasize an important point of logic. A proof consists of a logical sequence of statements in which the final statement is the statement to be proved.

We illustrate our point with a simple example. Suppose we wish "to prove" that $1 = 2$. If we are allowed to start with $1 = 2$ as the first step, then our "proof" could proceed as follows:

$$1 = 2,$$
$$0 \cdot 1 = 0 \cdot 2, \qquad \text{(multiply both sides by 0)}$$
$$0 = 0.$$

Since $0 = 0$ is an obvious equality, can we conclude that $1 = 2$? Clearly not; the only conclusion we can make from the sequence above is that "if $1 = 2$, then $0 = 0$," which is a true statement.

The important point this example illustrates is that it is not logically acceptable to begin a proof with the statement to be proved, perform algebraic manipulations on it, obtain an obvious equality, and then conclude that the starting statement is true. If such a procedure is followed *and if it can be shown that the steps are reversible,* then the proof is valid. However, the steps in reverse are a necessary part of the proof and should be included. What step or steps in the above faulty proof are not reversible?

Suggestion: As illustrated in Examples 1, 2, and 3 above, the best technique in communicating a proof, we believe, is to *work independently* with either or both of the left- and right-hand sides of the given equation to show that each reduces to the same expression. The final statement of LHS = RHS then follows from the transitive property of the equals relation.

Exercises 5.1

1. Which of the following are identities?

a) $x^3 + 1 = (x + 1)(x^2 - x + 1)$ $\qquad$ **b)** $xe^{\ln x} = x^2$ $\qquad$ **c)** $\dfrac{3x - 2}{x + 2} - \dfrac{1}{x - 1} = \dfrac{3x(x - 2)}{x^2 + x - 2}$

In problems 2 through 40, prove that the given equation is an identity.

2. $\sin \theta \cot \theta = \cos \theta$

3. $\dfrac{\tan \theta}{\sin \theta} = \sec \theta$

4. $\cot \theta = \csc \theta \cos \theta$

5. $\cos x \sec x = 1$

6. $\cos x \tan x = \sin x$

7. $1 - \cos^2 x = \cos^2 x \tan^2 x$

8. $\cot x \sec x = \csc x$

9. $\sin^2 x = (1 - \cos x)(1 + \cos x)$

10. $\dfrac{\cot x}{\sec x} = \csc x - \sin x$

11. $\dfrac{\sin x \csc x}{\cot x} = \tan x$

12. $\dfrac{\sin(-\theta)}{\cos \theta} = \tan(-\theta)$

13. $\sec \theta \csc \theta = \tan \theta + \cot \theta$

14. $\sec \theta(\csc \theta - \sin \theta) = \csc \theta \cos \theta$

15. $\dfrac{1 - \cos x}{1 + \cos x} = (\cot x - \csc x)^2$

16. $\dfrac{\sin \theta}{1 + \cos \theta} = \dfrac{1 - \cos \theta}{\sin \theta}$

17. $\tan x + \cot x = \dfrac{\csc x}{\cos x}$

18. $\dfrac{1 + \tan \theta}{\sec \theta} = \dfrac{1 + \cot \theta}{\csc \theta}$

19. $\cot \alpha \csc \alpha = \dfrac{1}{\sec \alpha - \cos \alpha}$

20. $\dfrac{1}{1 - \sin x} + \dfrac{1}{1 + \sin x} = 2 \sec^2 x$

21. $\sec^2 x + \csc^2 x = \sec^2 x \csc^2 x$

22. $\dfrac{\sin \theta}{1 + \cos \theta} + \dfrac{1 + \cos \theta}{\sin \theta} = \dfrac{2}{\sin \theta}$

23. $(\cos x + 1)(\sec x - 1) = \sec x - \cos x$

24. $\sec \theta - \cos \theta = \sin(-\theta)\tan(-\theta)$

25. $\sin^4 x - \cos^4 x = \sin^2 x - \cos^2 x$

26. $1 + \tan^2 x = \tan x \sec x \csc x$

27. $\dfrac{\tan \theta + \sec \theta}{\sin \theta \cot \theta} = \dfrac{1 + \sin \theta}{\cos^2 \theta}$

28. $\cot(-x)\cos(-x) = \sin x - \csc x$

29. $\dfrac{\cos \theta}{\sin \theta} + \dfrac{\sin \theta}{\cos \theta} = \sec \theta \csc \theta$

30. $\dfrac{1 - \sin(-x)}{\cos x} = \tan x + \sec x$

31. $\dfrac{1 - \cos x}{1 + \cos x} = \dfrac{\sec x - 1}{\sec x + 1}$

32. $1 - (\sin x - \cos x)^2 = 2 \sin x \cos x$

33. $\dfrac{\csc(-x)}{\cot(-x) + \tan(-x)} = \cos x$

34. $\dfrac{\cos x}{1 - \sin x} = \dfrac{1 + \sin x}{\cos x}$

35. $\sec^4 x - \tan^4 x = \sec^2 x(\sin^2 x + 1)$

36. $\tan^2 x - \sec^2 x = -1$

37. $\tan^4 x + \tan^2 x = \sec^4 x - \sec^2 x$

38. $\dfrac{1}{\sec\theta - \tan\theta} = \sec\theta + \tan\theta$

39. $\dfrac{\cot x + \tan x}{\sec x\,\csc x} = 1$

40. $\sin^2 x\,\tan^2 x + \sin^2 x = \tan^2 x$

In problems 41 through 45, determine whether the given functions are odd, even, or neither.

41. a) $f(x) = \sin x\cos x$ **b)** $g(x) = \sin x + \cos x$

42. a) $f(x) = x\cos x$ **b)** $g(x) = x + \cos x$

43. a) $f(x) = e^{\cos x}$ **b)** $g(x) = e^{\sin x}$

44. a) $f(x) = \sqrt{1 - \sin^2 x}$ **b)** $g(x) = \sqrt{\sec^2 x - 1}$

45. a) $f(x) = \dfrac{\sin x}{\csc x}$ **b)** $g(x) = \dfrac{x^2 + \cos x}{\sin x}$

In problems 46 through 50, determine the domain and range of each of the given functions.

46. $f(x) = \cos x\,\tan x$ **47.** $f(x) = \sin^2 x + \cos^2 x$

48. $f(x) = 1 - \tan x\,\cot x$ **49.** $f(x) = \sqrt{1 - \sin^2 x}$

50. $f(x) = e^{\sin^2 x} \cdot e^{\cos^2 x}$

5.2   SUM AND DIFFERENCE IDENTITIES

Expressions of the type $\sin(\alpha + \beta)$ occur frequently, and we might ask: Is $\sin(\alpha + \beta) = \sin\alpha + \sin\beta$ for all values of α and β? The answer is no. For instance, if $\alpha = \pi/2$ and $\beta = \pi/2$, then $\sin(\pi/2 + \pi/2) = \sin\pi = 0$, whereas $\sin\pi/2 + \sin\pi/2 = 1 + 1 = 2$. Hence the equation is not an identity. The next question is: Can we find a simple formula that gives $\sin(\alpha + \beta)$ in terms of trigonometric functions of α and of β? The answer to this is included in the following set of identities, which are called the *sum and difference formulas*.

(I.12) $\sin(\alpha + \beta) = \sin\alpha\cos\beta + \cos\alpha\sin\beta$

(I.13) $\sin(\alpha - \beta) = \sin\alpha\cos\beta - \cos\alpha\sin\beta$

(I.14) $\cos(\alpha + \beta) = \cos\alpha\cos\beta - \sin\alpha\sin\beta$

(I.15) $\cos(\alpha - \beta) = \cos\alpha\cos\beta + \sin\alpha\sin\beta$

(I.16)
$$\tan(\alpha + \beta) = \frac{\tan \alpha + \tan \beta}{1 - \tan \alpha \tan \beta}$$

(I.17)
$$\tan(\alpha - \beta) = \frac{\tan \alpha - \tan \beta}{1 + \tan \alpha \tan \beta}$$

We first prove identity **(I.14)** by using the diagrams of Fig. 5.1, in which α and β are shown as positive angles and points A and B are on the corresponding terminal sides, one unit from the origin. From the definitions of circular functions, the coordinates of A and B are given by

$$A\!:\!(\cos \alpha, \sin \alpha), \qquad B\!:\!(\cos(-\beta), \sin(-\beta)) = (\cos \beta, -\sin \beta).$$

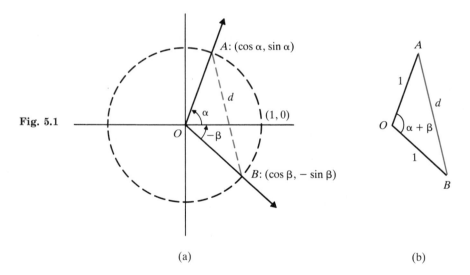

Fig. 5.1

(a) (b)

Let d be the distance between points A and B. Using the distance formula given by Eq. (1.2), we obtain

$$d^2 = (\cos \alpha - \cos \beta)^2 + (\sin \alpha + \sin \beta)^2.$$

After applying some simple algebra and using identity **(I.9)** twice, we get

$$d^2 = 2 - 2(\cos \alpha \cos \beta - \sin \alpha \sin \beta). \tag{5.1}$$

Now looking at triangle AOB of Fig. 5.1 (b), in which the points A, O, and B are taken from Fig. 5.1(a), and applying the Law of cosines, we get

$$d^2 = 1^2 + 1^2 - (2)(1)(1)\cos(\alpha + \beta) = 2 - 2\cos(\alpha + \beta). \tag{5.2}$$

From Eqs. (5.1) and (5.2) we have

$$2 - 2(\cos\alpha\cos\beta - \sin\alpha\sin\beta) = 2 - 2\cos(\alpha + \beta).$$

Thus $\cos(\alpha + \beta) = \cos\alpha\cos\beta - \sin\alpha\sin\beta$, and this is identity **(I.14)**.

Note: The diagrams of Fig. 5.1 show α and β as positive acute angles. A proof similar to that above could be given for α and β of any size.

Identity **(I.14)** along with identities established earlier can now be used to prove the remaining identities given above. The following is a proof of **(I.15)**, in which **(I.14)**, **(I.5)**, and **(I.4)** are used.

$$\cos(\alpha - \beta) = \cos(\alpha + (-\beta)) = \cos\alpha\cos(-\beta) - \sin\alpha\sin(-\beta)$$
$$= \cos\alpha\cos\beta + \sin\alpha\sin\beta.$$

This gives identity **(I.15)**: $\cos(\alpha - \beta) = \cos\alpha\cos\beta + \sin\alpha\sin\beta$.

Identity **(I.12)** can now be established by using **(I.15)** and two identities that are given in Chapter 4, namely,

$$\sin\left(\frac{\pi}{2} - \theta\right) = \cos\theta \qquad \text{and} \qquad \cos\left(\frac{\pi}{2} - \theta\right) = \sin\theta. \tag{5.3}$$

We proceed as follows:

$$\sin(\alpha + \beta) \underset{\substack{\uparrow \\ \text{by Eq. (5.3)}}}{=} \cos\left[\frac{\pi}{2} - (\alpha + \beta)\right] \underset{\substack{\uparrow \\ \text{algebra}}}{=} \cos\left[\left(\frac{\pi}{2} - \alpha\right) - \beta\right]$$

$$\underset{\substack{\uparrow \\ \text{by (I.15)}}}{=} \cos\left(\frac{\pi}{2} - \alpha\right)\cos\beta + \sin\left(\frac{\pi}{2} - \alpha\right)\sin\beta.$$

Now we apply Eq. (5.3) to get the desired identity:

$$\sin(\alpha + \beta) = \sin\alpha\cos\beta + \cos\alpha\sin\beta.$$

Identity **(I.16)** can now be proved as follows:

$$\tan(\alpha + \beta) \underset{\substack{\uparrow \\ \text{by (I.7)}}}{=} \frac{\sin(\alpha + \beta)}{\cos(\alpha + \beta)} \underset{\substack{\uparrow \\ \text{by (I.14), (I.15)}}}{=} \frac{\sin\alpha\cos\beta + \cos\alpha\sin\beta}{\cos\alpha\cos\beta - \sin\alpha\sin\beta} \underset{\substack{\uparrow \\ \text{algebra and (I.7)}}}{=} \frac{\tan\alpha + \tan\beta}{1 - \tan\alpha\tan\beta}.$$

The last step involves dividing numerator and denominator by $\cos\alpha\cos\beta$ and then applying **(I.7)**.

Proofs of **(I.13)** and **(I.17)** are left as exercises (see problem 2).

Example 1 Prove that $\tan\left(x - \dfrac{\pi}{4}\right) = \dfrac{\sin x - \cos x}{\sin x + \text{cox } x}$ is an identity.

Solution $$\text{LHS} = \underset{\substack{\uparrow \\ \text{given}}}{\tan\left(x - \dfrac{\pi}{4}\right)} = \underset{\substack{\uparrow \\ \text{by (I.17)}}}{\dfrac{\tan x - \tan(\pi/4)}{1 + \tan x \tan(\pi/4)}} = \underset{\substack{\uparrow \\ \text{since } \tan\frac{\pi}{4} = 1}}{\dfrac{\tan x - 1}{1 + \tan x}}$$

$$= \underset{\substack{\uparrow \\ \text{by (I.7)}}}{\dfrac{(\sin x/\cos x) - 1}{1 + (\sin x/\cos x)}} = \underset{\substack{\uparrow \\ \text{algebra}}}{\dfrac{\sin x - \cos x}{\cos x + \sin x}} = \underset{\substack{\uparrow \\ \text{given}}}{\text{RHS}}$$

Hence LHS = RHS, and the given equation is an identity. ■

Example 2 Evaluate $\sin 75°$ and express the answer in exact form.

Solution Using $\sin 75° = \sin(30° + 45°)$, applying **(I.12)**, and evaluating the result gives

$$\sin 75° = \sin(30° + 45°) = \sin 30° \cos 45° + \cos 30° \sin 45°$$

$$= \dfrac{1}{2}\dfrac{\sqrt{2}}{2} + \dfrac{\sqrt{3}}{2}\dfrac{\sqrt{2}}{2} = \dfrac{1}{4}(\sqrt{2} + \sqrt{6}).$$ ■

Example 3 Evaluate $\cos\dfrac{\pi}{12}$ and give the answer in exact form.

Solution Using $\cos \pi/12 = \cos\left(\dfrac{\pi}{4} - \dfrac{\pi}{6}\right)$, applying **(I.15)**, and evaluating the result gives

$$\cos\dfrac{\pi}{12} = \cos\left(\dfrac{\pi}{4} - \dfrac{\pi}{6}\right) = \cos\dfrac{\pi}{4}\cos\dfrac{\pi}{6} + \sin\dfrac{\pi}{4}\sin\dfrac{\pi}{6}$$

$$= \dfrac{\sqrt{2}}{2}\dfrac{\sqrt{3}}{2} + \dfrac{\sqrt{2}}{2}\dfrac{1}{2} = \dfrac{1}{4}(\sqrt{6} + \sqrt{2}).$$ ■

Example 4 Prove that

$$\sin x \cos y = \dfrac{1}{2}[\sin(x + y) + \sin(x - y)] \tag{5.4}$$

is an identity.

Solution Adding the two equations given in **(I.12)** and **(I.13)** with $\alpha = x$ and $\beta = y$, we have $\sin(x + y) + \sin(x - y) = 2 \sin x \cos y$. This is equivalent to the given equation. ■

An identity of the type given in Example 4 is useful in two types of problems: expressing a product as a sum, or expressing a sum as a product (factoring). This is illustrated in the next example.

Example 5   **a)** Express $\sin 5\alpha \cos 3\alpha$ as a sum.

                b) Express $\sin 4\alpha + \sin 2\alpha$ as a product.

Solution   **a)** Substituting $x = 5\alpha$, $y = 3\alpha$ into Eq. (5.4) gives

$$\sin 5\alpha \cos 3\alpha = \frac{1}{2}[\sin(5\alpha + 3\alpha) + \sin(5\alpha - 3\alpha)] = \frac{1}{2}[\sin 8\alpha + \sin 2\alpha].$$

Thus $\sin 5\alpha \cos 3\alpha = \dfrac{1}{2}\sin 8\alpha + \dfrac{1}{2}\sin 2\alpha$ is an identity.

b) Equation (5.4) can be written as

$$\sin(x + y) + \sin(x - y) = 2\sin x \cos y. \tag{5.5}$$

Let $x + y = 4\alpha$ and $x - y = 2\alpha$. Adding these two equations gives $2x = 6\alpha$, or $x = 3\alpha$. Similarly, subtracting gives $2y = 2\alpha$, or $y = \alpha$. Substituting $x = 3\alpha$, $y = \alpha$ into Eq. (5.5) gives

$$\sin 4\alpha + \sin 2\alpha = 2\sin 3\alpha \cos \alpha.$$

Thus the given expression has been written as a product.   ■

Exercises 5.2

1. Derive a formula for $\cot(\alpha + \beta)$ in terms of $\cot \alpha$ and $\cot \beta$.
2. Prove that the equations given in **(I.13)** and **(I.17)** are identities.
3. Establish each of the following cofunction identities.

   a) $\sin\left(\dfrac{\pi}{2} + \theta\right) = \cos\theta$ **b)** $\cos\left(\dfrac{\pi}{2} + \theta\right) = -\sin\theta$

   c) $\sin\left(\dfrac{3\pi}{2} - \theta\right) = -\cos\theta$ **d)** $\cos\left(\dfrac{3\pi}{2} - \theta\right) = -\sin\theta$

   e) $\sin\left(\dfrac{3\pi}{2} + \theta\right) = -\cos\theta$ **f)** $\cos\left(\dfrac{3\pi}{2} + \theta\right) = \sin\theta$

4. In each of the following, prove that the given equation is an identity.

   a) $\sin(180° - \theta) = \sin\theta$ **b)** $\cos(180° - \theta) = -\cos\theta$

   c) $\tan(180° - \theta) = -\tan\theta$ **d)** $\sin(180° + \theta) = -\sin\theta$

   e) $\cos(180° + \theta) = -\cos\theta$ **f)** $\tan(180° + \theta) = \tan\theta$

5. Evaluate each of the following. Give answers in exact form.

   a) $\cos 75°$ **b)** $\sin 195°$ **c)** $\tan 285°$

   d) $\cot 15°$ **e)** $\sec 255°$ **f)** $\csc(-75°)$

6. Evaluate each of the following. Give answers in exact form and then use your calculator to evaluate the result correct to two decimal places. As a check, evaluate directly by calculator (make certain it is in radian mode).

   a) $\tan\dfrac{7\pi}{12}$ **b)** $\sec\left(-\dfrac{5\pi}{12}\right)$ **c)** $\cos\dfrac{11\pi}{12}$

    d) $\sin \dfrac{23\pi}{12}$ **e)** $\sin \dfrac{13\pi}{12}$ **f)** $\csc \dfrac{25\pi}{12}$

7. Given that $\tan x = \dfrac{3}{4}$ and $x + y = \dfrac{\pi}{4}$, find $\tan y$.

8. Given that $\tan \alpha = 3$ and $\tan(\alpha + \beta) = -\dfrac{2}{3}$, find $\tan \beta$.

9. Given that $x - y = \dfrac{3\pi}{4}$ and $\tan y = 3$, find $\tan x$.

10. Given that $\tan(x - y) = -\dfrac{5}{4}$ and $\tan x = 0.4$, find $\tan y$.

In problems 11 through 17, determine whether or not the given equation is an identity.

11. $\tan\left(\dfrac{\pi}{4} + x\right) = \dfrac{1 + \tan x}{1 - \tan x}$ **12.** $\sin\left(\dfrac{\pi}{6} - x\right) = \dfrac{1}{2}(\cos x - \sqrt{3}\sin x)$

13. $\dfrac{\cos x - \sin x}{\cos x + \sin x} = \tan\left(\dfrac{\pi}{4} - x\right)$ **14.** $\sec(\alpha + \beta) = \sec \alpha + \sec \beta$

15. $\csc\left(\dfrac{\pi}{2} - x\right) = \sec x$ **16.** $\sin x + \sin 2x = \sin 3x$

17. $\cos\left(\dfrac{5\pi}{2} + x\right) = -\sin x$

18. Use $\cos 75° = \cos(30° + 45°)$ to get $\cos 75°$ in exact form. Similarly, express $\sin 75°$ in exact form.

19. Use $\cos 72° = \dfrac{1}{1 + \sqrt{5}}$ (see Exercises 4.7, problem 40) along with identity **(I.9)** to find $\sin 72°$ in exact form.

20. Use problems 18 and 19 and $\cos 3° = \cos(75° - 72°)$ to get $\cos 3°$ in exact form.

21. Prove that each of the following equations is an identity.

    a) $\cos x \cos y = \frac{1}{2}[\cos(x + y) + \cos(x - y)]$

    b) $\sin x \sin y = \frac{1}{2}[\cos(x - y) - \cos(x + y)]$.

22. Using Eq. (5.4) or the identities given in problem 21, express each of the following products as a sum or difference (see Example 5).

    a) $(\sin 3\theta)(\cos 5\theta)$ **b)** $(\cos 3\theta)(\cos 4\theta)$ **c)** $(\sin 2y)(\sin 4y)$

    d) $(\cos 3x)(\sin(-5x))$ **e)** $(\sin 2y)(\sin(-4y))$ **f)** $(\sin 3x)(\sin 2x)$

23. Using Eq. (5.5) or the identities given in problem 21, express each of the following as a product (see Example 5).

    a) $\sin 5\alpha + \sin 3\alpha$ **b)** $\cos 5\alpha + \cos 3\alpha$ **c)** $\cos 3\alpha - \cos \alpha$

24. In each of the following write the given expression in terms of $\sin x$ and $\cos x$.

    a) $\sin\left(x - \dfrac{\pi}{4}\right)$ **b)** $\sin\left(x - \dfrac{\pi}{2}\right)$ **c)** $\cos\left(x - \dfrac{\pi}{4}\right)$

    d) $\sin(2x)$ **e)** $\cos(2x)$ **f)** $\sin\left(2x - \dfrac{\pi}{3}\right)$

25. Evaluate each of the following. Express your answers in exact form.

a) $\sin \frac{\pi}{4} \cos \frac{\pi}{12} + \sin \frac{\pi}{12} \cos \frac{\pi}{4}$

b) $\cos 160° \cos 25° + \sin 160° \sin 25°$

c) $\cos^2 47° + \sin^2 47°$

d) $\dfrac{\tan 37° - \tan 67°}{1 + \tan 37° \tan 67°}$

26. If α, β, and γ are the angles of a triangle, prove that

a) $\sin \gamma = \sin \alpha \cos \beta + \cos \alpha \sin \beta$

b) $\cos \gamma = \sin \alpha \sin \beta - \cos \alpha \cos \beta$

5.3   DOUBLE-ANGLE AND HALF-ANGLE FORMULAS

Useful identities can be derived from the addition formulas given in Section 5.2. The following are called *double-angle identities*.

(I.18)
$$\sin 2\theta = 2 \sin \theta \cos \theta$$

(I.19)
$$\cos 2\theta = \cos^2\theta - \sin^2\theta = 1 - 2 \sin^2\theta = 2 \cos^2\theta - 1$$

(I.20)
$$\tan 2\theta = \frac{2 \tan \theta}{1 - \tan^2\theta}$$

These are special cases of **(I.12)**, **(I.14)**, and **(I.16)**, in which we take $\alpha = \theta$ and $\beta = \theta$ (see problem 1).

The double-angle identities are useful in simplifying certain trigonometric expressions, and the student should become familiar with them. We consider some examples in which these identities, along with **(I.1)** through **(I.17)**, are used.

Example 1   Prove that $\sin 2x = \dfrac{2 \tan x}{1 + \tan^2 x}$ is an identity.

Solution
$$\text{LHS} = \sin 2x = 2 \sin x \cos x.$$
$$\underset{\text{given}}{\uparrow} \qquad \underset{\text{by (I.18)}}{\uparrow}$$

$$\text{RHS} = \underset{\text{given}}{\frac{2 \tan x}{1 + \tan^2 x}} = \underset{\text{by (I.10)}}{\frac{2 \tan x}{\sec^2 x}} = \underset{\text{by (I.2), (I.7)}}{\left(\frac{2 \sin x}{\cos x} \right) \div \left(\frac{1}{\cos^2 x} \right)} = \underset{\text{algebra}}{2 \sin x \cos x}.$$

Therefore LHS = RHS, and the given equation is an identity.   ∎

Example 2   Suppose $\sin \theta = 3/5$ and $\cos \theta$ is negative. Evaluate in exact form.

a) $\sin 2\theta$

b) $\cos 2\theta$

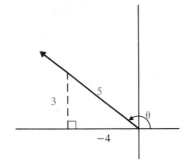

Fig. 5.2

If $\sin\theta > 0$ and $\cos\theta < 0$, then θ is an angle in the second quadrant, as shown in Fig. 5.2.

a) To find $\sin 2\theta$, use **(I.18)** to get $\sin 2\theta = 2\sin\theta\cos\theta = 2(\tfrac{3}{5})(-\tfrac{4}{5}) = -\tfrac{24}{25}$.

b) To find $\cos 2\theta$, use **(I.19)** to get $\cos 2\theta = \cos^2\theta - \sin^2\theta = (-\tfrac{4}{5})^2 - (\tfrac{3}{5})^2 = \tfrac{7}{25}$.

Example 3   Express $\sin 3x$ as a function of $\sin x$.

Solution
$$\sin 3x = \underset{\underset{\text{algebra}}{\uparrow}}{\sin(2x + x)} = \underset{\underset{\text{by (I.12)}}{\uparrow}}{\sin 2x \cos x + \cos 2x \sin x}$$

$$= \underset{\underset{\text{by (I.18), (I.19)}}{\uparrow}}{(2\sin x \cos x)\cos x + (\cos^2 x - \sin^2 x)\sin x}$$

$$= \underset{\underset{\text{algebra}}{\uparrow}}{3\sin x \cos^2 x - \sin^3 x} = \underset{\underset{\text{by (I.9)}}{\uparrow}}{3\sin x(1 - \sin^2 x) - \sin^3 x}$$

$$= \underset{\underset{\text{algebra}}{\uparrow}}{3\sin x - 4\sin^3 x}.$$

Therefore $\sin 3x = 3\sin x - 4\sin^3 x$ is an identity.

Example 4   Suppose $\sin\theta = 0.3487$ and $0° < \theta < 90°$. Using a calculator, evaluate each of the following correct to four decimal places.

a) $\sin 2\theta$ **b)** $\cos 2\theta$ **c)** $\tan 2\theta$

Solution   Enter 0.3487 into the display. Then with the calculator in either degree or radian mode, press $\boxed{\text{INV}}$ and $\boxed{\text{sin}}$ keys (or $\boxed{\text{sin}^{-1}}$ key), which gives θ in the display; multiply by 2 and store the result, using the $\boxed{\text{STO}}$ key. Using the $\boxed{\text{RCL}}$ key as needed, gives

a) $\sin 2\theta = 0.6536$ **b)** $\cos 2\theta = 0.7568$ **c)** $\tan 2\theta = 0.8637$

Note: On some calculators the store and recall keys may be labeled otherwise than $\boxed{\text{STO}}$ and $\boxed{\text{RCL}}$. ■

Half-angle Formulas

Writing identity **(I.19)** in the form $\cos 2x = 1 - 2\sin^2 x$ and replacing x by $\theta/2$, we get $\cos\theta = 1 - 2\sin^2(\theta/2)$. Solving for $\sin(\theta/2)$ gives

$$\sin\frac{\theta}{2} = \sqrt{\frac{1 - \cos\theta}{2}} \text{ for each } \theta \text{ for which } \sin\frac{\theta}{2} \geq 0,$$

$$\sin\frac{\theta}{2} = -\sqrt{\frac{1 - \cos\theta}{2}} \text{ for each } \theta \text{ for which } \sin\frac{\theta}{2} \leq 0.$$

These two equations are ordinarily written as

(I.21) $$\sin\frac{\theta}{2} = \pm\sqrt{\frac{1 - \cos\theta}{2}}$$

where the $\pm$ means not that we get two values for $\sin(\theta/2)$ but that we select the sign that is consistent with the sign of $\sin(\theta/2)$, depending on the quadrant in which $\theta/2$ is located.

In a similar manner, using identity **(I.19)** in the form $\cos 2x = 2\cos^2 x - 1$, replacing x by $\theta/2$, and solving for $\cos(\theta/2)$ gives

(I.22) $$\cos\frac{\theta}{2} = \pm\sqrt{\frac{1 + \cos\theta}{2}}$$

The $+$ or $-$ sign in **(I.22)** is selected to agree with the sign of $\cos(\theta/2)$.

We can now derive an identity for $\tan(\theta/2)$ by using **(I.21)** and **(I.22)** along with **(I.7)**:

(I.23) $$\tan\frac{\theta}{2} = \pm\sqrt{\frac{1 - \cos\theta}{1 + \cos\theta}}$$

Identity **(I.23)** can be expressed in a more desirable form that does not involve the $\pm$ sign. Instead of manipulating **(I.23)** directly, we can proceed as follows. Identities **(I.18)** and **(I.19)** can be written in the form

$$\sin \theta = 2 \sin \frac{\theta}{2} \cos \frac{\theta}{2} \quad \text{and} \quad 1 + \cos \theta = 2 \cos^2 \frac{\theta}{2},$$

respectively. Dividing the left sides and the right sides of these two equations gives

$$\frac{\sin \theta}{1 + \cos \theta} = \frac{2 \sin \frac{\theta}{2} \cos \frac{\theta}{2}}{2 \cos^2 \frac{\theta}{2}} = \frac{\sin \frac{\theta}{2}}{\cos \frac{\theta}{2}} = \tan \frac{\theta}{2}.$$

Thus

$$\tan \frac{\theta}{2} = \frac{\sin \theta}{1 + \cos \theta}.$$

An alternative form of this equation is

$$\tan \frac{\theta}{2} = \frac{1 - \cos \theta}{\sin \theta}$$

(see problem 16 of Exercises 5.1). Therefore the following are useful identities for $\tan(\theta/2)$:

(I.24) $$\tan \frac{\theta}{2} = \frac{\sin \theta}{1 + \cos \theta} = \frac{1 - \cos \theta}{\sin \theta}$$

Example 5   Evaluate each of the following and express the answer in exact form.

a) $\sin 22°30'$ **b)** $\cos 112.5°$ **c)** $\tan \dfrac{7\pi}{12}$

Solution   Using identities **(I.21)**, **(I.22)**, and **(I.24)** gives the following.

a) $\sin 22°30' = \sin \left(\dfrac{45}{2}\right)^\circ = \sqrt{\dfrac{1 - \cos 45°}{2}} = \dfrac{1}{2}\sqrt{2 - \sqrt{2}}.$

b) $\cos 112.5° = \cos \left(\dfrac{225}{2}\right)^\circ = -\sqrt{\dfrac{1 + \cos 225°}{2}} = -\dfrac{1}{2}\sqrt{2 - \sqrt{2}}.$

c) $\tan \dfrac{7\pi}{12} = \tan \dfrac{7\pi/6}{2} = \dfrac{1 - \cos \dfrac{7\pi}{6}}{\sin \dfrac{7\pi}{6}} = \dfrac{1 - \left(-\dfrac{\sqrt{3}}{2}\right)}{-\dfrac{1}{2}} = -(2 + \sqrt{3}).$

Example 6   Given that $\cos\theta = -3/5$ and $180° < \theta < 270°$, evaluate in exact form:

a) $\sin\dfrac{\theta}{2}$ **b)** $\cos\dfrac{\theta}{2}$ **c)** $\tan\dfrac{\theta}{2}$

Solution   First note that $90° < \theta/2 < 135°$, and so $\sin(\theta/2)$ is positive, and $\cos(\theta/2)$ is negative. From Fig. 5.3, $\sin\theta = -4/5$. Now use identities **(I.21)**, **(I.22)**, and **(I.24)** to get the following.

a) $\sin\dfrac{\theta}{2} = \sqrt{\dfrac{1-\cos\theta}{2}} = \sqrt{\dfrac{1-(-\frac{3}{5})}{2}} = \dfrac{2\sqrt{5}}{5}.$

b) $\cos\dfrac{\theta}{2} = -\sqrt{\dfrac{1+\cos\theta}{2}} = -\sqrt{\dfrac{1+(-\frac{3}{5})}{2}} = -\dfrac{\sqrt{5}}{5}.$

c) $\tan\dfrac{\theta}{2} = \dfrac{\sin\theta}{1+\cos\theta} = \dfrac{-\frac{4}{5}}{1+(-\frac{3}{5})} = -2.$

Fig. 5.3

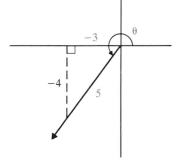

Example 7   Evaluate $\sin 15°$ in exact form in two ways:

a) By using **(I.13)**   **b)** By using **(I.21)**

Solution   **a)** $\sin 15° = \sin(45° - 30°) = \sin 45° \cos 30° - \cos 45° \sin 30° = (\sqrt{6} - \sqrt{2})/4.$ Therefore

$$\sin 15° = \frac{\sqrt{6} - \sqrt{2}}{4}.$$

b) $\sin 15° = \sin\left(\dfrac{30}{2}\right)° = \sqrt{\dfrac{1-\cos 30°}{2}} = \dfrac{1}{2}\sqrt{2 - \sqrt{3}}.$ Therefore

$$\sin 15° = \tfrac{1}{2}\sqrt{2 - \sqrt{3}}.$$

It appears that we get two different answers for $\sin 15°$. We leave it for the reader to evaluate each with a calculator to see if they both represent the same number (see problem 47 of Exercises 5.3).

Example 8   Suppose $\sin(\theta/2) = 0.6843$ and $0° < \theta < 180°$. Use a calculator to evaluate each of the following correct to four decimal places.

   a) $\sin\theta$ $\cos 2\theta$ **c)** $\tan\dfrac{\theta}{4}$

Solution   Enter 0.6843 into the display. Then with the calculator in degree (or radian) mode, press $\boxed{\text{INV}}$ and $\boxed{\text{sin}}$ (or $\boxed{\text{sin}^{-1}}$) keys, then multiply by 2 (this gives θ), and store into memory with the $\boxed{\text{STO}}$ key. Using the $\boxed{\text{RCL}}$ key as needed, gives the following.

   a) $\sin\theta = 0.9980$ **b)** $\cos 2\theta = -0.9919$ **c)** $\tan\dfrac{\theta}{4} = 0.3957$ ■

Exercises 5.3

1. By replacing α by θ and β by θ in identities **(I.12)**, **(I.14)**, and **(I.16)**, show that identities **(I.18)**, **(I.19)**, and **(I.20)** follow.

2. Given that $\cos\theta = -12/13$, and θ is in the second quadrant, find the following in exact form.

   a) $\sin 2\theta$ **b)** $\cos 2\theta$ **c)** $\tan 2\theta$

3. Given that $\sin\theta = -5/13$ and $\cos\theta = 12/13$, find the following in exact form.

   a) $\sin 2\theta$ **b)** $\cos 2\theta$ **c)** $\tan 2\theta$

4. Suppose $\cos\theta = 0.5873$ and $0° < \theta < 90°$. Using a calculator, evaluate the following correct to four decimal places.

   a) $\sin 2\theta$ **b)** $\cos 2\theta$ **c)** $\tan 2\theta$

5. Suppose $\sin\theta = 0.4385$ and $0 < \theta < \pi/2$. Using a calculator, evaluate correct to four decimal places each of the following.

   a) $\sin 2\theta$ **b)** $\cos 3\theta$ **c)** $\cot 3\theta$

In problems 6 through 9, evaluate the given expressions in exact form. Check your results by using a calculator.

6. **a)** $\sin 67°30'$ **b)** $\cos(-22.5°)$ **c)** $\sin 105°$ **d)** $\cos 105°$

7. **a)** $\tan 165°$ **b)** $\cos(247.5°)$ **c)** $\tan(-195°)$ **d)** $\cos 285°$

8. **a)** $\sin\dfrac{\pi}{12}$ **b)** $\cos\dfrac{5\pi}{8}$ **c)** $\sin\dfrac{11\pi}{8}$ **d)** $\tan\dfrac{13\pi}{12}$

9. **a)** $\cos\dfrac{19\pi}{8}$ **b)** $\sin\left(-\dfrac{7\pi}{8}\right)$ **c)** $\sin\dfrac{21\pi}{8}$ **d)** $\tan\left(-\dfrac{5\pi}{12}\right)$

10. Evaluate each of the following in exact form.

   a) $\sin 15° \cos 15°$ **b)** $\sin^2 105° - \cos^2 105°$ **c)** $1 - 2\sin^2\dfrac{5\pi}{12}$

11. Evaluate each of the given expressions, and give answers rounded off to five decimal places. In each case check to see if the two given expressions are equal.

   a) $2\sin 37° \cos 37°$ and $\sin 74°$ **b)** $\cos^2 21° - \sin^2 21°$ and $\cos 42°$ **c)** $1 - 2\sin^2 0.65$ and $\cos 1.3$

12. Given that $0 \leq \theta \leq \pi$ and $\cos \theta = -0.6$, evaluate in exact form $\sin \theta/2 + \sin 2\theta$.

13. Suppose $-90° \leq \theta \leq 90°$ and $\sin \theta = -0.6$. Evaluate in exact form $\sin \theta/2 + \sin 2\theta$.

In problems 14 through 33, prove that the given equations are identities.

14. $(\sin \theta + \cos \theta)^2 = 1 + \sin 2\theta$

15. $\dfrac{1}{\csc 2\theta} = 2 \sin \theta \cos \theta$

16. $\tan \dfrac{\theta}{2} = \csc \theta - \cot \theta$

17. $\left(\sin \dfrac{\theta}{2} + \cos \dfrac{\theta}{2} \right)^2 = 1 + \sin \theta$

18. $(\cos x + \sin x)(\cos x - \sin x) = \cos 2x$

19. $\cos 2x \tan 2x = \sin 2x$

20. $\cos^2 \dfrac{x}{2} - \sin^2 \dfrac{x}{2} = \cos x$

21. $\tan \dfrac{x}{2} = \dfrac{\sec x - 1}{\sin x \sec x}$

22. $\sin 2x \tan x = 2 \sin^2 x$

23. $(1 + \tan x) \tan 2x = \dfrac{2 \tan x}{1 - \tan x}$

24. $2 \sin^2 \dfrac{x}{2} = \sin x \tan \dfrac{x}{2}$

25. $2 \cos^2 \dfrac{x}{2} = \dfrac{\sin x + \tan x}{\tan x}$

26. $\tan \theta \sin 2\theta = 1 - \cos 2\theta$

27. $\sin 2\theta \sec \theta = 2 \sin \theta$

28. $\cot x - \tan x = 2 \cot 2x$

29. $2 \csc 2x = \tan x + \cot x$

30. $\cos^4 x - \sin^4 x = \cos 2x$

31. $\dfrac{1 - \tan x}{1 + \tan x} = \sec 2x - \tan 2x$

32. $\cos 3x = 4 \cos^3 x - 3 \cos x$

33. $\cos 4x = \cos^4 x - 6 \sin^2 x \cos^2 x + \sin^4 x$

In problems 34 through 40, determine whether or not the given equations are identities.

34. $\sin 2x + \sin 3x = \sin 5x$

35. $\sin^2 \dfrac{x}{2} = 1 - \cos^2 \dfrac{x}{2}$

36. $\cot \dfrac{x}{2} - \tan \dfrac{x}{2} = 2 \cot x$

37. $(\sin 6x + \cos 6x)^2 = 1$

38. $\sec 2x + \tan 2x = \tan \left(x + \dfrac{\pi}{4} \right)$

39. $(\cot x - \tan x) \tan 2x = 2$

40. $(\sin x - \cos x)^2 = 1 - \sin 2x$

41. Given that $\sin \dfrac{\theta}{2} = -\dfrac{3}{4}$, find $\cos \theta$. *Hint:* Use **(I.21)**.

42. Given that $\cos \dfrac{\theta}{2} = \dfrac{2}{5}$, find $\cos \theta$. *Hint:* Use **(I.22)**.

In problems 43 through 46, evaluate the given expressions in exact form, where α and β are given by

$$\sin \alpha = \frac{3}{5} \quad \text{and} \quad \frac{\pi}{2} \leq \alpha \leq \frac{3\pi}{2},$$

$$\tan \beta = -\frac{5}{12} \quad \text{and} \quad -\frac{\pi}{2} \leq \beta \leq \frac{\pi}{2}.$$

43. $\sin\left(\dfrac{\alpha + \beta}{2}\right)$ **44.** $\sin\left(\alpha + \dfrac{\beta}{2}\right)$ **45.** $\cos(\alpha + 2\beta)$ **46.** $\tan\left(\dfrac{\alpha + \beta}{2}\right)$

47. In Example 7 we concluded that $\dfrac{\sqrt{6} - \sqrt{2}}{4}$ and $\dfrac{1}{2}\sqrt{2 - \sqrt{3}}$ represent the same number.

    a) Use your calculator to check this conclusion (at least to the decimal-place capacity of the calculator).

    b) Prove that they actually are equal without using a calculator and without referring to the fact that both represent sin 15°.

48. Triangle ABC is inscribed in a circle, as shown in Fig. 5.4, where Q is the center of the circle, α is one angle, and a is the opposite side. Prove that the diameter d of the circle is given by $d = a/\sin \alpha$.

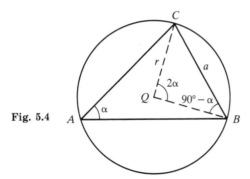

Fig. 5.4

Hint: Note that angle BQC is equal to 2α. (Why?) Now use the Law of sines on triangle BQC and identity **(I.18)** to get the result.

Note: This problem also appeared in Exercises 4.7 as problem 33. However, the solution suggested there is quite different.

5.4   INVERSE TRIGONOMETRIC FUNCTIONS

By looking at the graphs of the trigonometric functions in Section 4.5, we can easily see that the corresponding inverse relations are not functions, since horizontal lines intersect the graphs at more than one point (in fact, at infinitely many points). However, by restricting the domains, we can get functions whose inverse relations are functions. We proceed to do so for the sine, cosine, and tangent functions.

Inverse Sine Function

Consider the function Sin defined by

$$\text{Sin}(x) = \sin x \quad \text{and} \quad \mathcal{D}(\text{Sin}) = \left\{ x \,\middle|\, -\frac{\pi}{2} \le x \le \frac{\pi}{2} \right\}.$$

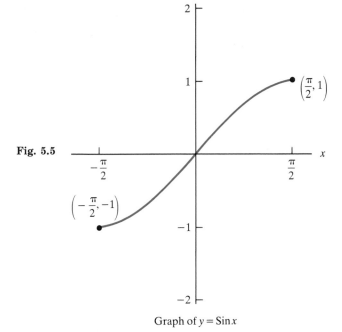

Graph of $y = \operatorname{Sin} x$

From the graph of $y = \operatorname{Sin} x$, as shown in Fig. 5.5, we see that for each value of y in $-1 \leq y \leq 1$ there is exactly one corresponding value of x associated with it on the curve. That is, the inverse of the Sin function is also a function; we denote it by Sin^{-1}. As suggested in our discussion of inverse functions in Section 1.8, interchanging the x and y variables gives $x = \operatorname{Sin} y$, and then "solving for y" we get $y = \operatorname{Sin}^{-1}x$. This result merely involves symbolism used to denote the inverse sine function. We now give a formal definition.

Definition 5.2   The *inverse sine function,* denoted by Sin^{-1}, is given by

$$\operatorname{Sin}^{-1} = \left\{(x, y)\,|\,-1 \leq x \leq 1, \ x = \sin y \ \text{and} \ -\frac{\pi}{2} \leq y \leq \frac{\pi}{2}\right\}.^{*}$$

---

*   We use the capital letter S in Sin^{-1} to distinguish the inverse sine function from the *inverse sine relation,* given by

$$\sin^{-1} = \{(x, y)\,|\,x = \sin y\}.$$

In some textbooks, for a given value of x in $-1 \leq x \leq 1$, $\operatorname{Sin}^{-1} x$ is referred to as the *principal value* of the inverse sine relation for that value of x.

In Section 1.8 we saw that the graph of an inverse function can be obtained by reflecting the graph of the function about the line $y = x$. Thus, if the graph shown in Fig. 5.5 is reflected about $y = x$, we get the graph of $y = \text{Sin}^{-1}x$, as shown in Fig. 5.6. The domain and range of the Sin^{-1} function are given by

$$\mathcal{D}(\text{Sin}^{-1}) = \{x \mid -1 \le x \le 1\}, \qquad \mathcal{R}(\text{Sin}^{-1}) = \left\{y \mid -\frac{\pi}{2} \le y \le \frac{\pi}{2}\right\}.$$

Fig. 5.6

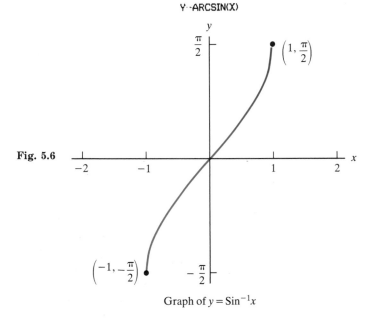

Y--ARCSIN(X)

Graph of $y = \text{Sin}^{-1}x$

Note on notation: In some textbooks the function Sin^{-1} is denoted by Arcsin. In this book we shall use Sin^{-1} and Arcsin interchangeably, so that the reader will become familiar with both notations in anticipation of notation that may be used in later calculus courses.

In the context in which trigonometric functions are introduced as functions on measures of angles, it is helpful to think of $\text{Sin}^{-1}x$ as that angle between or equal to $-\pi/2$ and $\pi/2$ whose sine is equal to x. For instance, $\text{Sin}^{-1}\frac{1}{2}$ is the angle $\pi/6$.

Example 1 Evaluate and give answers in exact form.

a) $\text{Sin}^{-1}\dfrac{\sqrt{2}}{2}$

b) $\text{Arcsin}\left(-\dfrac{1}{2}\right)$

Solution   **a)** Let $\mathrm{Sin}^{-1}(\sqrt{2}/2) = \alpha$. According to Definition 5.2, $\sin\alpha = \sqrt{2}/2$ and $-\pi/2 \le \alpha \le \pi/2$. Thus α is in the first quadrant (since $\sin\alpha > 0$), as shown in Fig. 5.7, from which we see that $\alpha = \pi/4$. Therefore, $\mathrm{Sin}^{-1}(\sqrt{2}/2) = \pi/4$.

   b) Let $\mathrm{Arcsin}(-\frac{1}{2}) = \beta$. According to Definition 5.2, $\sin\beta = -\frac{1}{2}$ and $-\pi/2 \le \beta \le \pi/2$. This tells us that β is an angle in the fourth quadrant, as shown in Fig. 5.8, from which we see that $\beta = -\pi/6$. Therefore $\mathrm{Arcsin}(-\frac{1}{2}) = -\pi/6$.

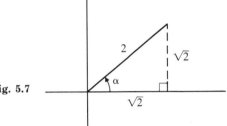

Fig. 5.7

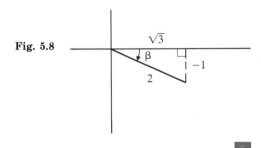

Fig. 5.8

In Example 1 we were able to evaluate the Sin^{-1} function at $\sqrt{2}/2$ and $-1/2$ and express the result in exact form. However, if we wish to evaluate $\mathrm{Sin}^{-1}0.47$, we see that we cannot proceed in a similar manner, since the resulting angle in the reference triangle is not one we can easily recognize (such as $\pi/6$, $\pi/4$, or $\pi/3$). However, we can use a calculator to get the result correct to several decimal places. *Calculators are programmed so that the values of inverse trigonometric functions are consistent with the definitions given in this section* (such as Definition 5.2).

Example 2   Evaluate and give results rounded off to four decimal places.

   a) $\mathrm{Sin}^{-1}(0.47)$ **b)** $\mathrm{Arcsin}\left(\dfrac{1 - \sqrt{5}}{2}\right)$ **c)** $\mathrm{Sin}^{-1}\left(\dfrac{1 + \sqrt{5}}{2}\right)$

Solution   First place the calculator in radian mode.

   a) Enter 0.47 into the calculator display, and then press $\boxed{\text{INV}}$ and $\boxed{\text{sin}}$ (or $\boxed{\text{sin}^{-1}}$ on some calculators*). The display will give 0.4893. Thus $\mathrm{Sin}^{-1}(0.47) = 0.4893$.

   b) First evaluate $(1 - \sqrt{5})/2$; then, with the result in the display, press $\boxed{\text{INV}}$ and $\boxed{\text{sin}}$ (or $\boxed{\text{sin}^{-1}}$) to get $\mathrm{Arcsin}[(1 - \sqrt{5})/2] = -0.6662$.

---

*   On most calculators the keys $\boxed{\text{INV}}$ and $\boxed{\text{sin}}$ (or $\boxed{\text{sin}^{-1}}$) are used to evaluate the Sin^{-1} function. However, some calculators have a key labeled $\boxed{\text{arc}}$ that is used in place of $\boxed{\text{INV}}$.

c) As in (b), first evaluate $(1 + \sqrt{5})/2$ and then press (INV) and (sin) (or (sin⁻¹)). The calculator will indicate *Error* because $(1 + \sqrt{5})/2$ is greater than 1 and so is not in $\mathfrak{D}(\text{Sin}^{-1})$. That is, $\text{Sin}^{-1}[(1 + \sqrt{5})/2]$ is not defined. ■

Inverse Cosine Function

The inverse cosine function can be introduced in a manner similar to that used above for the inverse sine function.

Suppose the function Cos is defined by

$$\text{Cos}(x) = \cos x \quad \text{and} \quad \mathfrak{D}(\text{Cos}) = \{x \mid 0 \le x \le \pi\}.$$

From the graph of $y = \text{Cos } x$, shown in Fig. 5.9, we see that the inverse of Cos is also a function; we denote it by Cos^{-1}(or by Arccos) and define it as follows.

Definition 5.3   The *inverse cosine function,* denoted by Cos^{-1}, is given by

$$\text{Cos}^{-1} = \{(x, y) \mid -1 \le x \le 1, x = \cos y, \text{ and } 0 \le y \le \pi\}.$$

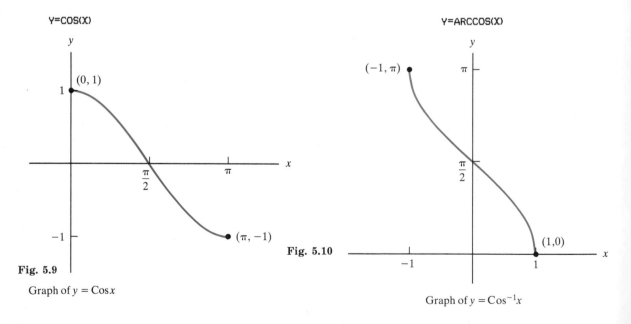

Y=COS(X)

Fig. 5.9

Graph of $y = \text{Cos } x$

Y=ARCCOS(X)

Fig. 5.10

Graph of $y = \text{Cos}^{-1}x$

The graph of $y = \text{Cos}^{-1}x$ is shown in Fig. 5.10. The domain and range of the Cos^{-1} function are given by

$$\mathfrak{D}(\text{Cos}^{-1}) = \{x \mid -1 \le x \le 1\}, \qquad \mathfrak{R}(\text{Cos}^{-1}) = \{y \mid 0 \le y \le \pi\}.$$

Example 3   **a)** Evaluate $\text{Cos}^{-1}(-\sqrt{3}/2)$ and give the result in exact form.

        b) Evaluate $\text{Arccos}(0.735)$ rounded off to four decimal places.

Solution   **a)** Let $\text{Cos}^{-1}(-\sqrt{3}/2) = \alpha$. According to Definition 5.3, $\cos\alpha = -\sqrt{3}/2$ and $0 \le \alpha \le \pi$. This tells us that α is an angle in the second quadrant, as shown in Fig. 5.11, from which we see that $\alpha = 5\pi/6$. Thus $\text{Cos}^{-1}(-\sqrt{3}/2) = 5\pi/6$.

        b) Place the calculator in radian mode. Enter 0.735 into the display, and then press $\boxed{\text{INV}}$ and $\boxed{\text{cos}}$ (or $\boxed{\text{cos}^{-1}}$) to get $\text{Arccos } 0.735 = 0.7451$.

Fig. 5.11

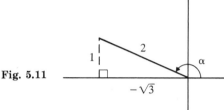

Inverse Tangent Function
Suppose function Tan is defined by

$$\text{Tan}(x) = \tan x \qquad \text{and} \qquad \mathfrak{D}(\text{Tan}) = \left\{ x \,\middle|\, -\frac{\pi}{2} < x < \frac{\pi}{2} \right\}.$$

From the graph of $y = \text{Tan } x$, shown in Fig. 5.12, we see that the inverse of Tan is also a function. We denote it by Tan^{-1}(or Arctan) and define it as follows:

Fig. 5.12

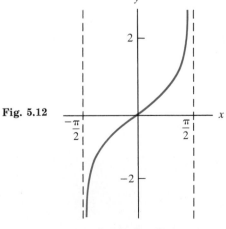

Graph of $y = \text{Tan } x$

Definition 5.4   The *inverse tangent function,* denoted by Tan^{-1}, is given by

$$\text{Tan}^{-1} = \left\{(x, y) \mid x \in \mathbf{R},\ x = \tan y,\ \text{and} -\frac{\pi}{2} < y < \frac{\pi}{2}\right\}.$$

The graph of $y = \text{Tan}^{-1}x$ is shown in Fig. 5.13. The domain and range of the Tan^{-1} function are given by

$$\mathfrak{D}(\text{Tan}^{-1}) = \mathbf{R}, \qquad \mathfrak{R}(\text{Tan}^{-1}) = \left\{y \mid -\frac{\pi}{2} < y < \frac{\pi}{2}\right\}.$$

Fig. 5.13

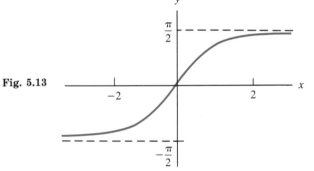

Y=ARCTAN(X)

Graph of $y = \text{Tan}^{-1}x$

Example 4   **a)** Evaluate $\text{Tan}^{-1}(-\sqrt{3})$ and give the answer in exact form.

b) Evaluate $\text{Arctan}(1.57)$ rounded off to four decimal places.

Solution   **a)** Let $\text{Tan}^{-1}(-\sqrt{3}) = \alpha$. According to Definition 5.4, $\text{Tan}\ \alpha = -\sqrt{3}$ and $-\pi/2 < \alpha < \pi/2$. This tells us that α is an angle as shown in Fig. 5.14. We note that $\alpha = -\pi/3$ and so $\text{Tan}^{-1}(-\sqrt{3}) = -\pi/3$.

Fig. 5.14

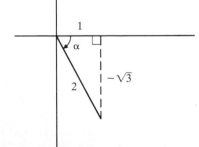

b) With the calculator in radian mode, enter 1.57 into the display; then press
INV and tan (or tan⁻¹) to get Arctan(1.57) = 1.0037. ▪

Example 5   Evaluate and give answers in exact form.

a) $\sin\left(\mathrm{Sin}^{-1}\dfrac{1}{4}\right)$ **b)** $\sin\left(2\,\mathrm{Sin}^{-1}\dfrac{1}{4}\right)$

Solution   **a)** Let $\mathrm{Sin}^{-1}(1/4) = \theta$. Then from Definition 5.2 we see that $\sin\theta = 1/4$ and $-\pi/2 \le \theta \le \pi/2$. Therefore θ is an angle as shown in Fig. 5.15, and so

$$\sin\left(\mathrm{Sin}^{-1}\frac{1}{4}\right) = \sin\theta = \frac{1}{4}.^{*}$$

b) Let θ be as in (a); we want to evaluate $\sin[2\,\mathrm{Sin}^{-1}(1/4)] = \sin 2\theta$. Applying the double-angle identity $\sin 2\theta = 2\sin\theta\cos\theta$ and reading the values of $\sin\theta$ and $\cos\theta$ from Fig. 5.15 gives

$$\sin\left(2\,\mathrm{Sin}^{-1}\frac{1}{4}\right) = 2\left(\frac{1}{4}\right)\left(\frac{\sqrt{15}}{4}\right) = \frac{\sqrt{15}}{8}.$$

Fig. 5.15

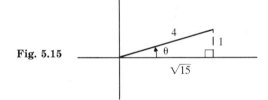

▪

Example 6   Evaluate and give answers in exact form.

a) $\sin[\mathrm{Tan}^{-1}(-\tfrac{5}{12}) + \mathrm{Tan}^{-1}(\tfrac{4}{3})]$ **b)** $\sin[\tfrac{1}{2}\,\mathrm{Tan}^{-1}(-\tfrac{3}{4})]$

Solution   **a)** Let $\mathrm{Tan}^{-1}(-5/12) = \alpha$ and $\mathrm{Tan}^{-1}(4/3) = \beta$. From Definition 5.4,

$$\tan\alpha = -\frac{5}{12}, \quad -\frac{\pi}{2} < \alpha < \frac{\pi}{2} \quad \text{and} \quad \tan\beta = \frac{4}{3}, \quad -\frac{\pi}{2} < \beta < \frac{\pi}{2}.$$

Thus α and β are angles as shown in Fig. 5.16. We want to evaluate $\sin(\alpha + \beta)$, and so we apply identity **(I.12)**, $\sin(\alpha + \beta) = \sin\alpha\cos\beta + \cos\alpha\sin\beta$. Using the diagrams in Fig. 5.16, we can evaluate the right-hand side to get

$$\sin\left[\mathrm{Tan}^{-1}\left(-\frac{5}{12}\right) + \mathrm{Tan}^{-1}\left(\frac{4}{3}\right)\right] = \left(-\frac{5}{13}\right)\left(\frac{3}{5}\right) + \left(\frac{12}{13}\right)\left(\frac{4}{5}\right) = \frac{33}{65}.$$

---

*   It might be helpful to state a problem such as (a) in words: "We want the sine of an angle whose sine is $\tfrac{1}{4}$." This is not so different from the popular quiz question: "Who is buried in Grant's tomb?"

Fig. 5.16
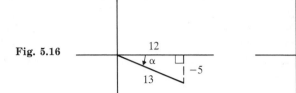

b) Let $\text{Tan}^{-1}(-3/4) = \theta$. By Definition 5.4, $\tan \theta = -3/4$ and $-\pi/2 < \theta < \pi/2$. Thus θ is an angle as shown in Fig. 5.17; we want to evaluate $\sin(\theta/2)$. Using the half-angle identity **(I.21)** with a minus sign before the square root, since $\theta/2$ is obviously in the fourth quadrant (look at Fig. 5.17), we get

$$\sin\left[\frac{1}{2}\text{Tan}^{-1}\left(-\frac{3}{4}\right)\right] = \sin\left(\frac{1}{2}\theta\right) = -\sqrt{\frac{1 - \cos\theta}{2}} = -\sqrt{\frac{1 - \frac{4}{5}}{2}} = -\frac{1}{\sqrt{10}}.$$

Fig. 5.17
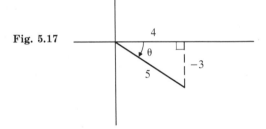

Example 7   Evaluate $\sec[\text{Tan}^{-1}0.348 - 2\,\text{Cos}^{-1}(-0.735)]$. Give answer rounded off to four decimal places.

Solution   Place the calculator in radian mode,* and first evaluate the angle $\text{Tan}^{-1}(0.348) - 2\,\text{Cos}^{-1}(-0.735)$. With the result in the display, press $\boxed{\text{cos}}$ and $\boxed{1/x}$ to get

$$\sec[\text{Tan}^{-1}(0.348) - 2\,\text{Cos}^{-1}(-0.735)] = -3.9742.$$

Example 8   Find all values of x that satisfy the inequality $5\,\text{Sin}^{-1}x - 4 \leq 0$.

Solution   The given inequality is equivalent to $\text{Sin}^{-1}x \leq 4/5$. To solve this inequality it is instructive to look at the graph of $y = \text{Sin}^{-1}x$, as shown in Fig. 5.18. The solution set consists of those values of x that correspond to points on the curve for which

---

*   Definitions 5.2, 5.3, and 5.4 imply that the inverse trigonometric function values are real numbers, that is, in radians if considered as angular measure. In this problem the correct answer would be obtained even if the calculator operated in degree mode.

$y \leq 4/5$; these points are those between Q and P. Thus the solution set is $\{x \mid -1 \leq x \leq c\}$, where $\text{Sin}^{-1}c = 0.8$. The value c is given by $c = \sin 0.8$. With the calculator in radian mode, evaluate $\sin 0.8$ to get $c = 0.717$ (to three decimal places). The solution set is $S = \{x \mid -1 \leq x \leq 0.717\}$.

Y=ARCSIN(X)

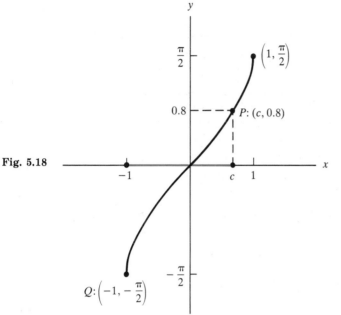

Fig. 5.18

Graph of $y = \text{Sin}^{-1}x$

Example 9   Is $\text{Sin}^{-1}\dfrac{3}{5} + \text{Sin}^{-1}\dfrac{8}{17} = \text{Sin}^{-1}\dfrac{77}{85}$?

Solution   As a first step, evaluating the left-hand side and the right-hand side by using a calculator in radian mode, we get

$$\text{LHS} = \text{Sin}^{-1}\frac{3}{5} + \text{Sin}^{-1}\frac{8}{17} = 1.133458435,$$

$$\text{RHS} = \text{Sin}^{-1}\frac{77}{85} = 1.133458435.$$

This is reasonably convincing evidence that the answer to the question is yes. However, two numbers could agree out to several decimal places and yet differ at some place beyond. Thus we give the following noncalculator proof.

Let $\text{Sin}^{-1}(3/5) = \alpha$ and $\text{Sin}^{-1}(8/17) = \beta$. Since $\alpha + \beta$ is an angle in the first quadrant (from the above computations of LHS), and $\text{Sin}^{-1}(77/85) = \gamma$ is also

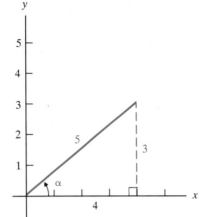

Fig. 5.19

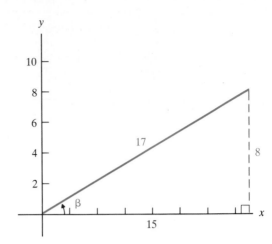

an angle in the first quadrant, it is sufficient to show that $\sin(\alpha + \beta) = \sin \gamma$. But $\sin \gamma = 77/85$, and so we wish to show that $\sin(\alpha + \beta) = 77/85$. Using **(I.12)** and reading off values from Fig. 5.19, we get

$$\sin(\alpha + \beta) = \sin \alpha \cos \beta + \cos \alpha \sin \beta = \left(\frac{3}{5}\right)\left(\frac{15}{17}\right) + \left(\frac{4}{5}\right)\left(\frac{8}{17}\right) = \frac{77}{85}.$$

Thus $\alpha + \beta$ and γ are in the first quadrant, and $\sin(\alpha + \beta) = \sin \gamma$, and so we can conclude that $\alpha + \beta = \gamma$.   ■

Exercises 5.4

In this exercise set there may be some problems in which the given expression is not defined. If a calculator is used, the display will show *Error*. Explain what part of the problem is responsible for such a result.

In problems 1 through 3, make a table of x, y values that satisfy the given equation, and then draw a graph.

1. $y = \text{Sin}^{-1}x$ **2.** $y = \text{Cos}^{-1}x$ **3.** $y = \text{Tan}^{-1}x$

In problems 4 through 24, evaluate the given expressions and give answers in exact form.

4. $\text{Sin}^{-1}1$ **5.** $\text{Cos}^{-1}(-1)$ **6.** $\text{Tan}^{-1}(-1)$

7. $\text{Arcsin}\left(-\dfrac{1}{\sqrt{2}}\right)$ **8.** $\text{Cos}^{-1}\left(\dfrac{\sqrt{3}}{2}\right)$ **9.** $\text{Arctan}\left(-\dfrac{\sqrt{3}}{3}\right)$

10. $\text{Sin}^{-1}\left(-\dfrac{3}{2\sqrt{3}}\right)$ **11.** $\text{Cos}^{-1}\left(\dfrac{2}{\sqrt{2}}\right)$ **12.** $\text{Arcsin}\dfrac{2}{\sqrt{3}}$

13. $\sin\left(\text{Sin}^{-1}\dfrac{3}{4}\right)$ **14.** $\tan\left(\text{Tan}^{-1}\dfrac{4}{3}\right)$ **15.** $\text{Cos}^{-1}\left(\sin\dfrac{3\pi}{2}\right)$

16. $\cos(2\ \text{Cos}^{-1}0)$

17. $\sin\left(\sin^{-1}\dfrac{4}{3}\right)$

18. $\cos\left[\dfrac{1}{2}\ \text{Tan}^{-1}\left(-\dfrac{3}{4}\right)\right]$

19. $\tan(2\ \text{Tan}^{-1}3)$

20. $\sec\left[\text{Cos}^{-1}\left(-\dfrac{3}{5}\right)\right]$

21. $\tan\left[\text{Cos}^{-1}\left(\dfrac{3}{5}\right)+\text{Sin}^{-1}\left(-\dfrac{3}{5}\right)\right]$

22. $\text{Sin}^{-1}\left(\cos\dfrac{\pi}{3}\right)$

23. $\cos\left(\dfrac{\pi}{2}+\text{Sin}^{-1}\dfrac{4}{5}\right)$

24. $\tan\left[\text{Tan}^{-1}(2)-\text{Cos}^{-1}\left(-\dfrac{3}{5}\right)\right]$

In problems 25 through 36, evaluate the given expressions and give answers rounded off to two decimal places.

25. $\text{Sin}^{-1}(0.3768)$

26. $\text{Arccos}(0.5732)$

27. $\text{Tan}^{-1}(-1.483)$

28. $\text{Cos}^{-1}\dfrac{\pi}{2}$

29. $\text{Arcsin}\left(\dfrac{\sqrt{47}-3}{4}\right)$

30. $\text{Arctan}\dfrac{\pi}{3}$

31. $\text{Cos}^{-1}(\sin 48°)$

32. $\text{Sin}^{-1}\left(\cos\dfrac{6}{\pi}\right)$

33. $\cos\left(2\ \text{Sin}^{-1}\dfrac{\sqrt{5}}{2}\right)$

34. $\sin\left(\text{Cos}^{-1}\dfrac{1+\sqrt{5}}{3}\right)$

35. $\cos(2\ \text{Sin}^{-1}0.4+\text{Cos}^{-1}0.6)$

36. $\sin(1.6-\text{Tan}^{-1}4)$

In problems 37 through 40, give reasons for your answers.

37. Is $\text{Sin}^{-1}\dfrac{3}{5}+\text{Sin}^{-1}\dfrac{5}{13}=\text{Sin}^{-1}\dfrac{56}{65}$?

38. Is $\text{Tan}^{-1}\dfrac{2}{3}+\text{Tan}^{-1}\dfrac{3}{2}=\dfrac{\pi}{2}$?

39. Is $\text{Tan}^{-1}(1)+\text{Tan}^{-1}(2)-\text{Tan}^{-1}(-3)=\pi$?

40. Is $\text{Sin}^{-1}\dfrac{3}{5}+\text{Sin}^{-1}\dfrac{5}{13}=\text{Cos}^{-1}\dfrac{33}{65}$?

In problems 41 through 47, determine all values of x that will satisfy the given equalities or inequalities. Give answers in exact form when it is reasonable to do so, otherwise rounded off to two decimal places.

41. a) $2\ \text{Sin}^{-1}x+1=0$ **b)** $2\ \text{Cos}^{-1}x-3=0$

42. a) $\text{Tan}^{-1}x=1$ **b)** $3\ \text{Sin}^{-1}x+4=0$

43. a) $2\ \text{Cos}^{-1}x+1\leq 0$ **b)** $1+\text{Tan}^{-1}x\leq 0$

44. a) $\cos(\text{Sin}^{-1}x)\leq 0$ **b)** $\tan(\text{Cos}^{-1}x)<0$

45. a) $\sin(\text{Cos}^{-1}x)\leq 0$ **b)** $\tan(\text{Sin}^{-1}x)<0$

46. a) $\sin(\text{Sin}^{-1}x)=x$ **b)** $\text{Sin}^{-1}(\sin x)=x$

47. a) $\cos(\text{Cos}^{-1}x)=x$ **b)** $\text{Cos}^{-1}(\cos x)=x$

48. A movie marquee on Main Street is 1.5 meters wide, and its bottom edge is 4 meters above the sidewalk, as shown in Fig. 5.20. A person with eye level h meters above the sidewalk and x meters from the point P directly below the edge of the marquee, is walking along Main Street and observes that the view of the marquee (as measured by angle θ) is small when far away (when x is large), but as the person gets closer, θ gets larger until it reaches a maximum, and then it begins to get smaller until it becomes $0°$ when the observer is directly underneath the edge of the marquee. For a given person, h is a fixed number, and θ can be considered as a function of x. Show

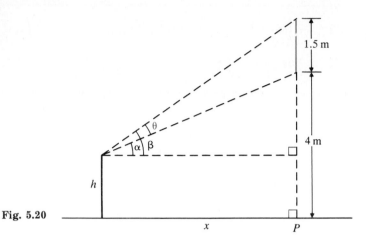

Fig. 5.20

that it is given by

$$\theta = \text{Tan}^{-1}\left[\frac{1.5x}{x^2 + (4 - h)(5.5 - h)}\right].$$

Hint: Note that $\theta = \beta - \alpha$. Use the two right triangles involving angles α and β and the identity

$$\tan(\beta - \alpha) = \frac{\tan \beta - \tan \alpha}{1 + \tan \beta \tan \alpha}$$

49. Suppose the person in problem 48 is Janet, whose eye level above the sidewalk is 1.5 meters.

a) Show that her "view" of the marquee is given by $\theta = \text{Tan}^{-1}\left(\dfrac{1.5x}{x^2 + 10}\right)$.

b) Use your calculator and the result in (a) to complete the following table, which gives her view for different values of x. The given values of x are in meters. Express θ in radians rounded off to three decimal places.

x	40	25	20	10	8	6	5	4	3.5	3.2	3.1	3.0	2.8	2.5	2.0	1.5	1.1	0.5
θ																		

c) Using the results in (b), make a reasonable estimate of how far from point P Janet should stand to get the "best view" (that is, the largest value of θ).

50. Suppose the person in problem 48 is Preston, whose eye level above the sidewalk is 2 meters.

a) Show that his "view" of the marquee is given by $\theta = \text{Tan}^{-1}\left(\dfrac{1.5x}{x^2 + 7}\right)$.

b) Make a table similar to that in problem 49.

c) How far from point P should Preston stand to get the "best view"?

5.5   **SOLUTION OF TRIGONOMETRIC EQUATIONS**

An equation is an identity if it is satisfied by all values of the variable (or variables) for which both sides are defined; otherwise, it is called a *conditional equation*. Techniques for solving conditional equations involving trigonometric functions are discussed in this section. These are illustrated by considering a variety of examples.

Example 1   Find the solution set for $2 \sin x - 1 = 0$. Give answer in terms of real numbers (radians).

Solution   This is a linear equation in $\sin x$. First solve for $\sin x$, obtaining $\sin x = \frac{1}{2}$. Thus we wish to determine all angles x whose sine is $\frac{1}{2}$. The sine function is positive for angles in the first or second quadrants. Two solutions, $\pi/6$ and $5\pi/6$, are shown in Fig. 5.21. All other solutions are angles that are coterminal with one of these two. Thus the solution set S is given by

$$S = \left\{ x \,\middle|\, x = \frac{\pi}{6} + k \cdot 2\pi \quad \text{or} \quad x = \frac{5\pi}{6} + k \cdot 2\pi, \; k \text{ is an integer} \right\}.$$

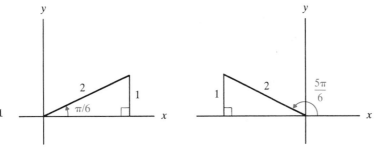

Fig. 5.21

Example 2   Find the solution set for equation $2 \sin^2 x - \cos^2 x - 5 \sin x - 1 = 0$. Express answer in terms of real numbers rounded off to four decimal places.

Solution   In the given equation use identity **(I.9)** to replace $\cos^2 x$ by $1 - \sin^2 x$. After simplifying, we get a quadratic equation in $\sin x$ that can be solved by factoring:

$$3 \sin^2 x - 5 \sin x - 2 = 0,$$
$$(\sin x - 2)(3 \sin x + 1) = 0.$$

Hence $\sin x = 2$ or $\sin x = -\frac{1}{3}$. Since $-1 \leq \sin x \leq 1$, there is no value of x for which $\sin x = 2$. To find solutions for $\sin x = -\frac{1}{3}$, use a calculator to get $\text{Sin}^{-1}(-\frac{1}{3}) = -0.3398$ as one solution. This angle is in quadrant IV, but there is another angle 3.4814 in quadrant III, whose sine is $-\frac{1}{3}$, as shown in Fig. 5.22. All other solutions will be coterminal with one of these two angles, and so the solution set S is given by

$$S = \{ x \,|\, x = -0.3398 + k \cdot 2\pi \quad \text{or} \quad x = 3.4814 + k \cdot 2\pi \}.$$

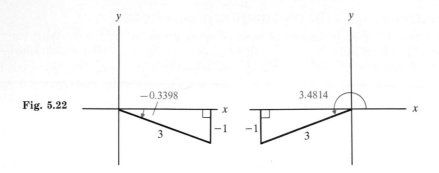

Fig. 5.22

Example 3   Find the solution set for the equation $\sin 2x - \sqrt{3} \cos 2x = 0$, where the replacement set is $\{x \mid 0 \leq x \leq 2\pi\}$. Give results in exact form.

Solution   The given equation involves both the sine and cosine functions, but we can express it in terms of a single function, as follows: Add $\sqrt{3} \cos 2x$ to both sides of the given equation, and then divide by $\cos 2x$ to get

$$\frac{\sin 2x}{\cos 2x} = \sqrt{3}.$$

Using identity **(I.7)** gives $\tan 2x = \sqrt{3}$. Recall that $\tan 2x > 0$ for $2x$ in the first or third quadrants. Thus two values of $2x$, $\pi/3$ and $4\pi/3$, are shown in Fig. 5.23; other values are given by angles coterminal to these. Thus

$$2x = \frac{\pi}{3} + k \cdot 2\pi \qquad \text{or} \qquad 2x = \frac{4\pi}{3} + k \cdot 2\pi.$$

Solving for x, we get

$$x = \frac{\pi}{6} + k \cdot \pi \qquad \text{or} \qquad x = \frac{2\pi}{3} + k \cdot \pi.$$

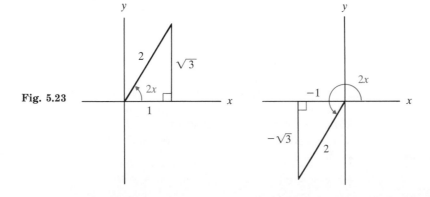

Fig. 5.23

The values of k that will give solutions in the interval $0 \le x \le 2\pi$ are given by $k = 0$ or $k = 1$, and so the solution set S is given by

$$S = \left\{ \frac{\pi}{6}, \frac{7\pi}{6}, \frac{2\pi}{3}, \frac{5\pi}{3} \right\}. \qquad \blacksquare$$

Example 4   Find all values of x that will make $3 \sin(2x - \frac{\pi}{4})$ a maximum. Give answers in exact form.

Solution   Since the maximum value of $\sin \theta$ is 1, the problem is to determine all values of x for which $\sin(2x - \frac{\pi}{4}) = 1$; this will make $3 \sin(2x - \frac{\pi}{4}) = 3$, which is its maximum value. Recall that $\sin \frac{\pi}{2} = 1$, and so all solutions will be given by

$$2x - \frac{\pi}{4} = \frac{\pi}{2} + k \cdot 2\pi.$$

Solving for x, we get

$$x = \frac{3\pi}{8} + k\pi.$$

Therefore $3 \sin(2x - \frac{\pi}{4})$ will assume a maximum value at each x in the set S given by

$$S = \left\{ x \mid x = \frac{3\pi}{8} + k\pi, \ k \text{ is an integer} \right\}. \qquad \blacksquare$$

Equations of the Form $a \sin x + b \cos x = c$
An equation of the form $a \sin x + b \cos x = c$ (where a, b, c are given numbers and a, b are not both zero) can be solved as follows: divide both sides of the equation by $\sqrt{a^2 + b^2}$ to get

$$\frac{a}{\sqrt{a^2 + b^2}} \sin x + \frac{b}{\sqrt{a^2 + b^2}} \cos x = \frac{c}{\sqrt{a^2 + b^2}} \qquad (5.6)$$

Figure 5.24 shows an angle α that has its terminal side passing through the point (a, b). (The diagram shown is for a negative and b positive.)

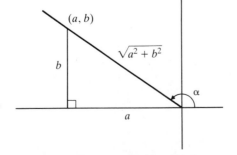

Fig. 5.24

Since a and b are given, angle α is determined. Note that $\cos\alpha = a/\sqrt{a^2 + b^2}$ and $\sin\alpha = b/\sqrt{a^2 + b^2}$, and so Eq. (5.6) can be written as

$$\cos\alpha\sin x + \sin\alpha\cos x = \frac{c}{\sqrt{a^2 + b^2}}.$$

The left-hand side of this equation reminds us of the identity for the sine of a sum of two angles, and indeed it can be replaced by $\sin(\alpha + x)$ (see identity **(I.12)** of Section 5.2). Therefore, the given equation can be written in equivalent form as

$$\sin(\alpha + x) = \frac{c}{\sqrt{a^2 + b^2}}, \tag{5.7}$$

and this is the form we can use to find the solution set.

Example 5   Find the solution set for the equation $3\sin x - 4\cos x = 5$, where the replacement set is $\{x \,|\, 0 \le x \le 2\pi\}$. Express results rounded off to three decimal places.

Solution   We first divide both sides of the given equation by $\sqrt{(3)^2 + (-4)^2} = \sqrt{25} = 5$:

$$\tfrac{3}{5}\sin x - \tfrac{4}{5}\cos x = \tfrac{5}{5}. \tag{5.8}$$

Plot the point $(3, -4)$, and let α be the angle shown in Fig. 5.25. We see that $\cos\alpha = 3/5$ and $\sin\alpha = -4/5$, and substituting these into Eq. (5.8) gives $\sin x\cos\alpha + \cos x\sin\alpha = 1$. This can be written as

$$\sin(x + \alpha) = 1. \tag{5.9}$$

Angle α is given by $\alpha = \mathrm{Sin}^{-1}(-4/5)$, and using a calculator, we get $\alpha = -0.9273$. Therefore Eq. (5.9) becomes

$$\sin(x - 0.9273) = 1. \tag{5.10}$$

Hence,

$$x - 0.9273 = \frac{\pi}{2} + k \cdot 2\pi \qquad \text{or} \qquad x = 0.9273 + \frac{\pi}{2} + k \cdot 2\pi.$$

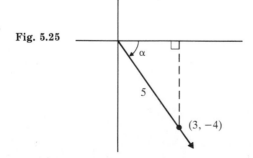

Fig. 5.25

The only value of k that will give a solution satisfying $0 \le x \le 2\pi$ is zero. Thus,

$$x = 0.9273 + \frac{\pi}{2} = 2.4981,$$

and the solution set is given by $S = \{2.498\}$.   ■

Example 6   Solve the equation $\sin 2x - \sin x = 0$. Express answer in terms of real numbers in exact form.

Solution   Using **(I.18)** to replace $\sin 2x$ in the given equation by $2 \sin x \cos x$ gives $2 \sin x \cos x - \sin x = 0$. This can be written as $\sin x(2 \cos x - 1) = 0$. Therefore the given equation is equivalent to $\sin x = 0$ or $\cos x = \frac{1}{2}$. From $\sin x = 0$ we get solutions of the form $x = k\pi$, and $\cos x = \frac{1}{2}$ gives

$$x = \frac{\pi}{3} + 2k\pi \qquad \text{or} \qquad x = -\frac{\pi}{3} + 2k\pi.$$

Therefore the solution set S is

$$S = \left\{ x \,|\, x = k\pi \quad \text{or} \quad x = \frac{\pi}{3} + 2k\pi \quad \text{or} \quad x = -\frac{\pi}{3} + 2k\pi, \ k \text{ is an integer} \right\}. \quad ■$$

Exercises 5.5

Express answers in exact form when it is reasonable to do so. Otherwise use a calculator and give results rounded off to two decimal places.

In problems 1 through 8, find all solutions and give answers in radians.

1. $2 \cos x + 1 = 0$
2. $2 \sin x + \sqrt{3} = 0$
3. $\sqrt{3} \tan x - 1 = 0$
4. $\sec x - 2 = 0$
5. $\tan^2 x - 1 = 0$
6. $\sqrt{3} \sin x - 4 = 0$
7. $\cos^2 x + 2 \cos x + 1 = 0$
8. $1 - 4 \sin^2 x = 0$

In problems 9 through 16, find all solutions and give answers in degrees.

9. $2 \sin x + 1 = 0$
10. $2 \cos x + \sqrt{3} = 0$
11. $\sqrt{3} \cot x + 1 = 0$
12. $\sqrt{3} \sec x - 2 = 0$
13. $\sqrt{3} \cos x - 4 = 0$
14. $2 \sin(x + 30°) - 1 = 0$
15. $2 \cos(x + 60°) + 1 = 0$
16. $2 \sin^2 x + 5 \sin x - 3 = 0$

In problems 17 through 41, find the solution set where the replacement set is $\{x \,|\, 0 \le x \le 2\pi\}$. *Note:* This implies that x is in radians.

17. $2 \sin x - \sqrt{3} = 0$
18. $2 \sin x - \sin^2 x = \cos^2 x$
19. $2 \sec x - \sqrt{3} = 0$
20. $\cot x + \sqrt{3} = 0$
21. $3.5 \sin x - 2.4 = 0$
22. $3 \cos x - 2 = 0$
23. $3 \sin x - 5 \cos x = 0$
24. $2 \sin x + 3 \cos x = 0$
25. $2 \sin 2x + \cos 2x = 0$
26. $4 - \tan^2 x = 0$

27. $\sin^2 x + 2 \sin x + 1 = 0$

28. $\cos^2 x - 3 \cos x - 2 = 0$

29. $2 \sec^2 x - 3 \sec x - 2 = 0$

30. $2 \sin^2 x + 2 \sin x - 1 = 0$

31. $3 \cos^2 x + 4 \cos x + 2 = 0$

32. $25 \sin^2 x = 30 \sin x + 7$

33. $\cos^2 x - 1.5 \cos x - 0.48 = 0$

34. $\sin^2 x - 2.4 \sin x - 1.8 = 0$

35. $9 \sin^2 x - 6 \sin x + 1 = 0$

36. $\tan^2 x - 4 \tan x + 3 = 0$

37. $\sqrt{3} \sin x + \cos x = 2$

38. $\sin x + \cos x = 1$

39. $2 \cos^2 x - \sin x \cos x = 0$

40. $\sin x + \cos x = 2$

41. $\sin^2 x + 2 \cos^2 x = 1$

In problems 42 through 45, for each of the given functions f find all values of x that will give the maximum value of $f(x)$.

42. $f(x) = 2 \sin\left(2x - \frac{\pi}{3}\right)$, $\mathfrak{D}(f) = \{x \,|\, 0 \leq x \leq 2\pi\}$

43. $f(\mathrm{x}) = 3 \cos\left(2x + \frac{\pi}{4}\right)$, $\mathfrak{D}(f) = \{x \,|\, \pi \leq x \leq 3\pi\}$

44. $f(x) = 4 - \cos 3x$, $\mathfrak{D}(f) = \{x \,|\, -\pi \leq x \leq 3\pi\}$

    Hint: $4 - \cos 3x$ is maximum when $\cos 3x$ is minimum.

45. $f(x) = 3 - 2 \sin 4x$, $\mathfrak{D}(f) = \{x \,|\, 0 \leq x \leq 2\pi\}$

    Hint: $3 - 2 \sin 4x$ is maximum when $\sin 4x$ is a minimum.

5.6   GRAPHS OF GENERAL SINE AND COSINE FUNCTIONS

In Section 4.5 we discussed graphs of $y = \sin x$ and $y = \cos x$; see Figs. 4.34 and 4.35. We observed that these functions are periodic with period 2π, and so it is sufficient to draw their graphs over an interval of one period.

In applications, particularly in engineering and physics, one frequently encounters the more general functions given by

$$y = a \sin(bx + c) \quad \text{and} \quad y = a \cos(bx + c), \tag{5.11}$$

where a, b, and c are called *parameters;* that is, they are specified real numbers in any particular case. We make the obvious exceptions that $a \neq 0$, $b \neq 0$.

We shall investigate properties of graphs of functions given by Eqs. (5.11) by considering a sequence of particular cases in order to determine what role each of the three parameters plays.

Example 1   Draw a graph of $y = 3 \sin x$.

Solution   In order to draw the graph of $y = 3 \sin x$, first draw the graph of $y = \sin x$. This is shown by the broken curve in Fig. 5.26. To draw the graph of $y = 3 \sin x$, note that for any given value of x the value of y is 3 times the corresponding value of y in $y = \sin x$. Thus we get the solid curve shown in Fig. 5.26.

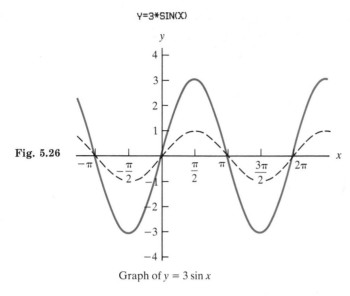

Fig. 5.26

Graph of $y = 3 \sin x$

From the graph in Fig. 5.26, we note that the function given by $y = 3 \sin x$ is periodic with period 2π. The graph is a "sine curve" that oscillates about the x-axis and reaches a maximum distance of three units above and below the axis. We describe this by saying that the *amplitude* of $y = 3 \sin x$ is 3. In general, then, we can conclude that $y = a \sin x$ is a sine curve with period 2π and amplitude $|a|$. *Parameter a plays the part of amplitude.*

Example 2   Draw a graph of $y = \sin 2x$.

Solution   Since the sine function has period 2π, it will be sufficient to draw a graph of the given function over the interval $0 \leq 2x \leq 2\pi$; that is, $0 \leq x \leq \pi$. The following table gives pairs of x, y values satisfying the given equation for x in this interval. These are used to draw the solid portion of the curve shown in Fig. 5.27. Since

x	0	$\dfrac{\pi}{8}$	$\dfrac{\pi}{4}$	$\dfrac{3\pi}{8}$	$\dfrac{\pi}{2}$	$\dfrac{5\pi}{8}$	$\dfrac{3\pi}{4}$	$\dfrac{7\pi}{8}$	π
y	0	$\dfrac{\sqrt{2}}{2}$	1	$\dfrac{\sqrt{2}}{2}$	0	$-\dfrac{\sqrt{2}}{2}$	-1	$-\dfrac{\sqrt{2}}{2}$	0

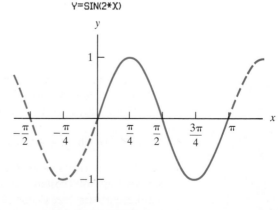

Fig. 5.27

Graph of $y = \sin 2x$

$\sin 2(x + \pi) = \sin(2x + 2\pi) = \sin 2x$, the graph can be continued as shown by the broken portion of the curve in Fig. 5.27.

We see from the graph of Fig. 5.27 that the function given by $y = \sin 2x$ is periodic with period π. ■

Example 2 suggests the following generalization: The function given by $y = \sin bx$ is a "sine curve" with period $2\pi/|b|$, where the absolute value of b is used because we want the period to be a positive number. *Parameter b plays the role of the period.*

Example 3   Draw a graph of $y = \sin\left(x + \dfrac{\pi}{4}\right)$.

Solution   Since the sine function has period 2π, it is sufficient to draw a graph for values of x satisfying $0 \le x + \pi/4 \le 2\pi$; that is, $-\pi/4 \le x \le 7\pi/4$. The following table gives pairs of x, y values satisfying the given equation for x in this interval. These are used to draw the solid portion of the curve shown in Fig. 5.28(a). Since

$$\sin\left[(x + 2\pi) + \frac{\pi}{4}\right] = \sin\left[\left(x + \frac{\pi}{4}\right) + 2\pi\right] = \sin\left(x + \frac{\pi}{4}\right),$$

the graph can be continued as shown by the broken portion of the graph in Fig. 5.28(a).

x	$-\dfrac{\pi}{4}$	0	$\dfrac{\pi}{4}$	$\dfrac{\pi}{2}$	$\dfrac{3\pi}{4}$	π	$\dfrac{5\pi}{4}$	$\dfrac{3\pi}{2}$	$\dfrac{7\pi}{4}$
y	0	$\dfrac{\sqrt{2}}{2}$	1	$\dfrac{\sqrt{2}}{2}$	0	$-\dfrac{\sqrt{2}}{2}$	-1	$-\dfrac{\sqrt{2}}{2}$	0

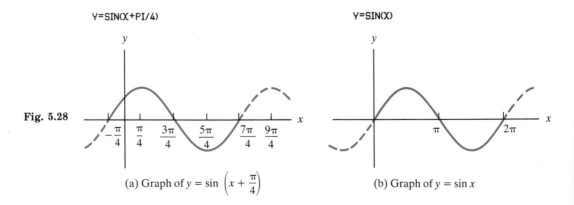

Fig. 5.28

(a) Graph of $y = \sin\left(x + \dfrac{\pi}{4}\right)$ (b) Graph of $y = \sin x$

In Fig. 5.28(b) the graph of the *standard sine function,* $y = \sin x$, is shown. Looking at the two graphs in Fig. 5.28, note that if the graph of the standard sine function is translated horizontally to the left $\pi/4$ units, we get the graph of $y = \sin(x + \pi/4)$. Such a horizontal translation is called a *phase shift*. We say that the graph of $y = \sin(x + \pi/4)$ has a phase shift of $\pi/4$ to the left.   ■

Example 3 suggests that in general the *parameter c* in $y = a \sin(bx + c)$ *plays the role of phase shift.*

The preceding three examples lead us to the following generalization (which can be proved although we are not interested in doing so here):

> The graph of $y = a \sin(bx + c)$, where $a \neq 0$ and $b \neq 0$, is a sine curve with period $2\pi/|b|$, amplitude $|a|$, and phase shift $|c/b|$. That is, to get the graph of $y = a \sin(bx + c)$, move the graph of $y = a \sin bx$ horizontally a distance of $|c/b|$ units to the left if $c/b > 0$ and to the right if $c/b < 0$.

If we replaced sine by cosine in the above discussion we would arrive at similar corresponding results. We shall merely state the following general conclusion.

> The graph of $y = a \cos(bx + c)$, where $a \neq 0$ and $b \neq 0$, is a cosine curve with period $2\pi/|b|$, amplitude $|a|$, and phase shift of $|c/b|$.

Example 4   Draw a graph of $y = -4 \sin(\pi/2 - 2x)$.

Solution   We can first write the given equation as $y = -4 \sin[-(2x - \pi/2)]$. Using the identity $\sin(-\theta) = -\sin\theta$ gives $y = 4 \sin(2x - \pi/2)$. Therefore the graph is a sine curve with period $2\pi/2 = \pi$, amplitude 4, and phase shift of $(\pi/2) \div 2$, or $\pi/4$ to the right. This is shown in Fig. 5.29, in which the graph of $y = 4 \sin 2x$ (shown as the broken curve) is moved to the right $\pi/4$ units to get the graph of the given equation shown as the solid curve.

As a check, we suggest locating a few "key points" on the graph, such as x-intercepts and highest or lowest points. These are given in the following table for the primary cycle given by $0 \leq 2x - \pi/2 \leq 2\pi$; that is, $\pi/4 \leq x \leq 5\pi/4$.

x	$\dfrac{\pi}{4}$	$\dfrac{\pi}{2}$	$\dfrac{3\pi}{4}$	π	$\dfrac{5\pi}{4}$	$\cdots$
y	0	4	0	-4	0	$\cdots$

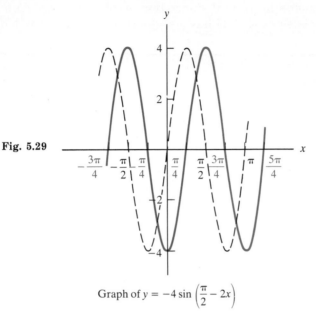

Y=−4*SIN(PI/2−2*X)

Fig. 5.29

Graph of $y = -4 \sin\left(\dfrac{\pi}{2} - 2x\right)$

Note: Since $\sin(\pi/2 - 2x) = \cos 2x$ is an identity, we could have written the given equation as $y = -4\cos 2x$ and used this equation to draw the graph.

Example 5   Draw a graph of $y = \sin x + \cos x$.

Solution   First apply a technique similar to that used in Section 5.5 (see Example 5), in which the given equation can be written as

$$y = \sqrt{2}\left(\frac{1}{\sqrt{2}}\sin x + \frac{1}{\sqrt{2}}\cos x\right)$$

Since $\sin \pi/4 = 1/\sqrt{2}$ and $\cos \pi/4 = 1/\sqrt{2}$, we can write

$$y = \sqrt{2}\left(\sin x \cos \frac{\pi}{4} + \cos x \sin \frac{\pi}{4}\right).$$

Using identity **(I.12)** gives $y = \sqrt{2}\sin(x + \pi/4)$. From this equation we see that the graph of the given function is a sine curve with period 2π, amplitude $\sqrt{2}$, and phase shift of $\pi/4$ to the left. This is shown as the solid curve in Fig. 5.30; the broken curve is the graph of $y = \sqrt{2}\sin x$.

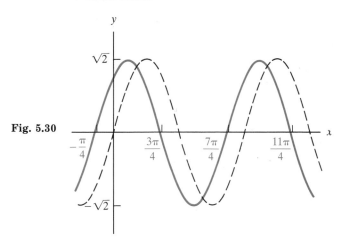

Y=SIN(X)+COS(X)

Fig. 5.30

Graph of $y = \sin x + \cos x$

Exercises 5.6 _____

In each of the following, give the period, amplitude, and phase shift. Then draw a graph of the given equation.

1. $y = 2 \sin x$ 2. $y = -3 \sin x$ 3. $y = -2 \cos x$ 4. $y = \cos 2x$

5. $y = 3 \sin(-2x)$ 6. $y = 3 \cos(-2x)$ 7. $y = 3 \sin(\pi x)$ 8. $y = -2 \cos(-\pi x)$

9. $y = 2 \sin\left(\frac{\pi}{2}x\right)$ 10. $y = \sin\left(x + \frac{\pi}{3}\right)$ 11. $y = 2 \cos\left(x - \frac{\pi}{2}\right)$ 12. $y = 3 \sin\left(x - \frac{\pi}{4}\right)$

13. $y = -3 \sin\left(2x - \frac{\pi}{3}\right)$ 14. $y = 2 \cos\left(\pi x - \frac{\pi}{4}\right)$ 15. $y = 4 \sin\left(2\pi x - \frac{\pi}{2}\right)$ 16. $y = 3 \sin\left(\frac{\pi}{2} - 2x\right)$

17. $y = \sin x - \cos x$ 18. $y = \sin x + \sqrt{3} \cos x$ 19. $y = \sqrt{3} \sin x - \cos x$ 20. $y = 2 \sin x + 2 \cos x$

5.7 Looking Ahead to Calculus

In Sections 2.7 and 3.6 we introduced ideas related to limits. Here we continue that discussion with examples involving trigonometric functions.

Example 1   Determine $\lim\limits_{x \to 0} \frac{\sin x}{x}$.

Solution   First we make a table giving values of $(\sin x)/x$ corresponding to values of x near zero both positive and negative.* However, since $f(x) = (\sin x)/x$ is an even func-

_____

* Note that we do not include extremely small values of x in the table. Calculators cannot handle such numbers without introducing substantial round-off errors.

tion, it is sufficient to use only positive values of x. Since x represents a real number, place the calculator in radian mode.

x	0.5	0.1	0.01	0.001
$\dfrac{\sin x}{x}$	0.959	0.998	0.99998	1.00000

From the values of $(\sin x)/x$ appearing in the table, we conlude that

$$\lim_{x \to 0} \frac{\sin x}{x} = 1.$$

Example 2   Determine $\displaystyle\lim_{x \to \infty} x\left(\frac{\pi}{2} - \text{Tan}^{-1}2x\right)$.

Solution   We make a table giving values of $x(\pi/2 - \text{Tan}^{-1}2x)$ corresponding to large values of x. First place the calculator in radian mode.

x	10	100	1000
$x(\pi/2 - \text{Tan}^{-1}2x)$	0.49958	0.499996	0.500000

From values appearing in the table, we conclude that

$$\lim_{x \to \infty} x\left(\frac{\pi}{2} - \text{Tan}^{-1}2x\right) = 0.5.$$

Example 3   Determine the slope of the line that is tangent to the curve $y = \cos x$ at the point $P{:}(\pi/3, 1/2)$. Draw a graph and show the tangent line.

Solution   The slope m of the tangent line is given by

$$m = \lim_{h \to 0} \frac{\cos\left(\dfrac{\pi}{3} + h\right) - \cos\dfrac{\pi}{3}}{h} = \lim_{h \to 0} \frac{\cos\left(\dfrac{\pi}{3} + h\right) - \dfrac{1}{2}}{h}.$$

First we make a table giving values of the difference quotient corresponding to values of h approaching zero. Place the calculator in radian mode.

h	0.5	0.1	0.01	0.001	$\ldots$	-0.5	-0.1	-0.01	-0.001
$\dfrac{\cos(\pi/3 + h) - 1/2}{h}$	-0.953	-0.890	-0.869	-0.8663	$\ldots$	-0.708	-0.840	-0.864	-0.8658

From the table of values of the difference quotients, we conclude that $m = -0.866$. The graph is shown in Fig. 5.31.

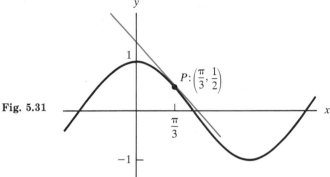

Fig. 5.31

Example 4   Find an equation of the line that is tangent to the curve $y = \text{Sin}^{-1}x$ at the point $P:(1/2, \pi/6)$. Give result with numbers rounded off to three decimal places.

Solution   The slope m of the tangent line is given by

$$m = \lim_{h \to 0} \frac{\text{Sin}^{-1}\left(\dfrac{1}{2} + h\right) - \text{Sin}^{-1}\dfrac{1}{2}}{h} = \lim_{h \to 0} \frac{\text{Sin}^{-1}\left(\dfrac{1}{2} + h\right) - \dfrac{\pi}{6}}{h}.$$

After placing the calculator in radian mode, we get the following table giving values of the difference quotient corresponding to values of h approaching zero.

h	0.5	0.1	0.01	0.001	...	-0.5	-0.1	-0.01	-0.001
$\dfrac{\text{Sin}^{-1}(1/2 + h) - \pi/6}{h}$	2.094	1.199	1.159	1.1551	...	1.047	1.121	1.151	1.1543

From the table of values we conclude that $m = 1.155$, and so an equation of the tangent line is given by $y - \dfrac{\pi}{6} = 1.155\left(x - \dfrac{1}{2}\right)$. This can be written as $y = 1.155x - 0.054$.

Exercises 5.7

In problems 1 through 10, evaluate the given limits. Give answers rounded off to two decimal places.

1. $\lim\limits_{x \to 0} \dfrac{\sin 2x}{x}$

2. $\lim\limits_{x \to 0} \dfrac{\sin 3x}{x}$

3. $\lim\limits_{x \to 0} \dfrac{1 - \cos x}{x}$

4. $\lim\limits_{x \to 0} \dfrac{1 - \cos^2 2x}{x^2}$

5. $\lim\limits_{x \to 1} \dfrac{\pi/4 - \text{Tan}^{-1}x}{\ln x}$

6. $\lim\limits_{x \to 0} \dfrac{\text{Sin}^{-1}2x}{x}$

7. $\lim\limits_{x \to \infty} x \sin \dfrac{1}{x}$

8. $\lim\limits_{x \to \infty} x \tan \dfrac{2}{x}$

9. $\lim\limits_{x \to \pi/3} \dfrac{2 \cos x - 1}{3x - \pi}$

10. $\lim\limits_{x \to \pi/6} \dfrac{2 \sin x - 1}{6x - \pi}$

In problems 11 through 14, a function f and a number c are given. Determine the slope of the line that is tangent to the curve $y = f(x)$ at the point $(c, f(c))$.

11. $f(x) = \sin x; \; c = \dfrac{\pi}{6}$

12. $f(x) = x \cos x; \; c = \dfrac{\pi}{2}$

13. $f(x) = \tan x; \; c = \dfrac{\pi}{4}$

14. $f(x) = x^2 \sin x; \; c = 1$

15. Suppose $f(x) = \sin x$. The slope of the line that is tangent to the curve $y = f(x)$ at the point $(c, f(c))$ depends on the given value of c. Denote this slope by $m(c)$. In each of the following give answers rounded off to three decimal places.

a) Determine $m(c)$ for each of the values of c in the following table.

c	0	1	1.5	3.4	-0.8
$m(c)$					

b) Complete the following table giving values of $\cos c$ corresponding to the values of c used in (a).

c	0	1	1.5	3.4	-0.8
$\cos c$					

c) Compare the results obtained in (a) and (b) and guess a formula that will give the value of $m(c)$ for any real number c. Use the formula to determine $m(2.5)$ and $m(-\sqrt{5})$.

Review Exercises

In problems 1 through 8, prove that the given equations are identities.

1. $\cos x \tan x = \sin x$

2. $\csc \theta \sin 2\theta = 2 \cos \theta$

3. $2 \sin^2 \dfrac{x}{2} = \dfrac{\sin^2 x}{1 + \cos x}$

4. $\sin x \tan \dfrac{x}{2} = 1 - \cos x$

5. $\left(\sin \dfrac{x}{2} - \cos \dfrac{x}{2} \right)^2 = 1 - \sin x$

6. $2 \sin \left(x + \dfrac{\pi}{6} \right) = \sqrt{3} \sin x + \cos x$

7. $\cos \left(\dfrac{\pi}{2} + x \right) = \cos x \tan(-x)$

8. $\cos^4 \dfrac{x}{2} - \sin^4 \dfrac{x}{2} = \cos x$

In problems 9 through 20, evaluate the given expressions in exact form, where α and β are angles that satisfy

$$\sin \alpha = \frac{3}{5} \quad \text{and} \quad \frac{\pi}{2} \leq \alpha \leq \pi,$$

$$\tan \beta = -\frac{5}{12} \quad \text{and} \quad -\frac{\pi}{2} < \beta < \frac{\pi}{2}.$$

9. $\sin 2\alpha$

10. $\sin \dfrac{\beta}{2}$

11. $\sin(\alpha + \beta)$

12. $\tan\left(\beta + \dfrac{\pi}{4}\right)$

13. $\cos 2\beta$

14. $\sec^2\alpha - \tan^2\alpha$

15. $\cos 2(\alpha - \beta)$

16. $\cos \dfrac{\beta}{2}$

17. $\tan 2\alpha$

18. $\sin^2\alpha + \cos^2\beta$

19. $\cos\left(\alpha + \dfrac{\pi}{3}\right)$

20. $\cos\left(\dfrac{\pi}{2} + 2\alpha\right)$

In problems 21 through 26, evaluate the given expressions in exact form. Give answers as real numbers (radians).

21. $\mathrm{Sin}^{-1}\left(-\dfrac{\sqrt{3}}{2}\right)$

22. $\mathrm{Cos}^{-1}\left(-\dfrac{1}{2}\right)$

23. $\mathrm{Cos}^{-1}(-1) - \mathrm{Tan}^{-1}(-1)$

24. $\mathrm{Tan}^{-1}\left(-\dfrac{1}{\sqrt{3}}\right)$

25. $\mathrm{Sin}^{-1}\left(\cos \dfrac{5\pi}{3}\right)$

26. $\mathrm{Cos}^{-1}\left(\cos \dfrac{4\pi}{3}\right)$

In problems 27 through 32, evaluate the given expressions in exact form, where α and β are given by

$$\alpha = \mathrm{Sin}^{-1}(-\tfrac{3}{5}), \qquad \beta = \mathrm{Tan}^{-1}\tfrac{5}{12}.$$

27. $\cos(\alpha + \beta)$

28. $\sin 2\alpha$

29. $\tan \dfrac{\beta}{2}$

30. $\tan(\alpha - \beta)$

31. $\cos\left(\alpha + \dfrac{\pi}{2}\right)$

32. $\sin\left(\dfrac{\alpha + \beta}{2}\right)$

In problems 33 through 40, evaluate the given expressions, and given results rounded off to three decimal places.

33. $\sin(\mathrm{Sin}^{-1}0.4362)$

34. $\mathrm{Sin}^{-1}0.4 + \mathrm{Cos}^{-1}0.5$

35. $\mathrm{Cos}^{-1}(\tan 123°)$

36. $\mathrm{Tan}^{-1}\left(\cot \dfrac{5\pi}{17}\right)$

37. $\cos\left(\dfrac{\pi}{4} - \mathrm{Cos}^{-1}0.41\right)$

38. $\mathrm{Cos}^{-1}\left(\dfrac{1}{2}\tan 0.75\right)$

39. $\sin(\mathrm{Sin}^{-1}0.45 - \mathrm{Cos}^{-1} - 0.32)$

40. $\sec\left(\mathrm{Sin}^{-1}\dfrac{1 + \sqrt{3}}{4}\right)$

In problems 41 through 48, assume that the replacement set is $\{x \mid 0 \leq x \leq 2\pi\}$ and solve the given equations. Express answers in exact form whenever it is reasonable to do so; otherwise give results rounded off to three decimal places.

41. $2\cos x - 1 = 0$

42. $2\sin \dfrac{x}{2} - \sqrt{3} = 0$

43. $3\sin x - 5\cos x = 0$

44. $\sin^2 x + 2\cos^2 x = 2$

45. $3\sin x - 4\cos x = 5$

46. $3\cos^2 x + \cos x - 1 = 0$

47. $\tan\left(\dfrac{3\pi}{2} - x\right) = \cos x$

48. $\cos^2 x - \sin 2x = 0$

49. $2 - \mathrm{Sin}^{-1}x = 0$

50. $1 - 2\mathrm{Sin}^{-1}x = 0$

In problems 51 through 56, evaluate the given expressions, where functions f, g, and h are given by

$$f(x) = \text{Sin}^{-1}x, \qquad g(x) = \cos x, \qquad h(x) = \sqrt{1 - x^2}.$$

Give answers in exact form whenever it is reasonable to do so, otherwise round off to two decimal places.

51. $(f \circ g)\left(\dfrac{\pi}{3}\right)$

52. $(g \circ f)\left(\dfrac{\pi}{3}\right)$

53. $(h \circ g)(1)$

54. $(f \cdot g)\left(\dfrac{\pi}{6}\right)$

55. $\left(\dfrac{f}{g}\right)(1)$

56. $(f \circ h)\left(\dfrac{1}{\sqrt{2}}\right)$

In problems 57 through 64, determine whether the given statement is true or false. Give reasons for your answers. Recall that $\mathcal{D}(f)$ and $\mathcal{R}(f)$ denote the domain and range of the function f, respectively.

57. $\sin\left(\dfrac{1 + \sqrt{3}}{2}\right)$ is not defined.

58. $\dfrac{\pi}{3}$ is in $\mathcal{D}(\text{Sin}^{-1})$

59. $\dfrac{\pi}{3}$ is in $\mathcal{R}(\text{Sin}^{-1})$

60. $\dfrac{\pi}{2} < \text{Tan}^{-1}(-1) < \pi$

61. $\text{Sin}^{-1}\left(\sin\dfrac{\pi}{3}\right) = \sin\left(\text{Sin}^{-1}\dfrac{\pi}{3}\right)$

62. $\text{Sin}^{-1}\left(\sin\dfrac{\pi}{4}\right) = \sin\left(\text{Sin}^{-1}\dfrac{\pi}{4}\right)$

63. $\text{Cos}^{-1}\left(\sin\dfrac{\pi}{6}\right) = \sin\left(\text{Cos}^{-1}\dfrac{\pi}{6}\right)$

64. $\sqrt{1 - \sin^2 x} = \cos x$ for every $x \geq 0$.

Systems of Equations and Inequalities

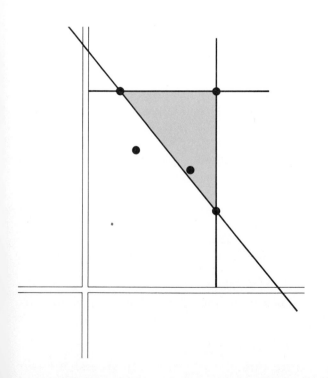

Suppose the hypotenuse of a right triangle is 17 and the perimeter is 40, and we wish to determine the lengths of the two sides. Here we are given two relations, and we want to find two "unknown" quantities.

Let x and y represent the lengths of the two sides, as shown in Fig. 6.1. From the given information, x and y must satisfy the two equations

$$x^2 + y^2 = 17^2,$$

$$x + y + 17 = 40.$$

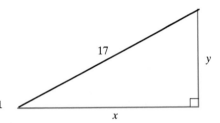

Fig. 6.1

Thus the problem becomes one of solving the system of equations

$$x^2 + y^2 = 289,$$

$$x + y = 23.$$

That is, we want to determine pairs of numbers x and y that will simultaneously satisfy these two equations.

For example, it is a simple matter to check that $x = 8$, $y = 15$ is a solution since $8^2 + 15^2 = 64 + 225 = 289$ and $8 + 15 = 23$. Similarly, we can check to see that $x = 15$, $y = 8$ is also a solution.

Developing techniques for determining solutions of systems of equations of this type is one of our primary concerns in this chapter. We shall delay discussion of details related to the solution of the system in this example until Example 1, Section 6.4.

In Sections 6.1 and 6.2, techniques for solving systems of linear equations are studied; Section 6.3 deals with systems of linear inequalities, and in Section 6.4 we discuss nonlinear systems.

6.1 SYSTEMS OF LINEAR EQUATIONS; MATRICES

A *linear equation in two variables* x and y is one that can be written in the form

$$ax + by = c,$$

where a, b, and c are given numbers with not both a and b equal to zero. Similarly,

$$ax + by + cz = d$$

represents a linear equation in three variables x, y, and z. In general, a *linear equation in n variables* $x_1, x_2, \ldots, x_n$ can be written as

$$a_1 x_1 + a_2 x_2 + \cdots + a_n x_n = b,$$

where $a_1, a_2, \ldots, a_n$ and b are given numbers and not all a_k are zero.

Suppose we are given two linear equations in two variables:

$$\begin{aligned} a_1 x + b_1 y &= c_1, \\ a_2 x + b_2 y &= c_2, \end{aligned} \tag{6.1}$$

and we wish to determine pairs of numbers x, y that satisfy both of these equations. Such pairs of numbers, if there are any, are called *solutions to the system of equations* given in (6.1).

Geometrically, each of the equations in (6.1) represents a line. A solution to the system will be a point (x, y) that is on both lines. Two lines intersect at exactly one point, or they are parallel and do not intersect, or they coincide, that is, are the same line, and thus have infinitely many points in common. Thus the system of equations will have

1. Exactly one solution, in which case the system is called *independent;* or
2. No solutions, in which case the system is called *inconsistent;* or
3. Infinitely many solutions, in which case the system is called *dependent.*

In general, a given system of n linear equations in m variables will have exactly one solution, or no solutions, or infinitely many solutions.

We now consider two techniques that can be used to solve a system of linear equations. These are referred to as the *method of substitution,* and the *method of elimination.* It will be helpful to introduce the idea of matrices in connection with the method of elimination.

Method of Substitution

In solving a system of two linear equations, we can solve one of the equations for one of the variables and substitute the result into the other equation. This gives an equation in one variable, the solution of which leads to the solution of the system. This procedure is illustrated in the following two examples.

Example 1   Use the method of substitution to solve the system of equations

$$\begin{aligned} -3x + y &= 5, \\ 2x - 3y &= -8. \end{aligned}$$

Solution   Solving the first equation for y gives $y = 3x + 5$, and substituting this into the second equation we get

$$\begin{aligned} 2x - 3(3x + 5) &= -8, \\ 2x - 9x - 15 &= -8, \\ -7x &= 7, \\ x &= -1. \end{aligned}$$

Thus the x value in the solution is -1, and the y value can be determined by replacing x by -1 in either of the given equations. Using the first equation gives

$$-3(-1) + y = 5,$$
$$y = 2.$$

Therefore the desired solution is given by $x = -1$, $y = 2$.   ■

Example 2   Solve the system of equations by substitution.

$$-3x + 6y = 5,$$
$$x - 2y = 4.$$

Solution   Solving the second equation for x gives $x = 2y + 4$, and substituting this into the first equation, we get

$$-3(2y + 4) + 6y = 5,$$
$$-6y - 12 + 6y = 5,$$
$$0y = 17.$$

Since $0 \cdot y = 0$ for every value of y, we conclude that the given system of equations has no solution. We say that the system is *inconsistent*.   ■

Method of Elimination

Two systems of linear equations are called *equivalent* if their solution sets are identical. Suppose S is a system of linear equations. Any of the following *elementary operations on S* will yield a system of equations that is equivalent to S:

a) Interchange any two equations of S;

b) Replace any equation E with an equation obtained by multiplying both sides of E by a nonzero number;

c) Replace any equation of S by the sum of that equation and a multiple of another equation of S.

The first goal in solving a system of n linear equations is to get an equivalent system in which one of the equations has zero coefficients for $n - 1$ of the variables (referred to as the elimination of $n - 1$ variables). If the coefficient of the remaining variable is not zero, we can solve for it; if it is zero and the constant term is not zero, then there are no solutions; if the constant term is zero, there are infinitely many solutions. We can achieve our first goal by a sequence of elementary operations. This and the concluding steps in solving a system of linear equations by elimination are illustrated in the following examples. Before considering examples, however, let us introduce some notation that will be helpful in describing the sequence of elementary operations.

Suppose the equations of a system of linear equations S are denoted symbolically by $E_1, E_2, \ldots, E_n$. The following illustrates the notation we shall use to indicate elementary operations on S.

1. $E_2 \leftrightarrow E_1$ indicates E_1 and E_2 are interchanged.

2. $4E_2 \rightarrow E_2$ tells us that E_2 is multiplied by 4 and the resulting equation becomes E_2 of the equivalent system.

3. $E_1 + 4E_2 \rightarrow E_1$ indicates that the sum of E_1 and 4 times E_2 becomes equation E_1 of the equivalent system.

4. $2E_1 + (-4)E_2 \rightarrow E_1$ tells us that the sum of 2 times E_1 and -4 times E_2 becomes equation E_1 of the equivalent system.

Example 3   Use the method of elimination to solve

$$E_1: \quad x + 3y = -3,$$
$$E_2: \quad 3x + 2y = \quad 5.$$

Solution   We can eliminate x by multiplying E_1 by -3 and adding the result to E_2. Thus elementary operation $E_2 + (-3)E_1 \rightarrow E_2$ gives the equivalent system

$$E_1: \quad x + 3y = -3,$$
$$E_2: \quad 0x - 7y = \quad 14.$$

Solving E_2 for y gives $y = -2$; substituting -2 for y in E_1 and solving for x gives $x = 3$. Therefore the solution to the given system is given by $x = 3$, $y = -2$. ∎

Example 4   Give a geometric interpretation of the solution of the system of equations given in Example 3.

Solution   The given equations E_1 and E_2 represent lines, as shown in Fig. 6.2. The two lines

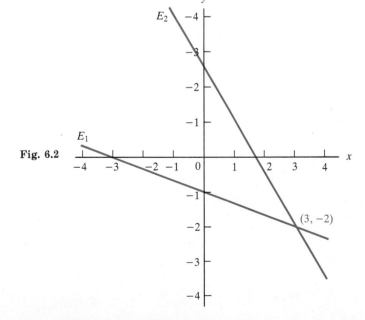

Fig. 6.2

intersect at the point $(3, -2)$. Since $(3, -2)$ is on both lines, $x = 3$ and $y = -2$ will satisfy both of the given equations, and those values are the desired solution. ▪

Example 5   Use the method of elimination to solve

$$E_1: \quad 2x - 5y + 3z = -4,$$
$$E_2: \quad x - 2y - 3z = \quad 3,$$
$$E_3: \quad -3x + 4y + 2z = -4.$$

Solution   Although we could first aim to eliminate any one of the three variables by applying elementary operations, the fact that the coefficient of x in E_2 is 1 suggests that we concentrate on eliminating x, since this will involve simple computations. In the following sequence of steps we first interchange E_1 and E_2, then in steps (2) and (3) we eliminate x in the second and third equations. In step (4) we eliminate y in the third equation.

1. $E_1 \leftrightarrow E_2$
$E_1: \quad x - 2y - 3z = \quad 3,$
$E_2: \quad 2x - 5y + 3z = -4,$
$E_3: \quad -3x + 4y + 2z = -4.$

2. $E_2 + (-2)E_1 \rightarrow E_2$
$E_1: \quad x - 2y - 3z = \quad 3,$
$E_2: \quad 0x - y + 9z = -10,$
$E_3: \quad -3x + 4y + 2z = \quad -4.$

3. $E_3 + 3E_1 \rightarrow E_3$
$E_1: \quad x - 2y - 3z = \quad 3,$
$E_2: \quad 0x - y + 9z = -10,$
$E_3: \quad 0x - 2y - 7z = \quad 5.$

4. $E_3 + (-2)E_2 \rightarrow E_3$
$E_1: \quad x - 2y - 3z = \quad 3,$
$E_2: \quad 0x - y + 9z = -10,$
$E_3: \quad 0x + 0y - 25z = \quad 25.$

Using the system of equations obtained in step (4), we get our solution as follows: Solve E_3 for z to get $z = -1$; substitute $z = -1$ into E_2, and solve for y to get $y = 1$; substitute $z = -1, y = 1$ into E_1, and solve for x to get $x = 2$. Therefore $x = 2$, $y = 1$, $z = -1$ constitute the solution to the given system of equations. ▪

Example 6   Solve the system of equations

$$E_1: \quad x + 2y - 2z = 3,$$
$$E_2: \quad 2x + 3y - 3z = 1,$$
$$E_3: \quad -4x - 5y + 5z = 3.$$

Solution   We first eliminate x by applying elementary operations shown in steps (1) and (2); then in (3) we eliminate y.

1. $E_2 + (-2)E_1 \rightarrow E_2$
$E_1: \quad x + 2y - 2z = \quad 3,$
$E_2: \quad 0x - y + z = -5,$
$E_3: \quad -4x - 5y + 5z = \quad 3.$

2. $E_3 + 4E_1 \rightarrow E_3$
$E_1: \quad x + 2y - 2z = \quad 3,$
$E_2: \quad 0x - y + z = -5,$
$E_3: \quad 0x + 3y - 3z = \quad 15.$

3. $E_3 + 3E_2 \rightarrow E_3$

$$E_1: \quad x + 2y - 2z = \quad 3,$$
$$E_2: \quad 0x - \ y + \ z = -5,$$
$$E_3: \quad 0x + 0y + 0z = \quad 0.$$

From the system of equations obtained in step (3) we conclude that E_3 is satisfied by all real numbers, and so the solution will consist of any x, y, z that satisfy E_1 and E_2. Suppose we let $z = t$, where t is any real number; then from E_2, $y = t + 5$. Substituting $z = t, y = t + 5$ into E_1 gives $x + 2(t + 5) - 2t = 3$, and so $x = -7$. Therefore, the given system of equations has infinitely many solutions given by

$$x = -7, \ y = t + 5, \ z = t,$$

where t is any real number. For example, each of the following will give a solution:

$$t = 0 \text{ gives } x = -7, \ y = 5, \ t = 0;$$
$$t = 3 \text{ gives } x = -7, \ y = 8, \ z = 3;$$
$$t = -\sqrt{2} \text{ gives } x = -7, \ y = 5 - \sqrt{2}, \ z = -\sqrt{2}.$$

We say that the given system of equations is *dependent*. ■

Systems of Linear Equations as Matrices

A close inspection of the solutions given in the preceding examples shows that the method of elimination involves a process of applying elementary operations along with careful bookkeeping. We can streamline the entire process by first noting that it is not necessary to carry along the variables in each step; second, the essential parts of the given system of linear equations are the coefficients of the variables and the numbers on the right-hand side of the equations.

Let us consider the solution given for Example 5. The given system is

$$2x - 5y + 3z = -4,$$
$$x - 2y - 3z = \quad 3, \qquad (6.2)$$
$$-3x + 4y + 2z = -4.$$

The coefficients of x, y, z and the constants on the right-hand side can be represented by rectangular arrays of numbers, as follows:

$$A = \begin{bmatrix} 2 & -5 & 3 \\ 1 & -2 & -3 \\ -3 & 4 & 2 \end{bmatrix}, \qquad B = \begin{bmatrix} -4 \\ 3 \\ -4 \end{bmatrix}. \qquad (6.3)$$

Such rectangular arrays of numbers are called *matrices* (plural for *matrix*). Matrix A, with *three rows* and *three columns,* is called a *three-by-three* (or 3×3) *matrix;* matrix B, with *three rows* and *one column,* is said to be a *three-by-one matrix*. Matrix A is called the *matrix of the coefficients,* and it is said to be a *square matrix* since it has the same number of rows and columns. The numbers inside the brackets are called *elements of the matrix;* for instance, -3 is the element in the second row and third column of A. Also, in a square matrix the sequence of numbers from upper left to lower right (that is, the $2, -2, 2$) constitutes the *main diagonal* of the matrix.

The system of equations given in (6.2) can be represented as a matrix as follows:

$$C = \begin{bmatrix} 2 & -5 & 3 & | & -4 \\ 1 & -2 & -3 & | & 3 \\ -3 & 4 & 2 & | & -4 \end{bmatrix}. \tag{6.4}$$

This is a three-by-four matrix, and it is called the *augmented matrix*. Note that matrix C completely describes the system of equations given in (6.2) if we remember that the first column is associated with the variable x, the second with y, and the third with z. We can now get a sequence of *equivalent matrices* in a manner similar to that in which we obtained a sequence of equivalent systems of equations. Here we show the operations on rows of matrices that are analogous to the elementary operations with equations in the solution for Example 5.

1. $R_1 \leftrightarrow R_2$

$$C_1 = \begin{bmatrix} 1 & -2 & -3 & | & 3 \\ 2 & -5 & 3 & | & -4 \\ -3 & 4 & 2 & | & -4 \end{bmatrix}.$$

2. $R_2 + (-2)R_1 \to R_2$

$$C_2 = \begin{bmatrix} 1 & -2 & -3 & | & 3 \\ 0 & -1 & 9 & | & -10 \\ -3 & 4 & 2 & | & -4 \end{bmatrix}.$$

3. $R_3 + 3R_1 \to R_3$

$$C_3 = \begin{bmatrix} 1 & -2 & -3 & | & 3 \\ 0 & -1 & 9 & | & -10 \\ 0 & -2 & -7 & | & 5 \end{bmatrix}.$$

4. $R_3 + (-2)R_2 \to R_3$

$$C_4 = \begin{bmatrix} 1 & -2 & -3 & | & 3 \\ 0 & -1 & 9 & | & -10 \\ 0 & 0 & -25 & | & 25 \end{bmatrix}.$$

In the solution of Example 5 we solved the system of equations corresponding to matrix C_4. However, we can continue performing elementary row operations to get the following equivalent matrices.

5. $(-1)R_2 \to R_2; \ -\frac{1}{25}R_3 \to R_3$

$$C_5 = \begin{bmatrix} 1 & -2 & -3 & | & 3 \\ 0 & 1 & -9 & | & 10 \\ 0 & 0 & 1 & | & -1 \end{bmatrix}.$$

6. $R_1 + 2R_2 \to R_1$

$$C_6 = \begin{bmatrix} 1 & 0 & -21 & | & 23 \\ 0 & 1 & -9 & | & 10 \\ 0 & 0 & 1 & | & -1 \end{bmatrix}.$$

7. $R_1 + 21R_3 \to R_1; \ R_2 + 9R_3 \to R_2$

$$C_7 = \begin{bmatrix} 1 & 0 & 0 & | & 2 \\ 0 & 1 & 0 & | & 1 \\ 0 & 0 & 1 & | & -1 \end{bmatrix}.$$

The system of equations corresponding to matrix C_7 is $x = 2, y = 1, z = -1$. Therefore the solution to the given system of equations is given by $x = 2, y = 1, z = -1$.

Matrix C_7 is an example of a *row-reduced echelon matrix*. In such a matrix the elements corresponding to the coefficients of the variables are zero except for those along the main diagonal, which are ones or zeros.

Elementary Row Operations on Matrices

The elementary operations on a system of linear equations described earlier translate into corresponding *elementary row operations on matrices*, as follows:

a) Interchange any two rows;

b) Replace any row by a nonzero multiple of itself;

c) Replace any row by the sum of itself and a multiple of some other row.

For instance, in the solution of the example above: in step (1)—$R_1 \leftrightarrow R_2$—we applied (a); in step (5)—$(-1)R_2 \to R_2$; $-(1/25)R_3 \to R_3$—we applied (b) twice; and in step (7)—$R_1 + 21R_3 \to R_1$; $R_2 + 9R_3 \to R_2$—we applied (c) twice.

We now summarize the essential features of solving a system of linear equations by using the technique of *row-reduction to echelon form*:

1. Write the augmented matrix associated with the given system of equations, remembering which columns correspond to which variables.

2. Use any of the elementary row operations (a) through (c) to get a sequence of equivalent matrices.

3. The ultimate goal is to obtain an equivalent augmented matrix in which the corresponding *coefficient matrix* has zeros for all of its elements except the elements of the main diagonal, each of which is a zero or a one. If the main diagonal elements of the row-reduced echelon matrix are all one, then the given system has a *unique solution;* otherwise the system is either *inconsistent* (no solutions) or *dependent* (infinitely many solutions).

Example 7   Solve the system of linear equations

$$3x - y = -1,$$
$$x + y = 3.$$

Solution   The augmented matrix corresponding to the given system is

$$\begin{bmatrix} 3 & -1 & \vdots & -1 \\ 1 & 1 & \vdots & 3 \end{bmatrix}.$$

We can get a sequence of equivalent matrices, as follows:

1. $R_1 \leftrightarrow R_2$

$$\begin{bmatrix} 1 & 1 & \vdots & 3 \\ 3 & -1 & \vdots & -1 \end{bmatrix}$$

2. $R_2 + (-3)R_1 \to R_2$

$$\begin{bmatrix} 1 & 1 & \vdots & 3 \\ 0 & -4 & \vdots & -10 \end{bmatrix}$$

3. $-\frac{1}{4}R_2 \to R_2$

$$\begin{bmatrix} 1 & 1 & | & 3 \\ 0 & 1 & | & \frac{5}{2} \end{bmatrix}$$

4. $R_1 + (-1)R_2 \to R_1$

$$\begin{bmatrix} 1 & 0 & | & \frac{1}{2} \\ 0 & 1 & | & \frac{5}{2} \end{bmatrix}$$

From the final matrix we see that the desired solution is given by $x = \frac{1}{2}$, $y = \frac{5}{2}$.

Example 8   Solve the system of linear equations

$$\begin{aligned} x - 3y + 2z &= 11, \\ -2x + y + 3z &= 4, \\ 3x + 2y - 4z &= -8. \end{aligned}$$

Solution   The augmented matrix associated with this system is

$$\begin{bmatrix} 1 & -3 & 2 & | & 11 \\ -2 & 1 & 3 & | & 4 \\ 3 & 2 & -4 & | & -8 \end{bmatrix}.$$

The following sequence of elementary row operations leads us to the solution.

1. $R_2 + 2R_1 \to R_2$; $R_3 + (-3)R_1 \to R_3$

$$\begin{bmatrix} 1 & -3 & 2 & | & 11 \\ 0 & -5 & 7 & | & 26 \\ 0 & 11 & -10 & | & -41 \end{bmatrix}$$

2. $-\frac{1}{5}R_2 \to R_2$

$$\begin{bmatrix} 1 & -3 & 2 & | & 11 \\ 0 & 1 & -\frac{7}{5} & | & -\frac{26}{5} \\ 0 & 11 & -10 & | & -41 \end{bmatrix}$$

3. $R_1 + 3R_2 \to R_1$; $R_3 + (-11)R_2 \to R_3$

$$\begin{bmatrix} 1 & 0 & -\frac{11}{5} & | & -\frac{23}{5} \\ 0 & 1 & -\frac{7}{5} & | & -\frac{26}{5} \\ 0 & 0 & \frac{27}{5} & | & \frac{81}{5} \end{bmatrix}$$

4. $\frac{5}{27}R_3 \to R_3$

$$\begin{bmatrix} 1 & 0 & -\frac{11}{5} & | & -\frac{23}{5} \\ 0 & 1 & -\frac{7}{5} & | & -\frac{26}{5} \\ 0 & 0 & 1 & | & 3 \end{bmatrix}$$

5. $R_1 + \frac{11}{5}R_3 \to R_1$; $R_2 + \frac{7}{5}R_3 \to R_2$

$$\begin{bmatrix} 1 & 0 & 0 & | & 2 \\ 0 & 1 & 0 & | & -1 \\ 0 & 0 & 1 & | & 3 \end{bmatrix}$$

Therefore, from the matrix in step (5) we see that the desired solution is $x = 2$, $y = -1$, $z = 3$.

Example 9   Solve the system of linear equations

$$\begin{aligned} 2x - 4y &= -3, \\ -3x + 6y &= 1. \end{aligned}$$

Solution   The augmented matrix associated with this system of equations is

$$\begin{bmatrix} 2 & -4 & | & -3 \\ -3 & 6 & | & 1 \end{bmatrix}.$$

We can get a sequence of equivalent matrices as follows:

1. $\frac{1}{2}R_1 \rightarrow R_1$ **2.** $R_2 + 3R_1 \rightarrow R_2$

$$\begin{bmatrix} 1 & -2 & | & -\frac{3}{2} \\ -3 & 6 & | & 1 \end{bmatrix} \qquad\qquad \begin{bmatrix} 1 & -2 & | & -\frac{3}{2} \\ 0 & 0 & | & -\frac{7}{2} \end{bmatrix}$$

At this point we see that the second row is associated with the equation

$$0 \cdot x + 0 \cdot y = -\tfrac{7}{2}.$$

Clearly, there are no values of x and y that will satisfy this equation. Thus the given system of equations has no solution, and we say that the system is inconsistent.   ■

Example 10   Solve the system of equations

$$\frac{3}{x} + \frac{2}{y} = \frac{5}{4},$$

$$-\frac{1}{x} + \frac{3}{y} = -\frac{9}{4}. \tag{6.5}$$

Solution   This is *not a system of linear equations in x and y*. However, it is a system of linear equations in $1/x$ and $1/y$. Suppose we let $u = 1/x$ and $v = 1/y$; then the given system can be written as

$$3u + 2v = \tfrac{5}{4},$$

$$-u + 3v = -\tfrac{9}{4}. \tag{6.6}$$

We can solve system (6.6) for u and v, and then the solution to system (6.5) will be given by $x = 1/u$, $y = 1/v$.

The augmented matrix for system (6.6) is

$$\begin{bmatrix} 3 & 2 & | & \frac{5}{4} \\ -1 & 3 & | & -\frac{9}{4} \end{bmatrix}$$

The following sequence of elementary row operations leads us to a row-reduced echelon matrix.

1. $R_1 \leftrightarrow R_2$ **2.** $(-1)R_1 \rightarrow R_1$

$$\begin{bmatrix} -1 & 3 & | & -\frac{9}{4} \\ 3 & 2 & | & \frac{5}{4} \end{bmatrix} \qquad\qquad \begin{bmatrix} 1 & -3 & | & \frac{9}{4} \\ 3 & 2 & | & \frac{5}{4} \end{bmatrix}$$

3. $R_2 + (-3)R_1 \rightarrow R_2$

$$\begin{bmatrix} 1 & -3 & | & \frac{9}{4} \\ 0 & 11 & | & -\frac{11}{2} \end{bmatrix}$$

4. $\frac{1}{11}R_2 \rightarrow R_2$

$$\begin{bmatrix} 1 & -3 & | & \frac{9}{4} \\ 0 & 1 & | & -\frac{1}{2} \end{bmatrix}$$

5. $R_1 + 3R_2 \rightarrow R_1$

$$\begin{bmatrix} 1 & 0 & | & \frac{3}{4} \\ 0 & 1 & | & -\frac{1}{2} \end{bmatrix}$$

Therefore, from step (5) we get $u = \frac{3}{4}$, $v = -\frac{1}{2}$ as the solution for system (6.6). The solution for system (6.5) is given by $x = \frac{4}{3}$, $y = -2$. ■

Example 11   A parts manufacturing firm finds that each week its machines are not used during a total of 18 machine hours, and there is a total of 34 surplus labor hours each week. In order to utilize these resources, the management decides to produce two additional products, A and B. To make each unit of A requires 2 machine hours and 4 labor hours. Each unit of B requires 3 machine hours and 5 labor hours. How many units of A and of B can be produced if the available resources are completely used?

Solution   Let x represent the number of units of A and y the number of units of B that can be produced. Then $2x + 3y$ represents the number of machine hours required to produce x units of A and y units of B; we want this number to be equal to 18. That is, $2x + 3y = 18$ (machine hours).

Similarly, to produce x units of A and y units of B requires $4x + 5y$ labor hours, and so we have $4x + 5y = 34$ (labor hours). Thus we have the following system of linear equations.

$$2x + 3y = 18,$$
$$4x + 5y = 34.$$

This system can be solved by the matrix row-reduction technique to get $x = 6$, $y = 2$ as the solution. Therefore the firm can produce 6 units of A and 2 units of B each week. ■

Exercises 6.1

Solve each of the given systems of linear equations. If the system is dependent (infinitely many solutions), describe all solutions and then give three specific solutions (see Example 6).

In problems 1 through 6,

a) Solve by using the method of substitution;

b) Draw graphs to show solutions graphically.

1. $x + y = 4,$
$3x - 2y = -3$

2. $-3x - y = 5,$
$-x + 2y = 4$

3. $2x + 4y = 3,$
$x + 2y = 1.5$

4. $\dfrac{x}{2} + \dfrac{y}{3} = 0,$

$\dfrac{3x}{4} - \dfrac{y}{2} = 3$

5. $\quad 4x - 2y = 3,$
$\quad -2x + \ y = 5$

6. $\quad 0.3x + 1.5y = 3.0,$
$\quad 1.4x - 0.2y = 3.2$

In problems 7 through 16, solve by using the method of elimination.

7. $x + 2y = 1,$
$\quad x - \ y = 4$

8. $\ 3x + 4y = -2,$
$\quad -x - \ y = \ \ 1$

9. $\quad 4x - 12y = 3,$
$\quad -x + \ 3y = 1$

10. $\dfrac{x}{2} + \dfrac{y}{4} = \quad 0,$

$\dfrac{5x}{2} - \dfrac{y}{4} = -12$

11. $0.3x - 0.4y = -0.2,$
$\quad -x + \quad y = \quad 1$

12. $\quad x + \ y + \ z = \quad 1,$
$\quad 2x - \ y - \ z = \quad 5,$
$\quad -x + 2y - 3z = -4$

13. $2x - y + 3z = 1,$
$\quad x + y - 5z = 2,$
$\quad 3x - \qquad 2z = 3$

14. $\quad x + 3y - \ z = 1,$
$\quad -2x + \ y + 3z = 0,$
$\quad -4x + 9y + 7z = 2$

15. $-x - y \qquad = 2,$
$\quad 3x + \qquad 4z = 5,$
$\quad 4x + y + 4z = 3$

16. $\dfrac{x}{2} + \dfrac{y}{3} - \dfrac{z}{4} = -1$

$\dfrac{x}{3} + \qquad \dfrac{z}{2} = \quad 8$

$\dfrac{2x}{3} + \dfrac{y}{3} - \dfrac{3z}{4} = -6$

In problems 17 through 30, solve the given systems of equations by using the method of augmented matrices and elementary row operations to reduce to echelon form.

17. $\ x + 3y = 5,$
$\quad 3x - \ y = 5$

18. $\dfrac{x}{3} - \dfrac{y}{2} = 4,$

$\dfrac{x}{2} - y = 7$

19. $6x - 12y = \quad 7,$
$\quad 4x - \ 8y = -5$

20. $\dfrac{x}{3} + \dfrac{y}{4} = -\dfrac{1}{2},$

$\dfrac{3x}{2} + \dfrac{5y}{2} = \quad \dfrac{1}{2}$

21. $\quad 4x - 8y = -5,$
$\quad -2x + 4y = \quad \dfrac{5}{2}$

22. $\quad 0.1x - 1.3y = 1.0,$
$\quad -0.4x + 0.5y = 0.7$

23. $\dfrac{2}{x} - \dfrac{4}{y} = -1,$

$\dfrac{3}{x} + \dfrac{5}{y} = \quad 2$

24. $\quad 2x + 6y + 3z = 1,$
$\quad 4x + 2y + \ z = 2,$
$\quad -2x + 3y - 6z = 9$

25. $\ x - 2y + 3z = -4,$

$\dfrac{x}{2} + 3y + \quad z = \quad 3,$

$\dfrac{x}{4} + \dfrac{y}{2} - \dfrac{z}{4} = \quad 2$

26. $\quad x + 2y - \ z = \quad 1,$
$\quad -2x + \ y + 2z = -2,$
$\quad -x + 8y + \ z = \quad 2$

27. $-x - \ y + 3z = \quad 1,$
$\quad 3x - \qquad 4z = -4,$
$\quad x - 2y + 2z = -2$

28. $3x + 4y + 4z = -1,$
$\quad 6x - 2y + 2z = -2,$
$\quad 2y + 6z = -6$

29. $2x - 4y + \ z = 4,$

$\dfrac{x}{3} + \qquad \dfrac{3z}{2} = 2,$

$\dfrac{y}{2} + 3z = 5$

30. $\quad 0.5x + 1.5y - 0.5z = \quad 2,$
$\quad -1.5x - 2.5y + 0.5z = -4,$
$\quad -0.5y + 1.5z = \quad 7$

31. A manufacturer has two different models of the same machine, model A and model B. If machine A operates for five hours and machine B for three hours, a total of 70 items can be produced. If each machine operates for four hours, a total of 80 items can be produced. How many items can be produced by each machine in one hour?

32. A 32-pound mixture of peanuts, cashew nuts, and walnuts contains four times as many pounds of peanuts as cashews and three times as many pounds of cashews as walnuts. How many pounds of each does the mixture contain?

33. The cost of a sandwich, a drink, and a piece of pie is $2.50. The sandwich costs a dollar more than the pie, and the pie costs twice as much as the drink. What is the cost of each?

34. A rectangular lot has a length-to-width ratio of 4 to 3. If it takes 168 meters of fence to enclose it, what are the dimensions of the lot?

35. Suppose x grams of food A and y grams of food B are mixed and the total weight is 2000 grams. Food A contains 0.25 units of vitamin D per gram, and food B contains 0.50 units of vitamin D per gram. Suppose the final mixture contains 900 units of vitamin D. How many grams of each type of food does the mixture contain?

36. A total of $3600 is invested in three different accounts; the first account earns interest at a rate of 8%, the second at 10%, and the third at 12%. The amount invested in the first account is twice as much as that in the second account. If the total amount of simple interest earned is $388, how much is invested in each account?

37. Suppose x grams of food A, y grams of food B, and z grams of food C are mixed together and the total weight is 2400 grams. The vitamin D and calorie content of each food is given in the table.

Food	Units of vitamin D per gram	Calories per gram
A	0.75	1.4
B	0.50	1.6
C	1.00	1.5

Suppose the 2400-gram mixture contains a total of 1725 units of vitamin D and 3690 calories. How many grams of each type of food does it contain?

38. Dessert consists of chocolate pudding with whipped cream, and we are interested in the energy (calorie) and vitamin A content. The necessary information is given in the table.

How much pudding (in cups) and whipped cream (in tablespoons) will give a dessert with 283 calories and 674 units of vitamin A?

Food	Energy (calories)	Vitamin A (units)
Chocolate pudding (1 cup; 260 g)	385	390
Whipped cream (1 tbsp)	26	220

39. Two pipelines A and B are used to fill a tank with water. The tank can be filled by having A run for 3 hours and B run for 6 hours, or it can be filled by having both of the supply lines open for 4 hours. How long would it take for A to fill the tank alone? How long would it take for B to fill the tank alone?

40. Suppose α, β, and γ represent the measures (in degrees) of three angles of a triangle. Angle α is twice as large as β, and γ is 54° more than three times α. Determine α, β, and γ.

41. Two circles are such that the circumference of the first is twice that of the second, and the radius of the first is 4 greater than that of the second. Find the area of each circle.

In problems 42 and 43, the given systems of equations have infinitely many solutions. In each case find the specific solutions for which

a) $z = 2$ **b)** $z = -3$

42. $\begin{aligned} x - 2y - z &= 1, \\ 2x + y - 2z &= -3, \\ x + 8y - z &= -9 \end{aligned}$ **43.** $\begin{aligned} 1.5x - 2.5y + 1.2z &= 0.4, \\ 2.1x + 0.4y - 3.2z &= 2.4, \\ 5.1x - 4.6y - 0.8z &= 3.2 \end{aligned}$

6.2   DETERMINANTS; CRAMER'S RULE; PARTIAL FRACTIONS

Solutions to systems of linear equations can be described in terms of determinants. Let us consider the general system of two linear equations in two variables.

$$\begin{aligned} E_1: \quad a_1 x + b_1 y &= c_1, \\ E_2: \quad a_2 x + b_2 y &= c_2, \end{aligned} \tag{6.7}$$

where $a_1, b_1, c_1, a_2, b_2, c_2$ are given numbers. We can "eliminate" y by the following operation on E_1 and E_2: $b_2 E_1 + (-b_1)E_2$ (that is, multiply E_1 by b_2 and E_2 by $(-b_1)$, and add the resulting equations). This gives

$$(a_1 b_2 - b_1 a_2)x = c_1 b_2 - b_1 c_2.$$

Solving for x, we get

$$x = \frac{c_1 b_2 - b_1 c_2}{a_1 b_2 - b_1 a_2}. \tag{6.8}$$

Similarly x can be eliminated by the operation $a_2 E_1 + (-a_1)E_2$: this gives

$$y = \frac{a_1 c_2 - c_1 a_2}{a_1 b_2 - b_1 a_2} \tag{6.9}$$

Note that the denominators of the results in Eqs. (6.8) and (6.9) are the same, and they are related to the coefficient matrix

$$A = \begin{bmatrix} a_1 & b_1 \\ a_2 & b_2 \end{bmatrix}.$$

In fact, this leads us to the definition of the *determinant of a two-by-two matrix*.

Definition 6.1   With each two-by-two matrix

$$A = \begin{bmatrix} a_1 & b_1 \\ a_2 & b_2 \end{bmatrix},$$

we associate a number called the *determinant of A*, denoted by $|A|$ and given by

$$|A| = a_1 b_2 - b_1 a_2. \tag{6.10}$$

Another notation that is commonly used in place of $|A|$ is $\det(A)$. Here "det" can be considered a function with domain the set of two-by-two matrices and range the set of real numbers. That is,

$$\mathcal{D}(\det) = \{A \mid A \text{ is a two-by-two matrix}\}, \qquad \mathcal{R}(\det) = \mathbf{R}.$$

Solutions to a system of two linear equations in two variables can now be stated in terms of determinants, as follows:

Cramer's Rule

If the determinant of the coefficient matrix is not zero, then the system of equations

$$a_1 x + b_1 y = c_1, \qquad a_2 x + b_2 y = c_2$$

has a unique solution given by

$$x = \frac{\begin{vmatrix} c_1 & b_1 \\ c_2 & b_2 \end{vmatrix}}{\begin{vmatrix} a_1 & b_1 \\ a_2 & b_2 \end{vmatrix}}, \qquad y = \frac{\begin{vmatrix} a_1 & c_1 \\ a_2 & c_2 \end{vmatrix}}{\begin{vmatrix} a_1 & b_1 \\ a_2 & b_2 \end{vmatrix}}. \tag{6.11}$$

Cramer's rule tells us that x and y are given by ratios of two determinants: The denominator is the determinant of the coefficient matrix, and the numerator is the determinant of the matrix obtained by replacing the coefficients of the variable in question by the numbers on the right-hand sides of the two equations.

Now let us consider some examples in which Cramer's rule is used.

Example 1   Solve the system of equations

$$\begin{aligned} x - 3y &= 6, \\ -2x + y &= -7. \end{aligned}$$

Solution   Applying Cramer's rule gives

$$x = \frac{\begin{vmatrix} 6 & -3 \\ -7 & 1 \end{vmatrix}}{\begin{vmatrix} 1 & -3 \\ -2 & 1 \end{vmatrix}} = \frac{(6)(1) - (-3)(-7)}{(1)(1) - (-3)(-2)} = \frac{6 - 21}{1 - 6} = \frac{-15}{-5} = 3,$$

$$y = \frac{\begin{vmatrix} 1 & 6 \\ -2 & -7 \end{vmatrix}}{\begin{vmatrix} 1 & -3 \\ -2 & 1 \end{vmatrix}} = \frac{(1)(-7) - (6)(-2)}{(1)(1) - (-3)(-2)} = \frac{-7 + 12}{1 - 6} = \frac{5}{-5} = -1.$$

Thus the solution is given by $x = 3$, $y = -1$.   ▬

Example 2   Solve the system of equations

$$2x - 6y = 1,$$
$$-3x + 9y = 5.$$

Solution   Applying Cramer's rule gives

$$x = \frac{\begin{vmatrix} 1 & -6 \\ 5 & 9 \end{vmatrix}}{\begin{vmatrix} 2 & -6 \\ -3 & 9 \end{vmatrix}} = \frac{(1)(9) - (-6)(5)}{(2)(9) - (-6)(-3)} = \frac{39}{0}.$$

Here we see that the determinant in the denominator is zero, and this suggests that there may be no solution. It is a simple matter to verify that there is indeed no solution by considering the operation $3E_1 + 2E_2 \rightarrow E_2$, which leads to the equivalent system

$$2x - 6y = 1,$$
$$0 \cdot x + 0 \cdot y = 13.$$

Clearly, there are no numbers x, y that will satisfy the second equation. Therefore we conclude that the given system of equations is inconsistent.   ▬

Example 3   Solve the system of equations

$$4x - 2y = 3,$$
$$-2x + y = -1.5.$$

Solution   Applying Cramer's rule gives

$$x = \frac{\begin{vmatrix} 3 & -2 \\ -1.5 & 1 \end{vmatrix}}{\begin{vmatrix} 4 & -2 \\ -2 & 1 \end{vmatrix}} = \frac{0}{0}, \qquad y = \frac{\begin{vmatrix} 4 & 3 \\ -2 & -1.5 \end{vmatrix}}{\begin{vmatrix} 4 & -2 \\ -2 & 1 \end{vmatrix}} = \frac{0}{0}.$$

Here we have a situation similar to that in Example 2 involving division by zero. However, in this case the numerators are also zero, and this suggests that we may have many solutions. Applying the elimination method with the elementary operation $E_1 + 2E_2 \rightarrow E_1$ gives the equivalent system

$$0 \cdot x + 0 \cdot y = 0,$$
$$-2x + y = -1.5.$$

The first equation is satisfied by any pair of numbers x, y, and so any pair that also satisfies the second equation will be a solution. Thus we can assign an arbitrary number to x and take $y = 2x - 1.5$ as the corresponding value of y to give us infinitely many solutions. For instance, $x_1 = 0$, $y_1 = -1.5$; $x_2 = 2$, $y_2 = 2.5$; $x_3 = -4$, $y_3 = -9.5$ are three particular examples of solutions. The given system is dependent. ◼

The solutions shown in Examples 2 and 3 suggest that we first evaluate the determinant of the coefficient matrix. If it is zero, we choose another method of solution, since Cramer's rule is not helpful for solving dependent or inconsistent cases.

System of Three Equations in Three Variables

The general system of three linear equations in three variables can be written as

$$a_1 x + b_1 y + c_1 z = d_1,$$
$$a_2 x + b_2 y + c_2 z = d_2, \qquad (6.12)$$
$$a_3 x + b_3 y + c_3 z = d_3,$$

where the a_i's, b_i's, c_i's, and d_i's are given numbers.

If we pursued a development similar to that of the two-by-two case given above, we would find that the solution can be expressed as ratios of determinants, where we must give an appropriate definition of the determinant of a three-by-three matrix. We shall omit the details, which require some careful algebraic manipulations, and merely present the results.

If we carried out the solution of the system in (6.12) by the method of elimination, we would find that the denominators for x, y, and z are all the same and are given by

$$D = a_1 b_2 c_3 - a_1 c_2 b_3 - b_1 a_2 c_3 + b_1 c_2 a_3 + c_1 a_2 b_3 - c_1 b_2 a_3. \qquad (6.13)$$

Note that D contains six terms, each of which is a product of three elements of the coefficient matrix,

$$A = \begin{bmatrix} a_1 & b_1 & c_1 \\ a_2 & b_2 & c_2 \\ a_3 & b_3 & c_3 \end{bmatrix}. \qquad (6.14)$$

Also note that each term of D is the product of an element from each row (each has 1, 2, 3 subscripts) and each column (each has a, b, c).

In a manner analogous to the two-by-two case, the *determinant of matrix A* is defined as the number D given in Eq. (6.13). This is denoted by $|A|$ or by $\det(A)$.

Definition 6.2

$$|A| = \det(A) = \begin{vmatrix} a_1 & b_1 & c_1 \\ a_2 & b_2 & c_2 \\ a_3 & b_3 & c_3 \end{vmatrix} \tag{6.15}$$

$$= a_1 b_2 c_3 - a_1 c_2 b_3 - b_1 a_2 c_3 + b_1 c_2 a_3 + c_1 a_2 b_3 - c_1 b_2 a_3.$$

Perhaps the first reaction is that this definition is hopelessly complicated to remember and to apply. However, a few observations will help considerably. The six terms of the right side of Eq. (6.15) can be grouped into three pairs so that a_1, b_1, c_1 (the first row of A) can be factored out of the respective pairs, as follows:

$$|A| = a_1(b_2 c_3 - c_2 b_3) - b_1(a_2 c_3 - c_2 a_3) + c_1(a_2 b_3 - b_2 a_3).$$

The three expressions within parentheses on the right remind us of determinants of two-by-two matrices, and indeed they are. Thus $|A|$ can be written as

$$|A| = a_1 \begin{vmatrix} b_2 & c_2 \\ b_3 & c_3 \end{vmatrix} - b_1 \begin{vmatrix} a_2 & c_2 \\ a_3 & c_3 \end{vmatrix} + c_1 \begin{vmatrix} a_2 & b_2 \\ a_3 & b_3 \end{vmatrix}. \tag{6.16}$$

The result given in Eq. (6.16) is considerably easier to remember by observing the following pattern: Each of the three terms on the right-hand side consists of an element from the first row of matrix A multiplied by the determinant obtained by crossing out the row and column of A in which that element appears. That is, a_1, b_1, and c_1 are multiplied respectively by the determinant of the remaining two-by-two matrices,

$$\begin{bmatrix} a_1 & b_1 & c_1 \\ a_2 & b_2 & c_2 \\ a_3 & b_3 & c_3 \end{bmatrix}, \quad \begin{bmatrix} a_1 & b_1 & c_1 \\ a_2 & b_2 & c_2 \\ a_3 & b_3 & c_3 \end{bmatrix}, \quad \begin{bmatrix} a_1 & b_1 & c_1 \\ a_2 & b_2 & c_2 \\ a_3 & b_3 & c_3 \end{bmatrix}.$$

The three two-by-two determinants in Eq. (6.16) are called the *minors of elements* a_1, b_1, c_1, respectively. This leads us to the following definition of a minor of any element of A.

Definition 6.3   The *minor of any element of the matrix A* is the determinant of the two-by-two matrix obtained by crossing out the row and column of A in which that element appears.

For example, the minor of element b_3 is

$$\begin{vmatrix} a_1 & c_1 \\ a_2 & c_2 \end{vmatrix} = a_1 c_2 - c_1 a_2.$$

Returning to Eq. (6.16), we must still associate the proper $+$ or $-$ sign with each of the three minors on the right. This leads to the idea of *cofactors,* which will allow us to state properties of $|A|$ in reasonably simple terms. The signs ($+$ or $-$) in front of the three terms of Eq. (6.16) suggest the following:

$$\text{Cofactor of } a_1 = (+1) \cdot (\text{Minor of } a_1),$$
$$\text{Cofactor of } b_1 = (-1) \cdot (\text{Minor of } b_1),$$
$$\text{Cofactor of } c_1 = (+1) \cdot (\text{Minor of } c_1).$$

Thus we have the following theorem, which is essentially a restatement of Definition 6.2.

Theorem 6.1   The determinant of the three-by-three matrix A is given by the sum of products of each element of the first row by its corresponding cofactor.

Let us now take another look at the right-hand side of Eq. (6.15). We can group the six terms into three pairs so that we can factor a_2, b_2, c_2 (the elements of the second row), respectively, out of the pairs, as follows:

$$|A| = -a_2 \begin{vmatrix} b_1 & c_1 \\ b_3 & c_3 \end{vmatrix} + b_2 \begin{vmatrix} a_1 & c_1 \\ a_3 & c_3 \end{vmatrix} - c_2 \begin{vmatrix} a_1 & b_1 \\ a_3 & b_3 \end{vmatrix}. \tag{6.17}$$

Eq. (6.17) suggests a result similar to that given in Theorem 6.1; that is, $|A|$ is equal to the sum of the products of each element of the second row of matrix A with its corresponding cofactor, where the cofactors of a_2, b_2, c_2 are respectively given by

$$(-1) \cdot \begin{vmatrix} b_1 & c_1 \\ b_3 & c_3 \end{vmatrix}, \qquad (+1) \cdot \begin{vmatrix} a_1 & c_1 \\ a_3 & c_3 \end{vmatrix}, \qquad (-1) \cdot \begin{vmatrix} a_1 & b_1 \\ a_3 & b_3 \end{vmatrix}.$$

This leads us to the definition of the *cofactor of any element of the matrix A.*

Definition 6.4   The *cofactor of any element of A* is the product of its minor and $(-1)^{i+j}$, where i is the row number and j is the column number in which the element appears.

For example,

$$\text{Cofactor of } b_3 = (-1)^{3+2} \cdot \begin{vmatrix} a_1 & c_1 \\ a_2 & c_2 \end{vmatrix} = (-1) \cdot (a_1 c_2 - c_1 a_2).$$

It should now be clear that there are other ways of combining the six terms of the right side of Eq. (6.15) that would lead to results similar to that given in Theorem 6.1. In fact, we can prove the following result, which tells us that we can evaluate $|A|$ by using any row or any column of A.

Theorem 6.2   The determinant of a three-by-three matrix A is the sum of the products of each element of *any row* (or *any column*) with its corresponding cofactor.

Although discussion above was limited to 2×2 or 3×3 square matrices, one can generalize Theorem 6.2 by considering matrix A to be a *square matrix of any size*. For example, the determinant of a 4×4 matrix is the sum of four terms, each of which involves the determinant of a 3×3 matrix.

Example 4   Find the determinant of the matrix

$$A = \begin{bmatrix} 4 & -2 & 0 \\ 3 & 5 & 1 \\ -2 & 1 & 3 \end{bmatrix}$$

by expanding (a) by the elements of the first row, (b) by the elements of the second column.

Solution   **a)** $|A| = (4)(-1)^{1+1} \begin{vmatrix} 5 & 1 \\ 1 & 3 \end{vmatrix} + (-2)(-1)^{1+2} \begin{vmatrix} 3 & 1 \\ -2 & 3 \end{vmatrix} + (0)(-1)^{1+3} \begin{vmatrix} 3 & 5 \\ -2 & 1 \end{vmatrix}$

$= (4)(15 - 1) + 2(9 + 2) = 78.$

b) $|A| = (-2)(-1)^{1+2} \begin{vmatrix} 3 & 1 \\ -2 & 3 \end{vmatrix} + (5)(-1)^{2+2} \begin{vmatrix} 4 & 0 \\ -2 & 3 \end{vmatrix} + (1)(-1)^{3+2} \begin{vmatrix} 4 & 0 \\ 3 & 1 \end{vmatrix}$

$= 2(9 + 2) + 5(12 - 0) - (4 - 0) = 78.$ ■

Let us now return to the problem of solving the system of linear equations given in (6.12). Cramer's rule, as stated for a system of two linear equations, can be generalized to systems of any number of linear equations. The following example illustrates its application to a system of three linear equations.

Example 5   Use Cramer's rule to solve the system of equations

$$3x - 4y + 2z = 1,$$
$$2x + \ y \quad\quad = 0,$$
$$-x + 2y - \ z = 3.$$

Solution   Let A represent the matrix of the coefficients:

$$A = \begin{bmatrix} 3 & -4 & 2 \\ 2 & 1 & 0 \\ -1 & 2 & -1 \end{bmatrix}.$$

Expanding by the elements of the third column (here we take advantage of the fact that one element is zero) gives

$$|A| = (2)(-1)^{1+3} \begin{vmatrix} 2 & 1 \\ -1 & 2 \end{vmatrix} + (-1)(-1)^{3+3} \begin{vmatrix} 3 & -4 \\ 2 & 1 \end{vmatrix}.$$

This gives $|A| = -1$.

Cramer's rule can be applied as follows: To find x, replace the coefficients of x in A (the first column) by the column of constants of the given equations; then divide the determinant of the resulting matrix by $|A|$ to get x. Proceed in a similar manner to find y and z:

$$x = \frac{\begin{vmatrix} 1 & -4 & 2 \\ 0 & 1 & 0 \\ 3 & 2 & -1 \end{vmatrix}}{|A|} = \frac{(1)(-1)^{2+2}\begin{vmatrix} 1 & 2 \\ 3 & -1 \end{vmatrix}}{-1} = \frac{-7}{-1} = 7;$$

$$y = \frac{\begin{vmatrix} 3 & 1 & 2 \\ 2 & 0 & 0 \\ -1 & 3 & -1 \end{vmatrix}}{|A|} = \frac{2(-1)^{2+1}\begin{vmatrix} 1 & 2 \\ 3 & -1 \end{vmatrix}}{-1} = \frac{14}{-1} = -14;$$

$$z = \frac{\begin{vmatrix} 3 & -4 & 1 \\ 2 & 1 & 0 \\ -1 & 2 & 3 \end{vmatrix}}{|A|} = \frac{2(-1)^{2+1}\begin{vmatrix} -4 & 1 \\ 2 & 3 \end{vmatrix} + (1)(-1)^{2+2}\begin{vmatrix} 3 & 1 \\ -1 & 3 \end{vmatrix}}{-1} = -38.$$

Thus the solution is given by $x = 7$, $y = -14$, $z = -38$.   ■

Example 6   Solve the equation $\begin{vmatrix} 2x & 0 & -1 \\ -2 & x & 3 \\ 1 & -1 & 2 \end{vmatrix} = 0.$

Solution   Evaluating the given determinant by the elements of the first row gives

$$(2x)\begin{vmatrix} x & 3 \\ -1 & 2 \end{vmatrix} + (-1)\begin{vmatrix} -2 & x \\ 1 & -1 \end{vmatrix} = 0.$$

That is,

$$4x^2 + 6x - 2 + x = 0,$$
$$4x^2 + 7x - 2 = 0.$$

Thus the given equation is equivalent to the quadratic equation $4x^2 + 7x - 2 = 0$. This can be solved by factoring: $(4x - 1)(x + 2) = 0$. Therefore $4x - 1 = 0$ or $x + 2 = 0$; and so $x = \frac{1}{4}$, $x = -2$ are two solutions.   ■

Partial Fractions

In dealing with fractions, we learn how to add, subtract, multiply, or divide two fractions. For example, to add $3/(x - 1)$ and $5/(x + 2)$ we first express each as equivalent fractions with a common denominator and then add the resulting fractions, as follows:

$$\frac{3}{x - 1} + \frac{5}{x + 2} = \frac{3(x + 2)}{(x - 1)(x + 2)} + \frac{5(x - 1)}{(x - 1)(x + 2)}$$

$$= \frac{3(x + 2) + 5(x - 1)}{(x - 1)(x + 2)} = \frac{8x + 1}{x^2 + x - 2}.$$

The following example illustrates the reverse problem.

Example 7   Given the fraction $\dfrac{8x + 1}{x^2 + x - 2}$, find two fractions whose denominators are linear expressions and whose sum is equal to the given fraction.

Solution   The given fraction can be written as $\dfrac{8x + 1}{(x - 1)(x + 2)}$, and it is reasonable to attempt to find two numbers a and b such that

$$\frac{a}{x - 1} + \frac{b}{x + 2} = \frac{8x + 1}{(x - 1)(x + 2)}. \tag{6.18}$$

Adding the two fractions on the left-hand side of Eq. (6.18) gives

$$\frac{a}{x - 1} + \frac{b}{x + 2} = \frac{a(x + 2) + b(x - 1)}{(x - 1)(x + 2)} = \frac{(a + b)x + (2a - b)}{(x - 1)(x + 2)}.$$

Thus Eq. (6.18) is equivalent to

$$\frac{(a + b)x + (2a - b)}{(x - 1)(x + 2)} = \frac{8x + 1}{(x - 1)(x + 2)}. \tag{6.19}$$

Note that the denominators of Eq. (6.19) are the same, and so it will be an identity if the numerators are identically equal. Thus a and b must satisfy

$$a + b = 8,$$
$$2a - b = 1.$$

Solving this system of linear equations gives $a = 3$ and $b = 5$. Substituting into Eq. (6.18), we get the identity

$$\frac{8x + 1}{(x - 1)(x + 2)} = \frac{3}{x - 1} + \frac{5}{x + 2}.$$

The technique for expressing a fraction as the sum of fractions with simpler denominators, as illustrated in this example, is called the *method of partial fractions*.

Example 8   Express $\dfrac{6x^2 + 3x + 1}{x^3 - x}$ as a sum of fractions with linear denominators.

Solution   It is reasonable to attempt to find numbers a, b, and c that will make the following an identity.

$$\frac{6x^2 + 3x + 1}{x^3 - x} = \frac{6x^2 + 3x + 1}{x(x + 1)(x - 1)} = \frac{a}{x} + \frac{b}{x + 1} + \frac{c}{x - 1}. \qquad (6.20)$$

Adding the three fractions on the right side of (6.20) and collecting like terms in the numerator, we get

$$\frac{6x^2 + 3x + 1}{x(x + 1)(x - 1)} = \frac{(a + b + c)x^2 + (-b + c)x - a}{x(x + 1)(x - 1)}.$$

This equation will be an identity if we can find a, b, and c such that

$$a + b + c = 6,$$
$$-b + c = 3,$$
$$-a \qquad\quad = 1.$$

Solving this system of linear equations gives $a = -1, b = 2, c = 5$. Therefore the solution is

$$\frac{6x^2 + 3x + 1}{x^3 - x} = \frac{-1}{x} + \frac{2}{x + 1} + \frac{5}{x - 1}.$$

Example 9   Express $\dfrac{x^2 + 11x + 5}{(x - 1)(x + 2)^2}$ as a sum of fractions.

Solution   Looking at the denominator of the given fraction, one can reasonably assume that it is a common denominator of three fractions whose denominators are $(x - 1)$, $(x + 2)$, and $(x + 2)^2$. Thus we look for three numbers a, b, and c such that

$$\frac{x^2 + 11x + 15}{(x - 1)(x + 2)^2} = \frac{a}{x - 1} + \frac{b}{x + 2} + \frac{c}{(x + 2)^2}.$$

Adding the fractions on the right and collecting terms in the numerator, we get

$$\frac{x^2 + 11x + 15}{(x - 1)(x + 2)^2} = \frac{(a + b)x^2 + (4a + b + c)x + (4a - 2b - c)}{(x - 1)(x + 2)^2}.$$

Equating corresponding coefficients of the two numerators gives

$$a + b = 1,$$
$$4a + b + c = 11,$$
$$4a - 2b - c = 15.$$

The solution to this system of linear equations is given by $a = 3$, $b = -2$, and $c = 1$. Therefore

$$\frac{x^2 + 11x + 5}{(x - 1)(x + 2)^2} = \frac{3}{x - 1} - \frac{2}{x + 2} + \frac{1}{(x + 2)^2}.$$

Exercises 6.2

In problems 1 through 6, find the determinants of the given matrices.

1. $A = \begin{bmatrix} 1 & -2 \\ 3 & 5 \end{bmatrix}$

2. $A = \begin{bmatrix} 0 & 1 \\ 3 & 2 \end{bmatrix}$

3. $B = \begin{bmatrix} 1 & 0 \\ 0 & 1 \end{bmatrix}$

4. $B = \begin{bmatrix} 1 & 0 \\ -1 & 1 \end{bmatrix}$

5. $B = \begin{bmatrix} -0.3 & 1.2 \\ -0.7 & 0.8 \end{bmatrix}$

6. $A = \begin{bmatrix} \sqrt{3} & -1 \\ 4 & \sqrt{12} \end{bmatrix}$

In problems 7 through 12, apply Cramer's rule to the given systems of equations. In cases where there are infinitely many solutions, describe all solutions, and give three particular solutions.

7. $\begin{aligned} x - 2y &= 1, \\ 3x + y &= 2 \end{aligned}$

8. $\begin{aligned} 5x - y &= -1, \\ -3x + 2y &= 4 \end{aligned}$

9. $\begin{aligned} 2x - 4y &= 1, \\ -3x + 6y &= -2 \end{aligned}$

10. $\begin{aligned} 3x - 2y &= 4, \\ -\frac{3}{2}x + y &= -2 \end{aligned}$

11. $\begin{aligned} \frac{3}{x} - \frac{5}{y} &= 2 \\ -\frac{2}{x} + \frac{1}{y} &= 3 \end{aligned}$

12. $\begin{aligned} \frac{2}{x} - \frac{1}{y} &= 1, \\ -\frac{1}{x} + \frac{1}{2y} &= -\frac{1}{2} \end{aligned}$

In problems 13 through 16, find the determinant of the given matrices. In each case, evaluate by using the elements of a row and then by using elements of a column to get a check on your answers.

13. $A = \begin{bmatrix} 1 & 0 & 2 \\ -1 & 2 & -3 \\ 0 & 1 & 4 \end{bmatrix}$

14. $A = \begin{bmatrix} -1 & 3 & 2 \\ 2 & -1 & 4 \\ 1 & 2 & 6 \end{bmatrix}$

15. $B = \begin{bmatrix} 2 & 4 & -3 \\ 1 & -1 & 4 \\ 2 & -5 & 1 \end{bmatrix}$ **16.** $B = \begin{bmatrix} 0.2 & -0.1 & 0 \\ 3.1 & 1.4 & -0.5 \\ -0.8 & 0 & 0.3 \end{bmatrix}$

In problems 17 through 20, use Cramer's rule to solve the given systems of equations.

17.
$$\begin{aligned} x - y + z &= 3, \\ 2x + y &= 1, \\ x - 3z &= -2 \end{aligned}$$

18.
$$\begin{aligned} 3x - y + 2z &= 0, \\ x + y &= 2, \\ y - 2z &= 1 \end{aligned}$$

19.
$$\begin{aligned} 2x - y &= -3, \\ x + y - 2z &= 2, \\ -x + 5y - 6z &= 1 \end{aligned}$$

20.
$$\begin{aligned} \frac{2}{x} - \frac{3}{y} + \frac{4}{z} &= -1, \\ \frac{1}{x} + \frac{4}{y} &= 4, \\ \frac{1}{x} - \frac{7}{y} + \frac{4}{z} &= -5 \end{aligned}$$

In problems 21 through 24, solve the given equations for x.

21. $\begin{vmatrix} 3 & x \\ -2 & -4 \end{vmatrix} = 2$ **22.** $\begin{vmatrix} x & 4 & 0 \\ 2 & 2 & -x \\ 1 & 1 & 1 \end{vmatrix} = 0$ **23.** $\begin{vmatrix} 1 & 0 & x \\ 3 & -1 & 2 \\ -5 & 3 & 0 \end{vmatrix} = -2$ **24.** $\begin{vmatrix} x & -2x & 1 \\ 3x & 1 & -2 \\ 2x & 2x+1 & -3 \end{vmatrix} = 0$

25. Determine three *integers* such that their sum is 48, the first is 4 more than twice the second, and the second is 52 less than the third.

26. A mixture of 50 pounds of peanuts, cashew nuts, and walnuts costs a total of $49. If peanuts cost $0.80 per pound, cashews cost $1.10 per pound, and walnuts cost $1.20 per pound, and if the mixture contains twice as many pounds of peanuts as of walnuts, how many pounds of each does the mixture contain?

27. A mixture of one-half cup of chocolate pudding and 3 tablespoons of whipped cream contains 330 calories; one-third cup of pudding and one tablespoon of cream contains 170 calories. How many calories are there in one cup of pudding? In one tablespoon of cream?

28. Suppose α, β, and γ represent the number of degrees in the three angles of a triangle, and α is 20° more than the sum of β and γ; also α is 60° greater than β. Find the measures of the three angles.

29. Two pipes, A and B, supply water to a reservoir, while pipe C (located at the bottom) drains the reservoir. When all three pipes are open, it takes 18 hours to fill the reservoir. If pipes A and B are open and C is closed, it takes 12 hours to fill the reservoir. If A and C are open and B is closed, it takes 24 hours to fill the reservoir. How many hours does it take to fill the reservoir if only pipe A is open?

30. A breakfast menu is to consist of oatmeal, whole milk, and fresh orange juice. We are interested in the protein-calcium-vitamin C content, and the following table gives the pertinent information.

Food	Protein	Calcium	Vitamin C
Oatmeal (1 cup; 245 g)	5 g	22 mg	0 mg
Milk (1 cup; 244 g)	8 g	291 mg	2 mg
Orange juice (1 cup; 248 g)	2 g	27 mg	124 mg

How many cups of each (oatmeal, milk, orange juice) are required to get a breakfast of 9 g of protein, 185.7 mg of calcium, and 125 mg of vitamin C?

In each of the following, use the method of partial fractions to express the given fraction as a sum of fractions with simple denominators.

31. $\dfrac{4x - 3}{(x - 2)(x + 3)}$ **32.** $\dfrac{8x - 7}{(2x - 1)(x + 1)}$ **33.** $\dfrac{-8}{3x^2 - 4x - 4}$ **34.** $\dfrac{14x}{3x^2 + 5x - 2}$

35. $\dfrac{-10x - 4}{x^3 - 4x}$ **36.** $\dfrac{5x^2 + x}{2x^3 + x^2 - 2x - 1}$ **37.** $\dfrac{2x^2 - 10x + 15}{(x + 1)(x - 2)^2}$ **38.** $\dfrac{9x^2 - 4x + 1}{2x^3 - x^2}$

6.3   SYSTEMS OF LINEAR INEQUALITIES

In the preceding sections of this chapter we discussed systems of linear equations; in most cases the problems encountered had exactly one solution. We now consider systems of linear inequalities in which generally there are infinitely many solutions. For cases involving only two variables, the best way to describe such results is graphically as sets of points in a certain region of the x-y plane. For this reason our discussion will be limited to linear inequalities in two variables. Problems involving more than two variables are studied in advanced courses.

In the following examples we first consider graphs of a single linear inequality; then problems involving systems of linear inequalities are discussed. Examples 6 and 7 illustrate how applied problems can be formulated mathematically in terms of a system of linear inequalities. This is essentially the first step in the solution of linear programming problems.

Graphs of Linear Inequalities

A point on a line divides the line into two half-lines. In a similar manner, a line L in a given plane partitions the plane into two half-planes H_1 and H_2, as shown in Fig. 6.3. Just as the set of points on line L can be described algebraically as a

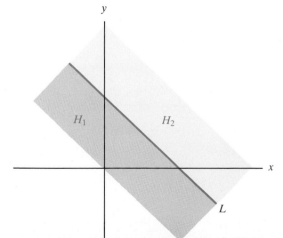

Fig. 6.3

linear equation in x and y (a linear equality), so the set of points of a half-plane can be described by a *linear inequality*. This is illustrated with an example.

Example 1   Discuss the relations given by

     a) $2x + y = 4$,   **b)** $2x + y < 4$,   **c)** $2x + y > 4$.

Solution   The graph of the given equation (a) is a line L, as shown in Fig. 6.4(a). This means that each pair of real numbers x, y satisfying the equation corresponds to the point (x, y) on the line, and the converse. For instance, $x = -1, y = 6$ satisfies the equation since $2(-1) + 6 = 4$ is a true statement, and so $(-1, 6)$ is on L; but $x = -1, y = 2$ does not satisfy the equation since $2(-1) + 2 = 4$ is a false statement, and so $(-1, 2)$ is not on L. However, $x = -1, y = 2$ satisfies the inequality given in (b) since $2(-1) + 2 < 4$ is a true statement. It should be clear after checking several points that any point below L will satisfy the inequality given in (b). Thus the graph of inequality $2x + y < 4$ is the half-plane H_1 (not including L), as shown in Fig. 6.4(b); note the broken line for L, indicating that L is not included. We shall call such a half-plane *an open half-plane*. In a similar manner, the graph of $2x + y > 4$ is the open half-plane H_2, as shown in Fig. 6.4(c).

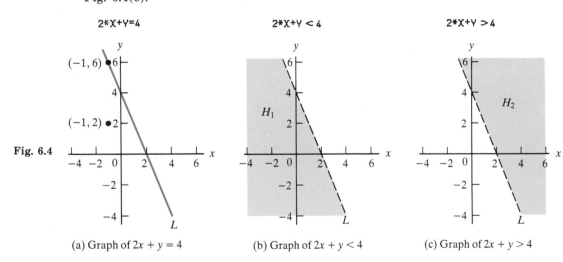

Fig. 6.4

(a) Graph of $2x + y = 4$ (b) Graph of $2x + y < 4$ (c) Graph of $2x + y > 4$

Thus line L partitions the plane into three disjoint subsets: L itself and the two half-planes H_1 and H_2. In terms of set notation, we have

$$\{(x, y) \mid 2x + y = 4\} \text{ is the set of points on } L,$$
$$\{(x, y) \mid 2x + y < 4\} \text{ is the set of points in } H_1,$$
$$\{(x, y) \mid 2x + y > 4\} \text{ is the set of points in } H_2.$$

More generally, we have the following: A linear equation $ax + by = c$ (where not both a and b are zero) describes a line L, and the inequalities

$ax + by < c$ and $ax + by > c$ describe open half-planes. To determine which of the two half-planes corresponds to which inequality, we can take any "test point" (x_1, y_1) not on L and check to see which of the two inequalities is satisfied when x is replaced by x_1 and y by y_1.

Example 2   Draw a graph of the inequality $x + 2y \geq 3$.

Solution   The $\geq$ relation means all pairs x, y that satisfy either $x + 2y > 3$ or $x + 2y = 3$. Thus the graph consists of all points in the half-plane $x + 2y > 3$ along with the points on the line $L: x + 2y = 3$. To determine which half-plane we want, take $(0, 0)$ as a test point; since $0 + 2 \cdot 0 > 3$ is not a true statement, $(0, 0)$ is not in the half-plane we want. The graph of the given inequality is shown in Fig. 6.5. Note that L is drawn as a solid line, which indicates that it is included in the graph. The graph is an example of a *closed half-plane,* meaning that the line is included.

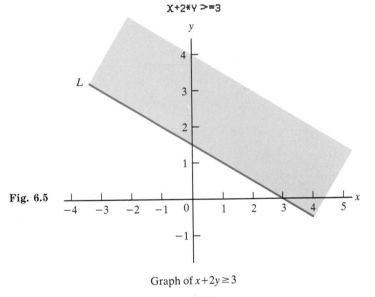

Fig. 6.5

Graph of $x + 2y \geq 3$

Systems of Linear Inequalities
Techniques for solving systems of linear equations were introduced in the preceding two sections. Methods for solving *systems of linear inequalities* are illustrated by the following examples.

Example 3   Solve the system of linear inequalities

$$x + 2y \leq 3,$$
$$-3x + y < 5.$$

Describe the solution graphically.

Solution   We want the set S of all points (x, y) that simultaneously satisfy both of the inequalities. That is, $S = A \cap B$, where

$$A = \{(x, y) \mid x + 2y \le 3\}, \qquad B = \{(x, y) \mid -3x + y < 5\}.$$

Geometrically, A and B are half-planes, and so S is the set of all points that are in both half-planes. This is shown in Fig. 6.6. Point P is called a *corner point* of S; the coordinates of P are determined by solving the system of linear equations

$$x + 2y = 3,$$
$$-3x + y = 5.$$

The solution to this system is given by $x = -1$, $y = 2$; thus P is the point $(-1, 2)$. Note that P is not in the solution set, since it is not in set B.

Fig. 6.6

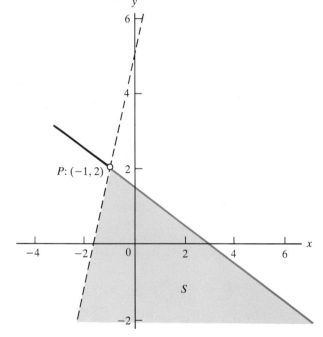

Graph of $x + 2y \le 3$ and $-3x + y < 5$

Example 4   Solve the system of inequalities

$$x + 2y \le 3,$$
$$-3x + y < 5,$$
$$-3x + 8y \ge -23.$$

Solution   Note that the first two inequalities are the same as those given in Example 3. Thus the solution set consists of all points that are in both set S (shown in Fig. 6.6) and the set

$$C = \{(x, y) \,|\, -3x + 8y \geq -23\}.$$

Set C is the closed half-plane above the line $-3x + 8y = -23$ (since test point $(0, 0)$ is in C). Thus the solution set is shown in Fig. 6.7. The solution set consists of all points inside triangle PMQ and the line segments PM and QM, excluding points P and Q. The corner points P, M, Q are determined by solving the systems of linear equations

$$P:\begin{cases} x + 2y = 3, \\ -3x + y = 5; \end{cases} \quad M:\begin{cases} x + 2y = 3, \\ 3x - 8y = 23; \end{cases} \quad Q:\begin{cases} -3x + y = 5, \\ 3x - 8y = 23. \end{cases}$$

The solutions of these give $P:(-1, 2)$, $M:(5, -1)$, and $Q:(-3, -4)$. In Fig. 6.7 points P and Q are shown with open circles to indicate that they do not belong to the solution set. Point M is shown with a solid circle to indicate that it is included in the solution set.

X+2*Y < =3 & -3*X+Y < 5 & -3*X+8*Y > =-23

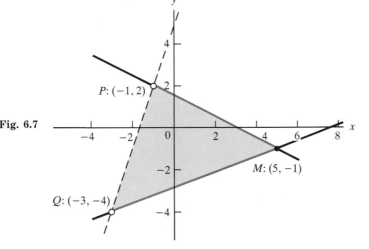

Fig. 6.7

Graph of $x + 2y \leq 3$, $-3x + y < 5$, and $-3x + 8y \geq -23$

Example 5   Solve the system of inequalities

$$\begin{aligned} x - 2y &\geq -2, \\ -2x + 4y &> 9. \end{aligned}$$

Solution   We first draw graphs of the two lines $L_1: x - 2y = -2$ and $L_2: -2x + 4y = 9$. These are parallel lines, and the half-planes H_1 and H_2 corresponding to the two inequalities are shown in Fig. 6.8. Since the two half-planes have no points in common, the solution set S is the empty set. This is indicated by writing $S = \emptyset$.

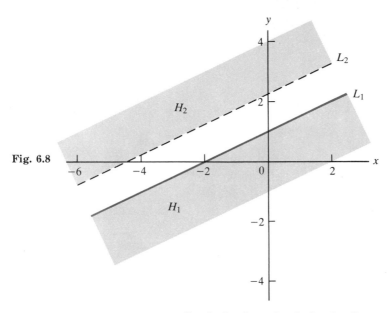

Fig. 6.8

Graph of $x - 2y \geq -2$ and $-2x + 4y > 9$

Example 6   A musical sponsored by the student association is to be held in the school auditorium, which has a seating capacity of 1500. Ticket prices are $5 each for the 500 reserved seats and $3 each for the remaining 1000 general-admission seats. The student council determines that the total cost of presenting the musical will be $3700. How many reserved-seat tickets and how many general-admission tickets must be sold if no financial loss is to be incurred?

Solution   Let x and y represent the numbers of reserved-seat and general-admission tickets, respectively, to be sold. We want all integers x and y that will satisfy the following inequalities: $x \geq 0$, $y \geq 0$, $x \leq 500$, $y \leq 1000$, and $5x + 3y \geq 3700$. The solution set for this system of inequalities is shown in Fig. 6.9.

Any point inside or on triangle ABC with integral coordinates will satisfy the given conditions and represent a no-loss situation. For example, 400 reserved tickets and 600 general-admission tickets sold will be a solution. In fact, the corresponding revenue would be $5(400) + 3(600) = \$3800$, a profit of $100. Clearly there are several other solutions. However, $x = 200$, $y = 700$ is not a solution; the revenue would be $5(200) + 3(700) = \$3100$, a loss of $600.

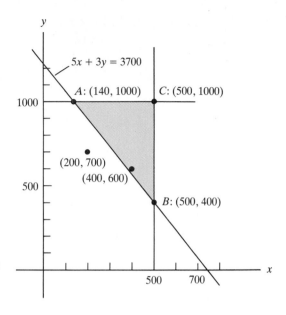

Fig. 6.9

Example 7   Diane wishes to have tuna salad and a chocolate milk shake for lunch. She plans her diet scientifically and is interested in the iron-protein-fat content of her lunch. The following constraints are to be satisfied:

a) At least 2 mg of iron;

b) At least 24 g of protein;

c) Not more than 30 g of fat.

The following table gives the pertinent nutrient values, where one unit means 100 grams. Determine how many units of tuna salad and how many units of milk shake will give a lunch that satisfies the desired constraints.

Food	Iron	Protein	Fat
Tuna salad 2 units (1 cup)	2 mg	30 g	20 g
Milk shake 3 units (1 cup)	1 mg	9 g	10 g

Solution   Let x and y denote the *number of units* of tuna salad and chocolate milk shake, respectively, that she can have for lunch. Translating the verbal statements of the problem into mathematical statements, we have the following.

a) Tuna salad contains $\frac{2}{2}$ mg/unit of iron, and the milk shake has $\frac{1}{3}$ mg/unit of iron. Thus "at least 2 mg of iron" translates into

$$x + \frac{1}{3}y \geq 2.$$

This is equivalent to $3x + y \geq 6$.

b) Similarly, "at least 24 g of protein" translates into

$$\frac{30}{2}x + \frac{9}{3}y \geq 24.$$

This is equivalent to $5x + y \geq 8$.

c) The constraint "not more than 30 g of fat" gives

$$\frac{20}{2}x + \frac{10}{3}y \leq 30.$$

This is equivalent to $3x + y \leq 9$.

   Therefore any pair of numbers x, y satisfying the following system of inequalities is acceptable:

$$3x + y \geq 6; \qquad 5x + y \geq 8; \qquad 3x + y \leq 9; \qquad x \geq 0; \qquad y \geq 0.$$

The solution set for this system is shown in Fig. 6.10.

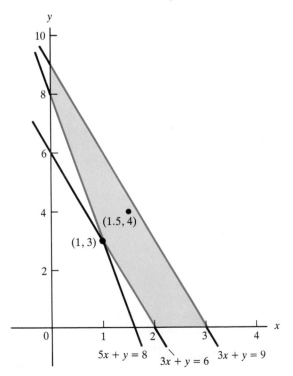

Fig. 6.10

Thus we see that Diane has considerable choice in how much of each she can have for lunch. For example, $x = 1, y = 3$ (that is 100 g of tuna salad and 300 g of milk shake) will satisfy the requirements. Similarly $x = 1.5, y = 4$ will also do it. However, $x = 2, y = 4$ is not acceptable, since it does not satisfy requirement (c).

Exercises 6.3

In problems 1 through 10, draw graphs of the given inequalities. In each case give two specific pairs of x, y values that satisfy the given inequality.

1. $x + 2y < 4$ 2. $-x + y > 3$ 3. $2x - 3y \geq 6$ 4. $-4x - 2y \leq 9$

5. $x + y + 4 < 0$ 6. $2x > y - 4$ 7. $y \geq 2x$ 8. $3x - 2y < 0$

9. $5x + 3y < 0$ 10. $4 > 3x - 2y$

In problems 11 through 14, draw graphs of the given sets.

11. $\{(x, y) | 2x + y < 0\}$ 12. $\{(x, y) | 2x + 3y > 6\}$ 13. $\{(x, y) | x + 2 > 0\}$ 14. $\{(x, y) | y + 3 \geq 0\}$

In problems 15 through 25, draw graphs of the regions described by the given systems of linear inequalities. In each case, determine the coordinates of any corner points, and label such points on your graph. Indicate whether or not these points belong to the solution set of the given system (see Examples 3 and 4).

15. $\quad x + y < \quad 4,$
$\quad 2x - y < -1$

16. $\quad 3x - 2y > \quad 5,$
$\quad -x - \quad y < -5$

17. $\quad x - 2y \geq 4,$
$\quad x > 2,$
$\quad y < 3$

18. $\quad 3x - 4y < 6,$
$\quad |x| < 2,$
$\quad y < 3$

19. $-x + 2y < 5,$
$\quad 2x + \quad y > 0,$
$\quad 3x - \quad y < 5$

20. $\quad 4x + 3y \leq \quad 16,$
$\quad -x + \quad y > -4,$
$\quad 6x + \quad y \geq \quad 10$

21. $\quad 3x - 2y \geq \quad -3,$
$\quad -3x + 2y > -14,$
$\quad 4x + 3y \leq \quad 13$

22. $y \leq 2x,$
$\quad |x| < 3,$
$\quad y \leq 2$

23. $x + y < 3,$
$\quad |x| \geq 2$

24. $\quad 2x - 4y > 5,$
$\quad -x + 2y > 4$

25. $\quad 3x - 6y > 8,$
$\quad -x + 2y > 3,$
$\quad x + \quad y \leq 4$

In problems 26 through 30, determine whether or not each of the pairs given in (a) and (b) belong to the given system of inequalities.

26. $\begin{cases} x - 3y < 4 \\ 2x + \quad y < 3 \end{cases}$ **a)** $x = 1, y = 1$
   b) $x = \sqrt{2}, y = -0.5$

27. $\begin{cases} -2x + \quad y > -3 \\ \quad 5x + 2y < \quad 1 \end{cases}$ **a)** $x = -1, y = 2$
   b) $x = 1, y = -5$

28. $\begin{cases} x - 3y \geq 1 \\ 4x - \quad y \leq \pi \end{cases}$ **a)** $x = 1, y = -1$
   b) $x = \sqrt{2}, y = \pi$

29. $\begin{cases} y \leq \quad 2x \\ y > -3x \end{cases}$ **a)** $x = 0, y = 0$
   b) $x = -1, y = 3$

30. $\begin{cases} x < 2y - 1 \\ |x| < \quad 4 \end{cases}$ **a)** $x = 0, y = 1$
   b) $x = -4, y = -\sqrt{2}$

In problems 31 through 35, draw graphs of the regions corresponding the given sets. Determine the coordinates of any corner points, and indicate whether or not they belong to the given set.

31. $\{(x, y) | x + y > 2, x + 2y < 5\}$ 32. $\{(x, y) | 2x - y \leq 4, |x| < 3, y + 1 > 0\}$

33. $\{(x, y) \mid x \geq 0,\ y \geq 0,\ x + 2y < 4\}$ **34.** $\{(x, y) \mid x \geq 0,\ y \geq 0,\ x + 2y \geq 4\}$

35. $\{(x, y) \mid x \geq 1,\ y \geq 2,\ x + y \leq 6,\ x - y \leq 0\}$

36. A concert is to be presented in an auditorium that has a seating capacity of 800. The price per ticket for 200 of the seats is $6, and for the remaining 600 seats it is $3. The management determines that the total cost for presenting the concert will be $2100. Draw a graph to show the various possible pairs of numbers of $6 and $3 tickets that must be sold so that the concert does not result in a financial loss.

37. A rancher who wants to purchase some lambs and goats cannot spend more than $800 and wants at least 5 lambs and at least 4 goats. The cost of each lamb is $80, and the cost of each goat is $50. How many of each can the rancher buy? Draw a graph to help you list all possible pairs, keeping in mind that lambs and goats come in whole numbers (assume they are live ones).

38. At a fish cannery two kinds of tuna are packed into cans, chunk style and solid pack. Limits on storage space and customer demand lead to the following conditions.

a) The total number of cases produced per day is to be not more than 3000.

b) The number of cases of chunk style is to be at least twice the number of cases of solid pack.

c) The number of cases of solid pack produced per day is to be at least 600.

How many cases of each type can be produced each day and still satisfy the given constraints? Draw a graph of the solution set, and show the coordinates of the corner points.

In problems 39 through 41, use information from the following table, which gives nutrient values of four foods A, B, C, and D, where one unit means 100 grams.

Food	Energy (calories/unit)	Vitamin C (mg/unit)	Iron (mg/unit)	Calcium (mg/unit)	Protein (g/unit)	Carbohydrates (g/unit)
A	200	2	0.5	10	2	15
B	100	3	1.5	4	3	30
C	300	0	2.0	20	9	10
D	400	1	0.0	5	3	10

39. In preparing a menu, determine how many units of A and of B can be included so that the combined nutrient values will satisfy the following constraints.

a) At least 8 mg of vitamin C

b) At least 18 mg of calcium

c) Not more than 800 calories

40. How many units of A and of C can be included in a menu to contribute the following?

a) At least 3 mg of vitamin C

b) At least 40 mg of calcium

c) Not more than 60 g of carbohydrates

41. How many units of C and of D will give a combined total satisfying the following constraints?

   a) At least 2 mg of vitamin C

   b) At least 15 g of protein

   c) Not more than 6 mg of iron

   d) Not more than 2100 calories

6.4 NONLINEAR SYSTEMS

Let us now consider systems of equations and inequalities that include quadratic expressions. Techniques for solving such systems will be illustrated through the following examples. The first example was used to introduce the discussion of this chapter.

Example 1 Suppose the hypotenuse of a right triangle is 17 and the perimeter is 40. Determine the lengths of the two sides.

Solution Let x and y denote the lengths of the two sides, as shown in Fig. 6.11. From the given information, x and y must satisfy the equations

$$x^2 + y^2 = 17^2 = 289 \text{ (hypotenuse is 17)}$$

and

$$x + y + 17 = 40 \text{ (perimeter is 40).}$$

Thus we wish to solve the system of equations

$$\begin{aligned} x^2 + y^2 &= 289, \\ x + y &= 23. \end{aligned} \tag{6.21}$$

The problem is similar to those discussed in preceding sections, except that the first equation of (6.21) is not linear. We can use the method of substitution as follows: Solving the second equation of (6.21) for y in terms of x, $y = 23 - x$, and substituting this into the first equation gives

$$x^2 + (23 - x)^2 = 289.$$

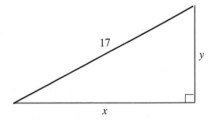

Fig. 6.11

This is a quadratic equation that can be solved as follows:

$$x^2 - 23x + 120 = 0,$$
$$(x - 8)(x - 15) = 0.$$

This gives $x = 8$ or $x = 15$. The corresponding values of y can be determined by substituting each of these values of x into the second equation of (6.21). Thus we get two solutions, given by $x = 8$, $y = 15$ or $x = 15$, $y = 8$. This tells us that the lengths of the two sides are 8 and 15. ■

Example 2 Solve the system of equations

$$2x + y = 10,$$
$$x^2 + y^2 = 25. \tag{6.22}$$

Draw a graph and give a geometric interpretation of the solution.

Solution Solving the first equation for y, $y = 10 - 2x$ and substituting this into the second equation gives $x^2 + (10 - 2x)^2 = 25$. This is equivalent to the quadratic equation $x^2 - 8x + 15 = 0$, which can be solved by factoring to get $x = 3$ or $x = 5$. Substituting each of these values of x into the first equation of (6.22) and solving for the corresponding values of y gives two solutions, $x = 3$, $y = 4$ or $x = 5$, $y = 0$.

Graphically, the first equation of (6.22) corresponds to a line and the second to a circle, as shown in Fig. 6.12. The two points of intersection of the line and circle correspond to the two solutions of the given system of equations.

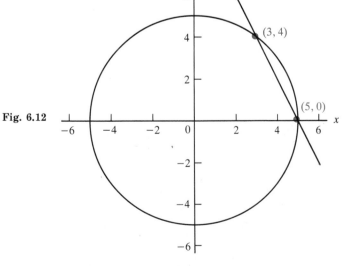

Fig. 6.12

Graph of $x^2 + y^2 = 25$ and $y = -2x + 10$

Example 3   Solve the system of equations

$$x + 2y = -4,$$
$$y = x^2 - 2x - 3. \tag{6.23}$$

Illustrate the solution graphically.

Solution   Substituting the expression given in the second equation of (6.23) into the first gives

$$x + 2(x^2 - 2x - 3) = -4.$$

This is equivalent to

$$2x^2 - 3x - 2 = 0,$$
$$(2x + 1)(x - 2) = 0.$$

Thus $x = -\frac{1}{2}$ or $x = 2$; the corresponding values of y can be found by using either of the equations of (6.23). Thus the solution to the given system is given by $x = -\frac{1}{2}$, $y = -\frac{7}{4}$ or $x = 2$, $y = -3$. The graphs of the given equations are shown in Fig. 6.13. The line and parabola intersect at points $(-\frac{1}{2}, -\frac{7}{4})$ and $(2, -3)$; these correspond to the solutions of the given system of equations.

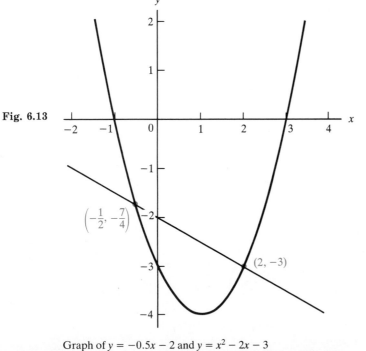

Fig. 6.13

Graph of $y = -0.5x - 2$ and $y = x^2 - 2x - 3$

Example 4   Solve the system of equations

$$2x + y = 3,$$
$$x^2 + y = 1.$$

(6.24)

Give a graphical interpretation of the result.

Solution   From the first equation of (6.24), $y = 3 - 2x$. Substituting this into the second equation gives

$$x^2 + (3 - 2x) = 1,$$
$$x^2 - 2x + 2 = 0.$$

Using the quadratic formula to solve this equation, we get

$$x = \frac{2 \pm \sqrt{4 - 8}}{2} = \frac{2 \pm \sqrt{-4}}{2} = \frac{2 \pm 2i}{2} = 1 \pm i.$$

Since our replacement sets are restricted to real numbers, we conclude that the solution set S for the given system is the empty set and write $S = \emptyset$.

The graphs corresponding to the two equations in (6.24) are given in Fig. 6.14. The line and parabola do not intersect.

2*X+Y=3 & X^2+Y=1

Fig. 6.14

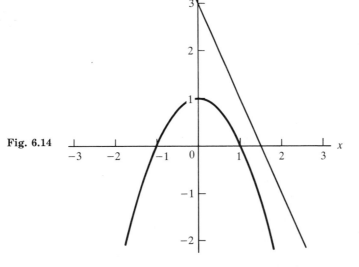

Graph of $y = -2x + 3$ and $y = 1 - x^2$

Example 5   Draw a graph of the inequality

$$y \geq x^2 - x - 6. \tag{6.25}$$

Solution   Just as a line partitions the plane into three disjoint sets, so the parabola $y = x^2 - x - 6$ will partition the plane into three disjoint sets: the parabola itself and two regions separated by the parabola. These regions are described by the two inequalities $y > x^2 - x - 6$ and $y < x^2 - x - 6$.

  The graph of $y = x^2 - x - 6 = (x - 3)(x + 2)$ is shown in Fig. 6.15. Using $(0, 0)$ as test point, we see that $0 \geq 0^2 - 0 - 6$ is a true statement, and so $(0, 0)$ belongs to the region we want. That is, the inequality given in (6.25) corresponds to all points inside or on the parabola, as shown in Fig. 6.15.

Fig. 6.15

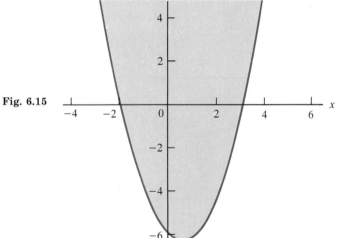

Graph of $y \geq x^2 - x - 6$

Example 6   Solve the system of inequalities

$$\begin{aligned} y &< x + 2, \\ y &\geq x^2 - x - 6. \end{aligned} \tag{6.26}$$

Show the solution graphically.

Solution   The second inequality corresponds to the region inside or on the parabola $y = x^2 - x - 6$ (see Example 5), and the first inequality corresponds to the half-plane below the line $y = x + 2$. Thus the system given in (6.26) corresponds to the intersection of these two sets, and it is shown in Fig. 6.16. The "corner points" are found by solving the system of equations $y = x + 2$, $y = x^2 - x - 6$. The solutions are: $x = -2$, $y = 0$ or $x = 4$, $y = 6$. These points are not in the solution set.

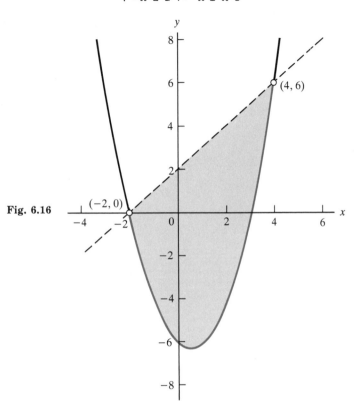

Y < X+2 & Y >= X^2-X-6

Fig. 6.16

Graph of $y < x + 2$ and $y \geq x^2 - x - 6$

Exercises 6.4

In problems 1 through 10, solve the given systems of equations, and draw graphs to illustrate the solutions geometrically.

1. $y = 3x + 4$,
$\quad y = x^2$

2. $2x - 3y + 26 = 0$,
$\quad x^2 + y^2 = 169$

3. $y = -3x$,
$\quad y = -2x^2 - 4x$

4. $y = 2x + 2$,
$\quad xy = 4$

5. $2x + 3y = -3$,
$\quad xy = -3$

6. $y = 5x - 10$,
$\quad y = x^2 + x - 6$

7. $y = 2x$,
$\quad y = x^2 + 3$

8. $x + y = 4$,
$\quad x^2 + y^2 = 1$

9. $y = x^2 - 4x + 4$,
$\quad y = -2x^2 + x + 16$

10. $3x - y = 5$,
$\quad x^2 + y^2 = 25$

In each of the problems 11 through 16, draw a graph of the given set.

11. $\{(x, y) \mid y \leq x^2 + 5x + 6\}$

12. $\{(x, y) \mid x^2 + y^2 \leq 25\}$

13. $\{(x, y) \mid x^2 + y^2 - 4 < 0\}$

14. $\{(x, y) \mid y > x^2 - 1\}$ **15.** $\{(x, y) \mid y^2 > 1 - x^2\}$ **16.** $\{(x, y) \mid x^2 + y^2 < 1\}$

In problems 17 through 21, draw graphs to illustrate the solution sets for the given system of inequalities.

17. $y \geq x^2 - 1,$ **18.** $y < x^2 + 2x - 3,$ **19.** $y \leq -x^2 - 2x + 3,$ **20.** $y \geq x^2 - 1,$ **21.** $y \geq x^2 - 4,$
     $x + 2y \leq 10$ $x + 2y < 1$ $x + y < 1,$ $y \leq x + 5,$ $y > x,$
                                          $x - y > 1$ $|x| < 2$ $y < 3x$

22. The perimeter of a rectangle is 40 cm and the area is 96 cm². Find the dimensions of the rectangle.

23. Find the dimensions of a rectangle that has a diagonal of length 13 cm and perimeter of 34 cm.

24. One side of a rectangle is 3 cm longer than twice the other side, and its area is 230 cm². Find the dimensions of the rectangle.

25. The altitude of a triangle is twice as long as the corresponding base. The area of the triangle is 36 cm². Determine the length of the altitude and base.

Review Exercises

In problems 1 through 10, solve the given systems of equations. If a system is dependent (has infinitely many solutions), describe all solutions, and then give two specific ones.

1. $3x - 2y = 5,$
    $x - y = -1$

2. $-2x + y = 3,$
     $5x - 3y = -4$

3. $\dfrac{x}{2} - \dfrac{y}{3} = 4,$
$\dfrac{x}{4} + \dfrac{y}{2} = -2$

4. $0.4x + 0.6y = 0,$
    $1.3x - 1.2y = 2.52$

5. $x - 2y + z = 3,$
   $-2x + y - z = 0,$
     $4x - 3y + 2z = 1$

6. $x + 2y = 2,$
   $3x - 4y + z = -2,$
     $x + 3z = -8$

7. $x + 2y - 5z = 1,$
   $3x + 2y + z = -2,$
   $3x - 2y + 17z = -7$

8. $x - y + z = 3,$
   $5x - 4y + 3z = 2,$
   $x - 2y + 3z = 16$

9. $\dfrac{1}{x} - \dfrac{2}{y} = \dfrac{1}{3},$
$\dfrac{2}{x} - \dfrac{5}{y} = -\dfrac{2}{5}$

10. $\dfrac{3}{x} + \dfrac{4}{y} = 2,$
$\dfrac{1}{x} + \dfrac{2}{y} = -1$

In problems 11 through 14, evaluate the determinants of the given matrices.

11. $A = \begin{bmatrix} 3 & -4 \\ 2 & 5 \end{bmatrix}$ **12.** $B = \begin{bmatrix} \sqrt{3} - 1 & 2 + \sqrt{5} \\ 2 - \sqrt{5} & \sqrt{3} + 1 \end{bmatrix}$ **13.** $A = \begin{vmatrix} 1 & -2 & 0 \\ 3 & 2 & -1 \\ 5 & 4 & 2 \end{vmatrix}$ **14.** $B = \begin{vmatrix} -3 & 2 & -1 \\ 2 & 2 & 3 \\ 5 & 4 & -1 \end{vmatrix}$

In problems 15 through 18, draw graphs of the set of points (x, y) satisfying the given inequalities.

15. $2x - y < 1$ **16.** $x + y \geq 1$

17. $y \leq x$ and $x - y < 2$ **18.** $2x + y < 4$ and $x - 2y \geq 1$

In problems 19 through 22, draw graphs of the regions described by the given systems of inequalities. In each case determine the coordinates of all corner points.

19. $x - y \leq 4,$
$\quad 2x + y \geq 2,$
$\quad x + 2y \leq 4$

20. $2x - y \geq 8,$
$\quad 2x + y \leq 4,$
$\quad x - y \leq 8$

21. $x \geq 0,$
$\quad y \geq 0,$
$\quad 2x + 3y \leq 18,$
$\quad 2x + y \leq 10$

22. $x \geq 0,$
$\quad y \geq 0,$
$\quad x \leq 3,$
$\quad y \leq 5,$
$\quad 3x + 2y \leq 13$

In problems 23 through 26, use the method of partial fractions to express the given fraction as sums of fractions with simple denominators.

23. $\dfrac{x + 5}{x^2 + x - 2}$

24. $\dfrac{6x}{(x + 1)(x + 4)}$

25. $\dfrac{2x - 3}{x^2 + x}$

26. $\dfrac{2x^2 - 2x - 3}{x^3 - x}$

27. A firm makes two models, A and B, of a product. The manufacture of each model requires a process in which two machines, I and II, are used. The number of hours each machine works on each model is given by the table.

Machine	Model A	Model B
I	1.5	3
II	3.5	2.5

       Machine I is used for a total of 36 hours per week and machine II is used for a total of 39 hours per week. How many of each model are manufactured per week?

28. The following table gives the protein-calcium content for oatmeal and milk.

Food	Protein	Calcium
Oatmeal (1 cup)	5 g	20 mg
Milk (1 cup)	8 g	300 mg

       Determine the amount (in cups) of oatmeal and of milk that will give a serving containing 12 g of protein and 383 mg of calcium.

Functions on Natural Numbers

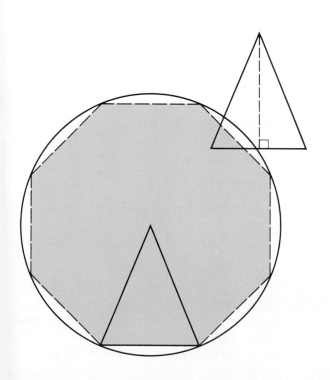

This chapter includes a variety of topics having a common theme in that each can be described in terms of functions defined on the set N, or a subset of N, where $N = \{1, 2, 3, \ldots\}$.

Sequences are discussed in the first three sections, followed by *mathematical induction* (involving sequences of statements). The *Binomial theorem* is introduced in Section 7.5 and the chapter concludes with topics from *mathematics of finance,* in which ideas from preceding sections are applied.

7.1   SEQUENCES

If a function f has domain N, then we say that f is a *sequence function*. The image values of f are given by

$$f(1), f(2), f(3), \ldots, f(n), \ldots. \tag{7.1}$$

This listing can be considered as an *ordered collection* with $f(1)$ the first term, $f(2)$ the second term, and so on; the nth term is $f(n)$. The *ordered* listing given in (7.1) is called the *sequence* associated with the function f.

Any sequence described by (7.1) is referred to as an *infinite sequence* because the domain of the function f is the infinite set N. There are situations in which the domain of f is a finite set $\{1, 2, 3, \ldots, m\}$, where m is a given positive integer. The ordered listing $f(1), f(2), f(3), \ldots, f(m)$ is called a *finite sequence*.

If the terms in (7.1) are real numbers, the sequence is called a *real-number sequence*. Unless there is an explicit statement to the contrary, *it will be assumed that we are dealing with sequences of real numbers*. It is customary to use subscript notation, in which $f(n)$ is denoted by, say a_n, and so (7.1) becomes

$$a_1, a_2, a_3, \ldots, a_n, \ldots. \tag{7.2}$$

We shall also indicate this sequence by $\{a_n\}_{n=1}^{\infty}$. Occasionally it is convenient to begin a sequence at some integer greater than 1, say m; in such cases the sequence $a_m, a_{m+1}, \ldots$ is denoted by $\{a_n\}_{n=m}^{\infty}$.

Let us now consider some examples illustrating these ideas.

Example 1   List the first four terms of the sequence described by

    a) $f(n) = 2n$ **b)** $g(n) = 2n - 1$

Solution   **a)** Let $a_n = f(n)$; then $a_1 = 2 \cdot 1 = 2$, $a_2 = 2 \cdot 2 = 4$,
                                $a_3 = 2 \cdot 3 = 6$, $a_4 = 2 \cdot 4 = 8$.

    Thus $\{2n\}_{n=1}^{\infty}$ is the sequence of positive even integers: $\{2, 4, 6, 8, \ldots, 2n, \ldots\}$.

    b) Let $b_n = g(n)$; then $b_1 = 2 \cdot 1 - 1 = 1$, $b_2 = 2 \cdot 2 - 1 = 3$,
                                  $b_3 = 2 \cdot 3 - 1 = 5$, $b_4 = 2 \cdot 4 - 1 = 7$.

    Thus $\{2n - 1\}_{n=1}^{\infty}$ is the sequence of positive odd integers: $\{1, 3, 5, \ldots, 2n - 1, \ldots\}$.   ■

Example 2   Suppose 2, 4, 6 are the first three terms of a sequence. Determine the next two terms.

Solution   Most of us would assume that the person proposing this problem is thinking of the sequence of positive even integers; that is, a_n is given by $a_n = 2n$. In this case the next two terms would be 8, 10. However, it is possible that a_n is given by

$$a_n = 2n + (n-1)(n-2)(n-3),$$

in which case $a_4 = 14$, and $a_5 = 34$. In fact, the proposer of this problem could have been thinking of many other sequence functions, each of which would give a sequence starting with 2, 4, 6. ∎

Example 2 illustrates an important fact: An infinite sequence is not completely described by listing the first few terms—indeed, not even by listing the first million terms. If one wishes to be unambiguous, then it is necessary to clearly understand what the *sequence function* is.

In most problems considered here, the sequence function will be given by a formula. However, there will be occasions when this is not possible and we must settle for a verbal description of $f(n)$, as illustrated by the following example.

Example 3   Find the first eight terms of the sequence given by "c_n is *the number* of prime numbers less than or equal to n."

Solution   The set of prime numbers is given by

$$\mathbf{P} = \{2, 3, 5, 7, 11, 13, 17, 19, 23, 29, 31, 37, \ldots\}.$$

Applying the function "rule" gives $c_1 = 0$, $c_2 = 1$, $c_3 = 2$, $c_4 = 2$, $c_5 = 3$, $c_6 = 3$, $c_7 = 4$, $c_8 = 4$. That is, the sequence starts as 0, 1, 2, 2, 3, 3, 4, 4, 4, 4, 5, 5, ... There is no simple formula that allows us to evaluate c_n for various values of n, even though c_n is unambiguously defined for each value of n.

The sequence of this example is somewhat more complicated than those given in Example 1. For instance, if we want the millionth term of the $\{c_n\}_{n=1}^{\infty}$ sequence, a rather extensive table of prime numbers would be required. ∎

In the following example we illustrate a finite sequence.

Example 4   List the terms of the sequence defined by $f(n) = \sqrt{25 - n^2}$.

Solution   Here we have the implicit assumption: $f(1), f(2), \ldots$ are to be real numbers. The domain of f is 1, 2, 3, 4, 5, and so we have the finite sequence $\sqrt{24}$, $\sqrt{21}$, 4, 3, 0. ∎

Sequences Described Recursively

In some cases it is not convenient or not possible to give the sequence function explicitly by a formula, but it is possible to describe it recursively. That is, the nth term can be defined as a function of preceding terms. This is illustrated in the following examples.

Example 5   The sequence $\{a_n\}_{n=1}^{\infty}$ is described by

$$a_n = \begin{cases} 1 & \text{if } n = 1, \\ 2a_{n-1} & \text{if } n > 1. \end{cases} \tag{7.3}$$

Write the first four terms of $\{a_n\}_{n=1}^{\infty}$, and determine another expression that can be used to find a_n.

Solution   We write $a_1 = 1$, $a_2 = 2 \cdot a_1 = 2$, $a_3 = 2 \cdot a_2 = 4$, $a_4 = 2 \cdot a_3 = 8$. These terms can be written as follows: $a_1 = 2^0$, $a_2 = 2^1$, $a_3 = 2^2$, $a_4 = 2^3$. Observing the pattern suggests that a_n is given by the formula

$$a_n = 2^{n-1} \text{ for } n = 1, 2, 3, \ldots. \tag{7.4}$$

Although this formula is based on observations of only four terms, a rigorous proof that it is valid for all values of n can be given by using mathematical induction (see Section 7.4).   ■

In Example 5 the same sequence is described in two different ways. In Eq. (7.3) the rule for a_n is given *recursively* whereas in Eq. (7.4) a_n is given by a formula. The following example illustrates a sequence described recursively that cannot be described easily by a simple formula.

Example 6   Determine the first six terms of the sequence $\{b_n\}_{n=1}^{\infty}$ described by $b_1 = 1$, $b_2 = 1$, and $b_n = b_{n-1} + b_{n-2}$ for $n > 2$.

Solution   Here $b_1 = 1$, $b_2 = 1$, $b_3 = b_2 + b_1 = 1 + 1 = 2$, $b_4 = b_3 + b_2 = 2 + 1 = 3$, $b_5 = b_4 + b_3 = 3 + 2 = 5$, $b_6 = b_5 + b_4 = 5 + 3 = 8$.   ■

Fibonacci Sequence

The recursive formula describing b_n in Example 6 tells us that after the first two terms each subsequent term is the sum of the preceding two terms. Thus the first several terms of $\{b_n\}_{n=1}^{\infty}$ are:

$$1, 1, 2, 3, 5, 8, 13, 21, \ldots$$

This is a well-known sequence that was first introduced by Leonardo of Pisa (who was also called Fibonacci) around the year 1200. In his honor, any sequence in which each term is the sum of the preceding two is called a *Fibonacci sequence*.

Partial Sums of a Sequence

Associated with a sequence $\{a_n\}_{n=1}^{\infty}$ of real numbers is another sequence, denoted by $\{S_n\}_{n=1}^{\infty}$, in which the terms are sums defined by

$$S_1 = a_1; \qquad S_2 = a_1 + a_2; \qquad S_3 = a_1 + a_2 + a_3; \ldots$$

In general S_n is given by

$$S_n = a_1 + a_2 + \cdots + a_n. \tag{7.5}$$

Sums of the type indicated in Eq. (7.5) are called *partial sums*. They occur frequently, and it is convenient to introduce a shorthand notation using the Greek letter Σ (sigma). Then S_n is denoted by $\Sigma_{k=1}^{n} a_k$, that is,

$$S_n = \sum_{k=1}^{n} a_k = a_1 + a_2 + \cdots + a_n. \tag{7.6}$$

Note the use of the letter k as a subscript variable; $\Sigma_{k=1}^{n} a_k$ means add the terms obtained by replacing k with 1, 2, 3, ... until you reach the number n. The sequence $\{S_n\}_{n=1}^{\infty}$ is called the *sequence of partial sums* associated with the sequence $\{a_n\}_{n=1}^{\infty}$.

An important topic in advanced mathematics is the study of "infinite sums," indicated by $a_1 + a_2 + a_3 + \cdots$. Such a sum is called an *infinite series* and is denoted by $\Sigma_{k=1}^{\infty} a_k$.

Example 7   Find the partial sum sequence associated with $\{2n - 1\}_{n=1}^{\infty}$.

Solution   The first few terms of the given sequence are 1, 3, 5, 7, 9, ... The partial sum sequence is given by

$$S_1 = 1; \; S_2 = 1 + 3 = 4; \; S_3 = 1 + 3 + 5 = 9; \; S_4 = 1 + 3 + 5 + 7 = 16; \; \ldots.$$

These can be written as $S_1 = 1^2$, $S_2 = 2^2$, $S_3 = 3^2$, $S_4 = 4^2$, ... In this form we observe that there *appears* to be a simple pattern giving a formula for S_n; that is, $S_n = n^2$. In summation notation this is written as

$$S_n = \sum_{k=1}^{n} (2k - 1) = 1 + 3 + 5 + \cdots + (2n - 1) = n^2.$$

From the above computations we see that this formula is valid for $n = 1, 2, 3,$ and 4. After the discussion of mathematical induction in Section 7.4, a proof will be given that this is a valid formula for *every* natural number n.   ■

Example 8   Find the first four terms of the partial sum sequence associated with the sequence given by $a_n = 1/n^2$.

Solution   The first four values of S_n are given by

$$S_1 = \sum_{k=1}^{1} \frac{1}{k^2} = \frac{1}{1^2} = 1; \qquad S_2 = \sum_{k=1}^{2} \frac{1}{k^2} = \frac{1}{1^2} + \frac{1}{2^2} = \frac{5}{4};$$

$$S_3 = \sum_{k=1}^{3} \frac{1}{k^2} = \frac{1}{1^2} + \frac{1}{2^2} + \frac{1}{3^2} = \frac{49}{36};$$

$$S_4 = \sum_{k=1}^{4} \frac{1}{k^2} = \frac{1}{1^2} + \frac{1}{2^2} + \frac{1}{3^2} + \frac{1}{4^2} = \frac{205}{144}.$$ ■

Note: In Example 7 we were able to observe a simple formula to describe S_n. However, there is no such formula in Example 8, and each sum must be added; there is no simple formula for S_n other than

$$S_n = \frac{1}{1^2} + \frac{1}{2^2} + \frac{1}{3^2} + \cdots + \frac{1}{n^2}.$$

Example 9   Find a sequence that approximates the area of a circle of radius 1.

Solution   Inscribe a regular polygon of n sides in a circle of radius 1, as shown in Fig. 7.1. We can get an approximation to the area of the circle by the area, a_n, of the polygon; a_n is given by the sum of the areas of n triangles such as $\triangle AOB$. Let the central angle be θ, as shown, where $\theta = 360°/n$. In Fig. 7.1(b) note that $\overline{AB} = 2\overline{MB} = 2\sin(\theta/2)$ and $h = \cos(\theta/2)$.

$$\text{Area of } \triangle AOB = \frac{1}{2}(\overline{AB})h = \frac{1}{2}\left(2\sin\frac{\theta}{2}\right)\left(\cos\frac{\theta}{2}\right) = \frac{1}{2}\sin\theta = \frac{1}{2}\sin\frac{360°}{n}.$$

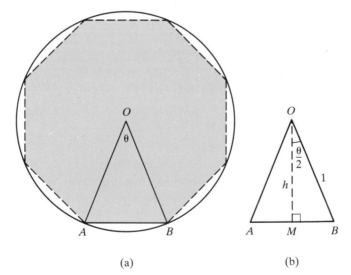

Fig. 7.1

(a)   (b)

Therefore $a_n = n \cdot \frac{1}{2}\sin(360°/n) = (n/2)\sin(360°/n)$.   Thus   the   sequence $\{a_n\}_{n=3}^{\infty}$ is given by

$$a_n = \frac{n}{2}\sin\frac{360°}{n}.$$

To get an approximation to the area of the circle, we can calculate a_n for large values of n. For example, $a_{100} = 3.13953$; $a_{500} = 3.14151$; $a_{1000} = 3.14157$; $a_{10000} = 3.14159$.

Exercises 7.1

In problems 1 through 4, find the first four terms of the given sequences $\{a_n\}_{n=1}^{\infty}$ and $\{b_n\}_{n=1}^{\infty}$.

1. a) $a_n = n^2$ **b)** $b_n = 2^n$ **2. a)** $a_n = 2^n - 1$ **b)** $b_n = 2^{n-1}$

3. a) $a_n = (-1)^n$ **b)** $b_n = \dfrac{(-1)^{n+1}}{2^n}$ **4. a)** $a_n = \left(\dfrac{1}{2}\right)^{n-1}$ **b)** $b_n = \left(-\dfrac{1}{2}\right)^{n-1}$

In problems 5 through 8, sequence functions are given. Find the first four terms of the corresponding sequences. Give answers in exact form.

5. $f(n) = \sqrt{n}$ **6.** $f(n) = \sqrt{n^2 + 1}$ **7.** $g(n) = \left(1 + \dfrac{1}{n}\right)^n$ **8.** $g(n) = \dfrac{1}{2^n - 1}$

In problems 9 through 12, write each of the given sums in expanded form; then evaluate the sum. Give answers in exact form.

9. $S_4 = \displaystyle\sum_{k=1}^{4} (2k + 1)$ **10.** $S_5 = \displaystyle\sum_{k=1}^{5} \dfrac{1}{k(k+1)}$ **11.** $S_6 = \displaystyle\sum_{k=1}^{6} k[1 - (-1)^k]$ **12.** $S_6 = \displaystyle\sum_{k=1}^{6} \left(-\dfrac{1}{2}\right)^{k=1}$

In each of the problems 13 through 16, the first four terms of a sequence $\{a_n\}_{n=1}^{\infty}$ are given. As illustrated in Example 2, this is not sufficient information to determine a unique sequence. Find a formula that will give *a sequence* in which the first four terms agree with the given sequence.

13. 3, 7, 11, 15, ...   **14.** $\dfrac{1}{2}, \dfrac{2}{3}, \dfrac{3}{4}, \dfrac{4}{5}, \ldots$ **15.** $\dfrac{2}{3}, -\dfrac{4}{9}, \dfrac{8}{27}, -\dfrac{16}{81}, \ldots$ **16.** 3, −5, 7, −9, ...

In problems 17 through 20, write the given sums in sigma notation.

17. $s_4 = 1^3 + 2^3 + 3^3 + 4^3$ **18.** $s_n = \dfrac{1}{1^2} + \dfrac{1}{2^2} + \dfrac{1}{3^2} + \cdots + \dfrac{1}{n^2}$

19. $s_n = 1 - \dfrac{1}{2} + \dfrac{1}{4} - \cdots + \dfrac{(-1)^{n-1}}{2^{n-1}}$ **20.** $s_5 = 2 + 6 + 10 + 14 + 18$

21. Consider the decimal expansion of $\frac{1}{7} = 0.14285\ldots$.

    a) Suppose a_n is defined to be the nth decimal digit of this decimal expansion (for instance $a_3 = 2$). List the first seven terms of $\{a_n\}_{n=1}^{\infty}$. What are a_{10}, a_{12}, and a_{20}? What is a_{6k}, where k is any positive integer?
    Hint: The decimal expansion for $\frac{1}{7}$ is a repeating decimal.

    b) Suppose b_n is given by

$$b_n = \begin{cases} 0 & \text{if the } n\text{th decimal digit of } \frac{1}{7} \text{ is odd,} \\ 1 & \text{if the } n\text{th decimal digit of } \frac{1}{7} \text{ is even.} \end{cases}$$

    List the first seven terms of $\{b_n\}_{n=1}^{\infty}$. What are b_8, b_{17}, and b_{24}?

22. Consider the sequences given in Problem 21.

    a) The sum of the first six terms of $\{a_n\}_{n=1}^{\infty}$ is 27. Use this to find the sum of the first 72 terms. What is the sum of the first $6N$ terms, where N is any positive integer?

    b) Suppose $S_n = \sum_{k=1}^{n} b_k$. Find a formula for S_{6N} where N is any positive integer.

23. Consider the decimal expansion of $\pi/10 = 0.31415926535897932384\ldots$.

    a) Suppose a_n is defined to be the nth decimal digit of $\pi/10$; for instance, $a_3 = 4$. List the first six terms of the sequence $\{a_n\}_{n=1}^{\infty}$. Give a_{16}. Is there a formula for a_n in general?

    b) Suppose b_n is given by

$$b_n = \begin{cases} 0 & \text{if the } n\text{th digit of } \pi/10 \text{ is odd,} \\ 1 & \text{if the } n\text{th digit of } \pi/10 \text{ is even.} \end{cases}$$

    List the first six terms of $\{b_n\}_{n=1}^{\infty}$. Give b_{12}. Is there a formula for b_n in general?

24. Find all of the terms of the sequence given by $a_n = \sqrt{16 - n^2}$,

25. Find all of the terms of the sequence given by $b_n = \sqrt{5n - n^2}$.

26. Suppose a sum of $100 is invested at 6 percent interest compounded n times per year. At the end of one year the value, A_n, of the investment is given by $A_n = 100(1 + 0.06/n)^n$. Find the first five terms in this sequence. Give answers rounded off to two decimal places.

27. Find the first six terms of $\{S_n\}_{n=1}^{\infty}$ for the sequence $\{1/n(n + 1)\}_{n=1}^{\infty}$. Use these terms to guess a simple formula for S_n.

28. Find the first six terms of $\{S_n\}_{n=1}^{\infty}$ for the sequence $\{1/2^n\}_{n=1}^{\infty}$. Use these terms to guess a simple formula for S_n.

29. Suppose the sequence $\{a_n\}_{n=1}^{\infty}$ is given recursively by

$$a_n = \begin{cases} 3 & \text{if } n = 1, \\ 2a_{n-1} & \text{if } n > 1. \end{cases}$$

    a) Find the first five terms of the sequence.

    b) Find a formula that gives a_n in closed form.

30. Suppose the sequence $\{b_n\}_{n=1}^{\infty}$ is given recursively by

$$b_n = \begin{cases} 1 & \text{if } n = 1, \\ b_{n-1} + 3 & \text{if } n > 1. \end{cases}$$

    a) Find the first five terms of the sequence.

    b) Determine a formula in closed form for b_n.

31. Approximate the circumference of a circle of radius 1 by finding a sequence giving the perimeter of inscribed polygons. See Example 9.

32. Sequence $\{a_n\}_{n=1}^{\infty}$ is defined as follows: a_1 is any given *positive integer,* and for $n \geq 1$ the terms are given recursively by

$$a_{n+1} = \begin{cases} \dfrac{a_n}{2} & \text{if } a_n \text{ is even,} \\ 3a_n + 1 & \text{if } a_n \text{ is odd.} \end{cases}$$

This rule tells us that whenever a term in the sequence is even, divide it by 2 to get the next term; if a term is odd, multiply it by 3 and add 1 to get the next term. For each of the following starting values, determine several terms of the sequence; in each case continue until you see something interesting happening.

    a) $a_1 = 4$ **b)** $a_1 = 5$ **c)** $a_1 = 10$ **d)** $a_1 = 21$ **e)** $a_1 = 34$

In problems 33 and 34, a function f and a number c are given. Let sequence $\{a_n\}_{n=1}^{\infty}$ be defined by $a_1 = f(c)$, $a_2 = f(a_1)$, $a_3 = f(a_2)$, ... In general,

$$a_{n+1} = f(a_n) \quad \text{for} \quad n = 1, 2, 3, \dots.$$

Give the first eight terms of the sequence; record results rounded off to two decimal places in each case. Observe the results and arrive at interesting conjectures.

33. $f(x) = e^{-x}$

    a) $c = 1$ **b)** $c = 0.5$ **c)** $c = 2$

34. $f(x) = 3^{-x}$

    a) $c = 0.5$ **b)** $c = 0.6$ **c)** $c = 1.5$

7.2   ARITHMETIC SEQUENCES

There are two special types of sequences that occur frequently and have many applications. *Arithmetic sequences* are studied in this section and *geometric sequences* are discussed in the next section.

    An *arithmetic sequence* is characterized by the following property: After the first term, each term is obtained from the preceding term by adding a fixed number. That is, $a_{n+1} = a_n + d$, where d is a constant; d is called the *common difference* between any two consecutive terms of the sequence. Arithmetic sequences are also commonly referred to as *arithmetic progressions*.

    The following are examples of arithmetic sequences:

$$3, 7, 11, 15, \dots; \ d = 4$$
$$8, 3, -2, -7, \dots; \ d = -5$$
$$1, \tfrac{5}{4}, \tfrac{3}{2}, \tfrac{7}{4}, \dots; \ d = \tfrac{1}{4}$$

    In general, suppose $\{a_n\}_{n=1}^{\infty}$ is an arithmetic sequence with common difference d. To determine a formula for a_n, let us list a few terms of the sequence and look for a pattern:

$$a_1 = a_1,$$
$$a_2 = a_1 + d,$$
$$a_3 = a_2 + d = (a_1 + d) + d = a_1 + 2d,$$
$$a_4 = a_3 + d = (a_1 + 2d) + d = a_1 + 3d.$$

Observe that each term is the sum of the first term and a multiple of d, where the multiple is 1 less than the number of the term. This suggests the following formula:

$$a_n = a_1 + (n - 1)d. \tag{7.7}$$

Our claim is that the formula given by Eq. (7.7) is valid for every positive integer n. The discussion above substantiates this for the first four values of n, but for the remaining positive integers it is still a guess. In order to prove our claim, a technique called mathematical induction is needed. This will be discussed in Section 7.4 (see Exercise set 7.4, problem 27). Let us now derive a relatively simple formula for the sum of the first n terms of an arithmetic sequence. First write S_n in normal order; then write it in reverse order, and add the resulting equations, term by term, as follows:

$$\begin{array}{rl} S_n = & a_1 \quad + (a_1 + d) + (a_1 + 2d) + \cdots + (a_n - d) + \quad a_n \\ S_n = & a_n \quad + (a_n - d) + (a_n - 2d) + \cdots + (a_1 + d) + \quad a_1 \\ \hline 2S_n = & (a_1 + a_n) + (a_1 + a_n) + (a_1 + a_n) + \cdots + (a_1 + {}_n) + (a_1 + a_n) \end{array}$$

Therefore, $2S_n = n(a_1 + a_n)$, and solving for S_n gives the desired formula:

$$S_n = \frac{n(a_1 + a_n)}{2}. \tag{7.8}$$

Equation (7.8) gives a formula for finding the sum of any number of terms of an arithmetic sequence. A convenient way to remember this formula is to think of the sum as the average of the first and last terms, $(a_1 + a_n)/2$, multiplied by the number of terms n; this seems intuitively reasonable for an arithmetic sequence.

An alternative expression for S_n can be derived by using Eq. (7.7) to replace a_n in Eq. (7.8) by $a_1 + (n - 1)d$:

$$S_n = \frac{n}{2}(a_1 + a_n) = \frac{n}{2}[a_1 + a_1 + (n - 1)d] = \frac{n}{2}[2a_1 + (n - 1)d],$$

$$S_n = \frac{n}{2}[2a_1 + (n - 1)d]. \tag{7.9}$$

Equation (7.9) gives a formula for S_n in terms of a_1, d, and n. It is more difficult to remember than Eq. (7.8), but occasionally it is convenient to use. We suggest deriving it from Eqs. (7.7) and (7.8) whenever it is needed.

Example 1   Determine whether or not the following can be considered as the first three terms of an arithmetic sequence.

a) 2, 6, 10   **b)** 5, 2, -1 **c)** 2, 4, 8

d) $x^2 - 2x$, $(x - 1)^2$, $x^2 - 2x + 2$

Solution   Here a_1, a_2, a_3 will be the first three terms of an arithmetic sequence if $d = a_2 - a_1$, and $d = a_3 - a_2$—that is, if $a_2 - a_1 = a_3 - a_2$.

a) $6 - 2 = 10 - 6$ **b)** $2 - 5 = -1 - 2$ **c)** $4 - 2 \neq 8 - 4$

d) $a_2 - a_1 = (x - 1)^2 - (x^2 - 2x) = 1; a_3 - a_2 = x^2 - 2x + 2 - (x - 1)^2 = 1;$ thus $a_2 - a_1 = a_3 - a_2$ for every real number x.

Therefore the sequences given in (a), (b), and (d) are the first three terms of arithmetic sequences, but that in (c) is not.   ■

Example 2   Find the tenth term and the sum of the first ten terms of the arithmetic sequence 15, 11, 7, 3, ...

Solution   Here $a_1 = 15$ and $d = -4$. Using Eq. (7.7) gives

$$a_{10} = 15 + (10 - 1)(-4) = 15 - 36 = -21.$$

Substituting into Eq. (7.8) gives

$$S_{10} = 10\left(\frac{a_1 + a_{10}}{2}\right) = 10\left[\frac{15 + (-21)}{2}\right] = 10(-3) = -30.   ■$$

Example 3   Evaluate the sum $\sum_{k=1}^{16} (3k - 4)$.

Solution $$\sum_{k=1}^{16} (3k - 4) = (3 \cdot 1 - 4) + (3 \cdot 2 - 4) + (3 \cdot 3 - 4) + (3 \cdot 4 - 4) + \cdots + (3 \cdot 16 - 4)$$

$$= -1 + 2 + 5 + 8 + 11 + \cdots + 44.$$

This is the sum of an arithmetic sequence where $a_1 = -1, a_n = 44, n = 16$, and $d = 3$. Applying Eq. (7.8) gives

$$\sum_{k=1}^{16} (3k - 4) = 16\left(\frac{-1 + 44}{2}\right) = 344.   ■$$

Example 4   The fourth term of an arithmetic sequence is 9, and the fifteenth term is 64. Find the sum of the first 20 terms.

Solution   Equation (7.8) gives $S_{20} = 20(a_1 + a_{20})/2 = 10(a_1 + a_{20})$; thus we first need to find a_1 and a_{20}. Using the given information and Eq. (7.7), we get

$$a_4 = 9 \quad \text{gives} \quad 9 = a_1 + 3d,$$
$$a_{15} = 64 \text{ gives } 64 = a_1 + 14d.$$

The solution to this system of equations is given by $a_1 = -6$ and $d = 5$. To find a_{20}, use Eq. (7.7) with $n = 20, a_1 = -6$, and $d = 5$: $a_{20} = -6 + 19 \cdot 5 = 89$. Therefore

$$S_{20} = 10(a_1 + a_{20}) = 10(-6 + 89) = 830.   ■$$

Example 5   A person accepts a job with a salary of $12 500 for the first year and is promised an increase of $800 for each subsequent year.

a) What will the salary be during the tenth year?

b) How much will this person earn during the ten years of employment?

c) How many years will it take this person to earn a total of more than one million dollars?

Solution   Here we have an arithmetic sequence with $a_1 = 12\,500$ and $d = 800$.

a) Using Eq. (7.7) gives $a_{10} = 12\,500 + 9(800) = \$19\,700$.

b) Substituting into Eq. (7.8) gives

$$S_{10} = 10\left(\frac{\$12\,500 + \$19\,700}{2}\right) = 161\,000.$$

c) Equation (7.9) can be used to get a formula that gives the total earnings S_n in n years.

$$S_n = \frac{n}{2}[2(12\,500) + (n-1)(800)] = 400\,n^2 + 12\,100\,n.$$

We want n such that

$$400\,n^2 + 12\,100\,n > 1\,000\,000$$
$$4\,n^2 + 121\,n - 10\,000 > 0.$$

The solution will be the smallest positive integer n such that

$$n > \frac{-121 + \sqrt{121^2 + 4(4)(10\,000)}}{2(4)} = 37.11.$$

Hence in 38 years the person's total earnings will surpass a million dollars and in fact will be

$$S_{38} = \$1\,037\,400 \text{ dollars.}$$ ■

Exercises 7.2

In problems 1 through 16, some information is given describing an arithmetic sequence $\{a_n\}_{n=1}^{\infty}$. Find the indicated quantities.

1. $a_1 = -4$, $d = 5$. Find

   a) a_3 **b)** a_6

3. $a_1 = -3$, $d = 4$. Find

   a) a_{15} **b)** S_{15}

5. $a_1 = 16$, $d = -5$. Find

   a) S_{16} **b)** S_{17}

7. $a_1 = 5$, $a_{40} = 122$. Find

   a) d **b)** S_{40}

2. $a_1 = 5$, $d = -2$. Find

   a) a_{12} **b)** a_{20}

4. $a_1 = 12$, $d = -3$. Find

   a) a_{12} **b)** S_{12}

6. $a_1 = -8$, $d = 3$. Find

   a) S_{15} **b)** S_{16}

8. $a_1 = -15$, $a_{32} = 109$. Find

   a) d **b)** S_{40}

9. $a_5 = 8$, $a_{33} = -48$. Find

  a) a_1 **b)** S_{33}

10. $a_6 = -5$, $a_{15} = -41$. Find

  a) d **b)** S_{20}

11. $a_4 = 16$, $S_4 = 22$. Find

  a) a_1 **b)** d

12. $a_5 = 0$, $S_5 = 30$. Find

  a) a_{16} **b)** S_{16}

13. $a_1 = 3$, $d = \frac{3}{5}$. Find

  a) a_8 **b)** S_8

14. $a_3 = 5$, $a_{12} = 11$. Find

  a) a_{15} **b)** S_{15}

15. $a_1 = 6$, $S_{17} = 0$. Find

  a) a_{17} **b)** a_{18}

16. $a_6 = -1$, $S_{16} = 8$. Find

  a) a_1 **b)** d

17. Find the sum of all integers between 200 and 500 that are divisible by 3.

18. Find the sum of all integers between 100 and 500 that are divisible by 7.

In problems 19 through 22, evaluate the given sums.

19. $\displaystyle\sum_{k=1}^{32} (2k - 15)$ **20.** $\displaystyle\sum_{k=1}^{24} (8 - 3k)$ **21.** $\displaystyle\sum_{k=1}^{10} \left(\frac{2}{3}k + 1\right)$ **22.** $\displaystyle\sum_{k=1}^{20} \left(5 - \frac{k}{3}\right)$

23. Given that $f(x) = 2x - 5$, find $\sum_{k=1}^{30} f(k)$.   **24.** Given that $g(x) = 3 - 2x$, find $\sum_{k=1}^{30} g(k)$.

25. Find the value of x such that the numbers $x + 2$, $2x - 4$, $5x - 4$ will be three consecutive terms of an arithmetic sequence.

26. Determine x such that the numbers $x - 1$, $2x + 5$, $5x - 1$ will be the first three terms of an arithmetic sequence; then find the tenth term.

27. Determine x such that the numbers $x + 1$, $3x - 5$, $6x + 6$ will be the first, third, and fifth terms of an arithmetic sequence; then find the eighth term.

28. Find x such that the numbers x^2, $2x + 1$, $2x - 1$ will be the first three terms of an arithmetic sequence.

29. A ball is dropped from the Goodyear blimp 2000 meters above the ground. During the first second it falls 4.9 m, during the second second it falls $3(4.9)$ m, during the third second it falls $5(4.9)$ m, and so on.

  a) Find how far it falls during the twentieth second.

  b) Find the total distance it falls during the first 20 seconds.

30. Suppose $\{a_n\}_{n=1}^{\infty}$ and $\{b_n\}_{n=1}^{\infty}$ are arithmetic sequences with $a_1 = 10$, $a_2 = 13$ and $b_1 = 16$, $b_2 = 21$. Consider the first 100 terms of each sequence; how many numbers do they have in common?

31. Suppose $\{a_n\}_{n=1}^{\infty}$ and $\{b_n\}_{n=1}^{\infty}$ are arithmetic sequences, with $a_1 = 10$, $a_2 = 13$ and $b_1 = 16$, $b_2 = 21$. Let $\{c_n\}_{n=1}^{\infty}$ be defined by $c_n = a_n + b_n$ for $n = 1, 2, 3, \ldots$. Is $\{c_n\}_{n=1}^{\infty}$ an arithmetic sequence? If so, what is the common difference?

32. If $\{a_n\}_{n=1}^{\infty}$ and $\{b_n\}_{n=1}^{\infty}$ are arithmetic sequences and $\{c_n\}_{n=1}^{\infty}$ is a sequence defined by $c_n = a_n + b_n$, is $\{c_n\}_{n=1}^{\infty}$ an arithmetic sequence? If so, what is the common difference?

33. Let $\{b_n\}_{n=1}^{\infty}$ be an arithmetic sequence given by $b_1 = 3$, $b_2 = 7$.

  a) Give the first five terms of $\{b_n\}_{n=1}^{\infty}$, and of $\{b_n^2\}_{n=1}^{\infty}$.

  b) Is $\{b_n^2\}_{n=1}^{\infty}$ an arithmetic sequence?

34. If $\{a_n\}_{n=1}^{\infty}$ is an arithmetic sequence, is the sequence $\{a_n^2\}_{n=1}^{\infty}$ always, sometimes, or never an arithmetic sequence? Give reasons.

35. Upon graduation from college a 23-year-old engineering student is offered a job with a starting salary of $24,000 during the first year and an increase of $1200 in each subsequent year.

a) What will the engineer's salary be during the eighth year?

b) What will total earnings be during the first eight years of employment.

c) How old will this engineer be when total earnings amount to more than a million dollars?

7.3   GEOMETRIC SEQUENCES

In the preceding section we studied arithmetic sequences, in which each term after the first is obtained by *adding* a fixed number d to the preceding term. Geometric sequences are characterized by the following property: Each term after the first is obtained by *multiplying* the preceding term by a fixed number. This is equivalent to saying that the ratio of any term to the preceding term is a fixed number. The *common ratio* will be denoted by r; consecutive terms are related by the formula

$$a_{n+1} = ra_n \quad \text{for} \quad n = 1, 2, 3, \ldots.$$

Let us derive a formula for a_n by listing a few terms and looking for a pattern: $a_1 = a_1$; $\quad a_2 = a_1 \cdot r$; $\quad a_3 = a_2 \cdot r = (a_1r) \cdot r = a_1r^2$; $\quad a_4 = a_3 \cdot r = a_1r^2 \cdot r = a_1r^3$; $\ldots$. Thus the first four terms of $\{a_n\}_{n=1}^{\infty}$ are

$$a_1 = a_1r^0; \qquad a_2 = a_1r^1; \qquad a_3 = a_1r^2; \qquad a_4 = a_1r^3.$$

The pattern here suggests the following formula for the general term,

$$a_n = a_1r^{n-1} \quad \text{for} \quad n = 1, 2, 3, \ldots. \tag{7.10}$$

Mathematical induction can be used to give a formal proof that the result stated in Eq. (7.10) is valid (see Exercise set 7.4, problem 28).

The sum S_n of the first n terms of a geometric sequence is given by

$$S_n = a_1 + a_1r + a_1r^2 + \cdots + a_nr^{n-1}. \tag{7.11}$$

A formula for S_n can be derived as follows: Multiply both sides of Eq. (7.11) by r and then subtract the resulting equation from Eq. (7.11).

$$S_n = a_1 + a_1r + a_1r^2 + \cdots + a_1r^{n-2} + a_1r^{n-1}$$
$$rS_n = \qquad a_1r + a_1r^2 + \cdots \qquad + a_1r^{n-1} + a_1r^n.$$

Subtracting the second equation from the first gives $S_n - rS_n = a_1 - a_1 r^n$. This is equivalent to $(1 - r)S_n = a_1 - a_1 r^n$, and solving for S_n gives the desired formula.

$$S_n = \frac{a_1 - a_1 r^n}{1 - r} = \frac{a_1(1 - r^n)}{1 - r}. \tag{7.12}$$

This can also be written as

$$S_n = \frac{a_1 r^n - a_1}{r - 1} = \frac{a_1(r^n - 1)}{r - 1}. \tag{7.13}$$

Example 1   Which of the following can be considered as the first three terms of a geometric sequence?

a) 2, 6, 18   **b)** $\frac{1}{2}, -\frac{1}{4}, \frac{1}{8}$ **c)** x, x^3, x^5 **d)** 4, 2, 0

Solution   Each of the sequences in (a), (b), and (c) constitute the first three terms of a geometric sequence since for (a), $r = 3$; for (b), $r = -\frac{1}{2}$; for (c), $r = x^2$. For (d) we have

$$\frac{a_2}{a_1} = \frac{2}{4} = \frac{1}{2} \quad \text{and} \quad \frac{a_3}{a_2} = \frac{0}{2} = 0.$$

Thus, $a_2/a_1 \neq a_3/a_2$, and there is no common ratio.   ▇

Example 2   Find the tenth term and the sum of the first ten terms of a geometric progression in which the first three terms are 2, 1, $\frac{1}{2}$.

Solution   From the given information, $a_1 = 2$ and $r = \frac{1}{2}$. Substituting into Eqs. (7.10) and (7.12) gives

$$a_{10} = a_1 r^{10-1} = a_1 r^9 = (2)(\tfrac{1}{2})^9 = \tfrac{1}{256};$$

$$S_{10} = \frac{a_1 - a_1 r^{10}}{1 - r} = \frac{2 - 2(1/2)^{10}}{1 - 1/2} = \frac{1023}{256}. \quad ▇$$

Example 3   Suppose $\{a_n\}_{n=1}^{\infty}$ is a geometric sequence with $a_4 = 1/27$ and $a_7 = -1/729$. Find the first term, the common ratio, and the sum of the first seven terms.

Solution   We wish to find a_1, r, and S_7. Using Eq. (7.10) with $n = 4$ and $n = 7$ gives $a_4 = a_1 r^3$ and $a_7 = a_1 r^6$. Use the given information to get

$$a_1 r^3 = \tfrac{1}{27} \quad \text{and} \quad a_1 r^6 = -\tfrac{1}{729}. \tag{7.14}$$

Solving this system of equations for a_1 and r by dividing the second by the first, we get

$$\frac{a_1 r^6}{a_1 r^3} = \left(-\frac{1}{729}\right) \div \left(\frac{1}{27}\right).$$

Simplifying gives $r^3 = -\frac{1}{27}$, and so $r = \sqrt[3]{-\frac{1}{27}} = -\frac{1}{3}$. Substituting $r = -\frac{1}{3}$ into either of the equations in (7.14)—say, the first—we get $a_1(-\frac{1}{3})^3 = \frac{1}{27}$, and so $a_1 = -1$.

We can now find S_7 by using Eq. (7.12),

$$S_7 = \frac{a_1 - a_1 r^7}{1 - r} = \frac{-1 - (-1)(-1/3)^7}{1 - (-1/3)} = -\frac{547}{729}.$$

Thus we have

$$a_1 = -1, \qquad r = -\tfrac{1}{3}, \qquad S_7 = -\tfrac{547}{729}. \qquad\blacksquare$$

Infinite Geometric Series

Although it is not possible to add infinitely many numbers, in some cases meaning can be given to such sums by considering a limit process. This is illustrated with the geometric sequence $\{a_n\}_{n=1}^{\infty}$, for which $a_1 = 1$ and $r = \frac{1}{2}$. First get a formula for the sum of n terms by using Eq. (7.12).

$$S_n = 1 + \frac{1}{2} + \frac{1}{4} + \cdots + \frac{1}{2^{n-1}}$$

$$S_n = \frac{1 - (1/2)^n}{1 - 1/2} = 2\left[1 - \frac{1}{2^n}\right] = 2 - \frac{1}{2^{n-1}}.$$

Now consider what happens to S_n when n becomes large. The term $1/2^{n-1}$ approaches zero, and so S_n approaches 2. We say that the sum of the infinite series $S = 1 + \frac{1}{2} + \frac{1}{4} + \cdots$ is 2 and write

$$S = 1 + \frac{1}{2} + \frac{1}{4} + \cdots + \frac{1}{2^{n-1}} + \cdots = 2.$$

In Σ notation this is written as

$$S = \sum_{n-1}^{\infty} \frac{1}{2^{n-1}} = 2.$$

In general, suppose $\{a_n\}_{n=1}^{\infty}$ is a geometric sequence in which $-1 < r < 1$. In considering the sum

$$S_n = \frac{a_1 - a_1 r^n}{1 - r},$$

as n becomes large, we see that n occurs only in $a_1 r^n$. Since $-1 < r < 1$, $a_1 r^n$

approaches zero as n becomes large, and so S_n approaches $a_1/(1-r)$. Therefore S is given by the formula

$$S = a_1 + a_1 r + a_1 r^2 + \cdots = \frac{a_1}{1-r} \quad \text{for} \quad -1 < r < 1. \qquad (7.15)$$

This defines the *sum of an infinite geometric series*. We say the series *converges* to $a_1/(1-r)$. If $r \geq 1$ or $r \leq -1$ and $a_1 \neq 0$, the geometric series has no sum because S_n does not approach any number as n becomes large (see problem 52). Such an infinite series is said to *diverge*.

Example 4   **a)** Write the infinite geometric series $3 - 1 + \frac{1}{3} - \frac{1}{9} + \cdots$ using Σ notation.

   b) Find its sum.

Solution   **a)** For the given series, $a_1 = 3$ and $r = -\frac{1}{3}$. The nth term is given by

$$a_n = a_1 r^{n-1} = 3\left(-\frac{1}{3}\right)^{n-1} = \frac{(-1)^{n-1}}{3^{n-2}}.$$

Therefore the given series can be written as

$$\sum_{n=1}^{\infty} \frac{(-1)^{n-1}}{3^{n-2}}.$$

b) Since $r = -\frac{1}{3}$ satisfies $-1 < r < 1$, Eq. (7.15) can be applied to get

$$\sum_{n=1}^{\infty} \frac{(-1)^{n-1}}{3^{n-2}} = \frac{a_1}{1-r} = \frac{3}{1-(-1/3)} = \frac{9}{4}. \qquad \blacksquare$$

Rational Numbers And Repeating Decimals

A *rational number* is one that can be expressed as a quotient (or a ratio) of two integers. For example, $\frac{3}{4}, \frac{1}{3}, \frac{4097}{3300}, -\frac{2}{7}$ are rational numbers. Numbers can also be represented in decimal notation. For instance,

$$\tfrac{3}{4} = 0.7500\ldots; \qquad \tfrac{1}{3} = 0.333\ldots; \qquad \tfrac{4097}{3300} = 1.24151515\ldots.$$

Note that in each of these cases we have a "repeating decimal" in which the decimal digits, after a certain point, repeat in a cycle. The notation commonly used to denote repeating decimals is a bar above the repeating portion. Thus

$$\tfrac{3}{4} = 0.75\overline{0}; \qquad \tfrac{1}{3} = 0.\overline{3}; \qquad \tfrac{4097}{3300} = 1.24\overline{15}; \qquad -\tfrac{2}{7} = -0.\overline{285714}.$$

The repeating-decimal feature characterizes rational numbers. Irrational numbers can also be represented as infinite decimals, but the digits do not follow a repeating pattern.

Example 5   Express the repeating decimal number $1.3\overline{54}$ as a quotient of two integers.

Solution   The given number can be written as follows:

$$1.3\overline{54} = 1.3545454 \cdots = 1.3 + 0.054 + 0.00054 + 0.0000054 + \cdots.$$

Observe that after the 1.3 we have an infinite geometric series with $a_1 = 0.054$ and $r = 0.01$. Substituting these numbers into Eq. (7.15) gives

$$1.3\overline{54} = 1.3 + \frac{0.054}{1 - 0.01} = \frac{13}{10} + \frac{54}{990} = \frac{149}{110}.$$

Exercises 7.3

1. Which of the following can be considered the first three terms of a geometric sequence?

   a) 3, 6, 12   **b)** 1, -2, 4   **c)** 3, -1.2, 0.48   **d)** 2, 4, 6

2. Which of the following can be considered three consecutive terms of a geometric sequence?

   a) 5, 10, 20   **b)** 0.1, 0.01, 0.001   **c)** 16, 10, 4   **d)** 3, -3, 3

In problems 3 through 6, the first three terms of a geometric sequence are given. Find the eighth term and the sum of the first eight terms.

3. $\frac{1}{9}$, $-\frac{1}{3}$, 1   **4.** 1, $-\frac{1}{2}$, $\frac{1}{4}$ **5.** 5, 2.5, 1.25   **6.** 5.1, 0.51, 0.051

In problems 7 through 12, suppose $\{a_n\}_{n=1}^{\infty}$ is a geometric sequence. Using the given information, determine the indicated quantities.

7. $a_1 = 8$, $r = 2$; find a_4 and a_5.   **8.** $a_1 = -4$, $r = -\frac{1}{2}$; find a_8 and a_{12}.

9. $a_1 = \frac{2}{3}$, $a_6 = \frac{1}{48}$; find r and a_8.   **10.** $a_1 = -\frac{8}{5}$, $a_8 = \frac{1}{80}$; find r and a_3.

11. $a_3 = -\frac{8}{5}$; $a_{10} = \frac{1}{80}$; find a_1, r, and S_{10}.   **12.** $a_5 = \frac{2}{3}$, $a_{10} = -\frac{1}{48}$; find a_1, r, and S_6.

In problems 13 through 16, a geometric series is described by the first term and common ratio. Determine the number of terms n that correspond to the given sum S_n.

13. $a_1 = \frac{2}{3}$, $r = \frac{1}{2}$, $S_n = \frac{31}{24}$ **14.** $a_1 = 2$, $r = 3$, $S_n = 6560$

15. $a_1 = -16$, $r = -\frac{1}{2}$, $S_n = -\frac{255}{24}$ **16.** $a_1 = 27$, $r = -\frac{1}{3}$, $S_n = \frac{61}{3}$

In problems 17 through 20, evaluate the indicated sums.

17. $\displaystyle\sum_{k=1}^{5} 2^k$ **18.** $\displaystyle\sum_{k=1}^{5} \frac{1}{3^{k-1}}$ **19.** $\displaystyle\sum_{k=1}^{6} 3(0.1)^k$ **20.** $\displaystyle\sum_{k=1}^{10} \frac{16}{2^{k-1}}$

In problems 21 through 24, evaluate the sum $\sum_{k=1}^{5} f(k)$ for the given function f.

21. $f(x) = \dfrac{3}{2^{x-1}}$ **22.** $f(x) = \dfrac{-5}{3^x}$ **23.** $f(x) = 3(0.1)^x$ **24.** $f(x) = (0.2)^{x-1}$

In problems 25 through 28, find x such that the given expressions are the first three terms of a geometric sequence.

25. $2x - 1$; $2x - 3$; $4x + 1$ **26.** $x - 1$; $x + 1$; $x + 4$

27. $x^2 + 1$; x; 2 **28.** $x + 4$; $x - 2$; $2x + 1$

In problems 29 through 32, (a) write the given infinite geometric series using Σ notation, and (b) find its sum.

29. $8 + 4 + 2 + \cdots$ **30.** $6 - 2 + \dfrac{2}{3} - \dfrac{2}{9} + \cdots$

31. $1.8 + 0.018 + 0.00018 + \cdots$ **32.** $2.4 - 0.024 + 0.00024 - \cdots$

In problems 33 through 36, determine whether or not the given infinite series converges. If it does, find the number to which it converges.

33. $\displaystyle\sum_{k=1}^{\infty} 4\left(-\frac{3}{4}\right)^k$ **34.** $\displaystyle\sum_{k=1}^{\infty} \left(\frac{3}{2}\right)^{k-1}$ **35.** $\displaystyle\sum_{k=1}^{\infty} 8\left(-\frac{3}{2}\right)^{k-1}$ **36.** $\displaystyle\sum_{k=1}^{\infty} 5(0.1)^{k-1}$

In problems 37 through 42, find the repeating decimal expansions for the given rational numbers.

37. $\dfrac{5}{12}$ **38.** $\dfrac{9}{11}$ **39.** $\dfrac{1}{7}$ **40.** $\dfrac{10}{7}$ **41.** $\dfrac{1}{13}$ **42.** $\dfrac{2}{13}$

In problems 43 through 50, express the given repeating decimal number as a quotient of two integers, and simplify answers.

43. $1.2\overline{3}$ **44.** $3.1\overline{55}$ **45.** $1.5\overline{81}$ **46.** $1.50\overline{45}$

47. $0.36\overline{36}$ **48.** $0.3\overline{63}$ **49.** $0.\overline{142857}$ **50.** $0.\overline{857142}$

51. The sequence of partial sums $\{S_n\}_{n=1}^{\infty}$ associated with the geometric sequence $\{(\frac{4}{3})^n\}_{n=1}^{\infty}$ is an increasing sequence; that is, $S_{n+1} > S_n$ for each $n = 1, 2, 3, \ldots$. Since $r = \frac{4}{3} > 1$, S_n becomes large when n is large. What is the smallest value of n such that $S_n > 8$? *Hint:* Find a formula for S_n and then, using a calculator, evaluate S_n for various values of n until you find a value of n for which $S_n > 8$ and $S_{n-1} \le 8$. Or you may wish to use logarithms.

52. A city has a population of 50,000 at the end of 1980, and the population is increasing at a rate of 10% each year. What will its population be at the end of 1990?

53. The first swing of a bob on a pendulum is 20 cm. On each subsequent swing it travels $\frac{4}{5}$ as far as the preceding swing. What is the total distance it travels before essentially coming to rest?

54. A rubber ball is dropped from the top of the Washington monument (170 meters high). Suppose each time it hits the ground it rebounds $\frac{2}{3}$ of the distance of the preceding fall.

    a) What is the total distance it travels up to the instant it hits the ground the fifth time?

    b) What is the total distance it travels before it essentially comes to rest?

55. Prove that the infinite geometric series does not converge if $r \ge 1$ or $r \le -1$, and $a_1 \ne 0$. *Hint:* Consider the following cases: (1) $r = 1$; (2) $r = -1$; (3) $r > 1$, and see what happens to $a_1 r^n$ in $S_n = (a_1 - a_1 r^n)/(1 - r)$ when n becomes large; (4) $r < -1$; follow suggestion of (3).

7.4 MATHEMATICAL INDUCTION

In mathematics, as well as in any area of communication, one of our primary concerns is to determine whether a given statement is true or false. For instance, the following three statements are true:

$$\sqrt{3} > 1.5; \qquad \frac{\sqrt{6} - \sqrt{2}}{4} = \frac{1}{2}\sqrt{2 - \sqrt{3}};$$

$$x = 3 \text{ is a root of } x^2 - 2x - 3 = 0;$$

and the following three are false:

$$4 < 2; \qquad 3.5 - 10 = 1;$$

$$4 \text{ is in the set } \{x \mid x^2 - 2x < 0\}.$$

Thus each of these can be viewed as a *sentence* to which we can assign a true or a false value. Such a sentence is said to have a *truth value* and is called a *statement*.

If we consider a sentence such as $x^2 - 2x - 3 = 0$, we cannot say it is true or false unless we replace x by a specific number. For instance, replacing x by 4 yields $4^2 - 2 \cdot 4 - 3 = 0$, which is a false statement, whereas replacing x by -1 gives $(-1)^2 - 2(-1) - 3 = 0$, which is a true statement.

The sentence $x^2 - 2x - 3 = 0$ is an example of an *open sentence*, that is, a sentence which involves a variable and to which we do not assign a truth value.

Let us consider still another example. The sentence "The sum of the first n positive odd integers is equal to n^2 for each positive integer n" is a statement that happens to be true. This statement can be expressed in mathematical terms as:

$$\sum_{k=1}^{n} (2k - 1) = n^2 \quad \text{for} \quad n = 1, 2, 3, \ldots. \tag{7.16}$$

In expanded notation this is

$$1 + 3 + 5 + \cdots + (2n - 1) = n^2 \quad \text{for} \quad n = 1, 2, 3, \ldots.$$

To better understand what this means, we denote Eq. (7.16) by $P(n)$. Thus

$$P(1) \text{ represents the statement } \sum_{k=1}^{1} (2k - 1) = 1^2, \text{ or } 1 = 1^2;$$

$$P(2) \text{ is } \sum_{k=1}^{2} (2k - 1) = 2^2, \text{ or } 1 + 3 = 2^2;$$

$$P(3) \text{ is } \sum_{k=1}^{3} (2k - 1) = 3^2, \text{ or } 1 + 3 + 5 = 3^2; \text{ and so on.}$$

We see that $P(1)$, $P(2)$, and $P(3)$ are true statements. To say that the statement given in (7.16) is true means that each statement in the sequence $P(1)$, $P(2)$,

$P(3), \ldots$ is a true statement. We have verified the truth of the first three, but it is impossible to directly verify the truth of all the remaining statements—$P(4)$, $P(5), P(6), \ldots$; there are too many of them. The primary goal in this section is to introduce a technique of proof, called *mathematical induction,* that can be used to prove that statements such as (7.16) are true. See Example 2 for a proof of (7.16). Note that the expansion of (7.16) involves the open sentence

$$1 + 3 + 5 + \cdots + (2n - 1) = n^2,$$

to which we do not assign a true or false value by itself. But when the quantifier "for $n = 1, 2, 3, \ldots$" is attached, as in (7.16), then the result is a statement. That is, $P(n)$ is either true for each positive integer $n = 1, 2, 3, \ldots$, or there is at least one value of n for which $P(n)$ is false.

Before stating the Principle of mathematical induction, we present a brief discussion to illustrate the need for proving statements that *appear* to be obviously true.

Inductive Inference

The common usage of the word *induction* implies a process of arriving at a general statement from particular cases. In mathematics as well as in all areas of science, many important discoveries are made by observing particular instances, and by drawing general conclusions from these observations. Consider the following simple example.

Example 1   The algebraic expression $n^2 - n + 41$ is evaluated for several values of n, where $n \in \mathbf{N}$, with the results shown in the table. Formulate general conclusions suggested by this table.

n	1	2	3	4	5	6	....
$n^2 - n + 41$	41	43	47	53	61	71	....

Solution   Our first observation is that all of the numbers in the second row are odd integers. Hence a general conclusion would be: "They are all odd." However, such a statement is not mathematically precise. We can say it better as follows:

$$n^2 - n + 41 \text{ is an odd number for each } n = 1, 2, 3, \ldots. \qquad (7.17)$$

From the given table we can see that the statement in (7.17) is true for the first six values of n, but we are claiming considerably more than that when we say that statement (7.17) is true. Of course, we could extend the table to include a few more values of n, but that still would not be sufficient. To *prove* that statement (7.17) is true requires an argument other than a direct verification for *all* values of n (see problem 21).

Our second observation is that the numbers in the second row of the table are prime numbers. Thus we arrive at a generalization and claim that

$$n^2 - n + 41 \text{ is a prime number for each } n = 1, 2, 3, \ldots. \qquad (7.18)$$

Our claim is that the statement (7.18) is true. From the numbers in the table, the best we can be certain of is that $n^2 - n + 41$ is a prime number for $n = 1, 2, 3, 4, 5$, and 6. We could get more evidence by extending the table. After $n = 6$ the next four values in the second row of the table are 83, 97, 113, 131. We can check to see that each of these is a prime number, (see Appendix D), and this gives us a stronger feeling that the statement in (7.18) is true. However, there are still many values of n we have not yet tried. Suppose we try $n = 41$; the value of $n^2 - n + 41$ is $41^2 - 41 + 41 = 41^2$, and this clearly is not a prime number. Suddenly our strong feeling that (7.18) is a true statement is shattered. Thus we have proved that statement (7.18) is false of finding a *counterexample*.

This example is interesting because $n^2 - n + 41$ is a prime number for the first 40 cases.*

Another conclusion concerning the second row of the table is that they differ by 2, 4, 6, and so on. Again we want to state this in precise mathematical language. Here is a good example of the value of introducing appropriate notation. Let $f(n) = n^2 - n + 41$. Then we can say that

$$f(n + 1) - f(n) = 2n \text{ for each } n = 1, 2, 3, \ldots \text{ is a true statement.} \quad (7.19)$$

We leave it to the reader to prove by evaluating $f(n + 1) - f(n)$ that statement (7.19) is true. The reader may also be interested in stating other claims (generalizations) suggested by the given table. ∎

We now introduce a principle that will be useful in proving the truth of many statements of the type: $P(n)$ for $n = 1, 2, 3, \ldots$. The situation is that we have a sequence of statements $P(1), P(2), P(3), \ldots$, each of which is either true or false. Suppose we let S denote the set of *all values of n* for which $P(n)$ is a true statement. That is,

$$S = \{n \mid P(n) \text{ is true}\}.$$

To claim that $P(n)$ is true for every natural number n is equivalent to claiming that S contains all the natural numbers; that is, $S = \mathbf{N}$. Suppose we can prove that (a) 1 is in S, and (b) $k + 1$ is in S whenever k is in S. Then it is intuitively reasonable to conclude that S contains all the natural numbers. For instance, from (a) we know $1 \in S$; now using $k = 1$ in (b), we conclude that $2 \in S$; now using $k = 2$ in (b), we can say that $3 \in S$; and so on.† That is the essence of the Principle of mathematical induction, which we now state.

---

*   Discovered by the Swiss mathematician Leonhard Euler (1707–1783).

†   In popular terms we might describe this as the *domino effect:* when dominoes are lined up vertically and the first one falls, it knocks down the second domino, which in turn knocks down the third; and so on.

> *Principle of Mathematical Induction*
>
> Let S be a subset of the set of natural numbers **N**. If set S has the following two properties: (A) $1 \in S$ and (B) $k \in S$ implies that $(k + 1) \in S$, then set S contains all the natural numbers; that is, $S = \mathbf{N}$.

Let us now see how this principle can be applied to prove statements similar to those discussed earlier in this chapter.

Example 2   Prove that the statement $\sum_{j=1}^{n} (2j - 1) = n^2$ for each $n = 1, 2, 3, \ldots$ is true.

Solution   Let S represent the set of natural numbers for which the given equality yields a true statement; that is,

$$S = \left\{ n \,\middle|\, \sum_{j=1}^{n} (2j - 1) = n^2 \right\}.$$

We wish to prove that S contains all the natural numbers. The Principle of mathematical induction provides us with a technique for doing this. We must show that S has the two properties stated in (A) and (B).

Property (A) says that the number 1 is in S. To verify this we must show that $\sum_{j=1}^{1} (2j - 1) = 1^2$ is a true statement. This follows from the fact that both sides are equal to 1. Hence S has property (A).

To prove that set S has property (B) it is necessary to show that if k is any fixed (but unspecified) positive integer in S, then its successor $(k + 1)$ in also in S. Thus our problem can be stated as follows:

Suppose $k \in S$, so $\sum_{j=1}^{k} (2j - 1) = k^2$. Using this supposition, show that $(k + 1) \in S$; that is, $\sum_{j=1}^{k+1} (2j - 1) = (k + 1)^2$. In expanded notation we can say this as follows: Given

$$1 + 3 + 5 + \cdots + (2k - 1) = k^2. \tag{7.20}$$

Using (7.20), we show that

$$1 + 3 + 5 + \cdots + (2k - 1) + (2k + 1) = (k + 1)^2. \tag{7.21}$$

We can accomplish this by adding $(2k + 1)$ to both sides of (7.20):

$$[1 + 3 + 5 + \cdots + (2k - 1)] + (2k + 1) = k^2 + (2k + 1).$$

Clearly this is equivalent to the equation in (7.21), since the left sides are identical and the right sides are equal because

$$k^2 + (2k + 1) = k^2 + 2k + 1 = (k + 1)^2.$$

Hence set S has property (B), and by the Principle of mathematical induction, $S = \mathbf{N}$. This is equivalent to saying that the given statement is true. ∎

Example 3   Prove that $3^{2n} - 1$ is divisible by 8, where n is any natural number.

Solution   Let $S = \{n \mid 3^{2n} - 1 \text{ is divisible by } 8\}$. We wish to show that $S = \mathbf{N}$; this can be done by mathematical induction. To show that set S has property (A) we must verify that $3^{2 \cdot 1} - 1$ is divisible by 8. This is immediate since $3^{2 \cdot 1} - 1 = 9 - 1 = 8$. To show that set S has property (B) we must argue the following: Given that

$$3^{2k} - 1 \text{ is divisible by } 8, \tag{7.22}$$

show that

$$3^{2(k+1)} - 1 \text{ is divisible by } 8. \tag{7.23}$$

The definition of "divisible by 8" and (7.22) tell us that there is *an integer m* such that

$$3^{2k} - 1 = 8m. \tag{7.24}$$

Using this, we wish to prove that there is *an integer q* such that

$$3^{2(k+1)} - 1 = 8q. \tag{7.25}$$

We can start with the left side of (7.25) and manipulate it into a form in which we can apply the information given in (7.24), as follows:

$$3^{2(k+1)} - 1 = 3^{2k+2} - 1 = 3^{2k} \cdot 3^2 - 1 = 3^{2k} \cdot 9 - 1$$
$$= (3^{2k} \cdot 9 - 9) + (9 - 1) \qquad \text{by adding and subtracting 9}$$
$$= 9(3^{2k} - 1) + 8 \qquad\qquad \text{and grouping}$$
$$= 9(8m) + 8 \qquad\qquad\qquad \text{by using (7.24)}$$
$$= 8(9m + 1) \qquad\qquad\qquad \text{by factoring.}$$

Therefore $3^{2(k+1)} - 1 = 8(9m + 1)$. Since m is an integer, $(9m + 1)$ is an integer, and so this is the value of q that can be used in (7.25). Thus set S has properties (A) and (B), and so $S = \mathbf{N}$.   ■

Exercises 7.4

In each of the problems 1 through 4, an open sentence $P(n)$ is defined, with n assuming positive integer values $1, 2, 3, \ldots$. Give the statements that correspond to $P(1)$, $P(2)$, $P(3)$, and $P(4)$; in each case determine whether the corresponding statement is true or false.

1. $P(n)$: $n^3 + n \geq 2n^2 + 2$ **2.** $P(n)$: $n^3 - n^2 = 4n - 4$

3. $P(n)$: $3^n \geq 3n$ **4.** $P(n)$: $n^2 - n + 5$ is an odd integer

In problems 5 through 8, determine the smallest positive integer n that will yield a false statement when it is substituted into the given open sentence.

5. $n^3 + 11n = 6(n^2 + 1)$ **6.** $n^2 - n + 17$ is a prime number. Use table in Appendix D.

7. $n^2 < 2n$ **8.** $2^n \leq (n + 1)^2$

In each of the problems 9 through 20, an open sentence $P(n)$ is given. Use mathematical induction to prove that the corresponding statement, "$P(n)$ for each $n = 1, 2, 3, \ldots$," is true.

9. $1 + 2 + 3 + \cdots + n = \dfrac{n(n + 1)}{2}$

10. $2 + 4 + 6 + \cdots + 2n = n(n + 1)$

11. $\displaystyle\sum_{m=1}^{n} m^2 = \tfrac{1}{6}n(n + 1)(2n + 1)$

12. $\displaystyle\sum_{m=1}^{n} m^3 = \tfrac{1}{4}n^2(n + 1)^2$

13. $\displaystyle\sum_{m=1}^{n} 2^{m-1} = 2^n - 1$

14. $\displaystyle\sum_{m=1}^{n} 3^{m-1} = \tfrac{1}{2}(3^n - 1)$

15. $1 + \dfrac{1}{2} + \dfrac{1}{4} + \cdots + \dfrac{1}{2^{n-1}} = 2 - \dfrac{1}{2^{n-1}}$

16. $1^2 + 3^2 + 5^2 + \cdots + (2n - 1)^2 = \tfrac{1}{3}n(2n + 1)(2n - 1)$

17. $1 \cdot 2 + 2 \cdot 3 + 3 \cdot 4 + \cdots + n(n + 1) = \tfrac{1}{3}n(n + 1)(n + 2)$

18. $n^3 + 2n$ is divisible by 3   **19.** $4^n - 1$ is divisible by 3   **20.** $2^n > n$

21. Prove that $n^2 - n + 41$ is an odd integer for each $n = 1, 2, 3, \ldots$

   a) by mathematical induction,

   b) by algebraic technique using $n^2 - n + 41 = n(n - 1) + 41$.

22. Suppose f is a function with domain the set of positive integers and defined as a sum

$$f(n) = \sum_{j=1}^{n} \frac{1}{j(j + 1)}.$$

   a) Make a table giving the values of $f(n)$ for $n = 1, 2, 3$, and 4.

   b) From the table in (a), extended if necessary, formulate a general conclusion that gives $f(n)$ in terms of a simpler formula. Prove that your formula is valid for each $n = 1, 2, 3, \ldots$.

In problems 23 through 26, determine whether the given statement is true or false. Give reasons for your answer.

23. $n(n + 1)(n + 2)$ is divisible by 6 for each $n = 1, 2, 3, \ldots$.

24. $\displaystyle\sum_{m=1}^{n} (2m - 3) = n(n - 2)$ for each $n = 1, 2, 3, \ldots$.

25. $\displaystyle\sum_{m=1}^{n} (2^{m-1} - 1) = 2^n - n - 1$ for each $n = 1, 2, 3, \ldots$.

26. $2\left(1 + \dfrac{1}{n}\right)^n < 5$ for each $n = 1, 2, 3, \ldots$.

27. Suppose $\{a_n\}_{n=1}^{\infty}$ is an arithmetic sequence, as defined in Section 7.2. Prove that the result stated in Eq. (7.7), $a_n = a_1 + (n - 1)d$, is valid.

28. Suppose $\{a_n\}_{n=1}^{\infty}$ is a geometric sequence, as defined in Section 7.3. Prove that the result stated in Eq. (7.10), $a_n = a_1 r^{n-1}$, is valid.

29. Suppose sequence function f is given by $f(n) = n^2 - 39n + 421$, $\mathfrak{D}(f) = \mathbf{N}$.

    a) Evaluate $f(n)$ for $n = 1, 2, 3, \ldots, 8$.

    b) Are all values of $f(n)$ in (a) prime numbers? Use table of prime numbers in Appendix D.

    c) What is the smallest value of n for which $f(n)$ is not a prime number?

30. For the sequence function given in problem 29,

    a) Prove that $f(60 + k^2) = (k^2 + k + 41)(k^2 - k + 41)$ for $k = 0, 1, 2, \ldots$ This gives a sequence of composite numbers for $f(n)$.

    b) Prove that $f(60 + 41k) = 41(41k^2 + 81k + 41)$ for $k = 0, 1, 2, \ldots$

    c) Is $f(61 + 41k)$ a composite number for every $k = 0, 1, 2, \ldots$?

7.5　BINOMIAL EXPANSION FORMULA

In Chapter 3 we noted that a product raised to a power follows the simple formula $(a \cdot b)^n = a^n \cdot b^n$. However, raising a sum to a power does not follow a similarly simple formula; that is, in general $(a + b)^n \neq a^n + b^n$. Expanding $(a + b)^n$ by direct multiplication for several values of n will lead us to a formula for the general case:

$$n = 1: \quad (a + b)^1 = a + b,$$
$$n = 2: \quad (a + b)^2 = a^2 + 2ab + b^2,$$
$$n = 3: \quad (a + b)^3 = (a + b)(a + b)^2 = a^3 + 3a^2b + 3ab^2 + b^3,$$
$$n = 4: \quad (a + b)^4 = (a + b)(a + b)^3 = a^4 + 4a^3b + 6a^2b^2 + 4ab^3 + b^4,$$
$$n = 5: \quad (a + b)^5 = (a + b)(a + b)^4$$
$$= a^5 + 5a^4b + 10a^3b^2 + 10a^2b^3 + 5ab^4 + b^5.$$

From these special cases we arrive at some generalizations. Expanding $(a + b)^n$ and collecting like terms yields the following.

    1. We get $(n + 1)$ terms;

    2. Each term is of the type $c \cdot a^s b^t$, where c is an appropriate coefficient and $s + t = n$;

    3. The expansion can be written so that for consecutive terms the exponent of a decreases by 1 and the exponent of b increases by 1.

    It is not easy to guess what the coefficients are before the multiplication is done, but we can get some insight by looking at how like terms collect in the multiplication for $(a + b)^5 = (a + b)(a + b)^4$. For this we first write $(a + b)^4$ in expanded form:

$$(a + b)^4 = a^4 + 4a^3b + 6a^2b^2 + 4ab^3 + b^4.$$

Then we multiply both parts of the above equation first by a and then by b, and add the results collecting like terms:

$$a(a + b)^4 = a^5 + 4a^4b + 6a^3b^2 + 4a^2b^3 + ab^4$$
$$b(a + b)^4 = a^4b + 4a^3b^2 + 6a^2b^3 + 4ab^4 + b^5$$
$$\overline{(a + b)(a + b)^4 = a^5 + 5a^4b + 10a^3b^2 + 10a^2b^3 + 5ab^4 + b^5}$$

This suggests the following pattern for the coefficients, where we indicate how the last row is obtained from the preceding row:

$(a + b)^0$ 1

$(a + b)^1$ 1   1

$(a + b)^2$ 1   2   1

$(a + b)^3$ 1   3   3   1

$(a + b)^4$ 1   4   6   4   1

$(a + b)^5$ 1   5   10   10   5   1

The above array (continued) is known as *Pascal's triangle*. The coefficients of the expansion of $(a + b)^n$ can be found for any value of n by continuing the array to the appropriate number of lines.

In order to describe the binomial coefficients in general without relying on Pascal's triangle, it is helpful to introduce some notation. We need a symbol to denote the product of several consecutive positive integers; for instance, $1 \cdot 2 \cdot 3 \cdot 4 \cdot 5 \cdot 6$ is denoted by 6!, which is read "six factorial." In general, for any positive integer n we define n *factorial* as the product of all positive integers 1 through n. This is denoted by

$$n! = 1 \cdot 2 \cdot 3 \cdots n. \tag{7.26}$$

We can also express $n!$ by the recursive formula for $n \geq 2$:

$$n! = n(n - 1)! \tag{7.27}$$

It is convenient to have (7.27) hold also for $n = 1$:

$$1! = 1 \cdot (1 - 1)! \quad \text{or} \quad 1 = 1 \cdot 0!$$

Thus we shall define 0! to be equal to 1.

$$0! = 1.$$

Binomial Coefficients

The symbol $\binom{n}{k}$ is commonly used to denote the coefficients of the binomal expansion. This is defined by

$$\binom{n}{k} = \frac{n(n-1)(n-2)\cdots(n-k+1)}{k!}, \tag{7.28}$$

where n is any positive integer and k is a *positive* integer with $k \leq n$. For $k = 0$ we define

$$\binom{n}{0} = 1.$$

Equation (7.28) can be written in compact form, as follows: Multiplying numerator and denominator by $(n-k)!$ gives

$$\begin{aligned}\binom{n}{k} &= \frac{n(n-1)\cdots(n-k+1)}{k!} \cdot \frac{(n-k)!}{(n-k)!} \\ &= \frac{n(n-1)\cdots(n-k+1)(n-k)(n-k-1)\cdots 2\cdot 1}{k!(n-k)!} \\ &= \frac{n!}{k!(n-k)!}\end{aligned}$$

Thus an equivalent form of Eq. (7.28) is

$$\binom{n}{k} = \frac{n!}{k!(n-k)!}. \tag{7.29}$$

The following examples illustrate factorial and binomial coefficient notations.

Example 1 Evaluate

    a) $4!$ **b)** $\dfrac{6!}{3!}$ **c)** $\dfrac{(n+1)!}{(n-1)!}$

Solution **a)** $4! = 1 \cdot 2 \cdot 3 \cdot 4 = 24$

b) $\dfrac{6!}{3!} = \dfrac{1 \cdot 2 \cdot 3 \cdot 4 \cdot 5 \cdot 6}{1 \cdot 2 \cdot 3} = 4 \cdot 5 \cdot 6 = 120$

c) $\dfrac{(n+1)!}{(n-1)!} = \dfrac{1 \cdot 2 \cdots (n-1) \cdot n(n+1)}{1 \cdot 2 \cdots (n-1)} = n(n+1) = n^2 + n$ ◼

Example 2   Evaluate

a) $\dbinom{5}{3}$ **b)** $\dbinom{8}{6}$ **c)** $\dbinom{12}{4}$

Solution   **a)** To apply Eq. (7.28) with $n = 5$ and $k = 3$, we first have $(n - k + 1) = (5 - 3 + 1) = 3$ (this gives the last factor of the numerator). Thus

$$\binom{5}{3} = \frac{5 \cdot 4 \cdot 3}{1 \cdot 2 \cdot 3} = 10.$$

b) Using Eq. (7.28), again we first calculate $(n - k + 1) = (8 - 6 + 1) = 3$, and so

$$\binom{8}{6} = \frac{8 \cdot 7 \cdot 6 \cdot 5 \cdot 4 \cdot 3}{1 \cdot 2 \cdot 3 \cdot 4 \cdot 5 \cdot 6} = \frac{8 \cdot 7}{1 \cdot 2} = 28.$$

c) Since either (7.28) or (7.29) will give $\binom{12}{4}$, we choose here to illustrate the use of (7.29):

$$\binom{12}{4} = \frac{12!}{4!(12-4)!} = \frac{1 \cdot 2 \cdot 3 \cdots 8 \cdot 9 \cdot 10 \cdot 11 \cdot 12}{(1 \cdot 2 \cdot 3 \cdot 4)(1 \cdot 2 \cdot 3 \cdots 8)} = \frac{9 \cdot 10 \cdot 11 \cdot 12}{1 \cdot 2 \cdot 3 \cdot 4} = 495.$$

◼

Example 3   Evaluate

a) $\dbinom{4}{0}; \dbinom{4}{1}; \dbinom{4}{2}; \dbinom{4}{3}; \dbinom{4}{4};$

b) $\dbinom{5}{0}; \dbinom{5}{1}; \dbinom{5}{2}; \dbinom{5}{3}; \dbinom{5}{4}; \dbinom{5}{5}.$

Solution   **a)** $\dbinom{4}{0} = 1; \dbinom{4}{1} = 4; \dbinom{4}{2} = \dfrac{4 \cdot 3}{2!} = 6; \dbinom{4}{3} = \dfrac{4 \cdot 3 \cdot 2}{3!} = 4; \dbinom{4}{4} = \dfrac{4 \cdot 3 \cdot 2 \cdot 1}{4!} = 1$

b) $\dbinom{5}{0} = 1; \dbinom{5}{1} = 5; \dbinom{5}{2} = 10; \dbinom{5}{3} = 10; \dbinom{5}{4} = 5; \dbinom{5}{5} = 1.$ ◼

Observation of the results in Example 3 leads to several generalizations. First, we see that $\binom{4}{1} = \binom{4}{3}$, $\binom{5}{2} = \binom{5}{3}$, and so on. This suggests the important symmetric relationship for the binomial coefficients:

$$\binom{n}{k} = \binom{n}{n-k}. \tag{7.30}$$

We can easily prove this relationship by applying Eq. (7.29) to both sides of Eq. (7.30) to get

$$\binom{n}{k} = \frac{n!}{k!(n-k)!}; \qquad \binom{n}{n-k} = \frac{n!}{(n-k)![n-(n-k)]!} = \frac{n!}{(n-k)!k!}$$

Therefore, $\binom{n}{k} = \binom{n}{n-k}$.

Looking at the results in Example 3, we also see that the sequences of coefficients in a and b are precisely the numbers in the last two rows of Pascal's triangle (see p. 391). Actually Pascal's triangle can be written as follows:

$(a + b)^0$:   1

$(a + b)^1$: $\binom{1}{0}$ $\binom{1}{1}$

$(a + b)^2$: $\binom{2}{0}$ $\binom{2}{1}$ $\binom{2}{2}$

$(a + b)^3$: $\binom{3}{0}$ $\binom{3}{1}$ $\binom{3}{2}$ $\binom{3}{3}$

$(a + b)^4$: $\binom{4}{0}$ $\binom{4}{1}$ $\binom{4}{2}$ $\binom{4}{3}$ $\binom{4}{4}$

$(a + b)^5$: $\binom{5}{0}$ $\binom{5}{1}$ $\binom{5}{2}$ $\binom{5}{3}$ $\binom{5}{4}$ $\binom{5}{5}$

  ...   ...

The two rows of Pascal's triangle associated with $(a + b)^{n-1}$ and $(a + b)^n$ are

$(a + b)^{n-1}$: $\binom{n-1}{0}$ $\binom{n-1}{1}$ $\binom{n-1}{2}$ $\cdots$ $\binom{n-1}{k-1}$ $\binom{n-1}{k}$ $\cdots$ $\binom{n-1}{n-1}$

$(a + b)^n$: $\binom{n}{0}$ $\binom{n}{1}$ $\binom{n}{2}$ $\cdots$ $\binom{n}{k}$ $\cdots$ $\binom{n}{n}$

This suggests the following identity for binomial coefficients.

Suppose n and k are positive integers with $n \geq 2$ and $k < n$. Then

$$\binom{n-1}{k-1} + \binom{n-1}{k} = \binom{n}{k}. \tag{7.31}$$

Verification that Eq. (7.31) is indeed an identity is left to the reader (see problem 38).

Finally, the row of Pascal's triangle associated with $(a + b)^n$, along with our earlier observations, suggests a formula for $(a + b)^n$.

Binomial Expansion Formula
Suppose n is a positive integer. Then

$$(a + b)^n = \sum_{k=0}^{n} \binom{n}{k} a^{n-k} b^k. \tag{7.32}$$

We arrived at formula (7.32) by observing several particular instances (a process of inductive inference), and at this point it should be considered a conjecture. The result is precisely the type of statement discussed in the preceding section: Formula (7.32) holds for $n = 1, 2, 3, \ldots$. This suggests proof by mathematical induction. Such a proof can indeed be given, but since it is somewhat lengthy, it is not included here. We urge the reader to give details of the proof and suggest the use of identity (7.31) in the process.

Example 4   Expand $(x - 2y)^4$.

Solution   Using (7.32) with $n = 4$, $a = x$, and $b = -2y$ gives

$$(x - 2y)^4 = \sum_{k=0}^{4} \binom{4}{k} x^{4-k} (-2y)^k$$

$$= \binom{4}{0} x^4 + \binom{4}{1} x^3 (-2y) + \binom{4}{2} x^2 (-2y)^2 + \binom{4}{3} x (-2y)^3 + \binom{4}{4} (-2y)^4$$

$$= x^4 - 8x^3 y + 24x^2 y^2 - 32xy^3 + 16y^4. \qquad \blacksquare$$

Example 5   Find the fourth term in the expansion of $(x^2 - 2y)^{12}$.

Solution   Using (7.32) with $n = 12$, $a = x^2$, $b = -2y$, we get the expansion of $(x^2 - 2y)^{12}$ as

$$(x^2 - 2y)^{12} = \sum_{k=0}^{12} \binom{12}{k} (x^2)^{12-k} (-2y)^k.$$

With the expansion in this form, the fourth term is given by $k = 3$. Thus we have as the fourth term:

$$\binom{12}{3}(x^2)^{12-3}(-2y)^3 = \frac{12 \cdot 11 \cdot 10}{3!}(x^2)^9(-8y^3)$$

$$= (220)(x^{18})(-8y^3) = -1760x^{18}y^3. \qquad \blacksquare$$

Example 6   If $(x^2 - 2/x)^{10}$ is expanded and each term is simplified, we get terms involving x to various powers. Find the term that involves x^8.

Solution   Applying (7.32) with $n = 10$, $a = x^2$, $b = -2/x$ and then performing some algebraic simplification gives

$$\left(x^2 - \frac{2}{x}\right)^{10} = \sum_{k=0}^{10} \binom{10}{k}(x^2)^{10-k}\left(-\frac{2}{x}\right)^k$$

$$= \sum_{k=0}^{10} \binom{10}{k}(x^{20-2k})(-2)^k x^{-k}$$

$$= \sum_{k=0}^{10} (-2)^k \binom{10}{k}x^{20-3k}.$$

In this form we can first find the value of k that gives 8 as the exponent of x. That is, k is determined by $20 - 3k = 8$. Solving this equation gives $k = 4$. Therefore, the term involving x^8 is the fifth term in the expansion and is given by

$$(-2)^4 \binom{10}{4}x^{20-3\cdot4} = 16\left(\frac{10 \cdot 9 \cdot 8 \cdot 7}{4!}\right)x^8 = 3360x^8. \qquad \blacksquare$$

Exercises 7.5

In problems 1 through 16, evaluate the given expressions.

1. $5!$

2. $7!$

3. $\dfrac{12!}{10!}$

4. $\dfrac{10!}{7!}$

5. $\dfrac{6!}{2!4!}$

6. $\dfrac{8!}{3!5!}$

7. $\dfrac{6!}{0!(6-0)!}$

8. $\dfrac{5!}{3!(5-2)!}$

9. $\dfrac{6! + 4!}{3!}$

10. $\dfrac{8! - 5!}{3!}$

11. $\dbinom{8}{3}$

12. $\dbinom{5}{4}$

13. $\dbinom{12}{12}$

14. $\dbinom{16}{14}$

15. $\dbinom{3}{0}$

16. $\dbinom{17}{3}$

In problems 17 through 20, verify the given equations by evaluating each side. Then show that each is a particular case of Eq. (7.30) or Eq. (7.31).

17. $\dbinom{12}{3} = \dbinom{12}{9}$

18. $\dbinom{16}{4} = \dbinom{16}{12}$

19. $\binom{8}{3} + \binom{8}{4} = \binom{9}{4}$ **20.** $\binom{6}{2} + \binom{6}{3} = \binom{7}{3}$

In problems 21 through 28, expand the given expressions and simplify the results.

21. $(2x + y)^4$ **22.** $(2x - y)^4$ **23.** $(x^2 - 1)^6$ **24.** $(x^2 + 1)^6$

25. $(x^2 - y)^5$ **26.** $(x^2 - 3y)^5$ **27.** $\left(x - \dfrac{2}{x}\right)^4$ **28.** $\left(x + \dfrac{2}{x}\right)^6$

In problems 29 through 32, find the indicated term in the expansion of the given expression.

29. $(x - y)^{12}$; sixth term   **30.** $(x^2 - 2y)^{10}$; seventh term

31. $\left(x^2 - \dfrac{2}{x}\right)^7$; fourth term   **32.** $\left(x - \dfrac{3}{x^2}\right)^8$; third term

For each of the problems 33 through 36, in the expansion of the given binomial expression with terms simplified, find the term that involves the given power of x.

33. $\left(x - \dfrac{2}{x^2}\right)^{16}$; the term involving x^7

34. $\left(x^3 - \dfrac{1}{x}\right)^{15}$; the term involving x^{25}

35. $\left(2x - \dfrac{3}{x}\right)^8$; the constant term (that is, involving x^0)

36. $\left(2x^3 - \dfrac{3}{x}\right)^{12}$; the constant term (that is, involving x^0)

In problems 37 through 40, n and k are positive integers with $k < n$. Prove that the given equations are identities.

37. $\binom{n}{k} = \dfrac{n - k + 1}{k}\binom{n}{k - 1}$ **38.** $\binom{n-1}{k-1} + \binom{n-1}{k} = \binom{n}{k}$

39. $\binom{n}{0} + \binom{n}{1} + \binom{n}{2} + \cdots + \binom{n}{n} = 2^n$. *Hint:* Expand $(1 + 1)^n$.

40. $\binom{n}{0} - \binom{n}{1} + \binom{n}{2} - \binom{n}{3} + \cdots + (-1)^n \binom{n}{n} = 0$. *Hint:* Expand $(1 - 1)^n$.

7.6 MATHEMATICS OF FINANCE

Concepts studied in this chapter can be applied to problems involving finance. Let us consider a variety of problems involving *simple interest, compound interest,* and *amortization.*

When you borrow money from a bank, you are expected not only to repay the amount borrowed within a specified period of time but also to pay a fee for the use of the money for that period. This fee is called *interest.* Similarly, if you deposit money in the bank, the bank pays you interest for the use of your money. The amount invested or borrowed is called *principal,* denoted by P. The *interest* rate, r, determines the percentage of the principal that will be paid as interest when the principal has been kept for a specified time (*interest period*). Thus the amount of interest earned in *one interest period* is given by

$$I = r \cdot P,$$

where r *is the decimal form* of the interest rate, which is usually stated as a percentage.

If the principal is kept for longer than one interest period, the total amount of interest may be calculated in two different ways. For instance, suppose $1000 is invested at 10 percent interest per year for two years. The interest earned during the first year is $(0.10)(1000) = 100$ dollars; suppose interest for the second year is also determined in the same way, $(0.10)(1000) = 100$ dollars. Then the total interest earned for the two years is $200, and the value of the investment at the end of two years is $1000 + 200 = 1200$. In this case we say that interest has been computed as *simple interest*. Now suppose interest is computed as follows: Interest for the first year is $(0.10)(1000) = 100$ dollars, and the value of the investment at the end of the first year is $1100. If interest for the second year is computed on this amount, then it is $(0.10)(1100) = 110$ dollars. The total interest for the two years is $210 and the value of the investment at the end of two years is $1210. In this case we call it *compound interest;* it includes interest on interest.

Simple Interest
Simple interest is defined by

$$I = P \cdot r \cdot n. \tag{7.33}$$

where P is the amount invested (or borrowed) for n interest periods at interest rate r per period. The accumulated value of P at the end of n periods of time is given by

$$A = P + Prn = P(1 + rn). \tag{7.34}$$

The following two examples illustrate applications of simple interest.

Example 1   What is the total repayment of a loan of $500 for four months if the interest rate is 15% per year?

Solution   Using formula (7.34) with $P = 500$, $r = 0.15$, $t = 4/12$ gives

$$A = 500 + 500(0.15)(\tfrac{4}{12}) = 500 + 25 = 525 \text{ dollars.}$$

Here we are considering r as the rate per year (interest period is one year) and $n = 4/12 = 1/3$ of a year.

We could consider the interest period as one month, in which case the rate would be $15/12 = 1.25$ percent per month. Then in formula (7.34) we would use $P = 500$, $n = 4$, $r = 0.0125$, which would yield the same result as above,

$$A = 500 + 500(0.0125)(4) = 525 \text{ dollars.}$$

Example 2 Suppose a home improvement loan of $3600 is to be repaid in 12 monthly payments of $300 plus the interest on the unpaid balance for that month. If the rate of interest is 18% per year, what is the total sum to be repaid?

Solution Let a_1 represent the amount to be paid at the end of the first month; that is, a_1 is the sum of $300 and the interest on $3600 for one month. Thus

$$a_1 = 300 + 3600\left(\frac{0.18}{12}\right)(1) = 300 + 54 = \$354.$$

Let a_2 be the amount to be repaid at the end of the second month; that is, a_2 is equal to $300 plus the interest on $3600 - $300 = $3300 for one month. Thus

$$a_2 = 300 + 3300\left(\frac{0.18}{12}\right)(1) = 300 + 49.50 = 349.50 \text{ dollars.}$$

Continuing with a_3 (the amount repaid at the end of the third month) and so on to a_{12} (the amount repaid at the end of the twelfth month), we get $a_3 = 345.00$; $a_4 = 340.50$; $\dots$; $a_{12} = 304.50$.

The sequence $\{a_n\}_{n=1}^{12}$ *is an arithmetic sequence* with $a_1 = 354$, $a_{12} = 304.50$, and $d = -4.50$. From Eq. (7.8)

$$S_{12} = 12\left(\frac{354 + 304.50}{2}\right) = 3951.$$

Therefore the amount to be repaid is $3951. ■

Compound Interest

Compound interest is similar to simple interest, except that interest is paid not only on the principal but also on the previously accumulated interest. This can be illustrated as follows:

Suppose a sum (or principal) of P dollars is invested, and the *interest rate per year* is r. If interest is *compounded annually,* what is the value of the investment at the end of n years? It is customary to state the interest rate as a percentage; for instance, interest at 12% means that $r = 0.12$. Let us denote the value of the investment at the end of n years by A_n. The value of the investment at the end of the first year consists of P dollars invested for one year, and so

$$A_1 = P + P \cdot r \cdot 1 = P(1 + r).$$

The value at the end of the second year, A_2, is equivalent to an investment of A_1 dollars for one year at interest rate r. That is,

$$A_2 = A_1 + A_1 \cdot r \cdot 1 = A_1(1 + r) = P(1 + r)(1 + r) = P(1 + r)^2.$$

In a similar manner, A_3 is equivalent to an investment of A_2 dollars for one year at interest rate r, and so we have

$$A_3 = A_2 + A_2 r = A_2(1 + r) = P(1 + r)^2(1 + r) = P(1 + r)^3.$$

In general, using inductive inference, we get

$$A_n = P(1 + r)^n. \tag{7.35}$$

Now suppose interest is *compounded semiannually*. What modifications are necessary in formula (7.35) to get the corresponding A_n? The rate of interest r is usually given as the *rate per year*, so the rate for half a year is $r/2$. The number of interest periods in n years is $2n$. If we follow a procedure similar to that above, we get

$$A_n = P\left(1 + \frac{r}{2}\right)^{2n}. \tag{7.36}$$

In general, if interest is *compounded m times per year*, then the rate of interest for each interest period is r/m, and the number of interest periods in n years is $m \cdot n$. Therefore

$$A_n = P\left(1 + \frac{r}{m}\right)^{mn}.$$

Thus we have the important result for computing compound interest:

> Suppose a sum of P dollars is invested at an interest rate per year of r. If interest is *compounded m times per year*, and A_n represents the amount of the investment at the end of n years, then
>
> $$A_n = P\left(1 + \frac{r}{m}\right)^{mn}. \tag{7.37}$$

Note that the sequence of numbers

$$A_1 = P\left(1 + \frac{r}{m}\right)^{m}, \qquad A_2 = P\left(1 + \frac{r}{m}\right)^{2m}, \qquad A_3 = P\left(1 + \frac{r}{m}\right)^{3m}, \ldots$$

is a geometric sequence with a common ratio of $(1 + r/m)^m$.

Example 3   Suppose a sum of $1000 is invested at 8% interest compounded quarterly. What is the value of this investment at the end of 10 years?

Solution   Applying formula (7.37) with $P = 1000$, $r = 0.08$, $m = 4$, and $n = 10$ gives

$$A_{10} = 1000\left(1 + \frac{0.08}{4}\right)^{4 \cdot 10} = 1000(1.02)^{40} = 2208.04.$$

Therefore the value of the investment at the end of 10 years is $2208.04.   ■

Interest Compounded Continuously

In Eq. (7.37) m represents the number of times per year interest is compounded. Compounding interest continuously implies that we are interested in what happens to A_n *as m becomes large*. Let us concentrate on $A_n = P(1 + r/m)^{mn}$, where P, r, and n remain fixed, and only m varies. Suppose we let $x = r/m$; then $m = r/x$, and Eq. (7.37) can be written as

$$A_n = P\left(1 + \frac{r}{m}\right)^{mn} = P(1 + x)^{nr/x} = P[(1 + x)^{1/x}]^{nr}. \qquad (7.38)$$

As m becomes large, $x = r/m$ approaches zero. Thus we can ask what happens to $(1 + x)^{1/x}$ as x approaches zero? In Section 3.2 the number e was introduced as the limiting value of $f(x) = (1 + x)^{1/x}$ as x approaches zero (see p. 172). Thus, in Eq. (7.38) we conclude that A_n approaches Pe^{nr} as m becomes large. This gives the following result for *interest compounded continuously:*

> If the sum of P dollars is invested at a yearly rate of r *compounded continuously,* the value A_n of the investment at the end of n years is given by
>
> $$A_n = Pe^{nr}. \qquad (7.39)$$

Example 4   Suppose $1000 is invested at 8 per cent. What is the value of the investment at the end of 10 years if interest is compounded in each of the following ways?

a) Annually   **b)** Daily   **c)** Continuously

Solution   **a)** Substituting $P = 1000$, $r = 0.08$, $m = 1$, $n = 10$ into Eq. (7.37) gives $A_{10} = 1000(1 + 0.08)^{10} = 2{,}158.93$.

b) Using Eq. (7.37) as in part (a) and changing m to 365 gives

$$A_{10} = 1000(1 + 0.08/365)^{365 \cdot 10} = 2{,}225.35.$$

c) Using Eq. (7.39) with $P = 1000$, $r = 0.08$, and $n = 10$ gives

$$A_{10} = 1000e^{10(0.08)} = 2{,}225.54.$$

The amount at the end of 10 years is $2,158.93 if interest is compounded annually, $2,225.35 if compounded daily, and $2,225.54 if compounded continuously. ■

Note on calculator round off error: When calculations involve small or large numbers, one must be aware of possible round off errors. Suppose in Example 4 we are interested in determining the value of the investment when interest is compounded each minute. This is given by

$$A_{10} = 1000\left(1 + \frac{0.08}{525600}\right)^{5256000}$$

Using a calculator that operates with 10-digit capacity gives $A_{10} = 2223.12$, while one with 13-digit capacity gives $A_{10} = 2225.54$. The correct answer is 2225.54. If interest is compounded each second, the 13-digit machine gives

$A_{10} = 2224.99$, while the correct result is 2225.54. Round off error is a topic of major concern in numerical-analysis courses.

Example 5　What rate of interest compounded annually is equivalent to 6% compounded continuously?

Solution　Let r be the desired rate of interest. At the end of n years, P dollars will be worth $P(1 + r)^n$ if interest is compounded annually at rate r and $Pe^{0.06n}$ if compounded continuously at 6%. We want to find r such that these two are equal. That is

$$P(1 + r)^n = Pe^{0.06n}.$$

Dividing by P and then taking the nth root of both sides gives

$$1 + r = e^{0.06} \quad \text{or} \quad r = e^{0.06} - 1 = 0.0618 = 6.18\%.$$

Thus money compounded annually at 6.18% would yield the same amount as it would if invested at 6% compounded continuously.　　　　　■

Present Value

In planning for future expenditure, whether for education, recreation, or retirement, we often ask the question "What amount should be invested now in order to accumulate a given sum at the end of n years?". This amounts to solving for P in Eq. (7.37) or Eq. (7.39), where A_n is the given amount. Thus, if interest is compounded m times per year,

$$P = A_n\left(1 + \frac{r}{m}\right)^{-mn}. \tag{7.40}$$

If interest is compounded continuously,

$$P = A_n e^{-nr}. \tag{7.41}$$

We call P the *present value* of the final amount A_n over n years.

Example 6　How much should be invested now, so that the investment will be worth $10 000 at the end of 8 years if the interest rate is 6% compounded

a) Quarterly　　　　　　　　　　**b)** Continuously

Solution　**a)** Substituting $A_n = 10{,}000$, $r = 0.06$, $m = 4$, and $n = 8$ into Eq. (7.40) gives

$$P = 10{,}000\left(1 + \frac{0.06}{4}\right)^{-4(8)} = 6209.93.$$

Thus $6209.93 invested now would be worth $10,000 in 8 years if interest is compounded quarterly.

b) Applying Eq. (7.41), we get

$$P = 10000e^{-8(0.06)} = 6187.83.$$

Therefore, $6187.83 invested now would be worth $10 000 in 8 years if interest is compounded continuously. ▄

Amortization refers to the elimination of a debt through periodic payments. An amortization schedule shows the division of the payments into interest charges and the reduction of principal. Let us illustrate by considering the problem of repaying a home loan, a topic that is or will be of interest to most of us.

Repaying a Home Loan

Suppose you get a loan from a bank for the purchase of a home, and you agree to repay the loan in monthly payments over a period of years. If the agreed interest rate is r, then in current banking practice this means you will pay interest at the rate $i = r/12$ each month on the unpaid balance of the loan. This is equivalent to a yearly interest rate of r compounded monthly. For instance, if the bank advertises a rate of 15 per cent, it means that you pay interest at a rate $i = 0.15/12 = 0.0125$ per month on the unpaid balance of the loan. The determining factor for most people in this situation is not the amount of interest paid over a period of time but the amount the bank expects to be repaid each month. Let us now derive a formula that will give the amount of the monthly payment.

For a loan of A dollars at an interest *rate of i per month* to be repaid over a period of n months, what should each monthly payment R be? The bank looks at the situation as follows: The sum of A dollars now would be worth $A(1 + i)^n$ at the end of n months. The total value of n monthly payments, each of R dollars, should also be worth $A(1 + i)^n$. That is, the total amount repaid, T_n, at the end of n months should be

$$T_n = A(1 + i)^n.$$

Figure 7.2 illustrates the contribution of each payment of R dollars to T_n, as shown in the last column. Thus the total value of the n payments is given by

$$T_n = R + R(1 + i) + R(1 + i)^2 + \cdots + R(1 + i)^{n-2} + R(1 + i)^{n-1}.$$

This is the sum of n terms of a *geometric sequence,* in which the first term is $a_1 = R$ and the common ratio is $r = 1 + i$. Using formula (7.13) gives

$$T_n = \frac{a_1 r^n - a_1}{r - 1} = \frac{R(1 + i)^n - R}{(1 + i) - 1} = \frac{R[(1 + i)^n - 1]}{i}.$$

Equating this value of T_n with that given by $T_n = A(1 + i)^n$, we get

$$\frac{R[(1 + i)^n - 1]}{i} = A(1 + i)^n.$$

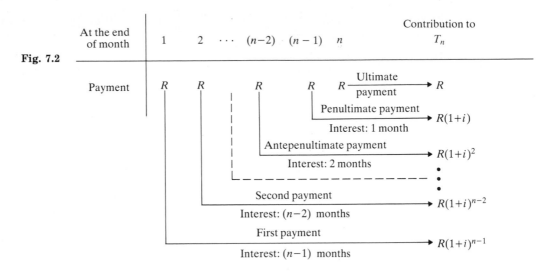

Fig. 7.2

Solving for R gives

$$R = \frac{A(1 + i)^n i}{(1 + i)^n - 1}.$$

By multiplying numerator and denominator by $(1 + i)^{-n}$, we get

$$R = \frac{Ai}{1 - (1 + i)^{-n}}.$$

Therefore the monthly payment R on a loan of A dollars for n months at a monthly interest rate of i is given by

$$R = \frac{A(1 + i)^n i}{(1 + i)^n - 1} = \frac{Ai}{1 - (1 + i)^{-n}}. \tag{7.42}$$

Example 7   What monthly payment is required by a bank on a 30-year loan of $48 000 at 12 per cent interest? What is the total amount paid on this loan at the end of 30 years?

Solution   Using formula (7.42) with $A = 48\,000$, $i = 0.12/12 = 0.01$, and $n = 30(12) = 360$, we get

$$R = \frac{48000(1.01)^{360}(0.01)}{(1.01)^{360} - 1}.$$

Evaluating by a calculator gives $R = \$493.73$ as the amount to be paid each month for 30 years. The total amount paid at the end of 30 years is

$$(493.73)(12)(30) = \$177\,742.80.$$

Example 8   *Amortization schedule*. For the loan discussed in Example 7, suppose we wish to determine how much of each of the first three monthly payments contributes toward reducing the principal and how much is interest.

Solution   At the end of the first month the interest due on $48,000 at 12% is

$$I_1 = 48\,000(0.12)(\tfrac{1}{12}) = 480.$$

Thus $480 of the first payment of $493.73 is interest, and $13.73 is the amount the principal is reduced. At the end of the first month we still owe $P_1 = 48\,000 - 13.73 = 47\,986.27$ dollars.
    The interest paid during the second month is

$$I_2 = 47986.27(0.12)(\tfrac{1}{12}) = 479.86.$$

Therefore $479.86 of the second payment is interest and $13.87 goes toward payment of the loan. At the end of the second month the amount remaining to be paid on the principal is

$$P_2 = 47986.27 - 13.87 = 47\,972.40.$$

During the third month,

$$I_3 = (47\,972.40)(0.12)(\tfrac{1}{12}) = 479.72,$$

and so the monthly payment consists of $479.72 interest and $14.01 toward reduction of the debt.
    Summarizing the above results in tabular form gives the first three entries of an amorization schedule.

Month n	Amount of R that is interest ($)	Amount of R that reduces debt ($)	Outstanding principal ($)
1	480.00	13.73	47,986.27
2	479.86	13.87	47,972.40
3	479.72	14.01	47,958.39
⋮	⋮	⋮	⋮

We see that the amount of monthly payment that goes toward the reduction of the debt is discouragingly small. However, 25 years later the situation has reversed, and the amount that is interest is very small.   ▬

Exercises 7.6

In the following problems interest rates are given as rates *per year*. For instance a rate of 18% is 18% per year; the corresponding rate per month is (18/12)%, or 1.5%.

Problems 1 through 4 refer to simple interest.

1. What total payment is due on a loan of $800 for six months if the interest rate is 10%?

2. What is the interest on $2000 for 9 months at a rate of 12%?

3. A home improvement loan of $4800 is to be repaid in 24 monthly installments, each of $200 plus interest on the unpaid balance. If the interest rate is 15%, what is the total sum to be repaid? (See Example 2.)

4. Banks usually discount (deduct interest from the principal) before a short-term loan is made. Borrowing $1000 for 6 months means that the bank will give you P dollars now and you repay $1000 in six months. If the interest rate is 15%, what is the value of P?

In problems 5 through 8, determine the value of the investment at the end of the given number of years if interest is compounded in each of the following ways

a) Annually   **b)** Quarterly   **c)** Monthly   **d)** Continuously

5. Principal of $2000 at 8% for 10 years.   6. Principal of $1000 at 12% for 20 years.

7. Principal of $2500 at $7\frac{3}{4}$% for 16 years.   8. Principal of $5000 at $8\frac{1}{2}$% for 24 years.

9. Suppose you perform a service for which you expect to be paid, and you are given the choice of being paid (a) $1500 immediately or (b) by a note for $2000 payable in three years. Assume that you do not need the money immediately, and that the $1500 could be invested at 12% compounded monthly. Which would be the better choice?

10. Suppose in problem 9 you are given the choice between (a) immediate payment of $1500 or (b) a note for $2000 payable in 2 years. If you can invest the $1500 at 10% compounded continuously, which would be the better choice?

11. What rate of interest compounded annually is equivalent to 8% compounded continuously?

12. What rate of interest compounded annually is equivalent to 8% compounded monthly?

13. Suppose you purchase a home and need a loan of $40,000. If the bank advertises a rate of 13%, what will the monthly payments be if you get

   a) a 20-year loan?   **b)** a 30-year loan?

14. Suppose you wish to borrow $25,000 from a bank at a rate of 12%. If your financial situation is such that you can afford no more than $260 for your monthly payments, will you be able to afford

   a) a 30-year loan?   **b)** a 25-year loan?

15. If $1000 is deposited at the beginning of each year for 20 years at 6% interest compounded annually, how much will be in the account at the end of 20 years?

16. Suppose $100 is deposited at the beginning of each month in an account that pays 8% interest compounded monthly. How much will be in the account at the end of 10 years?

7.7 Looking Ahead to Calculus

Limit concepts were introduced intuitively in Sections 2.7, 3.6, and 5.7. Let us continue that discussion as it relates to functions with domain **N**, that is, sequences.

A sequence $\{a_n\}_{n=1}^{\infty}$ of real numbers is said to *converge* to a number c if for large values of n, the corresponding sequence numbers approach c.* The number c is called the *limit of the sequence,* and we write

$$\lim_{n\to\infty} a_n = c.$$

As an illustration, consider the sequence given by $a_n = n/(n+1)$. The first few terms are given by

$$\tfrac{1}{2}, \tfrac{2}{3}, \tfrac{3}{4}, \tfrac{4}{5}, \ldots.$$

Note that the sequence values are increasing as n increases, but they never become greater than 1. Intuition leads us to conclude that this sequence converges to 1, and we write

$$\lim_{n\to\infty} \frac{n}{n+1} = 1.$$

As another example, consider the sequence described by $b_n = n^2/(n+1)$. The first few terms are given by

$$\tfrac{1}{2}, \tfrac{4}{3}, \tfrac{9}{4}, \tfrac{16}{5}, \tfrac{25}{6}, \ldots.$$

Here the sequence values increase as n increases. Note that they become indefinitely large and do not approach any fixed number. We say that sequence $\{b_n\}_{n=1}^{\infty}$ *diverges.*

In the above examples, the sequence functions are relatively simple, and it was an easy matter to decide intuitively whether or not the sequence converges. In the following examples that is not so, but we can get some insights by evaluating the sequence functions for several large values of n and from these arrive at conjectures concerning convergence. Proofs that our conjectures are valid must be deferred to courses in calculus.

Example 1   Does the sequence $\{n^{1/n}\}_{n=1}^{\infty}$ converge?

Solution   The first few terms of the sequence are given by $1, \sqrt{2}, \sqrt[3]{3}, \sqrt[4]{4}, \sqrt[5]{5}, \ldots.$ Let $f(n) = n^{1/n}$ and make a table giving values of $f(n)$ corresponding to large values of n:

n	10	100	1000	5000	100000
$f(n)$	1.259	1.047	1.0069	1.0017	1.000115

From the values of $f(n)$ in the table we conjecture that the given sequence converges to the number 1, and we write $\lim_{n\to\infty} n^{1/n} = 1.$ ■

---

*   The idea of convergence of sequences will be defined more precisely in calculus.

Example 2   A sequence function is given by $f(n) = n \sin \pi/n$. Does the sequence converge?

Solution   The first few terms of the sequence are given by

$$f(1) = \sin \pi = 0, \quad f(2) = 2 \sin \frac{\pi}{2} = 2, \quad f(3) = 3 \sin \frac{\pi}{3} = 2.598, \ldots$$

The following table will give some insight concerning the behavior of $f(n)$ for large values of n:

n	10	100	1000	4000	1000000
$f(n)$	3.090	3.141	3.1416	3.14159	3.141592654

Looking at the values of $f(n)$ given in the table, we conjecture that the sequence converges to a number we recognize as being π. Therefore

$$\lim_{n \to \infty} n \sin \frac{\pi}{n} = \pi. \qquad \blacksquare$$

Example 3   A sequence is described recursively by $a_1 = \sqrt{2}$ and $a_{n+1} = \sqrt{2 + a_n}$ for $n = 1$, 2, 3, ... Build a table giving several values of a_n, and make a conjecture about the convergence of $\{a_n\}_{n=1}^{\infty}$.

Solution   The first few terms of the sequence are given by

$$a_1 = \sqrt{2}, \quad a_2 = \sqrt{2 + \sqrt{2}}, \quad a_3 = \sqrt{2 + \sqrt{2 + \sqrt{2}}}, \ldots$$

These are expressed in approximate decimal form and included in the table:

n	1	2	3	4	5	6	7	8
a_n	1.414	1.848	1.962	1.990	1.998	1.9994	1.9998	1.99996

From the values of a_n in the table it appears reasonable to conjecture that the sequence converges to the number 2. $\qquad \blacksquare$

Exercises 7.7

In problems 1 through 10, sequence functions are given. Make a table giving values of $f(n)$ corresponding to large values of n. Use your table to arrive at conjectures about convergence of the sequence $\{a_n\}_{n=1}^{\infty}$, where $a_n = f(n)$.

1. $f(n) = (2n)^{1/n}$ **2.** $f(n) = \left(1 + \frac{1}{n}\right)^n$ **3.** $f(n) = \left(1 - \frac{1}{n}\right)^n$ **4.** $f(n) = \left(1 + \frac{0.08}{n}\right)^n$

5. $f(n) = \frac{\sin n}{2n}$ **6.** $f(n) = \frac{1 - \cos n}{n^2}$ **7.** $f(n) = n \sin \frac{2\pi}{n}$ **8.** $f(n) = n \tan \frac{\pi}{4n}$

9. $f(n) = \sqrt{n}(\sqrt{n + 1} - \sqrt{n})$ **10.** $f(n) = (\ln n)^{1/n}$

In problems 11 and 12, sequences are described recursively. Make a table giving the first eight terms. Using the values of a_n in your table, arrive at a conjecture concerning $\lim_{n \to \infty} a_n$.

11. $a_1 = \sqrt{3}$ and $a_{n+1} = \sqrt{3 + a_n}$ for $n = 1, 2, 3, \ldots$

12. $a_1 = \sqrt{5}$ and $a_{n+1} = \sqrt{5 + a_n}$ for $n = 1, 2, 3, \ldots$

Review Exercises

In problems 1 through 4, a formula is given for the nth term of a sequence.

a) Find the first four terms.

b) Evaluate $\sum_{k=1}^{4} a_k$.

1. $a_n = 1 - \dfrac{1}{2^n}$

2. $a_n = \dfrac{1}{2^{n-1}}$

3. $a_n = 3n - 1$

4. $a_n = \dfrac{1}{n(n+2)}$

5. The first three terms of an arithmetic sequence are 3, 8, 13. Find the following.

   a) The 24th term

   b) The sum of the first 24 terms

6. In an arithmetic sequence $a_4 = 16$ and $a_{13} = -2$, find the following.

   a) a_{20}

   b) $\sum_{k=1}^{20} a_k$

   c) The number of terms n such that $\sum_{k=1}^{n} a_k = -140$.

7. Find the values of x such that $x^2, x, -3$ are three consecutive terms of an arithmetic sequence.

8. The first three terms of a geometric sequence are $3, \frac{3}{2}, \frac{3}{4}$. Find the following.

   a) The fifth term

   b) The sum of the first five terms.

9. Suppose a sequence $\{a_n\}_{n=1}^{\infty}$ is given by $a_n = 1 + 1/2^n$.

   a) Write out the first four terms.

   b) Is this a geometric sequence?

   c) Find the sum of the first four terms.

10. In a geometric sequence, $a_1 = \frac{2}{3}$ and $r = \frac{1}{3}$. Find the number of terms n such that the sum S_n is $\frac{6560}{6561}$.

11. Find the repeating decimal expansion for the following.

   a) $\frac{4}{15}$

   b) $\frac{18}{11}$

   c) $\frac{3}{14}$

12. Express the repeating decimal $0.727272\ldots$ (that is, $0.\overline{72}$) as a fraction of two integers.

13. Prove that $3 + 9 + 15 + \cdots + (6n - 3) = 3n^2$ for $n = 1, 2, 3, \ldots$.

14. Is $3n^3 + 6n$ divisible by 9 for each $n = 1, 2, 3, \ldots$? Give reasons for your answer.

15. Is "$3^n \le (n+3)^2$ for $n = 1, 2, 3, \ldots$" a true statement? If so, give a proof. If not, give a counterexample.

16. Use the binomial expansion formula to expand $(2x - 1/x)^5$.

17. Evaluate:

a) $\dfrac{6!}{2!4!}$
 b) $\dbinom{15}{3}$
 c) $\dbinom{8}{2} + \dbinom{8}{3}$

18. Find the fourth term in the expansion of $(x + 2y)^8$.

19. Suppose $(x^2 + 1/x)^{15}$ is expanded and the resulting terms are simplified. Find the term that involves x^6.

20. Find the sum of all positive integers less than 400 that are divisible by both 2 and 3.

21. Suppose sequence $\{a_n\}_{n=1}^{\infty}$ is given by $a_n = (1 - 1/n)^n$. Find the following and give answers rounded off to five decimal places.

a) a_{10}
 b) a_{100}
 c) a_{1000}
 d) $a_{10,000}$

22. Suppose $1000 is invested at 8% interest compounded quarterly. The accumulated value A_n of the investment at the end of n years is given by

$$A_n = 1000 \left(1 + \frac{0.08}{4}\right)^{4n} \text{ dollars.}$$

Find the first four terms of the sequence $\{A_n\}_{n=1}^{\infty}$. Give results rounded off to the nearest cent.

23. Suppose $2400 is invested at $7\frac{3}{4}$ per cent interest. Determine the value of the investment at the end of 12 years if interest is compounded in each of the following ways.

a) Annually
 b) Quarterly
 c) Continuously

24. What rate of interest compounded continuously is equivalent to 6 per cent compounded quarterly?

25. Suppose you get a 30-year loan of $50,000 for the purchase of a house. If the rate of interest is 12 per cent, how much are the monthly payments?

In problems 26 through 30, evaluate the given sums.

26. $\displaystyle\sum_{k=1}^{15} (2k - 1)$
  27. $\displaystyle\sum_{k=1}^{50} (3k + 2)$
  28. $\displaystyle\sum_{k=1}^{\infty} \left(\frac{1}{3}\right)^k$
  29. $\displaystyle\sum_{k=1}^{\infty} 3\left(\frac{1}{4}\right)^k$
  30. $\displaystyle\sum_{k=1}^{5} (2^k - k)$

Analytic Geometry: Conics and Parametric Equations

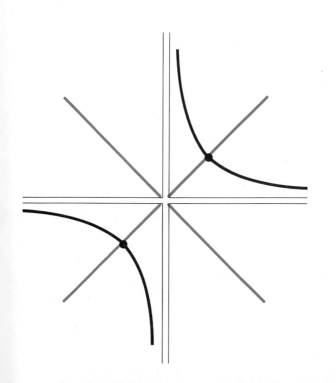

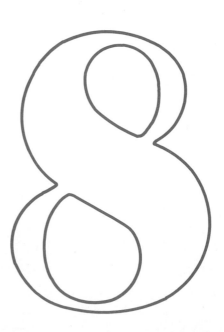

One of the significant contributions to the development of mathematics was the introduction of coordinate systems, which provides a link between algebra and geometry. In many cases, solution of a complex algebraic (or analytic) problem can be achieved by considering it in geometric terms. Geometric insight provides the key that is frequently not available in the algebraic setting. Similarly, many geometric problems can be more easily analyzed by algebraic methods. The marriage of algebra and geometry is referred to as *analytic geometry*.

We have already used methods of analytic geometry in earlier sections of this book. Let us now explore some properties of well-known plane curves called *conic sections.* These were studied as early as the third century B.C. by Apollonius, who became known to his contemporaries as "The Great Geometer."

Consider a surface in the shape of a vertical double right circular cone, as shown in Fig. 8.1. The intersection of such a cone by a plane will be a curve, the shape of which depends on the angle at which the plane is inclined. The cases of interest to us are those illustrated in Fig. 8.2. In (a) the plane is horizontal, and the resulting intersection is a *circle;* in (b) the plane is parallel to a line generating the cone, and the curve of intersection is a *parabola;* in (c) the plane is inclined at an angle between those in (a) and in (b), and the corresponding curve is an *ellipse;* in (d) the plane is vertical, and the intersection, which is in two parts, is called a *hyperbola.*

The definitions that we shall use to describe these curves will be given in terms of points, lines, and distances, rather than intersections of planes and cones. We shall derive equations of the conic sections and see that in each case we get an equation of the type

$$Ax^2 + Bxy + Cy^2 + Dx + Ey + F = 0, \tag{8.1}$$

where A, B, and C will not all be equal to zero. Just as we saw in Section 1.6 that *linear equations* in x and y, such as $ax + by + c = 0$, *are associated with lines,*

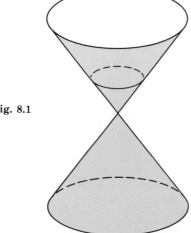

Fig. 8.1

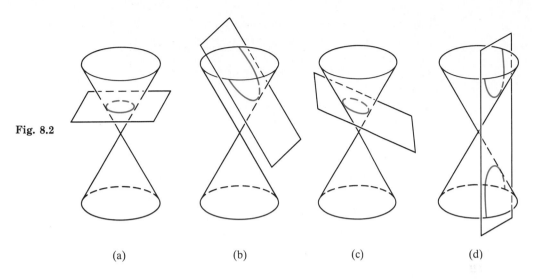

Fig. 8.2

(a)   (b)   (c)   (d)

so we shall see that *quadratic equations* in x and y, given by Eq. (8.1), *correspond to conic sections*. It is possible to show that our definitions are equivalent to the corresponding ones described in terms of intersecting cones by planes, but it is not of interest to do so here.

8.1   CIRCLE; PARABOLA

Definition 8.1   A circle is the set of all points in a plane, each of which is a given distance, called the *radius,* from a fixed point, called the *center*.

Suppose the fixed point is given as the origin $(0, 0)$ and the given distance is denoted by r (Fig. 8.3). Then the set of points (x, y), each of which is a distance r

Fig. 8.3

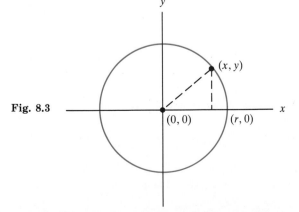

from $(0, 0)$, must satisfy the equation

$$\sqrt{(x - 0)^2 + (y - 0)^2} = r. \qquad (8.2)$$

Squaring both sides of Eq. (8.2) gives

$$x^2 + y^2 = r^2. \qquad (8.3)$$

Equation (8.3) represents a circle with center at the origin and radius r.

Now suppose we consider the general situation with the center at point (h, k) and radius r (Fig. 8.4). The corresponding circle is represented algebraically by the equation

$$(x - h)^2 + (y - k)^2 = r^2. \qquad (8.4)$$

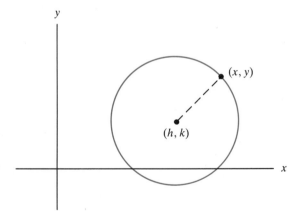

Fig. 8.4

Equation (8.4) is called the *standard form* of the equation of a circle. The *expanded form* of Eq. (8.4) is

$$x^2 + y^2 - 2hx - 2ky + h^2 + k^2 - r^2 = 0. \qquad (8.5)$$

Note that this is a special case of Eq. (8.1) in which $A = 1$, $B = 0$, $C = 1$, $D = -2h$, $E = -2k$, $F = h^2 + k^2 - r^2$.

Here we have an instance of a geometrical figure (a circle) being represented algebraically by Eq. (8.5). The process can be reversed. Suppose we have an algebraic equation of the form

$$x^2 + y^2 + ax + by + c = 0. \qquad (8.6)$$

The corresponding geometrical figure is a circle (or in a degenerate case it may represent a single point or no points; see Example 4).

Example 1   Draw a graph of the curve corresponding to the equation $x^2 + y^2 = 4$.

Solution   The curve is a circle with center at $(0, 0)$ and radius $r = 2$. The graph is shown in Fig. 8.5.

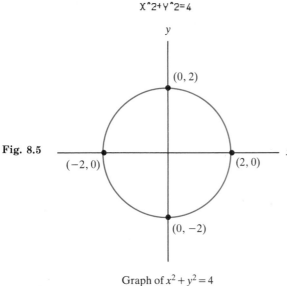

X^2+Y^2=4

Fig. 8.5

Graph of $x^2 + y^2 = 4$

Example 2   Find an equation for the circle with center at $(-2, 1)$ and radius $r = 3$.

Solution   Replacing h by -2, k by 1, and r by 3 in Eq. (8.4) gives $(x + 2)^2 + (y - 1)^2 = 3^2$. This can be written as $x^2 + y^2 + 4x - 2y - 4 = 0$.

Example 3   Describe the curve given by the equation $4x^2 + 4y^2 + 16x - 12y - 39 = 0$.

Solution   We can write the given equation in the form of Eq. (8.4) by completing the square on the x and y terms, as follows: First divide through by 4 and rearrange terms:

$$x^2 + 4x + y^2 - 3y = \tfrac{39}{4}.$$

Completing the square on the x terms and on the y terms gives

$$(x^2 + 4x + 4) + (y^2 - 3y + \tfrac{9}{4}) = \tfrac{39}{4} + 4 + \tfrac{9}{4},$$
$$(x + 2)^2 + (y - \tfrac{3}{2})^2 = 4^2.$$

Comparing this with Eq. (8.4), we see that $h = -2$, $k = \tfrac{3}{2}$, $r = 4$; and so the given equation represents a circle with center at $(-2, \tfrac{3}{2})$ and radius 4. This is shown in Fig. 8.6.

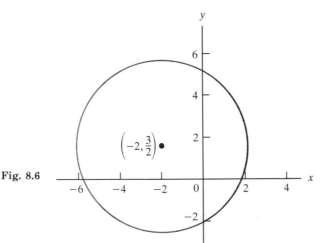

Graph of $4x^2 + 4y^2 + 16x - 12y - 39 = 0$

Example 4   What geometrical figure is associated with the following algebraic equation?

$$x^2 + y^2 + 2x - 4y + 8 = 0.$$

Solution   Our initial impulse is to conclude that the given equation represents a circle. Completing the square on the x and y terms gives

$$(x + 1)^2 + (y - 2)^2 = -3.$$

It is easy to see that there are no real numbers x and y that will satisfy this equation since the left side will always be a nonnegative number and cannot equal -3. Since the given equation is equivalent to this one, we conclude that there is no associated geometrical figure.

Example 5   Find an equation for a semicircle with center at $(1, -3)$, radius 1, and points A and B as endpoints of a diameter, where $A:(0, -3)$ and $B:(2, -3)$.

Solution   The point midway between A and B is $(1, -3)$. Using Eq. (8.4) with $h = 1$, $k = -3$, and $r = 1$ gives $(x - 1)^2 + (y + 3)^2 = 1$ as an equation for the given circle. There are two semicircles with diameter AB (see Fig. 8.7), arc ACB and arc ADB. In arc ACB, $y \geq -3$ and in arc ADB, $y \leq -3$. The equation of the circle can be written in the form

$$(y + 3)^2 = 1 - (x - 1)^2$$
$$y + 3 = \pm\sqrt{1 - (x - 1)^2}$$
$$y = -3 \pm \sqrt{2x - x^2}.$$

Fig. 8.7

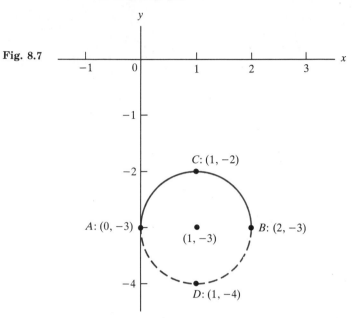

Graph of $(x-1)^2 + (y+3)^2 = 1$

Thus we have two equations: $y = -3 + \sqrt{2x - x^2}$ corresponds to arc ACB, and $y = -3 - \sqrt{2x - x^2}$ corresponds to arc ADB. Each of these describes a function with domain equal to $\{x \mid 0 \leq x \leq 2\}$. ∎

Parabola

Our definition of a parabola is in terms of a given line D, called the *directrix,* and a given point F, called the *focus.*

Definition 8.2

For a given line D and point F not on D, the corresponding parabola is the set of all points P having the property that P is equidistant from F and D; that is, $\overline{PF} = \overline{PQ}$, as shown in Fig. 8.8, where Q is a point on D with PQ perpendicular to D.

The line through F and perpendicular to D is called the *axis of the parabola.* The point V, which is on this line and midway between F and D, is called the *vertex of the parabola.*

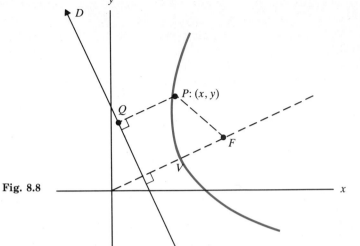

Fig. 8.8

Special case   Suppose we consider the special case in which F is the point $(p, 0)$, where $p \neq 0$, and D is given by the equation $x = -p$, as shown in Fig. 8.9 (which illustrates the case where $p > 0$). If we apply Definition 8.2, the equation $\overline{PF} = \overline{PQ}$ becomes

$$\sqrt{(x - p)^2 + (y - 0)^2} = \sqrt{[x - (-p)]^2 + (y - y)^2}.$$

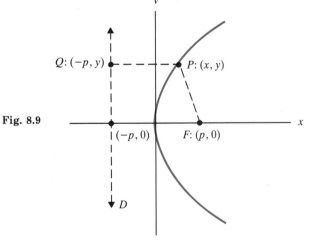

Fig. 8.9

Squaring both sides and simplifying gives

$$y^2 = 4px. \tag{8.7}$$

Thus we have the following result: the equation $y^2 = 4px$ represents a parabola with vertex V:(0, 0), focus F:(p, 0), directrix D: $x = -p$, and axis $y = 0$ (the x-axis). If $p > 0$, then the parabola opens to the right; if $p < 0$, the parabola opens to the left.

Another special case If in the above discussion we had taken F as (0, p) and D as $y = -p$, then the resulting equation of the parabola would have become

$$x^2 = 4py. \tag{8.8}$$

Equation (8.8) represents a parabola with vertex V:(0, 0), focus F:(0, p), directrix D: $y = -p$, and axis $x = 0$ (the y-axis).

If $p > 0$, the parabola opens upward, as shown in Fig. 8.10(a); if $p < 0$, then the parabola opens downward as shown in Fig. 8.10(b). Parabolas represented by Eqs. (8.7) and (8.8) are said to be in *standard position*.

Fig. 8.10

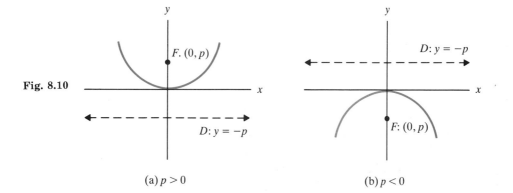

(a) $p > 0$ (b) $p < 0$

Parabolas not in standard position Suppose F and V are not on the coordinates axes, say F:($h + p$, k) and D: $x = h - p$, where $p \neq 0$; this will give V:(h, k) as the vertex, as shown in Fig. 8.11 (which illustrates the case for $p > 0$). Also the axis of the parabola is $y = k$. If we apply the definition of a parabola, $\overline{PF} = \overline{PQ}$ becomes

$$\sqrt{[x - (h + p)]^2 + [y - k]^2} = \sqrt{[x - (h - p)]^2 + [y - y]^2}.$$

Squaring both sides and simplifying gives

$$(y - k)^2 = 4p(x - h). \tag{8.9}$$

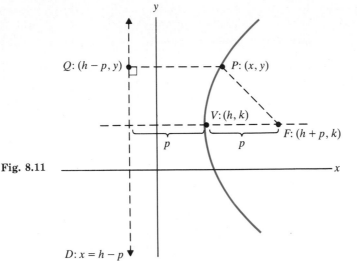

Fig. 8.11

Equation (8.9) represents a parabola with vertex $V:(h, k)$, focus $F:(h + p, k)$, directrix $D: x = h - p$, and axis $y = k$.

If $p > 0$, the parabola opens to the right, whereas if $p < 0$ it opens to the left. In an analogous manner, a parabola with a vertical axis is given by

$$(x - h)^2 = 4p(y - k). \tag{8.10}$$

Equation (8.10) represents a parabola with vertex $V:(h, k)$, focus $F:(h, k + p)$, directrix: $D: y = k - p$, and axis $x = h$.

If $p > 0$ the parabola opens upward; if $p < 0$ the parabola opens downward. Figure 8.12 illustrates the case for $p > 0$.

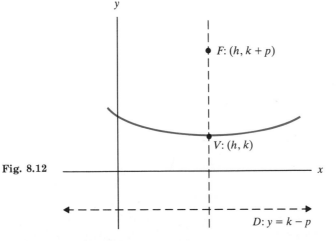

Fig. 8.12

Example 6   Draw a graph of $y^2 = -6x$. Give the coordinates of the vertex and focus and the equation of the directrix.

Solution   First note that the given equation is of the form $y^2 = 4px$, therefore $4p = -6$, and so $p = -\frac{3}{2}$. Thus the given equation represents a parabola with focus $F:(-\frac{3}{2}, 0)$, vertex $V:(0, 0)$, and directrix $D: x = \frac{3}{2}$. The graph is shown in Fig. 8.13.

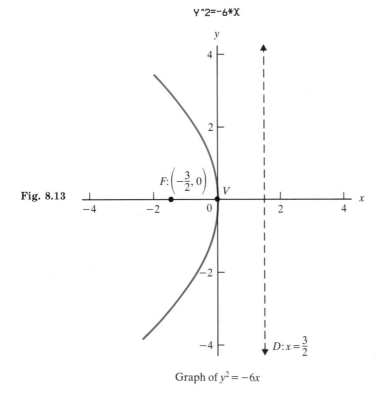

Fig. 8.13

Graph of $y^2 = -6x$

Example 7   Draw a graph of $y^2 + 2y - 8x + 25 = 0$. Determine the coordinates of the vertex and focus.

Solution   Completing the square on the y terms and rearranging terms gives $(y + 1)^2 = 8(x - 3)$. Comparing this equation with that given by Eq. (8.9), we conclude that the given equation represents a parabola with vertex $V:(3, -1)$ and focus $F:(5, -1)$, since $4p = 8$ and so $p = 2$. The graph is shown in Fig. 8.14.

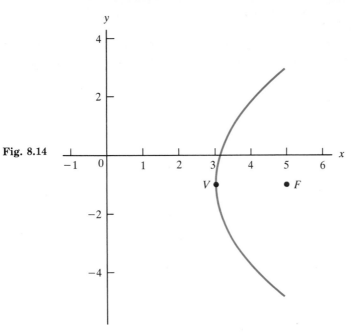

Fig. 8.14

Graph of $y^2+2y-8x+25=0$

Applications of Parabolas
There are numerous instances in which parabolas occur in applied problems. We mention a few of them here.

Parabolic mirrors   Parabolic mirrors are constructed by rotating part of a parabola about its axis, for example, rotating the arc of $y^2 = 4px$ from point 0 to point A about the x-axis as shown in Fig. 8.15(a).

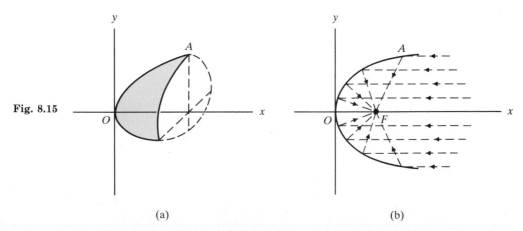

Fig. 8.15

(a)

(b)

Suppose the surface of a reflecting telescope is a parabolic mirror, and we take a cross section, as shown in Fig. 8.15(b), where F is the focus located at $(p, 0)$. If the telescope is pointed at a distant star, the incoming rays of light from the star are essentially parallel to the axis of the parabola. An interesting property of a *parabolic surface* is that each of the rays will be reflected through the focus. This is proved in calculus. Thus the eyepiece of a telescope is located at the focus to get a clear image of the star.

The same principle applies to parabolic radio telescopes. Also, in a spotlight the reflecting surface is a parabolic mirror with the light source at the focus; this produces a parallel beam of light.

Trajectories   When an object (such as a golf ball) moves near the surface of the earth under the influence of gravity only, its path is parabolic. Here we neglect secondary factors, such as air resistance.

Suspension bridges   A suspended cable hangs very nearly in the shape of a parabola if the weight of the cable is negligible in comparison with the weight it supports and the weight is uniformly distributed along the cable. Bridges supported by cables in this fashion are called suspension bridges. Most of the large bridges in the world are this type. An example is the San Francisco Golden Gate Bridge.

Exercises 8.1

In each of the problems 1 through 12, identify the type of curve associated with the given equation. If there is no curve, explain why not. If it is a circle, determine the coordinates of the center and give the radius. If it is a parabola, determine the coordinates of its vertex and focus and the equation of the directrix. Draw a graph of the curve.

1. $x^2 + y^2 - 9 = 0$ 2. $3x^2 + 3y^2 = 16$

3. $y^2 - 8x = 0$ 4. $x^2 + 8y = 0$

5. $x^2 + y^2 - 2x + 6y + 6 = 0$ 6. $x^2 + y^2 + 2x + 2 = 0$

7. $4x^2 + 4y^2 - 12x - 16y + 7 = 0$ 8. $3x^2 + 3y^2 + 8x - 6y + 3 = 0$

9. $4x + y^2 - 2y + 9 = 0$ 10. $x^2 + 2x - 8y + 9 = 0$

11. $12x^2 - 12x + 20y - 17 = 0$ 12. $9y^2 + 24y - 12x + 28 = 0$

In problems 13 through 16, determine the equation of a circle having the given center C and radius r. Give the answer both in standard form and in simplified expanded form.

13. $C:(2, 4)$; $r = 2$ 14. $C:(-3, -1)$; $r = 3$

15. $C:(-2, 0)$; $r = 2$ 16. $C:(\frac{3}{2}, -\frac{1}{2})$; $r = \frac{5}{2}$

In problems 17 through 24, determine the equation of the parabola having the given focus F, directrix D, or vertex V.

17. $F:(2, 0)$; $D: x = -2$ 18. $F:(0, -3)$; $D: y = 3$

19. $V:(0, 0)$; $F:(0, -1)$ 20. $V:(0, 0)$; $F:(-4, 0)$

21. F: $(1, 2)$; D: $x = 3$ **22.** F:$(-3, 1)$; D: $x = -5$

23. V:$(3, -4)$; F:$(3, 0)$ **24.** V:$(2, -1)$; F:$(0, -1)$

25. There are two semicircles having $(-1, 0)$ and $(3, 0)$ as diameter endpoints. Find the equation and draw a graph of each. Does each equation define a function with x as the independent variable?

26. There are two semicircles that have diameters with endpoints $(0, 0)$ and $(-4, 0)$. Find the equation and draw a graph of each of them.

27. The parabola $y^2 = -4x$ consists of an upper half and lower half. Draw a graph of each half, and find the corresponding equation.

28. The parabola $y^2 - 2y - 9x + 1 = 0$ consists of an upper half and a lower half. Draw a graph of each half, and find the corresponding equation.

29. Suppose a driven golf ball travels a distance of 200 meters as measured along the ground, and during its flight it reaches a maximum height of 50 meters. Consider the tee (point from which the ball leaves the ground) as the origin of a coordinate system with positive x-axis along the ground in the direction that the ball takes. Find an equation that describes the path of the ball, assuming that the path is parabolic.

30. The diameter of a parabolic mirror is 20 cm, and it is 10 cm deep at its center. How far is the focus from the vertex?

8.2   ELLIPSE

An ellipse is defined in terms of two given points and a distance, as follows:

Definition 8.3   Let F_1 and F_2 be two given points in the plane, and suppose k is a *number greater than* the distance between F_1 and F_2. The *ellipse* associated with these given quantities is the set of all points P, as shown in Fig. 8.16, such that

$$\overline{F_1P} + \overline{F_2P} = k. \tag{8.11}$$

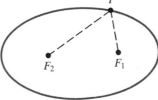

Fig. 8.16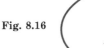

Points F_1 and F_2 are called *foci* of the ellipse. The definition of an ellipse states that the sum of the distances from a point P on the ellipse to the two foci is constant. For instance, if a piece of string k units long having endpoints anchored at F_1 and F_2 is stretched taut with a pencil and a curve is then traced, the result will be an ellipse. The point midway between F_1 and F_2 is referred to as the *center* of the ellipse. If the center is at the origin and the foci are on the x-axis or on the y-axis, we say the ellipse is in *standard position*.

Ellipse in Standard Position

Foci on the x-axis Suppose the two foci are on the x-axis with coordinates $F_1:(c, 0)$ and $F_2:(-c, 0)$, where $c > 0$, and for convenience take $k = 2a$, where $a > c$. Let $P:(x, y)$ be any point on the ellipse, as shown in Fig. 8.17. If we apply Definition 8.3 to this orientation, $\overline{F_1P} + \overline{F_2P} = k$ becomes

$$\sqrt{(x - c)^2 + (y - 0)^2} + \sqrt{(x + c)^2 + (y - 0)^2} = 2a. \qquad (8.12)$$

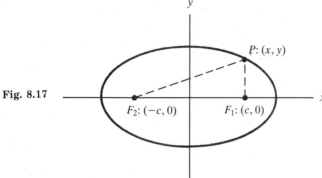

Fig. 8.17

This can be written as

$$\sqrt{(x - c)^2 + y^2} = 2a - \sqrt{(x + c)^2 + y^2}.$$

Squaring both sides of this equation and simplifying gives

$$a\sqrt{(x + c)^2 + y^2} = a^2 + cx.$$

Again squaring both sides, simplifying, and rearranging, we get

$$(a^2 - c^2)x^2 + a^2y^2 = a^2(a^2 - c^2). \qquad (8.13)$$

Since $a > c > 0$, $a^2 - c^2 > 0$. For convenience, denote $a^2 - c^2$ by b^2, where $b > 0$. That is,

$$b = \sqrt{a^2 - c^2} \qquad (8.14)$$

Then Eq. (8.13) can be written as $b^2x^2 + a^2y^2 = a^2b^2$. Dividing both sides by a^2b^2 gives the following standard form.

$$\frac{x^2}{a^2} + \frac{y^2}{b^2} = 1. \qquad (8.15)$$

Thus Eq. (8.15) represents an *ellipse in standard position with foci on the x-axis* having coordinates $F_1:(c, 0)$, $F_2:(-c, 0)$, where a, b, and c are related by

$$c^2 = a^2 - b^2. \tag{8.16}$$

The ellipse given by Eq. (8.15) is symmetric with respect to the x-axis, the y-axis, and the origin. The x-intercepts are $A:(a, 0)$ and $B:(-a, 0)$, and the y-intercepts are $C:(0, b)$ and $D:(0, -b)$, as shown in Fig. 8.18. The line segment AB is called the *major axis,* and each of the endpoints A and B is called a *vertex* of the ellipse; the line segment CD is called the *minor axis* of the ellipse. Note that a is always greater than b since $a^2 = b^2 + c^2$. Also note that the foci lie on the major axis.

Fig. 8.18

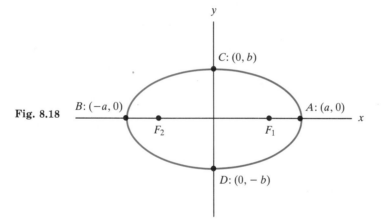

Foci on the y-axis   Suppose we take the foci on the y-axis as $F_1:(0, c)$, $F_2:(0, -c)$ and proceed as above; the resulting equation is

$$\frac{x^2}{b^2} + \frac{y^2}{a^2} = 1, \tag{8.17}$$

where again $c^2 = a^2 - b^2$.

The ellipse represented by Eq. (8.17) is shown in Fig. 8.19; the vertices are $A:(0, a)$ and $B:(0, -a)$. The major axis is line segment AB, and the minor axis is line segment CD. Note that Eqs. (8.15) and (8.17) are equations for ellipses in *standard form.* The right side is always 1, and a^2 is the larger of the two denomi-

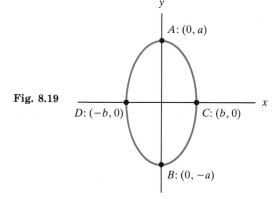

Fig. 8.19

nators on the left side. Also the *length of the major axis is 2a* and the *length of the minor axis is 2b*.

Example 1   Draw a graph of $x^2 + 4y^2 = 16$. Determine the coordinates of the vertices and foci. How long is the major axis?

Solution   First write the given equation in standard form. Divide both sides by 16 to get

$$\frac{x^2}{16} + \frac{y^2}{4} = 1.$$

The larger of the two denominators is 16, so take $a^2 = 16$ and $b^2 = 4$. Thus the vertices are given by $V_1:(4, 0)$, $V_2:(-4, 0)$. The major axis is on the x axis; hence the foci are on the x-axis at $(c, 0)$ and $(-c, 0)$. Substituting 4 for a and 2 for b into $c^2 = a^2 - b^2$ gives $c = 2\sqrt{3}$. Therefore $F_1:(2\sqrt{3}, 0)$ and $F_2:(-2\sqrt{3}, 0)$ are coordinates of the foci. The graph is shown in Fig. 8.20. The major axis is the length of the segment between V_1 and V_2; its length is 8.

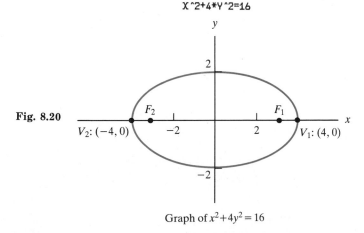

Fig. 8.20

Graph of $x^2 + 4y^2 = 16$

Ellipse Not in Standard Position

Suppose the ellipse shown in Fig. 8.18 is moved horizontally and vertically (this is called a *translation*) so that the center is located at a point (h, k) in the plane (see Fig. 8.21). In this position the equation of the ellipse becomes

$$\frac{(x - h)^2}{a^2} + \frac{(y - k)^2}{b^2} = 1. \qquad (8.18)$$

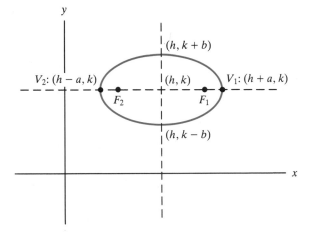

Fig. 8.21

Equation (8.18) represents an ellipse with center at (h, k); vertices at V_1:$(h + a, k)$, V_2:$(h - a, k)$; foci at F_1:$(h + c, k)$ and F_2:$(h - c, k)$. Again $c^2 = a^2 - b^2$. The major axis is the line segment V_1V_2.

Similarly, if the ellipse in Fig. 8.19 is translated so that its center is at (h, k), the corresponding equation is

$$\frac{(x - h)^2}{b^2} + \frac{(y - k)^2}{a^2} = 1. \qquad (8.19)$$

The graph of Eq. (8.19) is shown in Fig. 8.22. The vertices are V_1:$(h, k + a)$, V_2:$(h, k - a)$, and the foci are F_1:$(h, k + c)$, F_2:$(h, k - c)$, where once again c is given by $c^2 = a^2 - b^2$.

Equations (8.18) and (8.19) are called *standard form* equations of an ellipse *with center at* (h, k). If each of these equations is written in *expanded form,* the result is an equation of the type

$$Ax^2 + Cy^2 + Dx + Ey + F = 0,$$

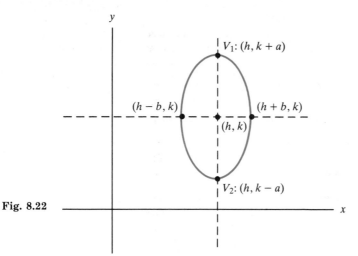

Fig. 8.22

where A and C have the same sign. If $A = C$, then the curve is a circle, which can be considered as a special case of an ellipse where the foci F_1 and F_2 coincide (at the center of the circle).

Example 2 Draw a graph of the ellipse whose equation is

$$9x^2 + 4y^2 - 18x + 16y - 11 = 0.$$

Solution Give the coordinates of the center, the vertices, and the foci.

The first step is to write the given equation in standard form. Thus we complete the square on the x terms and the y terms, as follows:

$$9(x^2 - 2x) + 4(y^2 + 4y) = 11$$
$$9(x^2 - 2x + 1) + 4(y^2 + 4y + 4) = 11 + 9 + 16$$
$$9(x - 1)^2 + 4(y + 2)^2 = 36.$$

Dividing both sides by 36 gives

$$\frac{(x - 1)^2}{4} + \frac{(y + 2)^2}{9} = 1$$

Since 9 is the larger of the two denominators on the left, we take $a^2 = 9$, $b^2 = 4$. This equation fits the standard form given by Eq. (8.19). Therefore the center is at $(1, -2)$; vertices are V_1:$(1, -2 + 3) = (1, 1)$ and V_2:$(1, -2 - 3) = (1, -5)$. To get the foci we need c, which is given by $c^2 = a^2 - b^2 = 9 - 4 = 5$. Thus $c = \sqrt{5}$, and so the foci are F_1:$(1, -2 + \sqrt{5})$ and F_2:$(1, -2 - \sqrt{5})$. The graph is shown in Fig. 8.23.

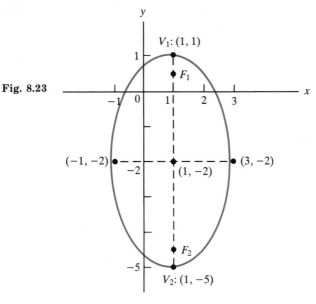

9*X^2+4*Y^2−18*X+16*Y−11=0

Fig. 8.23

Graph of $\dfrac{(x-1)^2}{4} + \dfrac{(y+2)^2}{9} = 1$

Example 3   Determine an equation of the ellipse with vertices $V_1\!:\!(1,2)$, $V_2\!:\!(-3,2)$ and length of minor axis equal to 3. Draw a graph and find the coordinates of the foci.

Solution   The given information suggests an equation of the type given by Eq. (8.18). We need only the given information to make a sketch of the graph, as shown in Fig. 8.24. This will help in finding h, k, a, and b. The center will be halfway between V_1 and V_2, which is the point $C\!:\!\left(\dfrac{1+(-3)}{2}, 2\right)$ or $(-1, 2)$. Therefore $h = -1$ and $k = 2$. Also the distance between the two vertices is 4, and this equals $2a$; thus $a = 2$. The length of the minor axis is given as 3, and so $2b = 3$ or $b = 3/2$. Substituting this information into Eq. (8.18) gives

$$\frac{(x+1)^2}{2^2} + \frac{(y-2)^2}{(3/2)^2} = 1$$

as the desired equation. To find the coordinates of the foci, first determine c as follows:

$$c = \sqrt{a^2 - b^2} = \sqrt{2^2 - \left(\frac{3}{2}\right)^2} = \sqrt{4 - \frac{9}{4}} = \frac{\sqrt{7}}{2}.$$

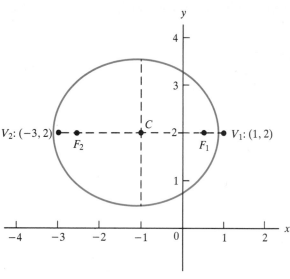

9*X^2+16*Y^2+18*X-64*Y+37=0

Graph of $\dfrac{(x+1)^2}{4} + \dfrac{4(y-2)^2}{9} = 1$

Fig. 8.24

Therefore the coordinates of the foci are

$$F_1\!\!:\!\left(-1 + \frac{\sqrt{7}}{2}, 2\right) \quad \text{and} \quad F_2\!\!:\!\left(-1 - \frac{\sqrt{7}}{2}, 2\right).$$

Applications of Ellipse

Elliptic domes Some buildings have domes that are elliptical in shape. The interior of the dome can be considered to be formed by an ellipse that is revolved about its major axis. Sound emanating from one focus is reflected from any point on the dome through the other focus; thus a whisper at one focus can be heard clearly by a person located at the other focus but not necessarily by others in the room. A room with an elliptical-shaped dome is commonly referred to as a "whispering gallery." Both the Capitol in Washington and the Mormon Tabernacle in Salt Lake City have whispering galleries. Historical rumor suggests that John C. Calhoun was aware of this property of the Statutory Hall, where the House of Representatives met in his time, and used it to eavesdrop on his adversaries.

Orbits Johannes Kepler discovered that the planets travel in (very nearly) an elliptical orbit around the sun, with the sun at one focus. Artificial satellites travel about the earth in elliptical orbits.

Exercises 8.2

In problems 1 through 12, find the coordinates of the vertices and the foci for the given ellipse, and sketch a graph.

1. $\dfrac{x^2}{9} + \dfrac{y^2}{4} = 1$

2. $\dfrac{x^2}{4} + \dfrac{y^2}{16} = 1$

3. $9x^2 + y^2 = 16$

4. $x^2 + 2y^2 = 16$

5. $\dfrac{(x-1)^2}{4} + \dfrac{(y+1)^2}{16} = 1$

6. $\dfrac{x^2}{8} + \dfrac{(y-1)^2}{2} = 1$

7. $(x + \frac{3}{2})^2 + 16(y - \frac{3}{2})^2 = 16$

8. $4(x + 1.4)^2 + (y + 2.5)^2 = 4$

9. $9x^2 + 4y^2 - 18x - 27 = 0$

10. $4x^2 + 9y^2 - 4x + 24y - 19 = 0$

11. $x^2 + 2y^2 + 2x - 8y + 5 = 0$

12. $x^2 + 4y^2 - 8y - 8 = 0$

In problems 13 through 20, determine an equation of the ellipse described by the given properties. In each case sketch the ellipse.

13. The foci are F_1:(3, 0), F_2:(−3, 0), and vertices are V_1:(5, 0), V_2:(−5, 0).

14. The foci are F_1:(0, 2), F_2:(0, −2), and vertices are V_1:(0, 4), V_2:(0, −4).

15. The foci are (3, 2) and (3, −2), and the length of the major axis is 6.

16. The vertices are (1, 5) and (1, 1), and the length of the minor axis is 3.

17. The vertices are (3, −1), (−1, −1), and the ellipse passes through (1, 0).

18. The ellipse is in standard position and passes through points (3, 1) and (1, $\sqrt{3}$).

19. The center is at (3, −1), the major axis is horizontal of length 4, and a focus is at (4.5, −1).

20. The center is at (−3, 0), one vertex is at (−3, 3), and the length of the minor axis is 4.

8.3   HYPERBOLA

The definition of a hyperbola is similar to that of the ellipse in that two points and a distance are given.

Definition 8.4   Let F_1 and F_2 be two given points in the plane, and suppose k is a positive *number less than* the distance between F_1 and F_2. The hyperbola associated with these quantities is the set of all points P such that

$$\overline{F_1 P} - \overline{F_2 P} = \pm k. \tag{8.20}$$

The diagram in Fig. 8.25 illustrates two points P_1 and P_2 on the hyperbola where $\overline{F_1 P_1} - \overline{F_2 P_1} = -k$ and $\overline{F_1 P_2} - \overline{F_2 P_2} = k$. As with the ellipse, each of the given points is called a *focus,* and the point C midway between F_1 and F_2 is called the

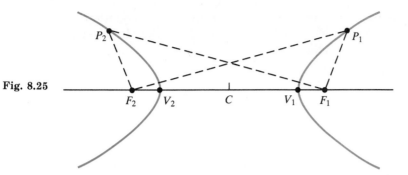

Fig. 8.25

center of the hyperbola. The line through F_1 and F_2 is called the *principal* axis of the hyperbola. The two points at which the hyperbola intersects the principal axis are called the *vertices* of the hyperbola. If the center is at the origin and the foci are on either the x- or the y-axis, then the hyperbola is said to be in *standard position*.

Hyperbola in Standard Position

Foci on the x-axis   Suppose the foci are the points F_1:$(c, 0)$ and F_2:$(-c, 0)$, where c is a given positive number, and the constant k in Definition 8.4 is taken for convenience to be $k = 2a$, where $0 < a < c$. If we apply the definition for this orientation (as shown in Fig. 8.26). Eq. (8.20) becomes

$$\overline{F_1P} - \overline{F_2P} = 2a \quad \text{(for points such as } P\text{),}$$
$$\overline{F_1Q} - \overline{F_2Q} = -2a \quad \text{(for points such as } Q\text{).}$$

(8.21)

That is, the hyperbola consists of all points (x, y) such that

$$\sqrt{(x - c)^2 + (y - 0)^2} - \sqrt{(x + c)^2 + (y - 0)^2} = \pm 2a. \qquad (8.22)$$

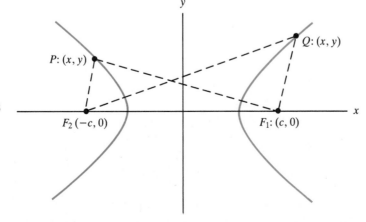

Fig. 8.26

In the preceding section we encountered an equation similar to this (see Eq. 8.12), which we simplified by squaring twice. In an analogous manner, Eq. (8.22) can be written as

$$\frac{x^2}{a^2} - \frac{y^2}{b^2} = 1, \tag{8.23}$$

where $b^2 = c^2 - a^2$. Remember the distance between F_1 and F_2, equal to $2c$, is greater than $2a$, so $c^2 - a^2 > 0$. When $y = 0$, $x = \pm a$, and so the vertices are given by $V_1:(a, 0)$ and $V_2:(-a, 0)$. It can be shown that the lines $y = (b/a)x$ and $y = -(b/a)x$ are *oblique asymptotes*. This fact is useful in drawing the graph of a hyperbola.

Foci on the y-axis   If the foci are $F_1:(0, c)$, $F_2:(0, -c)$, then the hyperbola is as shown in Fig. 8.27, and its equation is the same as Eq. (8.23) with x and y interchanged,

$$\frac{y^2}{a^2} - \frac{x^2}{b^2} = 1, \tag{8.24}$$

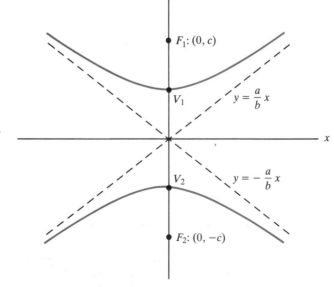

Fig. 8.27

where again $b^2 = c^2 - a^2$. The vertices are $V_1:(0, a)$ and $V_2:(0, -a)$. The lines $y = (a/b)x$ and $y = -(a/b)x$ are oblique asymptotes.

Example 1   Draw a graph of $y^2/4 - x^2/9 = 1$, and determine the coordinates of the vertices and the foci.

Solution   The given equation is a particular case of Eq. (8.24) with $a = 2$, $b = 3$, and so it represents a hyperbola with vertices at $V_1:(0, 2)$ and $V_2:(0, -2)$. To determine the coordinates of the foci, we need c; since $c^2 = a^2 + b^2$, $c = \sqrt{4 + 9} = \sqrt{13}$. The foci are given by $F_1:(0, \sqrt{13})$, $F_2:(0, -\sqrt{13})$. The graph can be drawn by plotting V_1 and V_2 along with several pairs of values of x, y that satisfy the equation. Also it is helpful to draw the asymptotes $y = (2/3)x$ and $y = -(2/3)x$, as shown in Fig. 8.28.

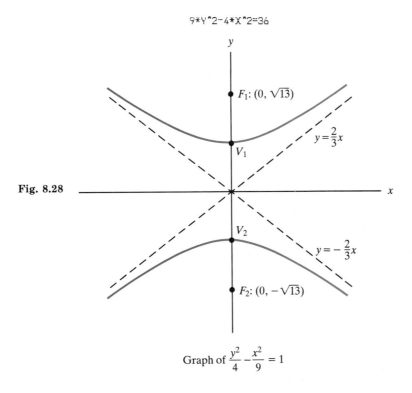

Fig. 8.28

Graph of $\dfrac{y^2}{4} - \dfrac{x^2}{9} = 1$

Hyperbola Not in Standard Position

Hyperbola with horizontal axis   If the hyperbola shown in Fig. 8.26 is translated in such a way that the center is located at (h, k), as shown in Fig. 8.29, an

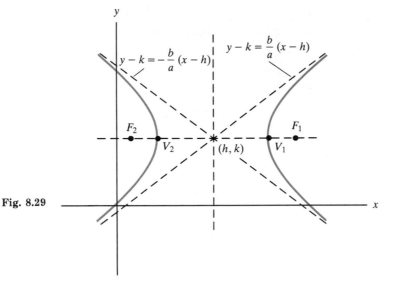

Fig. 8.29

equation for the resulting hyperbola is given by

$$\frac{(x-h)^2}{a^2} - \frac{(y-k)^2}{b^2} = 1. \tag{8.25}$$

The coordinates of the vertices are V_1:$(h + a, k)$, V_2:$(h - a, k)$, and the coordinates of the foci are F_1:$(h + c, k)$ and F_2:$(h - c, k)$, where c is given by $c^2 = a^2 + b^2$. The hyperbola has two oblique asymptotes; these are lines through (h, k) having slope b/a and $-b/a$, respectively.

Hyperbola with vertical axis   If the hyperbola shown in Fig. 8.27 is translated in such a way that its center is at (h, k), as shown in Fig. 8.30, the corresponding equation is

$$\frac{(y-k)^2}{a^2} - \frac{(x-h)^2}{b^2} = 1. \tag{8.26}$$

The coordinates of the vertices are V_1:$(h, k + a)$, V_2:$(h, k - a)$, and the coordinates of the foci are F_1:$(h, k + c)$ and F_2:$(h, k - c)$, where $c^2 = a^2 + b^2$. The hyperbola described here also has two *oblique asymptotes,* the lines through (h, k) with slope a/b and $-a/b$, respectively.

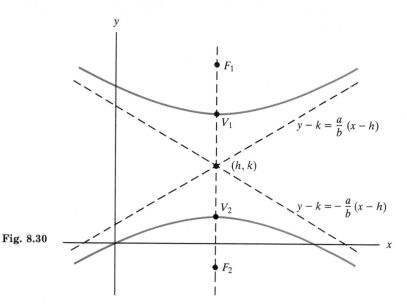

Fig. 8.30

Equations (8.25) and (8.26) are called *equations of hyperbolas in standard form* with center at (h, k). If each of these is written in expanded form, the result will be an equation of the type

$$Ax^2 + Cy^2 + Dx + Ey + F = 0,$$

where A and C have opposite signs.

Example 2   Draw a graph of $(x - 1)^2/4 - (y + 1)^2/9 = 1$. Give the coordinates of the center, the vertices, and the foci. Also give the equations of the asymptotes.

Solution   The given equation is a particular case of Eq. (8.25), and so the graph is a hyperbola with horizontal axis, where $h = 1$, $k = -1$, $a = 2$, and $b = 3$. The center is at $(1, -1)$; the vertices are $V_1:(3, -1)$, $V_2:(-1, -1)$. To determine the foci, we need c, which is given by $c = \sqrt{4 + 9} = \sqrt{13}$. Therefore the coordinates of the foci are $F_1:(1 + \sqrt{13}, -1)$, $F_2:(1 - \sqrt{13}, -1)$. The asymptotes are two lines through $(1, -1)$ having slopes $\frac{3}{2}$ and $-\frac{3}{2}$, respectively. The equations are given by

$$l_1: y - (-1) = (3/2)(x - 1),$$

which simplifies to $3x - 2y = 5$; and

$$l_2: 3x + 2y = 1.$$

The graph is shown in Fig. 8.31.

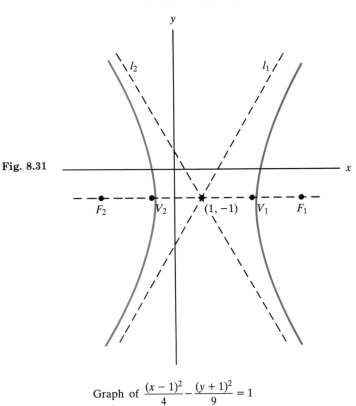

$$9*X\char94 2-4*Y\char94 2-18*X-8*Y-31=0$$

Fig. 8.31

Graph of $\dfrac{(x-1)^2}{4} - \dfrac{(y+1)^2}{9} = 1$

Example 3   Draw a graph of the hyperbola given by the equation $4x^2 - y^2 - 2y + 3 = 0$. Give the coordinates of the center, vertices, and foci. Also determine the equation of the asymptotes.

Solution   By completing the square in y, we get

$$4x^2 - (y^2 + 2y) = -3$$
$$4x^2 - (y^2 + 2y + 1) = -3 - 1$$
$$4(x - 0)^2 - (y + 1)^2 = -4.$$

Dividing both sides by -4 gives an equation of the hyperbola in standard form:

$$\frac{(y + 1)^2}{4} - \frac{(x - 0)^2}{1} = 1.$$

This is a particular case of Eq. (8.26) with $h = 0$, $k = -1$, $a = 2$, and $b = 1$. The hyperbola opens up and down (has a vertical principal axis) with center at $(0, -1)$ and vertices V_1:$(0, 1)$ and V_2:$(0, -3)$. To determine the foci, first find c, using $c^2 = a^2 + b^2$; $c = \sqrt{4+1} = \sqrt{5}$, and so the foci are $F_1:(0, -1 + \sqrt{5})$, $F_2:(0, -1 - \sqrt{5})$. The asymptotes are two lines through $(0, -1)$ having slopes $\pm\frac{2}{1}$; that is, $(y + 1)/x = 2$ or $(y + 1)/x = -2$. Thus the asymptotes are

$$l_1: y = 2x - 1 \qquad \text{and} \qquad l_2 = -2x - 1.$$

The graph is shown in Fig. 8.32.

4*X^2-Y^2-2*Y+3=0

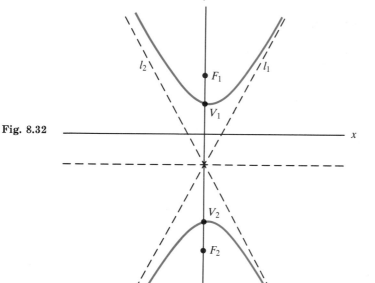

Fig. 8.32

Graph of $4x^2 - y^2 - 2y + 3 = 0$

Application of Hyperbola

The location of a distant source of sound can be determined by placing sound receiving devices at different locations. Suppose two receivers are located at points F_1 and F_2, and suppose sound waves emanating from a source such as thunder or a cannon arrive at F_1 and F_2 at different times. The difference in these times is sufficient to determine a hyperbola with foci at F_1 and F_2 on which the sound source lies. If still another receiver (say, F_3) is used, then in a similar manner its location along with that of one of the other receivers will determine a second hyperbola on which the source lies. The intersection of these hyperbolas will determine the location of the sound source, as illustrated in Fig. 8.33.

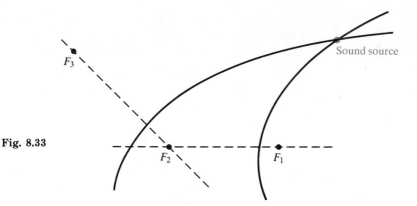

Fig. 8.33

Exercises 8.3

In problems 1 through 12, do the following.

a) Find the coordinates of the center, vertices, and foci of the given hyperbola;

b) Give equations of the asymptotes;

c) Sketch a graph.

1. $\dfrac{x^2}{9} - \dfrac{y^2}{4} = 1$

2. $\dfrac{x^2}{4} - \dfrac{y^2}{16} = 1$

3. $y^2 - 8x^2 = 16$

4. $x^2 - 2y^2 + 16 = 0$

5. $\dfrac{(x-1)^2}{4} - \dfrac{(y+1)^2}{9} = 1$

6. $\dfrac{x^2}{8} - \dfrac{(y-1)^2}{2} = 1$

7. $(x+2)^2 - (y-1)^2 + 4 = 0$

8. $4(x - \tfrac{1}{2})^2 - 2(y + \tfrac{3}{2})^2 = 4$

9. $x^2 - 4y^2 + 2x + 16y - 19 = 0$

10. $8x^2 - 4y^2 + 8x + 16y - 18 = 0$

11. $4x^2 - y^2 + 8x + 3y + \tfrac{11}{4} = 0$

12. $x^2 - 4y^2 - 12y + 7 = 0$

In problems 13 through 20, determine an equation of the hyperbola having the given properties. In each case sketch the hyperbola.

13. Foci $F_1:(3, 0)$; $F_2:(-3, 0)$; vertices $V_1:(2, 0)$; $V_2:(-2, 0)$

14. Foci $F_1:(0, 4)$; $F_2:(0, -4)$; vertices $V_1:(0, 2)$; $V_2:(0, -2)$

15. Vertices $V_1:(3, 2)$, $V_2:(3, -2)$ and passing through the point $(4, 4)$.

16. Center $(-2, 1)$; a vertex at $(-2, 3)$ and focus at $(-2, 4)$.

17. Center at $(3, -1)$; a vertex at $(1, -1)$ and a focus at $(0, -1)$.

18. Foci $(4, 2)$, $(-4, 2)$; a vertex $V_1:(2, 2)$.

19. Vertices $V_1:(3, 0)$, $V_2:(-3, 0)$ and asymptote $y = x$.

20. Vertices $V_1:(0, 1)$, $V_2:(0, -1)$; and asymptote $y = 2x$.

TRANSLATION AND ROTATION OF AXES

When a curve is given in the plane, the equation describing it will depend on the position of the coordinate system we choose. Ordinarily we are interested in using the coordinate system that will give us the simplest equation. On the other hand, for a given equation graphing is frequently made easier by changing from one set of axes to another. Transformations involving a translation or a rotation of coordinates are often helpful.

Translation of Coordinates

Suppose we have two systems of rectangular coordinates, the x, y system and the X, Y system, that are positioned so that the x-axis and the X-axis are parallel and the y-axis and the Y-axis are parallel, as shown in Fig. 8.34. A change from the x, y system to the X, Y system can be thought of as moving the axes horizontally a directed distance h and vertically a directed distance k. Such a transformation is called *translation of coordinates*.

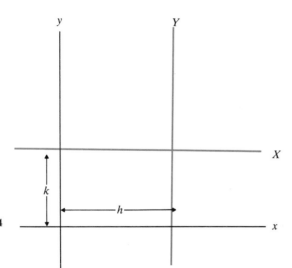

Fig. 8.34

Each point P in the plane has associated with it a pair of numbers relative to the x, y system and a different pair of numbers relative to the X, Y system. In order to avoid confusion as to which pair we mean, we shall use (x, y) to denote the coordinates of a point in the x, y system and $[X, Y]$ to denote the coordinates of the same point relative to the X, Y system. For instance, in Fig. 8.35, Q is the origin of the X, Y system and has coordinates (h, k) in the x, y system and $[0, 0]$ in the X, Y system. Similarly, the coordinates of O, the origin of the x, y system, are denoted either by $(0, 0)$ or $[-h, -k]$.

Let us first derive transformation equations that will enable us to determine the name of a point in one system if it is known in the other system. Let point P

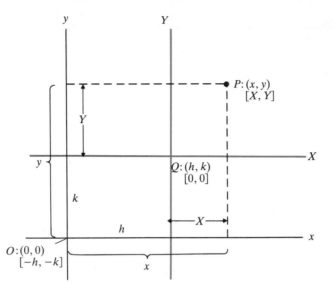

Fig. 8.35

be any point in the plane. From the diagram in Fig. 8.35 we see that

$$\begin{cases} x = X + h, \\ y = Y + k. \end{cases} \tag{8.27}$$

Solving these equations for X and Y gives

$$\begin{cases} X = x - h, \\ Y = y - k. \end{cases} \tag{8.28}$$

Equations (8.27) and (8.28) are called *transformation equations* relating the x, y system and X, Y system of coordinates.

Example 1   Suppose the x, y coordinate system is translated in such a way that the origin of the new system is at the point $(-2, 3)$.

a) Draw the new coordinates.

b) Give the transformation equations.

c) Given points A and B as $A:(3, 5)$ and $B:[-3, -1]$, locate these points in your figure and label each with names in both coordinate systems.

Solution   a) The two coordinates systems are shown in Fig. 8.36, where O is the origin of the original system and Q is the origin of the new system.

b) Substituting $h = -2$ and $k = 3$ into the transformation equation (8.27) and (8.28) gives

$$\begin{cases} x = X - 2, \\ y = Y + 3, \end{cases} \quad \text{or} \quad \begin{cases} X = x + 2, \\ Y = y - 3. \end{cases}$$

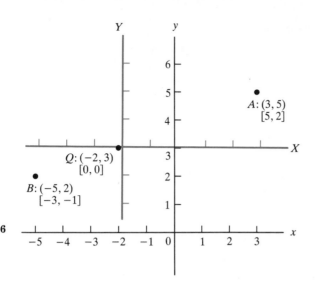

Fig. 8.36

c) For point A: $x = 3$, $y = 5$. Substituting these into the transformation equations $X = x + 2$, $Y = y - 3$ gives $X = 3 + 2 = 5$, $Y = 5 - 3 = 2$. Thus the point A:$(3, 5)$ can also be denoted by A:$[5, 2]$. Similarly, B:$[-3, -1]$ can be written as B:$(-5, 2)$ by using the transformation equations $x = X - 2$, $y = Y + 3$, where $X = -3$, $Y = -1$. Points A and B are shown in Fig. 8.36.

Example 2   In Section 8.2 we saw that an equation such as

$$9x^2 + 4y^2 + 36x - 24y + 36 = 0$$

represents an ellipse. By a translation of coordinates, find a simpler equation for this curve; then draw a graph of the equation.

Solution   First write the given equation in standard form by completing the square on the x terms and the y terms, as follows:

$$9(x^2 + 4x) + 4(y^2 - 6y) = -36$$
$$9(x^2 + 4x + 4) + 4(y^2 - 6y + 9) = -36 + 36 + 36$$
$$9(x + 2)^2 + 4(y - 3)^2 = 36$$
$$\frac{(x + 2)^2}{4} + \frac{(y - 3)^2}{9} = 1. \tag{8.29}$$

This form suggests that we should choose the X, Y system of coordinates such that

$$\begin{cases} X = x + 2, \\ Y = y - 3. \end{cases}$$

Then Eq. (8.29) becomes

$$\frac{X^2}{4} + \frac{Y^2}{9} = 1.$$

This is an equation for an ellipse in *standard position relative to the X, Y coordinate system.* The graph, relative to the X, Y system, can be quickly drawn. Now sketch in the x, y coordinates using $x = 0$ and $y = 0$ to find the coordinates of the origin of the x, y system relative to the X, Y system: $X = 0 + 2 = 2$ and $Y = 0 - 3 = -3$. Therefore the origin of the x, y system has coordinates $[2, -3]$, and so we sketch the x, y axes through this point parallel, respectively, to the X and Y axes. Figure 8.37 shows a completed graph. If the X, Y axes are erased (or ignored), the graph relative to the original coordinates remains.

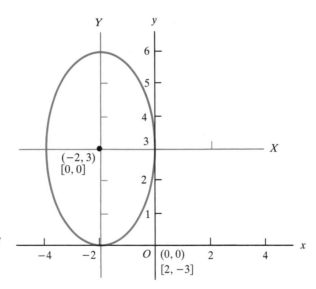

9*X^2+4*Y^2+36*X−24*Y+36=0

Fig. 8.37

Graph of $9x^2 + 4y^2 + 36x - 24y + 36 = 0$

In this and preceding sections all the equations we have encountered are of the form

$$Ax^2 + Cy^2 + Dx + Ey + F = 0, \tag{8.30}$$

where not both A and C are zero. The type of curve represented by Eq. (8.30), except for degenerate cases, is determined by the values of the coefficients A and C:

Parabola: if $A \cdot C = 0$ (one of A or C is zero)

Ellipse: if $A \cdot C > 0$ (A and C have the same sign)

Hyperbola: if $A \cdot C < 0$ (A and C have opposite signs)

At the beginning of this chapter we indicated that the general quadratic equation in x and y,

$$Ax^2 + Bxy + Cy^2 + Dx + Ey + F = 0,$$

represents a conic section, except for degenerate cases. In the remaining portion of this section we shall illustrate how a rotation of coordinates can be used to transform an equation of this form into an equation of the type given by Eq. (8.30). The presence of the Bxy term, with $B \neq 0$, will suggest that the equation represents a conic section with axes that are not horizontal or vertical. First let us introduce the idea of changing coordinates by rotation.

Rotation of Coordinates

Suppose we consider two systems of rectangular coordinates in which the x', y' system is obtained by rotating the x, y system about the origin counterclockwise through an acute angle θ, as shown in Fig. 8.38. Each point P in the plane can be given either by coordinates relative to the x, y system or the x', y' system. In order to avoid confusion, let (x, y) denote the coordinates of P relative to x, y and $\langle x', y' \rangle$ relative to x', y'.

Fig. 8.38

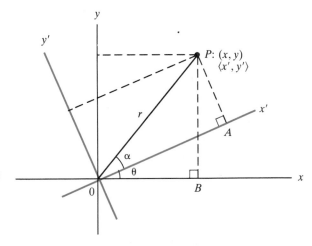

Let $r = \overline{OP}$ and α be as shown in Fig. 8.38, where $x' = \overline{OA}$, $y' = \overline{AP}$, and $x = \overline{OB}$, $y = \overline{BP}$. From right triangle OAP we have

$$x' = r \cos \alpha \qquad \text{and} \qquad y' = r \sin \alpha. \qquad (8.31)$$

Also from right triangle OBP we get

$$x = r \cos(\alpha + \theta) \qquad \text{and} \qquad y = r \sin(\alpha + \theta). \qquad (8.32)$$

Using trigonometric identities (see **I.12, I.14** on p. 279), we can write the equations in (8.32) as

$$x = r \cos \alpha \cos \theta - r \sin \alpha \sin \theta,$$
$$y = r \sin \alpha \cos \theta + r \cos \alpha \sin \theta.$$

Using (8.31), we can replace $r \cos \alpha$ by x' and $r \sin \alpha$ by y' to get

$$x = (\cos \theta)x' - (\sin \theta)y',$$
$$y = (\sin \theta)x' + (\cos \theta)y'. \tag{8.33}$$

The equations in (8.33) are *transformation equations* from x', y' to x, y associated with *rotating* the x, y coordinate axes counterclockwise through an acute angle θ.

Solving the system of equations in (8.33) for x and y gives the corresponding transformation equations from the x, y system to the x', y' system (see problem 20):

$$x' = (\cos \theta)x + (\sin \theta)y,$$
$$y' = (-\sin \theta)x + (\cos \theta)y. \tag{8.34}$$

Example 3 Suppose the x, y system of rectangular coordinates is rotated counterclockwise through an angle of 30°.

a) Draw a diagram showing the two systems of coordinates.

b) Locate points A and B, given by $A:(4, 3)$ and $B:\langle -3, 5 \rangle$, and label them with reference to each coordinate system; round off coordinates to two decimal places.

Solution **a)** The two systems of coordinates are shown in Fig. 8.39.

b) Locate points A and B by using $A:(4, 3)$ and $B:\langle -3, 5 \rangle$. The transformation equations given in (8.33) and (8.34) are

$$\begin{cases} x = (\cos 30°)x' - (\sin 30°)y' = \dfrac{\sqrt{3}}{2}x' - \dfrac{1}{2}y', \\[2mm] y = (\sin 30°)x' + (\cos 30°)y' = \dfrac{1}{2}x' + \dfrac{\sqrt{3}}{2}y'; \\[2mm] x' = (\cos 30°)x + (\sin 30°)y = \dfrac{\sqrt{3}}{2}x + \dfrac{1}{2}y, \\[2mm] y' = (-\sin 30°)x + (\cos 30°)y = -\dfrac{1}{2}x + \dfrac{\sqrt{3}}{2}y. \end{cases}$$

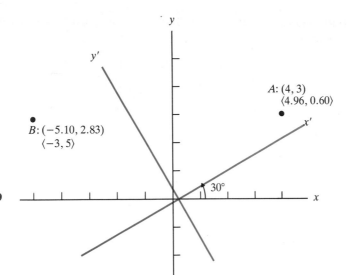

Fig. 8.39

For point A: $x = 4$, $y = 3$; so

$$x' = \left(\frac{\sqrt{3}}{2}\right) \cdot 4 + \left(\frac{1}{2}\right) \cdot 3 = 4.96,$$

$$y' = \left(-\frac{1}{2}\right) \cdot 4 + \left(\frac{\sqrt{3}}{2}\right) \cdot 3 = 0.60.$$

Thus A is also given by $A:\langle 4.96, 0.60 \rangle$ (rounded off to two decimals). Similarly, for B:

$$x = \left(\frac{\sqrt{3}}{2}\right)(-3) - \left(\frac{1}{2}\right)(5) = -5.10,$$

$$y = \left(\frac{1}{2}\right)(-3) + \left(\frac{\sqrt{3}}{2}\right)(5) = 2.83.$$

Thus the "name" of point B in the x, y system is $(-5.10, 2.83)$.   ▬

Example 4   Suppose the x, y coordinate system is rotated through an angle of $\theta = 45°$. Find the equation relative to the new system of coordinates that corresponds to $xy = 1$. Draw the graph of this equation.

Solution   Substituting $\theta = 45°$ into the equations in (8.33) gives

$$x = \frac{1}{\sqrt{2}}(x' - y'), \qquad y = \frac{1}{\sqrt{2}}(x' + y').$$

Thus the given equation, $xy = 1$, becomes

$$\frac{1}{\sqrt{2}}(x' - y') \cdot \frac{1}{\sqrt{2}}(x' + y') = 1.$$

Simplifying this gives $(x')^2 - (y')^2 = 2$. In standard form this becomes

$$\frac{(x')^2}{(\sqrt{2})^2} - \frac{(y')^2}{(\sqrt{2})^2} = 1.$$

We recognize this as the equation of a hyperbola with vertices at $\langle\sqrt{2},0\rangle$, $\langle -\sqrt{2},0\rangle$, having asymptotes $y' = \pm x'$. Note that $y' = x'$ is the y-axis, and $y' = -x'$ is the x-axis. Now sketch the curve relative to the x', y' coordinates, as shown in Fig. 8.40. Of course, this is also a graph of the original equation. That is, the graph of $xy = 1$ is a hyperbola with the line $y = x$ as the principal axis.

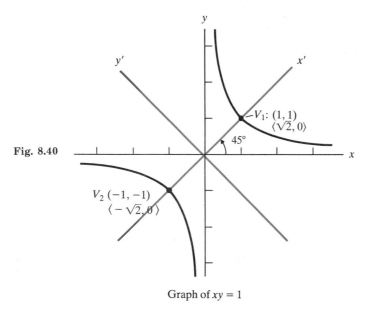

Fig. 8.40

Graph of $xy = 1$

Exercises 8.4

In problems 1 through 4, suppose the x, y coordinate system is translated in such a way that the origin Q of the new system is the given point.

a) Draw the two coordinate systems.

b) Give transformation equations corresponding to Eqs. (8.27) and (8.28).

c) Locate the points $A:(3, 4)$ and $B:[3, -2]$, and label each with names in both coordinate systems.

1. $Q:(1, 3)$	**2.** $Q:(5, 3)$	**3.** $Q:(2, -3)$	**4.** $Q:(-3, -1)$

In problems 5 through 10, by a suitable translation of coordinates express the given equation in terms of new coordinates giving an equation of the conic in reduced standard form. Use your result to draw a graph.

5. $x^2 + 2y^2 + 4x - 8y - 4 = 0$

6. $x^2 + 6x - y + 4 = 0$

7. $y^2 + 4y - 3x + 1 = 0$

8. $x^2 - y^2 - 4x + 2y + 8 = 0$

9. $x^2 + y^2 + 6x - 2y + 6 = 0$

10. $x^2 + 4y^2 - 4x - 8y - 8 = 0$

In problems 11 through 14, the x, y axes are rotated counterclockwise through an angle of $45°$ to give a new set of axes (see Eqs. (8.33) and (8.34)). Points are labeled (x, y) and $\langle x', y' \rangle$ relative to the original and new axes, respectively. In each case a given point is identified by a name relative to one of the coordinate systems. Find the name of the point relative to the other axes. Round off results to one decimal place.

11. a) $(0, 0)$ **b)** $\langle 0, 0 \rangle$ **12. a)** $(1, 2)$ **b)** $\langle 3, 5 \rangle$
13. a) $(-3, 2)$ **b)** $\langle 4, -1 \rangle$ **14. a)** $(3, -5)$ **b)** $\langle 3, -5 \rangle$

In each of the problems 15 through 19, an equation of a curve is given relative to the x, y system of coordinates. By a rotation of axes through the given angle θ, find the equation of the curve relative to the new system of coordinates. Draw the original and the new systems of coordinates, and sketch a graph of the curve.

15. $xy = 4$; $\theta = 45°$ **16.** $xy = -9$; $\theta = 45°$
17. $2x^2 + \sqrt{3}xy + y^2 = 5$; $\theta = 30°$ **18.** $2x^2 - \sqrt{3}xy + y^2 = 20$; $\theta = 60°$
19. $34x^2 - 24xy + 41y^2 = 200$; $\theta = \operatorname{Sin} \dfrac{-3}{5}$

20. Solve the system of equations given by (8.33) to get the system in (8.34).

8.5 PARAMETRIC EQUATIONS

In many applications that involve a moving object in the x, y plane it is natural to express the position of the object by giving the x and y values in terms of time t. That is, $x = f(t)$ and $y = g(t)$ determine the position (x, y) at any given time t. As the object moves in the plane, it traces out a curve (or a path), and we say that the curve is described by the equations

$$x = f(t), \qquad y = g(t), \tag{8.35}$$

where f and g are functions. These are called *parametric equations of the curve*, and t is referred to as a *parameter*.

In this setting, since t represents time, we would implicitly assume in most situations that $t \geq 0$. However, there are many cases in which we shall not think of a moving object and we want to talk about Eq. (8.35) in a broader sense; that is, t may assume values in an interval that could include negative as well as positive numbers.

Here we shall be content with illustrating through examples some of the ideas related to parametric equations.

Example 1   Draw a curve described by the parametric equations $x = 4 - t^2$, $y = t$, where $0 \leq t \leq 3$.

Solution   First, construct a table of x, y values for several values of t; then plot the (x, y) points and draw the curve, as shown in Fig. 8.41. Numbers in the table are rounded off to two decimal places. If the given parametric equations describe the path of a moving object, then it starts at point A, moves along the curve, and at the end of three units of time is at point B.

t	0	$\frac{1}{2}$	1	$\sqrt{2}$	2	$\sqrt{5}$	3
x	4	3.75	3	2	0	−1	−5
y	0	0.5	1	1.41	2	2.24	3

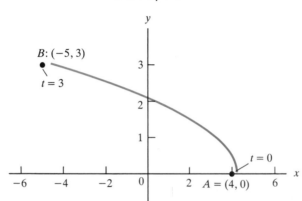

Fig. 8.41

Graph of $x = 4 - t^2$ and $y = t$

Example 2 The parametric equations

$$x = \cos \pi t, \qquad y = \sin \pi t,$$

where $0 \le t \le 4$, give the position of a moving particle in the x, y plane. Assume t is in seconds and x, y are in centimeters. Describe the motion.

Solution We draw the curve corresponding to the given parametric equations, using x, y values from the following table. The curve is shown in Fig. 8.42. The particle moves counterclockwise in a circular path (see problem 11) with center at the origin and radius one centimeter; it starts at A ($t = 0$), moves to B in 0.5 seconds, then to C ($t = 1$), and so on until it has moved around the curve twice. It is at point A at the end of four seconds.

t	0	0.25	0.50	0.75	1.00	1.25	1.50	1.75	2.00	2.5	3	3.5	4
x	1	0.71	0	−0.71	−1	−0.71	0	0.71	1	0	−1	0	1
y	0	0.71	1	0.71	0	−0.71	−1	−0.71	0	1	0	−1	0

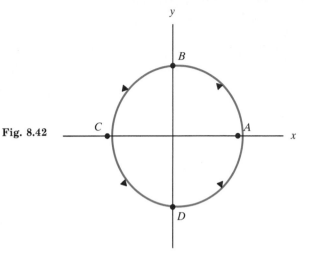

Fig. 8.42

Graph of $x = \cos \pi t$ and $y = \sin \pi t$

Example 3   Draw a graph of the curve given by the parametric equations

$$x = 1 + 2t, \qquad y = -1 + t,$$

where $-1 \leq t \leq 2$.

Solution   As in the preceding examples, we make a table giving x, y values to be used in plotting the curve. The graph is shown in Fig. 8.43. The curve described by the given parametric equations appears to be a line segment joining points A and B. In the following example we shall see that this is indeed so.

t	-1.0	-0.5	0	0.5	1.0	1.5	2.0
x	-1	0	1	2	3	4	5
y	-2	-1.5	-1	-0.5	0	0.5	1

Fig. 8.43

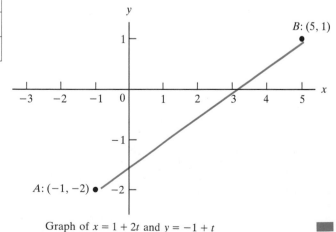

Graph of $x = 1 + 2t$ and $y = -1 + t$

Example 4   By eliminating t in the parametric equations

$$x = 1 + 2t \quad \text{and} \quad y = -1 + t, \quad -1 \le t \le 2,$$

determine the equation of the curve in terms of x and y.

Solution   Solving the second equation for t, $t = y + 1$, and substituting into the first equation gives $x = 1 + 2(y + 1)$. This is equivalent to $y = \frac{1}{2}x - \frac{3}{2}$. Since $-1 \le t \le 2$ and $t = (x - 1)/2$, then $-1 \le (x - 1)/2 \le 2$, and so $-1 \le x \le 5$. Thus we have the line segment shown in Fig. (8.43), given by

$$y = \tfrac{1}{2}x - \tfrac{3}{2}, \quad -1 \le x \le 5. \qquad \blacksquare$$

Example 5   Graph the curve given by the parametric equations

$$x = \sin t, \quad y = \cos^2 t,$$

where t is any real number. Determine an equation in terms of x and y that describes this curve.

Solution   We can eliminate the parameter t by squaring both sides of the first equation, $x^2 = \sin^2 t$, and adding this to the second equation:

$$x^2 + y = \sin^2 t + \cos^2 t.$$

Since $\sin^2 t + \cos^2 t$ is identically equal to 1, we get $x^2 + y = 1$. This is equivalent to $y = 1 - x^2$, but it is necessary to restrict the values of x since $x = \sin t$ and $-1 \le \sin t \le 1$. Therefore the points on the curve are given by

$$y = 1 - x^2, \quad -1 \le x \le 1.$$

This is an arc of a parabola, as shown in Fig. 8.44.

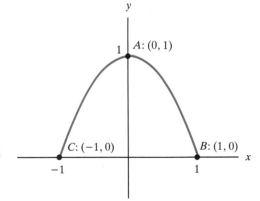

X=SIN(T), Y=(COS(T))^2

Fig. 8.44

Graph of $x = \sin t$ and $y = \cos^2 t$

If the given parametric equations described a moving particle with time $t \geq 0$, then the particle would start at point A $(t = 0)$, move to point B $(t = \pi/2)$, then back to A $(t = \pi)$, then to C $(t = 3\pi/2)$, then to A $(t = 2\pi)$, then to B $(t = 5\pi/2)$, and continue moving along the arc of the parabola indefinitely.

Exercises 8.5

In problems 1 through 5, draw the curve described by the given parametric equations. In each case make a table (as illustrated in examples of this section) to get several points (x, y) on the curve.

1. $x = t,\ y = 4 - t^2,\ 0 \leq t \leq 3$

2. $x = t,\ y = \sqrt{4 - t^2},\ -2 \leq t \leq 2$

3. $x = \sqrt{t},\ y = \sqrt{4 - t},\ 0 \leq t \leq 4$

4. $x = \cos \pi t,\ y = \sin \pi t,\ 0 \leq t \leq 1$

5. $x = 3 + t,\ y = 2 - t,\ -3 \leq t \leq 2$

In problems 6 through 10, the given parametric equations describe a moving particle where time t is in seconds and x, y are in centimeters. Draw graphs and use them to discuss the motion of the particle.

6. $x = 1 - t,\ y = t,\ 0 \leq t \leq 4$

7. $x = 4 \cos t,\ y = 3 \sin t,\ 0 \leq t \leq \pi$

8. $x = 1 + 4 \sin t,\ y = -3 + 2 \cos t,\ 0 \leq t \leq 2\pi$

9. $x = \cos^2 \pi t,\ y = \sin^2 \pi t,\ 0 \leq t \leq 4$

10. $x = \sin t,\ y = \cos 2t,\ 0 \leq t \leq \pi$

In problems 11 through 16, eliminate the parameter and get an equation in terms of x and y that describes the given curve. Be certain to give the restrictions on the x or y values that must accompany your equation. Draw a graph of the curve.

11. $x = \cos \pi t,\ y = \sin \pi t$

12. $x = 2 + \sin t,\ y = \cos t,\ 0 \leq t \leq \dfrac{\pi}{2}$

13. $x = 1 - t^2,\ y = 1 + 2t^2,\ -1 \leq t \leq 2$

14. $x = \cos t,\ y = \cos 2t,\ 0 \leq t \leq \pi$

15. $x = \tan t,\ y = \sec^2 t,\ 0 \leq t < \dfrac{\pi}{2}$

16. $x = \cos t,\ y = \sec t,\ 0 \leq t \leq \dfrac{\pi}{3}$

17. By eliminating the parameter, show that the parametric equations

$$x = x_0 + at, \qquad y = y_0 + bt, \quad t \text{ any real number,}$$

where a and b are given numbers, represent a straight line.

Review Exercises

In problems 1 through 8, sufficient information is given to determine a conic section. Write its equation. Give answers in

a) Standard form

b) Simplified expanded form

1. Circle with center at $(-2, 1)$ and radius 4.

2. Circle with center at $(0, -3)$ and radius $\sqrt{5}$.

3. Parabola with focus at $(3, 0)$ and vertex at $(0, 0)$.

4. Parabola with focus at $(1, -2)$ and directrix $y = 2$.

5. Ellipse with center at $(1, 4)$, focus at $(1, 2)$, and vertex at $(1, 0)$.

6. Ellipse with foci at $(4, -1)$ and $(0, -1)$, and vertex at $(5, -1)$.

7. Hyperbola with center at $(1, -1)$, focus at $(4, -1)$, and vertex at $(3, -1)$.

8. Hyperbola with vertices at $(1, 3)$ and $(1, -1)$ and focus at $(1, -2)$.

In problems 9 and 10, find the coordinates of the center and the radius of the circle corresponding to the given equation.

9. $x^2 + y^2 + 2x - 4y + 1 = 0$ **10.** $x^2 + y^2 - 2y - 2 = 0$

11. For the parabola whose equation is $y = x^2 - 2x$, find the coordinates of the (a) vertex and (b) focus, and (c) draw a graph of the parabola.

12. For the parabola whose equation is $x^2 - 2x + 2y - 5 = 0$, draw a graph and label the coordinates of the vertex and focus.

13. Determine the coordinates of the (a) center and (b) foci, and (c) draw a graph of the ellipse given by

$$9x^2 + 4y^2 - 8y - 32 = 0.$$

14. Determine the coordinates of the (a) center and (b) vertices, and (c) draw a graph of the hyperbola whose equation is

$$x^2 - 9y^2 - 4x - 5 = 0.$$

15. Draw a graph of $y = \sqrt{4 - x^2}$. What type of curve is this?

16. Draw a graph of $y = 1 + \sqrt{4 - x^2}$. What type of curve is this?

17. Assuming that each of the given equations represents a nondegenerate conic section, identify the type of curve.

a) $x^2 + 2y - 3 = 0$ **b)** $2x^2 + 4y^2 - x + y = 0$

c) $x^2 - y^2 + 2x - 3y + 1 = 0$ **d)** $4 - x^2 = y^2$

18. Transformation equations corresponding to a translation of coordinates are given by

$$x = X - 1, \qquad y = Y + 2.$$

a) Draw a diagram showing the two sets of coordinate axes relative to each other.

b) Label each of the origins relative to the (x, y) system and the $[X, Y]$ system.

19. In each of the following, the name of a point P is given relative to one of the coordinate systems described in problem 18. Locate P in your diagram, and determine its name relative to the other coordinate system.

a) $P{:}(3, 4)$ **b)** $P{:}[-1, 2]$ **c)** $P{:}[0, 3]$

20. Each of the following is an equation of a curve in the x, y or the X, Y system of coordinates described in problem 18. Determine the equation of the same curve relative to the other coordinate system.

a) $x^2 + y^2 + 2x - 4y + 1 = 0$ **b)** $X^2 - 2X - 2Y - 3 = 0$

21. Draw a graph of the curve given in parametric equations by $x = 2t$, $y = \sqrt{4 - 4t^2}$, $0 \le t \le 1$.

Polar Coordinates

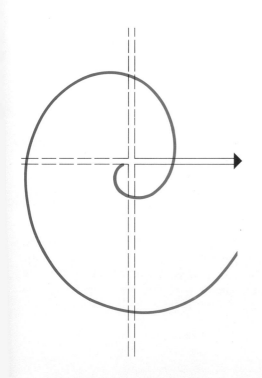

9.1   INTRODUCTION

Many problems involve equations relating two variables. We have seen that it is helpful to have geometrical representations of such relationships, since these can frequently provide insights that are not readily apparent from the equation itself. In some problems the situation is reversed in that we have a problem described geometrically and it becomes useful to consider it in an algebraic setting, which usually means an equation relating two variables. The form of the equation we get depends to a large degree on the reference (or coordinate) system we decide to use. So far, all our geometrical representations have been relative to a rectangular (or cartesian) system of coordinates. This has served us well for most problems. However, there are situations in which a given geometrical problem translates into a cumbersome equation when rectangular coordinates are used. A system of coordinates known as polar coordinates can be particularly useful in many situations.

As indicated at the beginning of this book, our geometrical considerations are restricted to a given plane (in future courses the student will encounter problems requiring three-dimensional geometry). A rectangular system of coordinates begins with two perpendicular lines. It is customary to take these lines as horizontal and vertical and call them the x-axis and y-axis, respectively. On each axis we have a one-to-one correspondence between points and real numbers. This provides us with a system that has a one-to-one correspondence between pairs of real numbers (x, y) and points P in the plane.

For the *system of polar coordinates* we begin with a ray (half line), which we call the polar axis; its endpoint is called the *polar origin* (point O), as shown in Fig. 9.1.

Fig. 9.1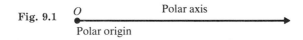

Let point P be any point (other than O) in the plane. Consider the ray $\overrightarrow{OP}$ (see Fig. 9.2(a)) as the terminal side of the directed angle θ obtained by rotating the polar axis about point O through the angle of measure θ. We call $\overrightarrow{OP}$ the θ *ray*. If the distance from O to P is denoted by r, where r is a positive number, then polar coordinates of P consist of the ordered pair r and θ, denoted by $[r, \theta]$.* This is shown in Fig. 9.2(b).

In many situations it is convenient to allow the first member of the ordered pair $[r, \theta]$ to be a negative number. Suppose we consider the ordered pair $[-r, \theta + \pi]$, where r is a positive number. The pair $[-r, \theta + \pi]$ represents the point that is a directed distance of $-r$ along the $(\theta + \pi)$ ray; we interpret this as

---

*   We use the bracket notation $[r, \theta]$ as the name of a point in polar coordinates corresponding name (x, y) in rectangular coordinates.

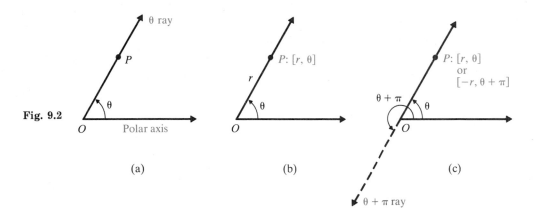

Fig. 9.2

(a)   (b)   (c)

meaning r units in the opposite direction, which is along the θ ray. This puts us at $P{:}[r, \theta]$. Therefore both $[r, \theta]$ and $[-r, \theta + \pi]$ are names in polar coordinates of the same point P, as shown in Fig. 9.2(c).

It is clear that the θ ray and the $(\theta + 2\pi)$ ray are the same; so $[r, \theta]$ and $[r, \theta + 2\pi)$ represent the same point. In fact, the point P shown in Fig. 9.2 can be represented by any of the ordered pairs $[r, \theta + 2k\pi]$ or $[-r, \theta + (2k + 1)\pi]$, where k is any integer.

The above discussion indicates how we name any point P in the plane in terms of polar coordinates. The special case where P is the *polar origin* is denoted by $[0, \theta]$, where θ can have any value.

Note that in polar coordinates we do not have the luxury we have in rectangular coordinates, in which there is a one-to-one correspondence between points in the plane and ordered pairs of real numbers. In polar coordinates each point P can be represented by infinitely many ordered pairs; however, a given ordered pair is associated with exactly one point. Although the lack of a one-to-one correspondence is an undesirable feature of polar coordinates, it does not create a serious problem.

We remind the reader that the *definition of equality of ordered pairs* is given by

$$(a, b) = (c, d) \quad \text{if and only if} \quad a = c \quad \text{and} \quad b = d.$$

We retain this definition for ordered pairs $[r, \theta]$, and *we do not say that* $[r, \theta]$ *equals* $[-r, \theta + \pi]$ even though they both represent the same point.

Example 1   For each of the following, draw a diagram to illustrate the given ray.

a) $30°$ ray   **b)** $480°$ ray   **c)** $-\dfrac{5\pi}{6}$ ray   **d)** $\dfrac{5\pi}{4}$ ray

Solution

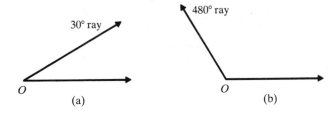

Fig. 9.3

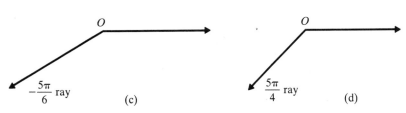

Example 2   In each of the following, give two other names for the given ray.

a) 45° ray   b) π ray   c) 2.5 ray   d) −2.5 ray

Solution   a) 405° ray; (−315°) ray

b) 3π ray; −3π ray

c) (2.5 + 2π) ray = 8.78 ray; (2.5 − 2π) ray = −3.78 ray

d) (−2.5 + 2π) ray = 3.78 ray; (−2.5 + 4π) ray = 10.07 ray

In (c) and (d) the results have been rounded off to two decimal places. ■

Example 3   Point P, shown in Fig. 9.4, is on the 30° ray at a distance 2 from the polar origin. Give four different names for P in polar coordinates.

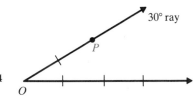

Fig. 9.4

Solution   Any of the following pairs can be used as the name of point P:

$$[2, 30°]; \qquad [2, 30° + 360°] = [2, 390°]; \qquad [2, 30° − 360°] = [2, −330°];$$
$$[−2, 30° + 180°] = [−2, 210°].$$ ■

Example 4   Suppose point P is 3 units from the polar origin on the $7\pi/6$ ray. Let Q be the point obtained by reflecting P about the line l perpendicular to the polar axis and passing through the polar origin. Give four different names for Q in polar coordinates.

Solution   From Fig. 9.5 we see that point Q is on the $11\pi/6$ ray and 3 units from O. Therefore Q can be represented by any of the following ordered pairs:

$$\left[3, \frac{11\pi}{6}\right]; \qquad \left[3, -\frac{\pi}{6}\right]; \qquad \left[-3, \frac{5\pi}{6}\right]; \qquad \left[3, -\frac{13\pi}{6}\right].$$

Fig. 9.5

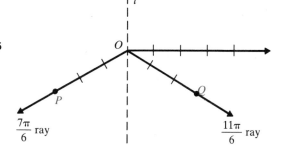

Example 5   In each of the following, draw a sketch to illustrate the point corresponding to the given ordered pairs in polar coordinates.

a) $[2, 40°]$ **b)** $[-3, 580°]$ **c)** $\left[3, \frac{3\pi}{4}\right]$ **d)** $[-4, -3\pi]$.

Solution   (Fig. 9.6)

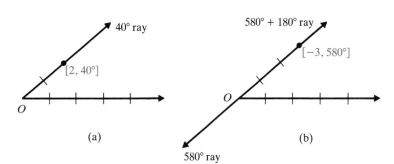

Fig. 9.6   (a)   (b)

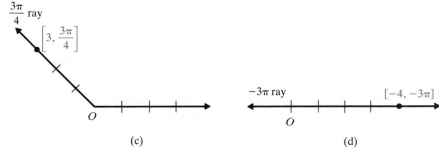

(c)   (d)

Exercises 9.1

1. In each of the following, a point is described relative to a polar axis with polar origin O. Draw a diagram showing the given point, and then give four different ordered pairs $[r, \theta]$ that name the point in polar coordinates.

   a) P is 3 units from O on the 50° ray.

   b) Q is 4 units from O on the $(-60°)$ ray.

   c) T is 2 units from O on the 540° ray.

2. In problem 1, suppose each of the points P, Q, and T is reflected about the polar origin O to get new points P_1, Q_1, and T_1, respectively. For each of these points, give an ordered pair $[r, \theta]$ that can be used to represent the point in polar coordinates.

3. In problem 1, suppose that each of the points P, Q, T is reflected about the line through the polar axis to get new points P_2, Q_2, T_2, respectively. For each of these points, give an ordered pair $[r, \theta]$ that corresponds to the point in polar coordinates.

4. In each of the following, a point is described relative to a polar axis with polar origin O. Draw a diagram showing the given point, and then give four different ordered pairs of real numbers $[r, \theta]$ that can be used to name the point in polar coordinates.

   a) P is 2 units from O on the $2\pi/3$ ray.

   b) Q is 3 units from O on the $-11\pi/12$ ray.

   c) T is 4 units from O on the $17\pi/6$ ray.

5. In problem 4, suppose that each of the points P, Q, T is reflected about the polar origin to get points P_1, Q_1, T_1, respectively. For each of these points, give an ordered pair $[r, \theta]$ of real numbers that is a name for the point in polar coordinates.

6. In problem 5, suppose each of the points P_1, Q_1, T_1 is reflected about the line through O perpendicular to the polar axis to get points P_2, Q_2, T_2, respectively. For each of these points, give an ordered pair $[r, \theta]$ of real numbers that can be used to represent the point in polar coordinates. How are P_2, Q_2, T_2 geometrically related to P, Q, T of problem 4?

7. In each of the following, draw a diagram that illustrates the point corresponding to the given ordered pairs.

   a) $[3, 60°]$ b) $[-4, 45°]$ c) $[-2, 180°]$ d) $[-3, -450°]$

8. In each of the following, draw a diagram showing the point that corresponds to the given ordered pairs.

   a) $\left[4, \dfrac{4\pi}{3}\right]$ b) $\left[-3, \dfrac{5\pi}{12}\right]$ c) $[2, 17\pi]$ d) $[-2, -2.36]$.

9. In each part of problem 7, the given point is reflected about the polar origin. Give an ordered pair of real numbers $[r, \theta]$ that represents the new point in polar coordinates.

10. In each part of problem 8, the given point is reflected about the line through the polar axis. Give an ordered pair $[r, \theta]$ of real numbers that can be used to represent the new point.

9.2   **GRAPHS IN POLAR COORDINATES**

In earlier parts of this book we encountered a variety of problems in which an equation was given in the form $y = f(x)$, and then by means of a system of rectangular coordinates, a graph (curve) corresponding to the given equation was drawn. The analogous problem in polar coordinates is: Given $r = f(\theta)$, draw a curve that corresponds to this equation.

Example 1   Sketch the curve whose equation in polar coordinates is $r = 2 \sin \theta$.

Solution   We first determine several ordered pairs $[r, \theta]$ that satisfy the given equation. These are shown in the following table. Note that it is not necessary to continue with larger values of θ, since $\sin(\theta + \pi) = -\sin \theta$ is an identity, and so

$$[r, \theta + \pi] = [2 \sin(\theta + \pi), \theta + \pi] = [-2 \sin \theta, \theta + \pi].$$

Therefore

$$[r, \theta + \pi] = [-2 \sin \theta, \theta + \pi] \quad \text{and} \quad [r, \theta] = [2 \sin \theta, \theta]$$

represent the same point.

In a similar manner we can show that negative values of θ produce no points that are not already included in the points given by $0 \le \theta \le \pi$.

We now plot the points given in the table and draw the curve shown in Fig. 9.7. The curve is a circle (see Exercise set 9.3, problem 11).

θ	0	$\dfrac{\pi}{6}$	$\dfrac{\pi}{4}$	$\dfrac{\pi}{3}$	$\dfrac{\pi}{2}$	$\dfrac{2\pi}{3}$	$\dfrac{3\pi}{4}$	$\dfrac{5\pi}{6}$	π
r	0	1	$\sqrt{2}$	$\sqrt{3}$	2	$\sqrt{3}$	$\sqrt{2}$	1	0

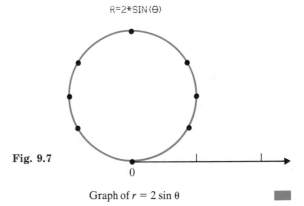

R=2*SIN(Θ)

Fig. 9.7

Graph of $r = 2 \sin \theta$

Example 2   Sketch the curve whose equation in polar coordinates is $r = 1 + \cos \theta$.

Solution   As in Example 1, we first make a table giving ordered pairs $[r, \theta]$ that satisfy the given equation. Values of r are given in decimal form to two places. Since $\cos(\theta + 2\pi) = \cos \theta$ is an identity, it is clear that we get no new points by considering values of θ that are outside the interval $0° \le \theta \le 360°$. Plot these points and draw the curve, as shown in Fig. 9.8. The curve is an example of a *cardioid*.

θ	0°	45°	90°	135°	180°	225°	270°	315°	360°
r	2	1.71	1	0.29	0	0.29	1	1.71	2

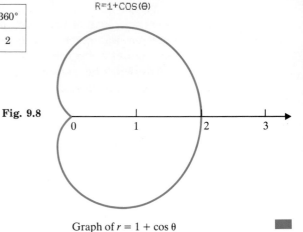

Fig. 9.8

Graph of $r = 1 + \cos\theta$

Example 3 Sketch the curve whose equation in polar coordinates is $r = 3$.

Solution As in the preceding two examples, first make a table of ordered pairs $[r, \theta]$. The variable θ does not appear explicitly in the given equation; if this causes any problems, we can write the equation in equivalent form as $r = 3 + 0 \cdot \theta$. The value of r is 3 for every value of θ, and so the corresponding points are on a circle with center at the polar origin and radius 3, as shown in Fig. 9.9.

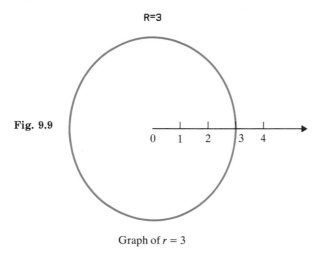

Fig. 9.9

Graph of $r = 3$

Example 4 Sketch the curve whose equation in polar coordinates is $r = \sin 3\theta$.

Solution First note that $\sin 3(\theta + \pi) = -\sin 3\theta$ is an identity. Thus

$$[r, \theta + \pi] = [\sin 3(\theta + \pi), \theta + \pi] = [-3\sin\theta, \theta + \pi].$$

Also $[r, \theta] = [3 \sin \theta, \theta]$. But $[-3 \sin \theta, \theta + \pi]$ and $[3 \sin \theta, \theta]$ represent the same point. Hence it is sufficient to use values of θ in the interval $0 \leq \theta \leq \pi$, as seen in the following table.

θ	0	$\dfrac{\pi}{12}$	$\dfrac{\pi}{6}$	$\dfrac{\pi}{4}$	$\dfrac{\pi}{3}$	$\dfrac{5\pi}{12}$	$\dfrac{\pi}{2}$	$\dfrac{7\pi}{12}$	$\dfrac{2\pi}{3}$	$\dfrac{3\pi}{4}$	$\dfrac{5\pi}{6}$	$\dfrac{11\pi}{12}$	π
r	0	0.71	1	0.71	0	-0.71	-1	-0.71	0	0.71	1	0.71	0

Plotting the points given in this table and connecting them in the appropriate manner gives the *three-leaf rose* shown in Fig. 9.10.

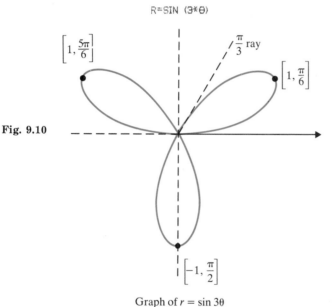

Fig. 9.10

Graph of $r = \sin 3\theta$

Note in Example 4 that $r = 0$ for values of θ such as 0, $\pi/3$, $2\pi/3$, and π. In each case the point is the origin, and the curve comes into the origin tangent to the corresponding θ ray (as shown in Fig. 9.10 for $\theta = \pi/3$). This illustrates a general situation: If $r = f(\theta)$ and $f(\theta_1) = 0$, then the curve comes into the origin tangent to the θ_1 ray.

Example 5 Sketch the curve whose equation in polar coordinates is given by $r = -\theta$, where $\theta \geq 0$.

Solution Note that the given equation implies that radian measure is to be used for θ since r is a real number. First make a table of ordered pairs $[r, \theta]$ that satisfy the equation; θ is given in exact form, and r is rounded off to two decimal places.

Plotting these points and drawing a curve through them gives a *spiral,* as shown in Fig. 9.11. The curve begins at the polar origin and, as θ increases, winds around in the counterclockwise direction, as illustrated.

θ	0	$\dfrac{\pi}{4}$	$\dfrac{\pi}{2}$	$\dfrac{3\pi}{4}$	π	$\dfrac{5\pi}{4}$	$\dfrac{3\pi}{2}$	$\dfrac{7\pi}{4}$	2π
r	0	-0.79	-1.57	-2.36	-3.14	-3.93	-4.71	-5.50	-6.28

Fig. 9.11

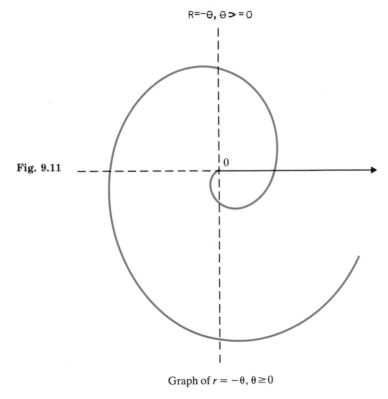

Graph of $r = -\theta,\ \theta \geq 0$

Exercises 9.2 _____

In each of the following, sketch the curve that corresponds to the given equation in polar coordinates.

1. $r = \cos\theta$ **2.** $r = 3\cos\theta$ **3.** $r = 2$ **4.** $r = -2\sin\theta$

5. $r = 1 + \sin\theta$ **6.** $r = 1 - \sin\theta$ **7.** $r = 1 - \cos\theta$ **8.** $r = 3 + \sin^2\theta + \cos^2\theta$

9. $r = \sin 2\theta$ **10.** $r = \cos 3\theta$ **11.** $r = \cos^2\theta - \sin^2\theta$ **12.** $r^2 = 4$

13. $r = \cos\theta\tan\theta$ **14.** $r = \sin^2\theta$ **15.** $r = \sin\left(\theta + \dfrac{\pi}{4}\right)$ **16.** $r = \cos(\theta + \pi)$

17. $r = 1 + 2\cos\theta$ **18.** $r = 2 - \sin\theta$ **19.** $r = \theta$, where $\theta \geq 0$ **20.** $r = \dfrac{3}{\theta}$, where $\theta \geq 1$

9.3　**RELATIONSHIP BETWEEN POLAR
AND RECTANGULAR COORDINATES**

Suppose the polar axis is taken in such a way that it coincides with the positive x-axis, as shown in Fig. 9.12, and let P be any point in the plane. The name of point P is (x, y) relative to the x, y coordinate system, and $[r, \theta]$ relative to the polar coordinate system.

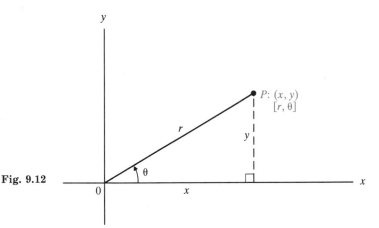

Fig. 9.12

The following equations give the relationship between rectangular and polar coordinates:

$$x = r \cos \theta, \qquad y = r \sin \theta. \tag{9.1}$$

$$r^2 = x^2 + y^2, \qquad \tan \theta = \frac{y}{x}. \tag{9.2}$$

The equations given in (9.1) are *transformation equations from polar to rectangular coordinates*. For each pair $[r, \theta]$, there is precisely one pair (x, y) corresponding to it.

The equations given in (9.2) are known as the *transformation equations from rectangular to polar coordinates*. Note that for a given pair (x, y) we can get multiple pairs $[r, \theta]$, each of which represents the same point. Since r can be taken as $\sqrt{x^2 + y^2}$ or as $-\sqrt{x^2 + y^2}$, and θ satisfying $\tan \theta = y/x$ is multiple-valued, we must be careful to match appropriate values of r and θ. This is illustrated in the following examples.

Example 1   In each of the following, find all ordered pairs $[r, \theta]$ that are associated with the given point in rectangular coordinates.

a) $(3, 4)$ **b)** $(-2, -1)$

Solution   **a)** We use Eq. (9.2) as follows (see Fig. 9.13): First find $[r, \theta]$ where $r > 0$; $r = \sqrt{3^2 + 4^2} = 5$, and θ satisfies $\tan \theta = 4/3$, where θ is in the first quadrant. Hence $\theta = 53.13°$. This gives the set of ordered pairs

$$A = \{[5, 53.13° + k \cdot 360°] \,|\, k \text{ is any integer}\}.$$

Now find $[r, \theta]$, where $r < 0$, $r = -\sqrt{3^2 + 4^2} = -5$, and θ satisfies $\tan \theta = 4/3$, where θ is in the third quadrant. This gives the set of ordered pairs

$$B = \{[-5, 233.13° + k \cdot 360°] \,|\, k \text{ is any integer}\}.$$

Therefore the name in polar coordinates of the point associated with $(3, 4)$ is given by any one of the ordered pairs in the union of sets A and B, where θ values are rounded off to two decimal places.

b) In a manner similar to (a), we have: For $r > 0$, $r = \sqrt{(-2)^2 + (-1)^2} = \sqrt{5}$, and θ satisfies $\tan \theta = \frac{1}{2}$, where θ is in the third quadrant (Fig. 9.14). That is, $r = \sqrt{5}$ and $\theta = 3.61 + k \cdot 2\pi$. For $r < 0$, $r = -\sqrt{5}$, and θ satisfies $\tan \theta = \frac{1}{2}$, where θ is in the first quadrant. That is, $r = -\sqrt{5}$ and $\theta = 0.46 + k \cdot 2\pi$. Therefore the point $(-2, -1)$ is represented in polar coordinates by any of the ordered pairs in the set

$$\{[\sqrt{5}, 3.61 + k \cdot 2\pi] \,|\, k \text{ any integer}\} \cup \{[-\sqrt{5}, 0.46 + k \cdot 2\pi] \,|\, k \text{ any integer}\},$$

where θ values are rounded off to two decimal places.

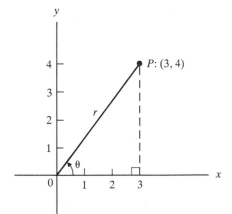

Fig. 9.13

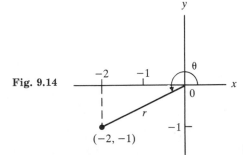

Fig. 9.14

Example 2   In each of the following, the given ordered-pair names a point P in polar coordinates. Find the corresponding name in rectangular coordinates.

    a) $[4, 60°]$ **b)** $[-3, 180°]$ **c)** $\left[4, \dfrac{-3\pi}{4}\right]$ **d)** $[-2, 2.48]$.

Solution   Use the equations in (9.1), which are valid for all values of θ and r.

    a) $x = 4 \cos 60° = 4(1/2) = 2$; $y = 4 \sin 60° = 4(\sqrt{3}/2) = 2\sqrt{3}$. Therefore the point in rectangular coordinates is given by $(2, 2\sqrt{3})$.

    b) $x = -3 \cos 180° = -3(-1) = 3$; $y = -3 \sin 180° = -3(0) = 0$. Hence the given point is $(3, 0)$ in rectangular coordinates.

    c) $x = 4 \cos(-3\pi/4) = -2\sqrt{2}$; $y = 4 \sin(-3\pi/4) = -2\sqrt{2}$. Thus the given point is denoted by $(-2\sqrt{2}, -2\sqrt{2})$ in rectangular coordinates.

    d) $x = -2 \cos 2.48 = 1.58$; $y = -2 \sin 2.48 = -1.23$ (to two decimal places). Therefore $[-2, 2.48]$ is represented by $(1.58, -1.23)$ in rectangular coordinates. ■

Example 3   Find an equation in polar coordinates that describes the same set of points (same curve) as $x^2 + y^2 - 2x = 0$ in rectangular coordinates.

Solution   Substituting $x = r \cos \theta$ and $y = r \sin \theta$ into the given equation gives

$$(r \cos \theta)^2 + (r \sin \theta)^2 - 2(r \cos \theta) = 0,$$
$$r^2[\cos^2\theta + \sin^2\theta] - 2r \cos \theta = 0.$$

This is equivalent to $r^2 - 2r \cos \theta = 0$. Thus $r(r - 2 \cos \theta) = 0$, and so $r = 0$ or $r = 2 \cos \theta$. Since $r = 0$ gives only the polar origin as a point, and from $r = 2 \cos \theta$ we get the point $[0, \pi/2]$, which is also the polar origin, we can ignore $r = 0$ in our solution. That is, $r = 2 \cos \theta$ will describe the same set of points as $x^2 + y^2 - 2x = 0$. ■

Example 4   Find an equation in rectangular coordinates that describes the same set of points in polar coordinates as

$$r = 2 \sin \theta + \cos \theta.$$

Solution   Since a direct substitution for r and θ from Eq. (9.2) would involve replacing r by $\sqrt{x^2 + y^2}$, it is simpler to first multiply both sides of the given equation by r:

$$r^2 = 2r \sin \theta + r \cos \theta.$$

Now replacing r^2 by $x^2 + y^2$, $r \sin \theta$ by y, and $r \cos \theta$ by x, we get

$$x^2 + y^2 = 2y + x.$$

Note: In this example we should check the possibility that we may have introduced some extraneous points by multiplying both sides of the given equation by

r. This can occur only if we have multiplied by the value of r equal to zero. Since $r = 0$ represents the origin, the only possible extraneous point is the origin. Thus we must check to see if the origin is also a point on the curve represented by the polar equation. We see that $2 \sin \theta + \cos \theta = 0$ for $\theta = \text{Tan}^{-1}(-\frac{1}{2}) = -0.46$; that is, $[0, -0.46]$ satisfies the given equation, and so the origin is on the given curve.   ■

Example 5 Draw a graph of the equation $\theta = 2$ in polar coordinates. Then find an equivalent equation in rectangular coordinates.

Solution The graph of $\theta = 2$ is a line through the origin, as shown in Fig. 9.15. Since $\tan \theta = y/x$, the corresponding equation in rectangular coordinates is $\tan 2 = y/x$, or $y = x(\tan 2)$. In decimal form this is $y = -2.19x$.

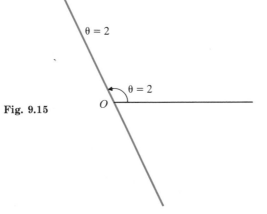

Fig. 9.15

Exercises 9.3

For each answer that is to be expressed in decimal form, give the result correct to two decimal places.

1. In each of the following, a point is given in rectangular coordinates. Find one name of the point in polar coordinates.

    a) $(-1, 1)$ **b)** $(-1, -\sqrt{3})$ **c)** $(\pi, 4)$ **d)** $(-1.57, 2.43)$

2. For each of the points given in problem 1, give the set of all possible ordered pairs $[r, \theta]$ that can be used as polar coordinates for the given points.

3. Express each of the following in polar coordinates with $r \geq 0$ and $0 \leq \theta \leq 2\pi$.

    a) $(-3, 3)$ **b)** $(1, -3)$ **c)** $\left(\pi, \dfrac{1 + \sqrt{5}}{2}\right)$

4. Express each of the following in polar coordinates, using the smallest positive angle θ and $r < 0$.

    a) $(4, -3)$ **b)** $(-\sqrt{3}, \sqrt{3})$ **c)** $(2.52, -2\pi)$

5. Express each of the following in rectangular coordinates.

a) $\left[2, \dfrac{\pi}{2}\right]$
 b) $\left[-3, -\dfrac{3\pi}{4}\right]$
 c) $[2.24, -0.37]$

6. Express each of the following in rectangular coordinates.

a) $[0, 30°]$
 b) $[4, -630°]$
 c) $[-2, 47°37']$

7. In each of the following, determine whether or not the given pair satisfies the equation $r^2 \sin \theta = 1$.

a) $\left[1, \dfrac{\pi}{2}\right]$
 b) $\left[-1, -\dfrac{\pi}{2}\right]$
 c) $\left[\sqrt{2}, \dfrac{5\pi}{6}\right]$
 d) $[0, 0]$
 e) $\left[1, \dfrac{3\pi}{2}\right]$

8. In each of the following, the coordinates of a point P are given in rectangular coordinates. Determine whether or not P lies on the curve whose equation in polar coordinates is $r = 1 + \cos \theta$.

a) $(0, 0)$
 b) $(0, 1)$
 c) $(2, 0)$
 d) $\left(\dfrac{1 + \sqrt{2}}{2}, \dfrac{1 + \sqrt{2}}{2}\right)$

9. Let $[r_1, \theta_1]$ be polar coordinates of point P and $[r_2, \theta_2]$ be polar coordinates of point Q. Let d represent the distance between P and Q. Show that d is given by

$$d = \sqrt{r_1^2 + r_2^2 - 2r_1 r_2 \cos(\theta_1 - \theta_2)}.$$

10. Use the result in problem 9 to find the distance between the given pairs of points.

a) $[3, 0], [\pi, \pi]$
 b) $\left[1, \dfrac{\pi}{3}\right], \left[-2, \dfrac{3\pi}{4}\right]$

c) $[-3.4, 32°], [1.6, 47°]$
 d) $[-2.4, 3.2], [3.7, -0.64].$

In problems 11 through 18, find an equation in rectangular coordinates that describes the same set of points (same curve) as the given equation in polar coordinates.

11. $r = 2 \sin \theta$
 12. $r = 4 \cos \theta$
 13. $3\theta = 4$
 14. $r \cos \theta = 3$

15. $r(1 - \sin \theta) = 2$
 16. $r(1 + \cos \theta) = 2$
 17. $r = 2 \cos(\theta + \pi)$
 18. $r = \cos 2\theta$

In problems 19 through 22, find an equation in polar coordinates that describes the same set of points (same curve) as the given equation in rectangular coordinates. Then sketch the curve, using the equation either in rectangular or in polar form.

19. $x^2 + y^2 = 1$
 20. $2xy = 3$
 21. $3x - y = 0$

22. $x^2 + y^2 + x = \sqrt{x^2 + y^2}$

23. Are all points on the curve whose equation is $r = \sin \theta$ also on the curve with equation $r \csc \theta = 1$? Give reason for your answer.

24. Express $r = \sin 2\theta$ as an equation in rectangular coordinates.

25. Suppose P is a point in the plane given in polar coordinates by $[-2, \pi]$. Is P on the curve whose equation is $r = 1 + \cos \theta$? *Hint: P is also given by* $[2, 0]$.

Review Exercises

In any problem in which both rectangular and polar coordinates are used, assume that the positive x-axis coincides with the polar axis.

1. In each of the following, the name of a point is given in rectangular coordinates. Give one name of the point in polar coordinates.

   a) $(1, 0)$ **b)** $(-3, 0)$ **c)** $(4, 4)$ **d)** $(-2, 2)$

   e) $(-\sqrt{3}, -1)$ **f)** $(\sqrt{2}, -\sqrt{2})$ **g)** $(0, 4)$ **h)** $(0, -3)$

2. Find the name in polar coordinates for the given points. Give r and θ (in radians) to two decimal places with $r > 0$ and $0 \le \theta \le 2\pi$.

   a) $(3, 4)$ **b)** $(-5, 1)$ **c)** $(3, -5)$ **d)** $(-2, -1)$

3. In each of the following, a name of a point is given in polar coordinates. Draw a diagram illustrating the point, and then give the name of the point in rectangular coordinates.

   a) $\left[4, \dfrac{\pi}{3}\right]$ **b)** $\left[-2, \dfrac{5\pi}{6}\right]$ **c)** $[4, \pi]$ **d)** $\left[-1, \dfrac{9\pi}{4}\right]$ **e)** $\left[-3, \dfrac{-3\pi}{4}\right]$

4. Follow the instructions of problem 3. Give answers to two decimal places.

   a) $\left[1, \dfrac{5\pi}{7}\right]$ **b)** $[-4, 3.47]$ **c)** $[2.3, 1.35]$ **d)** $\left[-2, \dfrac{17\pi}{5}\right]$ **e)** $[3, -4.32]$

In problems 5 through 12, an equation is given in polar coordinates. Draw a graph of the corresponding curve.

5. $r = \sin \theta$ **6.** $r^2 = 16$ **7.** $r = 2 \sin(-\theta)$ **8.** $r = \cos \theta - 1$

9. $r = 3 \sec \theta$ **10.** $r = \cos 2\theta$ **11.** $2r = \theta$, where $\theta \ge 0$ **12.** $r = \sin\left(\theta + \dfrac{\pi}{2}\right)$

13. Find an equation in polar coordinates that describes the same curve as $x^2 + y^2 = 4$. Draw a graph of the curve.

14. Find an equation in polar coordinates that describes the same curve as $x^2 + y^2 + y = \sqrt{x^2 + y^2}$. Draw a graph of the curve.

15. Draw a graph of $r(1 + \cos \theta) = 1$. Then find an equation in rectangular coordinates that describes the same curve.

16. Draw a graph of $r \sin \theta = 3$. Then find an equation in rectangular coordinates that describes the same curve.

Complex Numbers

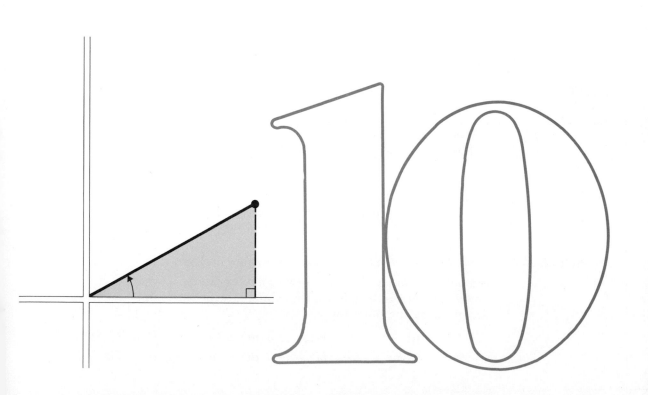

10.1 INTRODUCTION

The system of real numbers is essential in the development of pure mathematics, as well as in applications of mathematics. However, even such a simple problem as finding the roots of the equation $x^2 + 1 = 0$ has no solution in the set of real numbers. To remedy this situation we introduce a number denoted by i (also written as $\sqrt{-1}$) with the property $i^2 = -1$. Thus the solutions to $x^2 + 1 = 0$ are i and $-i$. Similarly, the quadratic equation $x^2 - 4x + 5 = 0$, which can be written as $(x - 2)^2 + 1 = 0$, has no solution in the set of real numbers. However, $x - 2 = i$ and $x - 2 = -i$ give solutions $2 + i$ and $2 - i$. The solutions in these examples, called complex numbers, lead us to the following definition.

Definition 10.1   The set $\mathbf{C}$ given by $\mathbf{C} = \{u + vi \mid u \text{ and } v \text{ are real numbers}\}$ is called the *set of complex numbers*.

Note that if we take $v = 0$ in Definition 10.1, the complex number $u + vi$ becomes simply u, a real number. Thus the set of real numbers $\mathbf{R}$ is a subset of $\mathbf{C}$. If we take $u = 0$ and $v \neq 0$, the resulting complex number is vi. Such a number is called an *imaginary number*. We shall refer to u as the *real part* and v as the *imaginary part* of the complex number $u + vi$.

Basic properties of real numbers related to the four binary operations $(+, -, \times, \div)$ and the order relations $(<$ and $>)$ are discussed in Section 1.0 and Appendix B. Now that the set $\mathbf{R}$ is extended to the set $\mathbf{C}$, it is of interest to define addition, subtraction, multiplication, and division of complex numbers. Since $\mathbf{R} \subset \mathbf{C}$, we want these definitions to be such that when they are applied to real numbers, the properties stated in Appendix B are still valid. In the system of complex numbers, it is not possible to define an order relation similar to that of "less than" for the real numbers; that is, we do not talk about one complex number being less than a second unless both are real numbers.

First let us define equality of two complex numbers.

Definition 10.2   Suppose a, b, c, and d are real numbers. We say that the complex numbers $a + bi$ and $c + di$ are equal if and only if $a = c$ and $b = d$.

Definition 10.3   *Binary operations on complex numbers*

Suppose $a + bi$ and $c + di$ are two complex numbers, where a, b, c, and d are real numbers. Their sum, difference, product, and quotient are given by the following.

Addition: $(a + bi) + (c + di) = (a + c) + (b + d)i$

Subtraction: $(a + bi) - (c + di) = (a - c) + (b - d)i$

$$\text{Multiplication:} \quad (a + bi) \cdot (c + di) = (ac - bd) + (ad + bc)i$$

$$\text{Division:} \quad \frac{a + bi}{c + di} = \left(\frac{ac + bd}{c^2 + d^2}\right) + \left(\frac{bc - ad}{c^2 + d^2}\right)i,$$

where c and d are not both zero.

Addition and subtraction as stated in Definition 10.3 appear to be natural, but the definitions of multiplication and division require some explanation. These are motivated by thinking of $a + bi$ and $c + di$ as algebraic expressions to which we can apply the familiar rules of algebra, except that we replace i^2 by -1. Thus for multiplication we have

$$\begin{aligned}
(a + bi) \cdot (c + di) &= ac + adi + bci + bdi^2 \\
&= ac + (ad + bc)i + bd(-1) \\
&= (ac - bd) + (ad + bc)i.
\end{aligned}$$

For division the first step in the following sequence involves multiplying the numerator and denominator by $c - di$; this gives the real number $c^2 + d^2$ in the denominator, as seen in the third step.

$$\begin{aligned}
\frac{a + bi}{c + di} &= \frac{(a + bi)(c - di)}{(c + di)(c - di)} = \frac{ac + bci - adi - bdi^2}{c^2 - d^2 i^2} \\
&= \frac{(ac + bd) + (bc - ad)i}{c^2 + d^2} = \left(\frac{ac + bd}{c^2 + d^2}\right) + \left(\frac{bc - ad}{c^2 + d^2}\right)i.
\end{aligned}$$

Actually, we shall follow the pattern above for multiplying or dividing two complex numbers, rather than substitute into Definition 10.3.

In the process of division described above, the numerator and denominator were multiplied by $c - di$. We call $c - di$ the *conjugate* of $c + di$.

Definition 10.4   Suppose $z = x + yi$, where x and y are real numbers. The *conjugate* of z, denoted by $\bar{z}$, is given by $\bar{z} = x - yi$.

A complex number is in *standard form* if it is written as $a + bi$, where a and b are real numbers. For instance, $(1 + i)/i$ represents a complex number that can be written in standard form as follows:

$$\frac{1 + i}{i} = \frac{(1 + i)(-i)}{i(-i)} = \frac{-i - i^2}{-i^2} = \frac{-i + 1}{1} = 1 - i.$$

Square Roots

The square root of a nonnegative real number b is defined to be a number x satisfying $x^2 = b$. For instance, the square root of 4 is a number x satisfying $x^2 = 4$; there are two such numbers, 2 and -2. We choose 2 as the *principal square root* and write $\sqrt{4} = 2$.

In a similar manner, we can talk about the square root of a negative real number. For example, $\sqrt{-4}$ is a number z satisfying $z^2 = -4$. Since $(2i)^2 = 4i^2 = 4(-1) = -4$ and $(-2i)^2 = 4i^2 = 4(-1) = -4$, we see that $z = 2i$ or $z = -2i$. We choose $2i$ as the *principal square root* of -4 and write $\sqrt{-4} = 2i$.

In general, suppose b is a positive real number. Then

$$\sqrt{-b} = \sqrt{b}\, i.$$

Let us now recall the square root property for real numbers: If a and b are nonnegative real numbers, then $\sqrt{a}\sqrt{b} = \sqrt{ab}$. This can be generalized to the following.

Square root property

Suppose a and b are real numbers such that *not both are negative*. Then $\sqrt{a}\sqrt{b} = \sqrt{ab}$.

Note that the conclusion stated in the square root property is not valid if both a and b are negative numbers. For instance, if $a = -3$ and $b = -12$, then

$$\sqrt{-3}\sqrt{-12} = (\sqrt{3}i)(\sqrt{12}i) = \sqrt{3}\sqrt{12}i^2 = \sqrt{36}(-1) = -6;$$

whereas $\sqrt{(-3)(-12)} = \sqrt{36} = 6$.

Thus, whenever we have $\sqrt{-b}$, where $b > 0$, it is good practice to write it as $\sqrt{b}i$ before performing algebraic manipulations. Thus is illustrated in the following example.

Example 1   Evaluate $(2 + \sqrt{-3})(2 - \sqrt{-3})$.

Solution
$$(2 + \sqrt{-3})(2 - \sqrt{-3}) = (2 + \sqrt{3}i)(2 - \sqrt{3}i) = 2^2 - (\sqrt{3}i)^2$$
$$= 4 - 3i^2 = 4 + 3 = 7. \qquad ■$$

From the above discussion note that we can determine the *square root of any real number*. We could continue with the investigation of the square root of any complex number $a + bi$ where $b \neq 0$. For instance, it is a simple matter to show that $\left(\dfrac{\sqrt{2}}{2} + \dfrac{\sqrt{2}}{2}i\right)^2 = i$ (see problem 6a), and so we could define $\sqrt{i}$ as the number $\dfrac{\sqrt{2}}{2} + \dfrac{\sqrt{2}}{2}i$. However, it is not in our interest to pursue this matter further at this point. See Section 10.5 for a discussion of roots of complex numbers.

Example 2   Write each of the following as complex numbers in standard form.

a) $(3 + 4i) + (5 - 8i)$ **b)** $(2 - 3i) - (-4 + i)$

c) $(3 - 4i)(2 + i)$ **d)** $(1 - 3i) \div (3 + 4i)$

Solution   **a)** $(3 + 4i) + (5 - 8i) = (3 + 5) + (4 - 8)i = 8 - 4i.$

b) $(2 - 3i) - (-4 + i) = (2 + 4) + (-3 - 1)i = 6 - 4i.$

c) $(3 - 4i)(2 + i) = 6 + 3i - 8i - 4i^2 = 6 - 5i + 4 = 10 - 5i.$

d) $(1 - 3i) \div (3 + 4i) = \dfrac{1 - 3i}{3 + 4i} = \dfrac{(1 - 3i)(3 - 4i)}{(3 + 4i)(3 - 4i)} = \dfrac{3 - 13i + 12i^2}{9 - 16i^2}$

$$= \dfrac{3 - 13i - 12}{9 + 16} = \dfrac{-9 - 13i}{25} = \dfrac{-9}{25} - \dfrac{13}{25}i.$$

Example 3   Given that $f(z) = z^3 + 2z^2 - 3$, find $f(1 + i)$.

Solution $f(1 + i) = (1 + i)^3 + 2(1 + i)^2 - 3 = 1 + 3i + 3i^2 + i^3 + 2(1 + 2i + i^2) - 3$

$$= 1 + 3i - 3 - i + 2 + 4i - 2 - 3 = -5 + 6i.$$

Note that we used the familiar rules of algebra, treating i as though it were a variable and replacing i^2 by -1.

Example 4   Given that $z = 2 - i$, find the following.

a) $\bar{z}$ **b)** $z \cdot \bar{z}$ **c)** $\dfrac{\bar{z}}{z}$

Solution   **a)** $\bar{z} = 2 + i$

b) $z \cdot \bar{z} = (2 - i)(2 + i) = 4 - i^2 = 4 + 1 = 5$

c) $\dfrac{\bar{z}}{z} = \dfrac{2 + i}{2 - i} = \dfrac{(2 + i)(2 + i)}{(2 - i)(2 + i)} = \dfrac{4 + 4i + i^2}{4 - i^2} = \dfrac{4 + 4i - 1}{4 + 1} = \dfrac{3}{5} + \dfrac{4}{5}i$

Example 5   Find the roots of $2z^2 + 2iz - 1 = 0$.

Solution   We apply the quadratic formula* to get

$$z = \frac{-2i \pm \sqrt{(2i)^2 - 4(2)(-1)}}{2(2)} = \frac{-2i \pm \sqrt{-4 + 8}}{4} = -\frac{1}{2}i \pm \frac{1}{2}.$$

Therefore the roots are given by $z = \frac{1}{2} - \frac{1}{2}i$ and $z = -\frac{1}{2} - \frac{1}{2}i$. ■

Example 6   Is $1 + \sqrt{3}i$ a zero of the polynomial $P(z) = z^2 - 2z + 4$?

Solution   To answer this question, we evaluate $P(1 + \sqrt{3}i)$ to see whether the result is equal to zero:

$$\begin{aligned}
P(1 + \sqrt{3}i) &= (1 + \sqrt{3}i)^2 - 2(1 + \sqrt{3}i) + 4 \\
&= (1 + 2\sqrt{3}i + 3i^2) - 2 - 2\sqrt{3}i + 4 \\
&= 1 + 2\sqrt{3}i - 3 - 2 - 2\sqrt{3}i + 4 \\
&= (1 - 3 - 2 + 4) + (2\sqrt{3} - 2\sqrt{3})i \\
&= 0 + 0i = 0.
\end{aligned}$$

Therefore the answer to the question is yes. ■

Exercises 10.1

Express answers in $a + bi$ form, where a and b are real numbers.

1. Evaluate each of the following.

a) i^3 **b)** i^6 **c)** i^{32} **d)** i^{17}

e) $(-i)^3$ **f)** $(-i)^5$ **g)** $(-i)^8$ **h)** $(-i)^{17}$

2. Evaluate each of the following.

a) $\dfrac{1}{i^4}$ **b)** $\dfrac{3+i}{i^3}$ **c)** $2i^4 - 3i^{20}$ **d)** $\dfrac{1}{i(i-1)}$

3. Evaluate each of the following.

a) $\sqrt{9} \cdot \sqrt{16}$ **b)** $\sqrt{9}\sqrt{-16}$ **c)** $\sqrt{-9}\sqrt{-16}$

d) $\dfrac{\sqrt{9}}{\sqrt{-16}}$ **e)** $\dfrac{\sqrt{-9}}{\sqrt{16}}$ **f)** $\dfrac{\sqrt{-9}}{\sqrt{-16}}$

4. Evaluate each of the following for $z = 1 - i$.

a) z^2 **b)** $\dfrac{1}{z^2}$ **c)** $3z^2 - 2z^3$

d) $z \cdot \bar{z}$ **e)** $(\bar{z})^3$ **f)** $z \div \bar{z}$

---

*   It can be shown that the quadratic formula is valid for quadratic equations whose coefficients are complex numbers.

5. Given that $f(z) = 2 - 3z - z^2$, evaluate each of the following.

   a) $f(-2)$ **b)** $f(1 + i)$ **c)** $f\left(\dfrac{1}{\sqrt{2}} + \dfrac{1}{\sqrt{2}}i\right)$

6. Show that the following are true.

   a) $\left(\dfrac{1}{\sqrt{2}} + \dfrac{1}{\sqrt{2}}i\right)^2 = i$ **b)** $\left(\dfrac{1}{\sqrt{2}} - \dfrac{1}{\sqrt{2}}i\right)^2 = -i$

7. Show that the following are true.

   a) $\left(\dfrac{\sqrt{3}}{2} + \dfrac{1}{2}i\right)^3 = i$ **b)** $\left(\dfrac{1}{2} + \dfrac{\sqrt{3}}{2}i\right)^3 = -1$

8. Given that $f(z) = z^2 + iz - 3$, evaluate the following.
   a) $f(1 + i)$ **b)** $f(-3i)$

9. Express each of the following in standard $a + bi$ form.

   a) $\sqrt{-4} + (3 - 5\sqrt{-4})$ **b)** $(\sqrt{-48} + 2) - \sqrt{-27}$ **c)** $\sqrt{-8}(2 + \sqrt{-2})$

   d) $(1 + \sqrt{-8})(1 - \sqrt{-8})$ **e)** $\dfrac{1}{1 - \sqrt{-9}}$ **f)** $\dfrac{\sqrt{-2}}{3 + \sqrt{-8}}$

10. In each of the following, determine the roots of the given equation.
    a) $z^2 - 3z + 4 = 0$ **b)** $3z^2 + z - 1 = 0$ **c)** $z^2 + 16 = 0$

11. Determine the roots of the given equations.
    a) $2z^2 - 3iz + 2 = 0$ **b)** $z^2 + 2iz + 3 = 0$
    c) $iz^2 - 3z + i = 0$ **d)** $2iz^2 + z + i = 0$

12. Given that $z = x + iy$, where x and y are real numbers, prove the following.

    a) The real part of z is equal to $\dfrac{z + \bar{z}}{2}$.

    b) The imaginary part of z is equal to $\dfrac{z - \bar{z}}{2i}$.

13. Determine real numbers x and y that satisfy the equation

$$x - 3y - (3x + y)i = -7 + i.$$

14. Solve the equation $z - 3\bar{z} = 1 + i$ for z. *Hint:* let $z = x + iy$; then find x and y.

15. Determine all pairs of real numbers x, y such that $x^2 + 2x + yi = 2 + y + (8 - x)i$.

16. **a)** Is $1 + i$ a root of the equation $z^2 - z + 1 - i = 0$?
    b) Is $1 - i$ a root of the equation given in (a)?

17. Is $-3i$ a solution of the equation $2z^3 - z^2 + 18z - 9 = 0$?

18. Is $1 - \sqrt{5}i$ a zero of the polynomial given by $f(z) = z^3 - z^2 + 4z + 6$?

19. **a)** Is $1 - i$ a root of the equation $z^3 - 3z^2 + 2z - 1 - i = 0$?
    b) Is $1 + i$ a solution of the equation given in (a)?

20. a) Is $1 + \sqrt{3}i$ a solution of the equation $z^3 - 3z^2 + 6z - 4 = 0$?

　　b) Is $1 - \sqrt{3}i$ a root of the equation given in (a)?

In problems 21 through 25, evaluate the given expressions, where f and g are functions with domain **C** and defined by $f(z) = 2z - 1$, $g(z) = z^2 + z$.

21. $(f + g)(1 - i)$　　　　　　**22.** $(f \cdot g)(1 + \sqrt{-2})$　　　　　**23.** $\left(\dfrac{f}{g}\right)(i)$

24. $(f \circ g)(1 + i)$　　　　　　**25.** $(g \circ f)(1 + i)$

10.2　GEOMETRIC REPRESENTATION OF COMPLEX NUMBERS

The set of complex numbers **C** is given by

$$\mathbf{C} = \{x + iy \mid x \text{ and } y \text{ are real numbers and } i^2 = -1\}.$$

We can establish a correspondence between **C** and the set of points in the plane in a natural way: For each complex number $x + iy$, associate the point (x, y) in the plane, and indicate this correspondence by

$$x + iy \longleftrightarrow (x, y).$$

In this setting the plane is referred to as the *complex plane*, where points are labeled either by (x, y) or by $x + iy$. The real numbers are associated with points on the x-axis $(x \leftrightarrow (x, 0))$, and the imaginary numbers correspond to points on the y-axis $(yi \leftrightarrow (0, y))$. Thus the x-axis is called the *real axis*, and the y-axis is referred to as the *imaginary axis*. Some examples of this correspondence are illustrated in Fig. 10.1.

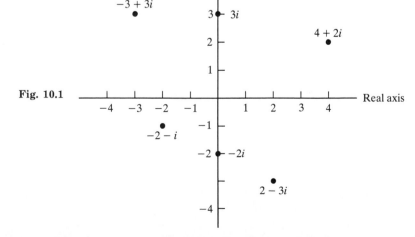

Fig. 10.1

In some problems it is useful to associate each complex number with a *geometric vector,* as shown in Fig. 10.2(a), in which the origin is the initial point and $x + iy$ is the terminal point. Figure 10.2(b) illustrates some examples of this correspondence.

Fig. 10.2

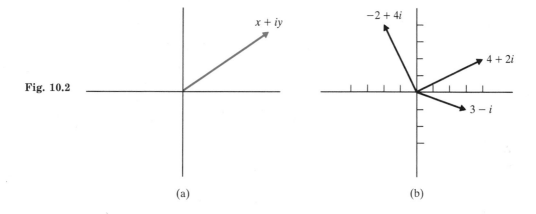

(a)   (b)

Representation of complex numbers by geometric vectors provides us with a convenient geometric interpretation of the sum of complex numbers. The sum $(a + bi) + (c + di)$ is associated with the geometric vector represented by the diagonal of the parallelogram illustrated in Fig. 10.3.

Fig. 10.3

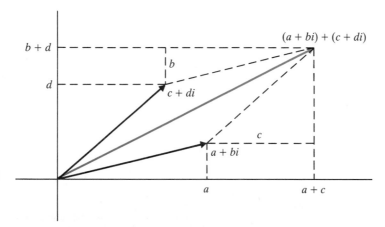

Example 1   For each of the given complex numbers, show the corresponding point (x, y) in the complex plane. Also, draw the corresponding geometric vector.

a) $5 + 3i$ **b)** $-\frac{5}{2} + 3i$ **c)** $\pi - 2i$ **d)** $3i$

Solution   (Fig. 10.4)

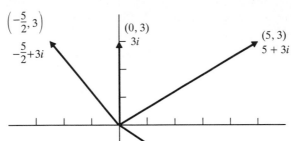

Fig. 10.4

Example 2   Illustrate each of the following by a diagram using geometric vectors.

a) $(4 + 2i) + (1 + 3i)$ **b)** $(1 - 4i) + (-2 + i)$ **c)** $(3 + i) - (1 + 3i)$

Solution   The solutions are shown in Fig. 10.5, where in (c) we use

$$(3 + i) - (1 + 3i) = (3 + i) + (-1 - 3i).$$

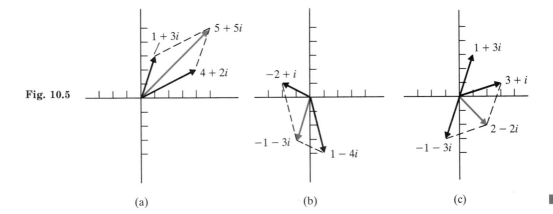

Fig. 10.5

(a)   (b)   (c)

Exercises 10.2

In problems 1 through 8, give the ordered pair of real numbers associated with the given complex number.

1. $3 + 5i$ **2.** $-3 + i$ **3.** $4i$ **4.** $\sqrt{5}$

5. $-\sqrt{-4} + 2i$ **6.** $1 - \pi i$ **7.** $i(1 - \sqrt{-4})$ **8.** $\dfrac{1}{1 - i}$

In problems 9 through 12, give the complex number associated with the given ordered pair.

9. $(0, -4)$ **10.** $(5, 2)$ **11.** $(-4, -3)$ **12.** $(\sqrt{2}, -\sqrt{3})$

In problems 13 through 16, illustrate the given complex number by drawing the associated geometric vector.

13. $-1 + 3i$ **14.** $-4 - 5i$ **15.** $-\sqrt{2} + i$ **16.** $\dfrac{1}{1 - 2i}$

In problems 17 through 20, illustrate geometrically the given sum or difference.

17. $(2 + 3i) + (5 + i)$ **18.** $(1 - 3i) + (4 + 2i)$ **19.** $(4 - i) - (3 + 5i)$ **20.** $(2 - 3i) - (5 + 2i)$

21. Given that $z = 3 - 4i$, on the same set of axes show the points associated with the following.

a) z **b)** $-z$ **c)** $\bar{z}$ **d)** $\dfrac{z + \bar{z}}{2}$

e) $\dfrac{z - \bar{z}}{2}$ **f)** $\sqrt{z \cdot \bar{z}}$

22. Given that $z = -1 + i$, give the ordered pairs corresponding to the following.

a) z^2 **b)** $(\bar{z})^2$ **c)** $\dfrac{1}{z}$ **d)** $z^2 + z + 1$

23. Given that $z = -1/2 + (\sqrt{3}/2)i$, draw the geometric vector associated with the following.

a) z **b)** z^2 **c)** $\dfrac{1}{(\bar{z})^2}$ **d)** $\sqrt{z \cdot \bar{z}}$

24. Let $z = 2(1 + \sqrt{3}i)$. Express each of the following in standard form.

a) z^2 **b)** z^3 **c)** z^4 **d)** z^5

25. Suppose point $P:(x, y)$ is associated with the complex number $x + iy$. State the conditions on x and y that characterize each of the following.

a) P is on the positive real axis.   **b)** P is on the imaginary axis.

c) P is in the first quadrant.   **d)** P is to the right of the imaginary axis.

e) P is below the real axis.

10.3   TRIGONOMETRIC FORM FOR COMPLEX NUMBERS

We continue the development of the preceding section, in which complex numbers are represented as points in the complex plane or as geometric vectors. Suppose $x + iy$ is associated with point $P:(x, y)$ in the complex plane, as shown in Fig. 10.6. Let r denote the distance from the origin O to P, and θ the directed angle that OP makes with the positive real axis. Since $\cos \theta = x/r$ and

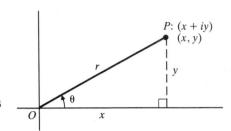

Fig. 10.6

$\sin \theta = y/r$, we have $x = r \cos \theta$ and $y = r \sin \theta$. Hence

$$x + iy = r \cos \theta + ir \sin \theta = r(\cos \theta + i \sin \theta).$$

The result, $r(\cos \theta + i \sin \theta)$, is called the *trigonometric form* or the *polar form* of the complex number $z = x + iy$. The real number r is given by

$$r = \sqrt{x^2 + y^2}$$

and is called the *absolute value* or the *modulus of z;* it is frequently denoted by $|z|$. Since r is the length of the geometric vector associated with z, it is sometimes referred to as the *length of z.*

The angle θ is called an *argument of z* and is denoted by $\theta = \arg z$. It is determined by the two equations

$$\sin \theta = \frac{y}{\sqrt{x^2 + y^2}} \quad \text{and} \quad \cos \theta = \frac{x}{\sqrt{x^2 + y^2}}$$

or by $\tan \theta = y/x$, with proper quadrant selection for θ.

Note that θ is not unique, since we can add or substract any integral multiple of 2π (or $360°$) to a given θ, and the resulting angle can be used in place of θ. The smallest nonnegative angle that can be used for θ is sometimes called the *principal argument of z.* Also, note that

$$z \cdot \bar{z} = (x + iy)(x - iy) = x^2 - i^2 y^2 = x^2 + y^2 = r^2,$$

and so

$$r = \sqrt{z \cdot \bar{z}}.$$

In the special case where P is the origin $(0, 0)$, we take $r = 0$ and do not specify any particular corresponding value of θ.

Representing complex numbers in trigonometric form is particularly useful in problems that involve multiplication or division.

Multiplication of Complex Numbers in Polar Form

Let $z_1 = r_1(\cos\theta_1 + i\sin\theta_1)$ and $z_2 = r_2(\cos\theta_2 + i\sin\theta_2)$ be complex numbers in polar form. Let us consider the product $z_1 \cdot z_2$, using the polar-form expressions.

$$
\begin{aligned}
z_1 \cdot z_2 &= r_1(\cos\theta_1 + i\sin\theta_1) \cdot r_2(\cos\theta_2 + i\sin\theta_2) \\
&= r_1 r_2[(\cos\theta_1\cos\theta_2 - \sin\theta_1\sin\theta_2) + i(\sin\theta_1\cos\theta_2 + \cos\theta_1\sin\theta_2)] \\
&= r_1 r_2[\cos(\theta_1 + \theta_2) + i\sin(\theta_1 + \theta_2)],
\end{aligned}
$$

where in the last step we used identities **I.12** and **I.14** of Chapter 5. Therefore

$$
z_1 \cdot z_2 = r_1 r_2[\cos(\theta_1 + \theta_2) + i\sin(\theta_1 + \theta_2)]. \tag{10.1}
$$

From Eq. (10.1) a geometric interpretation of the product of two complex numbers can be given: $z_1 \cdot z_2$ is a complex number and has length $r_1 r_2$ and argument $\theta_1 + \theta_2$. This is stated as follows:

$$
|z_1 z_2| = |z_1| \cdot |z_2| \qquad \text{and} \qquad \arg(z_1 z_2) = \arg z_1 + \arg z_2. \tag{10.2}
$$

Note: The addition of arguments in the product of complex numbers suggests that a complex number can be expressed in exponential form. This is indeed true. In advanced mathematics courses one learns that z can be expressed as $z = r \cdot e^{i\theta}$, where e is the irrational number $2.71828\ldots$ introduced in Section 3.2.

Division of Complex Numbers in Polar Form

Let z_1 and z_2 be complex numbers expressed in polar form, as above, and suppose $z_2 \neq 0$. Then

$$
\frac{z_1}{z_2} = \frac{r_1}{r_2}[\cos(\theta_1 - \theta_2) + i\sin(\theta_1 - \theta_2)]. \tag{10.3}
$$

The proof of Eq. (10.3), which is similar to that of (10.1), is left as problem 1. From Eq. (10.3) note that the modulus and argument of z_1/z_2 are given by

$$
\left|\frac{z_1}{z_2}\right| = \frac{|z_1|}{|z_2|} \qquad \text{and} \qquad \arg\left(\frac{z_1}{z_2}\right) = \arg z_1 - \arg z_2. \tag{10.4}
$$

In Examples 1, 2, and 3, complex numbers z_1, z_2, z_3, and z_4 are given by

$$z_1 = 1 + i, \qquad z_2 = \sqrt{3} - i, \qquad z_3 = -2 - 2\sqrt{3}\,i, \qquad z_4 = -3 + 4i.$$

Example 1   Express the following in polar form.

      a) z_1 **b)** z_2 **c)** z_3 **d)** z_4

Solution   **a)** $r_1 = |z_1| = \sqrt{1^2 + 1^2} = \sqrt{2}$, and $\theta_1 = \pi/4 = 45°$ (see Fig. 10.7(a)). Therefore

$$z_1 = \sqrt{2}\left(\cos\frac{\pi}{4} + i\sin\frac{\pi}{4}\right)$$

$$= \sqrt{2}(\cos 45° + i\sin 45°).$$

b) $r_2 = |z_2| = \sqrt{(\sqrt{3})^2 + (-1)^2} = \sqrt{4} = 2$, and $\theta_2 = 11\pi/6 = 330°$ (see Fig. 10.7(b)). Thus

$$z_2 = 2\left(\cos\frac{11\pi}{6} + i\sin\frac{11\pi}{6}\right)$$

$$= 2(\cos 330° + i\sin 330°).$$

c) From Fig. 10.7(c) we see that

$$z_3 = 4\left(\cos\frac{4\pi}{3} + i\sin\frac{4\pi}{3}\right)$$

$$= 4(\cos 240° + i\sin 240°).$$

Fig. 10.7

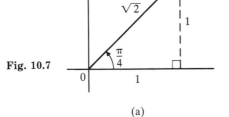

(a)

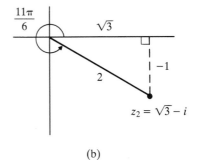

(b)

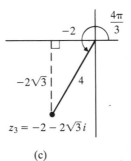

(c)

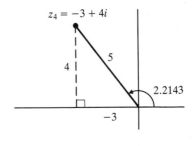

(d)

d) From Fig. 10.7(d) we see that $\theta_4 = \text{Cos}^{-1}(-\frac{3}{5}) = 2.2143 = 126.87°$. Therefore

$$z_4 = 5(\cos 2.2143 + i \sin 2.2143)$$
$$= 5(\cos 126.87° + i \sin 126.87°).$$ ▮

Example 2   Find the following. Express each answer both in polar form and in rectangular form. Use two-decimal place values for approximate answers.

a) $z_1 \cdot z_2$ **b)** $z_3 \cdot z_4$ **c)** $z_1 \cdot z_2 \cdot z_3$

Solution   In each case the formula given by Eq. (10.1) is used.

a) $z_1 \cdot z_2 = (\sqrt{2})(2)[\cos(45° + 330°) + i\sin(45° + 330°)]$
$$= 2\sqrt{2}[\cos 375° + i \sin 375°]$$
$$= 2\sqrt{2}(\cos 15° + i \sin 15°) \quad \text{(polar form)}$$
$$= 2.73 + 0.73i. \quad \text{(rectangular form)}.$$

b) $z_3 \cdot z_4 = (4)(5)[\cos(240° + 126.87°) + i\sin(240° + 126.87°)]$
$$= 20[\cos 366.87° + i \sin 366.87°]$$
$$= 20(\cos 6.87° + i \sin 6.87°) \quad \text{(polar form)}$$
$$= 19.86 + 2.39i. \quad \text{(rectangular form)}.$$

c) $z_1 \cdot z_2 \cdot z_3 = (\sqrt{2})(2)(4)\left[\cos\left(\frac{\pi}{4} + \frac{11\pi}{6} + \frac{4\pi}{3}\right) + i \sin\left(\frac{\pi}{4} + \frac{11\pi}{6} + \frac{4\pi}{3}\right)\right]$

$$= 8\sqrt{2}\left[\cos\frac{41\pi}{12} + i \sin\frac{41\pi}{12}\right]$$

$$= 8\sqrt{2}\left(\cos\frac{17\pi}{12} + i \sin\frac{17\pi}{12}\right) \quad \text{(polar form)}$$

$$= -2.93 - 10.93i. \quad \text{(rectangular form)}.$$ ▮

Example 3   Evaluate the following. Express each answer in both polar form and rectangular form.

a) $\dfrac{z_1}{z_2}$ **b)** $\dfrac{z_3}{z_4}$

Solution   We use Eq. (10.3).

a) $\dfrac{z_1}{z_2} = \dfrac{\sqrt{2}}{2}[\cos(45° - 330°) + i \sin(45° - 330°)]$

$$= \frac{\sqrt{2}}{2}[\cos(-285°) + i \sin(-285°)] \quad \text{(polar form)}$$

$$= \frac{\sqrt{2}}{2}[\cos 285° - i \sin 285°]$$

$$= 0.18 + 0.68i. \quad \text{(rectangular form)}.$$

b) $\dfrac{z_3}{z_4} = \dfrac{4}{5}\left[\cos\left(\dfrac{4\pi}{3} - 2.2143\right) + i\sin\left(\dfrac{4\pi}{3} - 2.2143\right)\right]$

$\qquad = \dfrac{4}{5}[\cos(1.9745) + i\sin(1.9745)]$ (polar form)

$\qquad = -0.31 + 0.74i.$ (rectangular form).   ■

Example 4   Express $3(\cos 60° - i\sin 60°)$ in polar form.

Solution   *Method 1.* For a complex number to be in polar form, it must be expressed as $r(\cos\theta + i\sin\theta)$, where $r \geq 0$. The given number is not in polar form because of the minus sign. However, since $\cos(-60°) = \cos 60°$ and $\sin(-60°) = -\sin 60°$, we can write

$$3(\cos 60° - i\sin 60°) = 3[\cos(-60°) + \sin(-60°)],$$

which is in polar form. Since $-60°$ and $300°$ are coterminal angles, this can also be written as $3(\cos 300° + i\sin 300°)$.

Method 2. Write the given number in rectangular form first.

$$3(\cos 60° - i\sin 60°) = \dfrac{3\sqrt{3}}{2} - \dfrac{3}{2}i.$$

From Fig. 10.8 we see that $3(\cos 300° + i\sin 300°)$ is a polar form of the given number.

Fig. 10.8

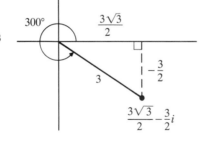

■

Example 5   Express $-4(\cos 120° + i\sin 120°)$ in polar form.

Solution   The given number is not in polar form, because the -4 is not an acceptable value for r (we require $r \geq 0$). The given number can be written as follows:

$$-4(\cos 120° + i\sin 120°) = 4[-\cos 120° + i(-\sin 120°)]$$
$$= 4[\cos(180° + 120°) + i\sin(180° + 120°)]$$
$$= 4(\cos 300° + i\sin 300°).$$

Note that we used identities $\cos(180° + \theta) = -\cos\theta$ and $\sin(180° + \theta) = -\sin\theta$

with $\theta = 120°$. Thus

$$-4(\cos 120° + i\sin 120°) = 4(\cos 300° + i\sin 300°),$$

which is in polar form.

We encourage the reader to follow Method 2 of Example 4 to get the solution.   ▪

Example 6   Express $3(\sin 47° - i\cos 47°)$ in polar form.

Solution $3(\sin 47° - i\cos 47°) = 3[\cos(270° + 47°) + i\sin(270° + 47°)]$
$$= 3(\cos 317° + i\sin 317°).$$

Here we used identities $\cos(270° + \theta) = \sin\theta$ and $\sin(270° + \theta) = -\cos\theta$ with $\theta = 47°$. Thus the polar form of the given number is

$$3(\cos 317° + i\sin 317°).$$ ▪

Exercises 10.3

In each of the problems of this exercise, give answers in exact form whenever it is reasonable to do so; otherwise use a calculator and state the results in decimal form (two places for degree measure, four places for radian measure).

1. Given that z_1 and z_2 are complex numbers expressed in polar form, prove that

$$\frac{z_1}{z_2} = \frac{r_1}{r_2}[\cos(\theta_1 - \theta_2) + i\sin(\theta_1 - \theta_2)].$$

2. Express each of the given numbers in polar form.
   a) -3 b) $1 - i$ c) $-i$ d) $1 + \sqrt{3}i$

3. Express the following in polar form.
   a) π b) $3 - 4i$ c) $i^5 - i^4$ d) $12 - 5i$

4. Express the following in polar form.

   a) $-3 - 3i$ b) $5i^2 - 2i - 3$ c) $\dfrac{1}{i}$ d) $\dfrac{1}{i - i^2}$

5. Express the following in rectangular form.

   a) $3(\cos 45° + i\sin 45°)$ b) $5(\cos 180° + i\sin 180°)$ c) $\cos\dfrac{4\pi}{3} + i\sin\dfrac{4\pi}{3}$

6. Express the following in rectangular form.

   a) $\cos\left(-\dfrac{7\pi}{6}\right) + i\sin\left(-\dfrac{7\pi}{6}\right)$ b) $\cos 450° + i\sin 450°$ c) $3(\cos 137° + i\sin 137°)$

7. Determine why the given number is not in polar form. Then express it in polar form.

   a) $4(\cos 45° - i\sin 45°)$ b) $-3(\cos 300° + i\sin 300°)$ c) $-\cos\dfrac{5\pi}{6} + i\sin\dfrac{5\pi}{6}$

8. Express the following in polar form.

   a) $3\left(-\cos\frac{\pi}{6} + i\sin\frac{\pi}{6}\right)$ **b)** $-5(\cos 40° - i\sin 40°)$ **c)** $-\cos 120° - i\sin 120°$

In problems 9 through 12, perform the indicated operations and express answers in (a) polar form, (b) rectangular form. *Hint:* Write numbers in polar form first and then use Eq. (10.1) or Eq. (10.3).

9. $(\cos 15° + i\sin 15°) \cdot (\cos 30° + i\sin 30°)$

10. $4(\cos 47° - i\sin 47°) \cdot (\cos 43° - i\sin 43°)$

11. $\dfrac{8(\cos 150° + i\sin 150°)}{4(\cos 30° + i\sin 30°)}$ **12.** $\dfrac{\cos 50° + i\sin 50°}{\cos 80° - i\sin 80°}$

In problems 13 through 15, let $z_1 = 3(\cos 210° - i\sin 210°)$, $z_2 = 6(\sin 60° + i\cos 60°)$. Evaluate the given expressions by using Eq. (10.1) or Eq. (10.3).

13. $z_1 \cdot z_2$ **14.** $z_2 \div z_1$ **15.** $\dfrac{1}{z_2}$

In problems 16 through 20, let $z_1 = \sqrt{3} + i$ and $z_2 = -2 + 2i$. Express each of the given numbers in polar form.

16. a) z_1 **b)** z_2 **17. a)** $\bar{z}_1$ **b)** $\bar{z}_2$ **18. a)** $z_1 \cdot z_2$ **b)** $\bar{z}_1 \cdot \bar{z}_2$

19. a) $z_1 \div z_2$ **b)** $\bar{z}_1 \div \bar{z}_2$ **20. a)** $\dfrac{1}{z_1}$ **b)** $\dfrac{1}{z_2}$

21. Given that $z = r(\cos\theta + i\sin\theta)$ represents a complex number in polar form, show that the following are true.

   a) $z^2 = r^2(\cos 2\theta + i\sin 2\theta)$ **b)** $z^3 = r^3(\cos 3\theta + i\sin 3\theta)$

22. Given that $z = r(\cos\theta + i\sin\theta)$ represents a complex number in polar form and $r \neq 0$, show that the following are true.

   a) $\dfrac{1}{z} = \dfrac{1}{r}[\cos(-\theta) + i\sin(-\theta)]$ **b)** $\dfrac{1}{z^2} = \dfrac{1}{r^2}[\cos(-2\theta) + i\sin(-2\theta)]$

23. Use problem 21 to evaluate the following.

   a) $(\sqrt{2} - \sqrt{2}i)^2$ **b)** $(1 + \sqrt{3}i)^3$

24. Use problem 22 to evaluate the following.

   a) $\dfrac{1}{1 + i}$ **b)** $\dfrac{1}{(\sqrt{3} - i)^2}$

    10.4   **DEMOIVRE'S THEOREM**

        Suppose z is a complex number in polar form, $z = r(\cos\theta + i\sin\theta)$. Applying Eq. (10.1) to the special case where both z_1 and z_2 are taken to be z gives

$$z \cdot z = r \cdot r[\cos(\theta + \theta) + i\sin(\theta + \theta)],$$
$$z^2 = r^2(\cos 2\theta + i\sin 2\theta).$$

If Eq. (10.1) is applied again with $z_1 = z$ and $z_2 = z^2$, we get

$$z^3 = r^3(\cos 3\theta + i \sin 3\theta).$$

This suggests that in general

$$z^n = r^n(\cos n\theta + i \sin n\theta) \tag{10.5}$$

for each positive integer n. This is indeed a true statement; the reader is asked to give a formal proof in problem 16.

Taking $r = 1$ in Eq. (10.5) gives the special case

$$(\cos \theta + i \sin \theta)^n = \cos n\theta + i \sin n\theta$$

for each positive integer n. This is known as *DeMoivre's theorem.*[*]

Equation (10.5) is stated for n a positive integer. For exponents that are not positive integers, we follow a pattern similar to that already encountered in algebra. We first define z^k, where k is zero, then for k a negative integer.

Definition 10.5

Zero exponent

If $z \neq 0$, then $z^0 = 1$.

Negative-integer exponent

If n is any positive integer and $z \neq 0$, then $z^{-n} = \dfrac{1}{z^n}$.

We now investigate z^{-n}, where n is a positive integer. Let $z = r(\cos \theta + i \sin \theta)$. Then

$$z^{-n} = \frac{1}{z^n} \qquad \text{(by Definition 10.5)}$$

$$= \frac{1}{r^n(\cos n\theta + i \sin n\theta)} \qquad \text{(by Eq. (10.5))}$$

$$= \frac{1}{r^n}\left(\frac{\cos 0 + i \sin 0}{\cos n\theta + i \sin n\theta}\right) \qquad \text{(since } 1 = \cos 0 + i \sin 0\text{)}$$

$$= r^{-n}[\cos(-n\theta) + i \sin(-n\theta)]. \qquad \text{(by Eq. (10.3))}$$

---

[*]   Named after the French-born English mathematician Abraham DeMoivre (1667–1754).

Thus we have

$$z^{-n} = r^{-n}[\cos(-n\theta) + i\sin(-n\theta)].$$

This is precisely Eq. (10.5) for negative integer exponents.
  Equation (10.5) also holds for $n = 0$ since $z^0 = 1$ and

$$r^0[\cos(0 \cdot \theta) + i\sin(0 \cdot \theta)] = 1 \cdot (\cos 0 + i\sin 0) = 1.$$

Therefore the formula given by Eq. (10.5) is generalized to

> If $z = r(\cos\theta + i\sin\theta)$ and n is *any integer,* then
> $$z^n = r^n(\cos n\theta + i\sin n\theta). \qquad (10.6)$$

Example 1   Express each of the following as a complex number in both polar form and rectangular form.

a) $(1 + i)^6$ **b)** $(-1 + \sqrt{3}i)^8$ **c)** $(3 - 4i)^4$

Solution   **a)** We first express $1 + i$ in polar form and then use the formula given by Eq. (10.6):

$$(1 + i)^6 = [\sqrt{2}(\cos 45° + i\sin 45°)]^6 = (\sqrt{2})^6[\cos(6 \cdot 45°) + i\sin(6 \cdot 45°)]$$
$$= 8(\cos 270° + i\sin 270°) \quad \text{(polar form)}$$
$$= 8[0 + i(-1)] = -8i. \quad \text{(rectangular form)}$$

b) $(-1 + \sqrt{3}i)^8 = \left[2\left(\cos\frac{2\pi}{3} + i\sin\frac{2\pi}{3}\right)\right]^8$

$$= 2^8\left[\cos\left(8 \cdot \frac{2\pi}{3}\right) + i\sin\left(8 \cdot \frac{2\pi}{3}\right)\right]$$

$$= 256\left[\cos\frac{16\pi}{3} + i\sin\frac{16\pi}{3}\right]$$

$$= 256\left[\cos\left(4\pi + \frac{4\pi}{3}\right) + i\sin\left(4\pi + \frac{4\pi}{3}\right)\right]$$

$$= 256\left[\cos\frac{4\pi}{3} + i\sin\frac{4\pi}{3}\right] \qquad \text{(polar form)}$$

$$= 256\left[-\frac{1}{2} + i\left(-\frac{\sqrt{3}}{2}\right)\right]$$

$$= -128 - 128\sqrt{3}i. \qquad \text{(rectangular form)}$$

c) $(3 - 4i)^4 = [r(\cos\theta + i\sin\theta)]^4 = r^4(\cos 4\theta + i\sin 4\theta)$, where $r = 5$ and $\theta = \text{Sin}^{-1}(-4/5)$ (see Fig. 10.9). Using a calculator, we evaluate

$$4\theta = 4\,\text{Sin}^{-1}(-\tfrac{4}{5}) = -212.52°.$$

Fig. 10.9

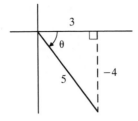

Therefore,

$$(3 - 4i)^4 = 625[\cos(-212.52°) + i\sin(-212.52°)] \quad \text{(polar form)}$$
$$= -527 + 336i. \quad \text{(rectangular form)}$$

Example 2   Evaluate each of the following, and express answers in both polar form and rectangular form.

a) $[2(\cos 22°30' + i\sin 22°30')]^4$ **b)** $(\cos 45° - i\sin 45°)^5$

Solution   **a)** Using Eq. (10.6) gives

$$[2(\cos 22°30' + i\sin 22'30')]^4 = 2^4[\cos 4(22°30') + i\sin 4(20°30')]$$
$$= 16(\cos 90° + i\sin 90°) \quad \text{(polar form)}$$
$$= 16\,i. \quad \text{(rectangular form)}$$

b) First express $\cos 45° - i\sin 45°$ in polar form as

$$\cos 45° - i\sin 45° = \cos(-45°) + i\sin(-45°).$$

Applying Eq. (10.6) we have

$$(\cos 45° - i\sin 45°)^5 = [\cos(-45°) + i\sin(-45°)]^5$$
$$= \cos 5(-45°) + i\sin 5(-45°)$$
$$= \cos(-225°) + i\sin(-225°) \quad \text{(polar form)}$$
$$= -\frac{\sqrt{2}}{2} + \frac{\sqrt{2}}{2}i. \quad \text{(rectangular form)}$$

Example 3   Express $\sin 4\theta$ and $\cos 4\theta$ as identities in terms of $\sin\theta$ and $\cos\theta$.

Solution   Substituting $n = 4$ into DeMoivre's theorem gives

$$(\cos\theta + i\sin\theta)^4 = \cos 4\theta + i\sin 4\theta.$$

Using the binomial expansion on the left-hand side of this equation, we get

$$\cos^4\theta + 4(\cos^3\theta \sin\theta)i + 6(\cos^2\theta \sin^2\theta)i^2 + 4(\cos\theta \sin^3\theta)i^3 + (\sin^4\theta)i^4$$
$$= \cos 4\theta + i\sin 4\theta.$$

Now use $i^2 = -1$, $i^3 = i^2 \cdot i = -i$ and $i^4 = i^2 \cdot i^2 = (-1)(-1) = 1$, and we collect real and imaginary terms to get

$$[\cos^4\theta - 6\cos^2\theta \sin^2\theta + \sin^4\theta] + [4\cos^3\theta \sin\theta - 4\cos\theta \sin^3\theta]i$$
$$= \cos 4\theta + i\sin 4\theta.$$

Using the definition of equality of two complex numbers (see Definition 10.2), we get

$$\sin 4\theta = 4\cos^3\theta \sin\theta - 4\cos\theta \sin^3\theta,$$
$$\cos 4\theta = \cos^4\theta - 6\cos^2\theta \sin^2\theta + \sin^4\theta.$$

These are identities.   ■

By using the technique illustrated in Example 3, we can solve the general problem of determining identities in which $\sin n\theta$ and $\cos n\theta$ are expressed in terms of $\sin \theta$ and $\cos \theta$.

Exercises 10.4

In the following problems, give answers in exact form whenever it is reasonable to do so; otherwise state results in decimal form, with numbers rounded off to two decimal places, angles to two places for degree measure and four places for radian measure. Express answers in both polar form and rectangular form.

In problems 1 through 8, perform the indicated operations.

1. a) $(\cos 30° + i\sin 30°)^5$ **b)** $[2(\cos - 45° + i\sin - 45°)]^4$ **c)** $(\cos 40° + i\sin 40°)^{-3}$

2. a) $(\cos 47° + i\sin 47°)^6$ **b)** $\left[3\left(\cos \frac{\pi}{3} + i\sin \frac{\pi}{3}\right)\right]^4$ **c)** $[\cos(-20°) + i\sin(-20°)]^{-6}$

3. a) $[2(\cos 150° - i\sin 150°)]^3$ **b)** $\dfrac{16}{[2(\cos 45° - i\sin 45°)]^4}$

4. a) $[-3(\cos 20° + i\sin 20°)]^4$ **b)** $\dfrac{81}{[-3(\cos \pi/12 + i\sin \pi/12)]^4}$

5. a) $(-1 + i)^8$ **b)** $(\sqrt{3} - i)^4$ **c)** $(1 + i)^{-3}$

6. a) $(\sqrt{2} + \sqrt{2}i)^4$ **b)** $\dfrac{1}{(1 - \sqrt{3}i)^6}$ **c)** $(2 + i)^6$

7. a) $(-1 + i)^4 \cdot (1 + \sqrt{3}i)^6$ **b)** $\dfrac{(2 + 2i)^4}{(\sqrt{3} + i)^3}$

8. a) $(1 - i)^{-3} \cdot (1 + i)^4$ **b)** $(2 - 3i)^2 \cdot (4 + 3i)^4$

In problems 9 through 12, $z = 1 - i$ and $w = -\sqrt{3} + i$. Evaluate the given expression.

9. $z^4 - z$ **10.** $z^3 \cdot w^4$ **11.** $z^4 - w^4$

12. $z^4 + z^3 + z^2 + z + 1$ *Hint:* The identity $(z - 1)(z^4 + z^3 + z^2 + z + 1) = z^5 - 1$ may be useful.

13. Given that $f(z) = z^4 - 2z^3 + z$, find the following.

    a) $f(i)$ **b)** $f(-1 + i)$

14. In Eq. (10.6) take $n = 2$, $r = 1$ and get identities **(I.18)** and **(I.19)** of Chapter 5.

15. Express $\sin 3\theta$ and $\cos 3\theta$ as identities in terms of $\sin \theta$ and $\cos \theta$ (see Example 3).

16. Prove that $z^n = [r(\cos \theta + i \sin \theta)]^n = r^n(\cos n\theta + i \sin n\theta)$ for each positive integer n. *Hint:* Use mathematical induction.

10.5   ROOTS OF COMPLEX NUMBERS

In Section 10.1 we discussed the problem of determining *the square root of any real number* and arrived at the following: Suppose c is a nonnegative real number. Then $\sqrt{c}$ is that nonnegative solution of $x^2 = c$, and $\sqrt{-c}$ is equal to the imaginary number $\sqrt{c}\,i$. In this section we are interested in the general problem of determining the nth roots of any complex number $z = a + bi$, where n is an integer greater than or equal to 2.

Definition 10.6   Suppose z is a given complex number. The *nth roots of z* are the solutions of the equation $w^n = z$.

Although we were able to talk about *the* square root of a real number, we shall make no attempt to define *the* nth root of z in general. Definition 10.6 refers to "the nth roots of z," and we do not select a particular solution of $w^n = z$ and call it the principal value or the nth root of z as we did for the square root of a real number.

Let us proceed with the problem of solving the equation $w^n = z$, where z is a given complex number. For the trivial case of $z = 0$, the solution is $w = 0$. Hence, in the following we shall assume that $z \neq 0$. Suppose z and w are expressed in polar form as

$$z = r(\cos \theta + i \sin \theta),$$
$$w = R(\cos \alpha + i \sin \alpha).$$

Then $w^n = z$ becomes

$$[R(\cos \alpha + i \sin \alpha)]^n = r(\cos \theta + i \sin \theta).$$

Applying Eq. (10.6) to the left side gives

$$R^n(\cos n\alpha + i \sin n\alpha) = r(\cos \theta + i \sin \theta).$$

From the definition of equality of two complex numbers (see Definition 10.2), it follows that

$$R^n \cos n\alpha = r \cos \theta \quad \text{and} \quad R^n \sin n\alpha = r \sin \theta.$$

Solving this pair of simultaneous equations for R and α (see problem 21) gives

$$R = r^{1/n} = \sqrt[n]{r} \quad \text{and} \quad \alpha = \frac{\theta + k \cdot 2\pi}{n} = \frac{\theta}{n} + \frac{2\pi k}{n},$$

where k is any integer.* Therefore $w^n = z$ has solutions given by

$$w_k = r^{1/n}\left[\cos\left(\frac{\theta}{n} + \frac{2\pi k}{n}\right) + i\sin\left(\frac{\theta}{n} + \frac{2\pi k}{n}\right)\right]. \qquad (10.7)$$

If we let k take on various integral values, we see that $w_0, w_1, w_2, \ldots, w_{n-1}$ will be n *distinct* complex numbers. These are given by

$$
\begin{aligned}
w_0 &= r^{1/n}\left[\cos\frac{\theta}{n} + i\sin\frac{\theta}{n}\right], \\[4pt]
w_1 &= r^{1/n}\left[\cos\left(\frac{\theta}{n} + \frac{2\pi}{n}\right) + i\sin\left(\frac{\theta}{n} + \frac{2\pi}{n}\right)\right], \\[4pt]
w_2 &= r^{1/n}\left[\cos\left(\frac{\theta}{n} + \frac{4\pi}{n}\right) + i\sin\left(\frac{\theta}{n} + \frac{4\pi}{n}\right)\right], \\
&\;\;\vdots \\
w_{n-1} &= r^{1/n}\left[\cos\left(\frac{\theta}{n} + \frac{2(n-1)\pi}{n}\right) + i\sin\left(\frac{\theta}{n} + \frac{2(n-1)\pi}{n}\right)\right].
\end{aligned}
\qquad (10.8)
$$

Suppose we evaluate w_n by replacing k by n in Eq. 10.7. This gives

$$
\begin{aligned}
w_n &= r^{1/n}\left[\cos\left(\frac{\theta}{n} + \frac{2\pi n}{n}\right) + i\sin\left(\frac{\theta}{n} + \frac{2\pi n}{n}\right)\right] \\[4pt]
&= r^{1/n}\left[\cos\left(\frac{\theta}{n} + 2\pi\right) + i\sin\left(\frac{\theta}{n} + 2\pi\right)\right] \\[4pt]
&= r^{1/n}\left[\cos\frac{\theta}{n} + i\sin\frac{\theta}{n}\right] = w_0.
\end{aligned}
$$

In a similar manner we can show that each value of k for which $k \geq n$ or $k < 0$ will give a w_k that is already included in Eqs. (10.8).

Geometrically, the n numbers given by (10.8) are located on the circle with center at the origin and radius $\sqrt[n]{r}$; they are equally spaced around the circle with the angle between any two consecutive values being $2\pi/n$. These are shown in Fig. 10.10.

---

* The solution for R involves $r^{1/n}$; here r is a positive real number, and $r^{1/n}$ has been defined in Section 3.1.

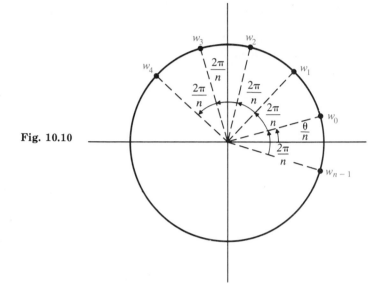

Fig. 10.10

Example 1   Find the four roots of $z^4 + 1 = 0$.

Solution   We wish to solve for the roots of the equation $z^4 = -1$. First we express -1 in polar form: $-1 = \cos \pi + i \sin \pi$. Substituting $n = 4$, $r = 1$, and $\theta = \pi$ into Eq. (10.7) gives

$$w_k = \cos\left(\frac{\pi}{4} + \frac{2\pi k}{4}\right) + i \sin\left(\frac{\pi}{4} + \frac{2\pi k}{4}\right)$$

$$= \cos\left(\frac{\pi}{4} + k \cdot \frac{\pi}{2}\right) + i \sin\left(\frac{\pi}{4} + k \cdot \frac{\pi}{2}\right).$$

Replacing k by 0, 1, 2, and 3 gives the four roots w_0, w_1, w_2, and w_3, respectively.

$$w_0 = \frac{\sqrt{2}}{2} + \frac{\sqrt{2}}{2}i, \qquad w_1 = -\frac{\sqrt{2}}{2} + \frac{\sqrt{2}}{2}i,$$

$$w_2 = -\frac{\sqrt{2}}{2} - \frac{\sqrt{2}}{2}i, \qquad w_3 = \frac{\sqrt{2}}{2} - \frac{\sqrt{2}}{2}i.$$

Example 2   Solve the equation $z^4 - 2z^2 + 2 = 0$.

Solution   The given equation is quadratic in z^2. Solving for z^2 by use of the quadratic formula gives

$$z^2 = \frac{-(-2) \pm \sqrt{(-2)^2 - 4(1)(2)}}{2(1)} = \frac{2 \pm \sqrt{-4}}{2} = 1 \pm i,$$

and

$$1 + i = 2(\cos 45° + i \sin 45°),$$
$$1 - i = 2(\cos 315° + i \sin 315°).$$

Using (10.8) with $n = 2$, we get the following solutions: $z^2 = 1 + i$ gives

$$w_0 = (\sqrt{2})^{1/2}\left[\cos\frac{45°}{2} + i \sin\frac{45°}{2}\right]$$

$$= \sqrt[4]{2}(\cos 22.5° + i \sin 22.5°) = 1.10 + 0.46i,$$

$$w_1 = (\sqrt{2})^{1/2}\left[\cos\left(\frac{45°}{2} + \frac{360°}{2}\right) + i \sin\left(\frac{45°}{2} + \frac{360°}{2}\right)\right]$$

$$= \sqrt[4]{2}(\cos 202.5° + i \sin 202.5°) = -1.10 - 0.46i;$$

$z^2 = 1 - i$ gives

$$w_0' = (\sqrt{2})^{1/2}\left[\cos\frac{315°}{2} + i \sin\frac{315°}{2}\right]$$

$$= \sqrt[4]{2}(\cos 157.5° + i \sin 157.5°) = -1.10 + 0.46i,$$

$$w_1' = (\sqrt{2})^{1/2}\left[\cos\left(\frac{315°}{2} + \frac{360°}{2}\right) + i \sin\left(\frac{315°}{2} + \frac{360°}{2}\right)\right]$$

$$= \sqrt[4]{2}(\cos 337.5° + i \sin 337.5°) = 1.10 - 0.46i.$$

Therefore the solution set for the given equation is

$$\{1.10 + 0.46i, \quad -1.10 + 0.46i, \quad -1.10 - 0.46i, \quad 1.10 - 0.46i\},$$

where the numbers are given to two decimal places. The numbers in the solution set are shown in Fig. 10.11, in which the radius of the circle is $\sqrt[4]{2}$.

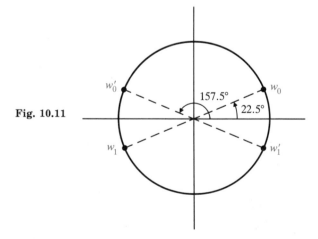

Fig. 10.11

Example 3   Find the square roots of $-3 - 4i$.

Solution   We want to solve the equation $w^2 = -3 - 4i$. We first express $-3 - 4i$ in polar form:

$$-3 - 4i = 5(\cos\theta + i\sin\theta),$$

where θ is the angle shown in Fig. 10.12. Substituting into (10.8) with $n = 2$ gives

$$w_0 = 5^{1/2}\left(\cos\frac{\theta}{2} + i\sin\frac{\theta}{2}\right),$$

$$w_1 = 5^{1/2}\left[\cos\left(\frac{\theta}{2} + \frac{2\pi}{2}\right) + i\sin\left(\frac{\theta}{2} + \frac{2\pi}{2}\right)\right]$$

$$= 5^{1/2}\left[\cos\left(\frac{\theta}{2} + \pi\right) + i\sin\left(\frac{\theta}{2} + \pi\right)\right]$$

$$= 5^{1/2}\left[-\cos\frac{\theta}{2} - i\sin\frac{\theta}{2}\right]$$

$$= -5^{1/2}\left(\cos\frac{\theta}{2} + i\sin\frac{\theta}{2}\right) = -w_0.$$

Fig. 10.12

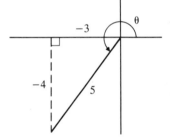

We can write w_0 in better form by using the half-angle identities. Since $\pi < \theta < 3\pi/2$, then $\pi/2 < \theta/2 < 3\pi/4$, and so $\theta/2$ is an angle in the second quadrant. Therefore $\cos\theta/2$ is negative, and $\sin\theta/2$ is positive. Since $\cos\theta = -3/5$ (see Fig. 10.12),

$$\cos\frac{\theta}{2} = -\sqrt{\frac{1 + \cos\theta}{2}} = -\sqrt{\frac{1 + (-3/5)}{2}} = -\frac{1}{\sqrt{5}},$$

$$\sin\frac{\theta}{2} = \sqrt{\frac{1 - \cos\theta}{2}} = \sqrt{\frac{1 - (-3/5)}{2}} = \frac{2}{\sqrt{5}}.$$

Thus we have

$$w_0 = \sqrt{5}\left(\cos\frac{\theta}{2} + i\sin\frac{\theta}{2}\right) = \sqrt{5}\left(-\frac{1}{\sqrt{5}} + \frac{2i}{\sqrt{5}}\right) = -1 + 2i,$$

$$w_1 = -w_0 = 1 - 2i.$$

Therefore the square roots of $-3 - 4i$ are $-1 + 2i$ and $1 - 2i$.   ■

Exercises 10.5

In the problems of this exercise, express answers in polar form. Then give answers in rectangular form as exact numbers when reasonable, otherwise to two decimal places.

1. Find the cube roots of 1.

2. Determine the fourth roots of i.

3. Find the fifth roots of $1 - \sqrt{3}i$.

4. Determine the roots of the equation $z^4 + 1 - i = 0$.

5. Find the sixth roots of -1, and show the results in a diagram.

6. Determine the sixth roots of $64(\cos 126° + i\sin 126°)$.

7. Find the fourth roots of $16(\sqrt{3} + i)$.

8. Determine the fourth roots of $(\sqrt{3} - i)^3$.

9. Determine the cube roots of $\left(\dfrac{1 - i}{\sqrt{2}}\right)^{-2}$.

In problems 10 through 13, solve the given quadratic equations.

10. $z^2 - (2 + 3i)z - 1 + 3i = 0$

11. $z^2 - 3z + 3 - i = 0$

12. $2z^2 + 2\sqrt{2}(-1 + i)z - 1 - 2i = 0$

13. $z^2 + z + 1 - i = 0$

14. Find the roots of the equation $z^4 + 1 = 0$.

15. Find the roots of the equation $z^3 + z^2 + iz + i = 0$. *Hint:* Factor first.

16. Find the roots of the equation $z^5 + 2z^3 - z^2 - 2 = 0$. *Hint:* Factor first.

17. Find the square roots of $3 - 4i$.

18. Find the square roots of $3 + 4i$.

19. Find the square roots of $-5 + 12i$.

20. Find the roots of $z^2 - iz - 1 + i = 0$.

21. In the derivation of Eq. (10.7) we encountered the problem of solving the following two equations simultaneously for R and α in terms of r and θ:

$$R^n \cos(n\alpha) = r\cos\theta, \qquad R^n \sin(n\alpha) = r\sin\theta.$$

Carry out the solution and show that $R = r^{1/n}$ and $\alpha = (\theta + k \cdot 2\pi)/n$.
Hint: First eliminate α by squaring each of the given equations and then adding the resulting equations. Use identity **(I.9)** twice. After getting R, substitute the result in either of the given equations and then solve for α.

Review Exercises

In problems 1 through 12, evaluate the given expressions and express results in standard form. Give answers in exact form whenever it is reasonable to do so; otherwise give them correct to two decimal places.

1. $(1 + i)^3$

2. $(3 - 2i)^2$

3. $(1 + 2i)^4$

4. $(\sqrt{3} + i)^6$

5. $(1 + i)^{-2}$

6. $625(3 + 4i)^{-4}$

7. $(1 + i)(\sqrt{3} - i)^{-4}$

8. $\dfrac{(3 + 4i)^5}{(4 + 3i)^4}$

9. $\dfrac{(1 + 2i)(3 + 4i)^3}{(1 - i)^4}$

10. $\left(\dfrac{\sqrt{3}}{2} - \dfrac{1}{2}i\right)^6$

11. $(1 + i)^3 - (1 - i)^5$

12. $\left(\dfrac{1}{2} + \dfrac{\sqrt{3}}{2}i\right)^{12}$

In problems 13 through 15, the function f is defined on the set of complex numbers and is given by $f(z) = 3 - 4i + z^2$, where z is any complex number. Evaluate the given expressions in exact form, and state answers in standard form.

13. $f(-3)$

14. $f(2 - 2i)$

15. $f(1 - \sqrt{3}i)$

In problems 16 through 20, give answers in standard form.

16. Solve the quadratic equation $z^2 + (2 - i)z - i = 0$.

17. Find the cube roots of $\frac{1}{2}(\sqrt{3} - i)$.

18. Find the fourth roots of $\frac{3}{5} - \frac{4}{5}i$.

19. Solve the equation $z^4 + (1 + i)z^2 + i = 0$.

20. Solve the equation $z^2 - 2iz - 2 = 0$.

Appendixes

Appendix A: Introduction to the Calculator

There are two types of scientific calculators suitable for studying a Precalculus course. One type involves algebraic entry while the other is based on the Reverse Polish Notation (RPN). The entry system depends upon the electronic circuitry installed in the calculator during its manufacture. The basic difference between using the algebraic entry and RPN is the order of pressing the four arithmetic-function keys. Calculators with algebraic entry place the binary operation between the two numbers, such as 2 ⊏+⊐3⊏=⊐, whereas in the RPN machines, the arithmetic operation follows both numbers after they are entered into the calculator; for instance, 2 ⊏ENT⊐3 ⊏+⊐is the sequence that evaluates the sum of 2 and 3.

One basic feature common to both entry systems is the use of real numbers in decimal form. Calculators operate with rational approximations of all real numbers correct to the capacity of the particular machine. Calculators cannot handle imaginary numbers directly. When an attempt is made to find $\sqrt{-4}$, the calculator will display *Error* in some way. In this case the error indication tells us that $\sqrt{-4}$ is not a real number. Another instance when the calculator indicates error occurs during an attempt to divide by zero. A good way to find the type of error indication a calculator displays is to press the keys 0 and ⊏1/x⊐. Whenever the error symbol is displayed, the user should be alerted to the fact that the calculator is being asked to perform an unacceptable operation.

Each entry system has its advantages and disadvantages. Students are urged to evaluate each system and choose the calculator that fits their interests and needs best. Appendix A is devoted to helping the student become proficient in using the calculator. We discuss separately algebraic calculators (Section A.1) and RPN calculators (Section A.2).

A.1 ALGEBRAIC CALCULATORS

Algebraic calculators can easily be identified by the presence of an ⊏=⊐ key on the keyboard. Some calculators with algebraic entry are preprogrammed to follow the conventional hierarchy of arithmetic operations, whereas others perform operations sequentially as entered into the calculator. To determine whether a calculator uses the hierarchy of arithmetic, calculate $2 + 3 \cdot 5$ by pressing 2 ⊏+⊐ 3 ⊏×⊐ 5 ⊏=⊐. If the display shows 17, the calculator is accepting the entire sequence of instructions and then performing the multiplication before the addition. In this case we say that addition is a *pending* operation. It is performed only after the entire sequence is entered, and the machine can then respond according to the conventional priority of multiplication and division over addition and subtraction. On the other hand, if the display shows 25, the machine is performing the operations in the order in which they are entered. That is, it is performing the calculation $(2 + 3) \cdot 5$.

Texas Instruments is a major manufacturer of calculators with algebraic entry. Some of their less sophisticated models do not follow arithmetical hierarchy; however, most of their scientific calculators use the so-called *algebraic operating system* (AOS) and are preprogrammed to follow the hierarchy of arithmetic in calculations. *In the instructions given here we assume that all algebraic calculators have AOS.* If this is not the case, the order of entry can be adjusted as necessary. For example, $2 + 3 \cdot 5$ can be calculated by pressing $3 \boxed{\times} 5 \boxed{+} 2 \boxed{=}$.

Using the Keys $\boxed{+}\ \boxed{-}\ \boxed{\times}\ \boxed{\div}\ \boxed{=}\ \boxed{+/-}\ \boxed{(}\ \boxed{)}\ \boxed{x^2}$
In order to use the calculator efficiently, it is helpful to know something about the operation of the machine. The series of examples given below is designed to help the reader make some important observations involving the order in which pending operations are carried out in an AOS calculator.

Example 1 Calculate $5 - 7 + 4$.

Solution Press the calculator keys corresponding to the numbers and operations, as written from left to right, carefully watching the display to see when a given command is executed. Press $5 \boxed{-} 7 \boxed{+} 4 \boxed{=}$. ■

Example 2 Calculate $5 - 7 + 4 \cdot 3$.

Solution Press $5 \boxed{-} 7 \boxed{+} 4 \boxed{\times} 3 \boxed{=}$. Observe how all pending operations are executed when the $\boxed{=}$ key is pressed. ■

Example 3 Calculate $\dfrac{5 - 7 + 4}{3}$.

Solution Press $\boxed{(} 5 \boxed{-} 7 \boxed{+} 4 \boxed{)} \boxed{\div} 3 \boxed{=}$. Note that the numerator is evaluated after the right-parenthesis key $\boxed{)}$ is pressed. As an alternative solution, press $5 \boxed{-} 7 \boxed{+} 4 \boxed{=} \boxed{\div} 3 \boxed{=}$. Thus, when the left-parenthesis key $\boxed{(}$ is not entered, one can use the $\boxed{=}$ key to compute the numerator before dividing by 3. ■

Example 4 Calculate $5 - 7 + 4 \cdot 3^2$.

Solution Press $5 \boxed{-} 7 \boxed{+} 4 \boxed{\times} 3 \boxed{x^2} \boxed{=}$. Note that pressing $\boxed{x^2}$ squares only the contents of the display. Pressing $\boxed{=}$ executes all pending operations. ■

Example 5 Calculate $5 - 7 + (4 \cdot 3)^2$.

Solution Press $5 \boxed{-} 7 \boxed{+} \boxed{(} 4 \boxed{\times} 3 \boxed{)} \boxed{x^2} \boxed{=}$. The problem requires that $4 \cdot 3$ be multiplied before squaring. Parentheses keys are used here to accomplish this. ■

Example 6 Calculate $5 \div (-7 + 4 \cdot 3)$.

Solution Press $5 \boxed{\div} \boxed{(} 7 \boxed{+/-} \boxed{+} 4 \boxed{\times} 3 \boxed{)} \boxed{=}$. The parentheses serve to compute the divisor before the division is carried out. Special note should be taken of the use of the change-sign key $\boxed{+/-}$. This key changes the sign of the number in the display. The calculator will not accept the sequence $5 \boxed{\div} \boxed{(} \boxed{-} 7 \ldots$. Such a sequence treats the $\boxed{-}$ 7 command as subtraction rather than a negative number, but the algebraic calculator cannot accept two operation commands in sequence (such as $\boxed{\div}$ and $\boxed{-}$). ▬

Clearing the Calculator

If the last key pressed is $\boxed{=}$, all pending operations have been executed, and the calculator is ready for a new problem without having the clear key pressed. Some calculators have a clear-entry key that clears only the number in the display, whereas a separate key is used to clear all pending operations. Other calculators have a key labeled $\boxed{\text{ON/C}}$ that serves three purposes. It is used to turn the calculator on; then, during computations, if it is pressed once, the number in the display *only* will be cleared; if it is pressed twice in succession, all pending operations are also cleared.

The clear-entry feature is especially useful since one of the most frequent mistakes is to key in an incorrect number after the calculator already has several pending operations. We illustrate this in the following example, where a 7 rather than an 8 was entered, and this mistake is corrected by using the clear-entry key.

Example 7   Evaluate $2 + 3 \cdot 5 - 24 \div 6 + 8$.

Solution   Press 2 $\boxed{+}$ 3 $\boxed{\times}$ 5 $\boxed{-}$ 24 $\boxed{\div}$ 6 $\boxed{+}$ 7 $\boxed{\text{ON/C}}$ 8 $\boxed{=}$. ▬

Exercises A.1

Calculations in problems 1 through 15 involve integers only. This is intended to allow the student to mentally follow the arithmetic and observe when the pending operations are performed by the calculator. Some important features of the calculator are illustrated in these problems; therefore the student is encouraged to consider each calculation carefully.

1. $5 + 3 \cdot 7$

2. $(5 + 3) \cdot 7$

3. $(5 + 3) - 7$

4. $(5 + 3)(-7)$

5. $2 + 12 \div 3 - 7$

6. $2 + 12 \div (3 - 7)$

7. $\dfrac{(15 - 4) \cdot 5}{2} + 3 \cdot 5 - 7$

8. $\dfrac{(15 - 4) \cdot 5}{2 + 3 \cdot 5 - 7}$

9. $\dfrac{(1/2) - 3}{4}$

10. $(1/2) - (3/4)$

11. $2 \cdot 3^2 + 4 \cdot 5^2$

12. $(2 \cdot 3)^2 + (4 \cdot 5)^2$

13. $(2 \cdot 3 + 4 \cdot 5)^2$

14. $\left(\dfrac{3 \cdot 4^2}{2}\right) \cdot 5^2$

15. $(3 \cdot 4^2) \div (2 \cdot 5^2)$

Use your calculator to solve problems 16 through 30. Answers correct to three decimal places are provided for a quick check.

16. $(1.87)(34.61) + 3.872$

17. $(45.9 - 29.76)^2 + 52.86$

18. $45.9 - 29.76^2 + 52.86$

19. $\dfrac{563 + 284}{18.7}$

20. $563 + \dfrac{284}{18.7}$

21. $\dfrac{52.9 \cdot 0.3876}{21.3}$

22. $12^2 + 5^2 - 2 \cdot 5 \cdot 12 \cdot 0.9848$

23. $(12^2 + 5^2 - 2 \cdot 5 \cdot 12)(0.9848)$

24. $(-37.48 + 59.32)^2 - 31.97$

25. $(37.48 - 59.32)^2 - 31.97$

26. $\dfrac{(15.39 - 4.72) \cdot 5}{2.3} + 3.78 \cdot 5.43$

27. $\dfrac{(15.39 - 4.72) \cdot 5}{2.3 + 3.78 \cdot 5.43}$

28. $\dfrac{21.8 + 4.32^2}{5.12} 5.39^2$

29. $\dfrac{2}{3} + \dfrac{3}{4} - \dfrac{7}{8}$

30. $\dfrac{(2/7) + (3/8)}{(1/6) + (1/7)}$

Answers to Exercises A.1

16. 68.593	**17.** 313.360	**18.** -786.898	**19.** 45.294
20. 578.187	**21.** 0.963	**22.** 50.824	**23.** 48.255
24. 445.016	**25.** 445.016	**26.** 43.721	**27.** 2.337
28. 229.593	**29.** 0.542	**30.** 2.135	

Using the Keys $\boxed{1/x}$ $\boxed{\sqrt{x}}$ $\boxed{y^x}$ $\boxed{\text{STO}}$ $\boxed{\text{RCL}}$

Scientific calculators have several keys in addition to the basic keys described in the preceding section. Here we shall consider the use of five more keys and defer discussion of others until the appropriate places in the text. The $\boxed{1/x}$ and $\boxed{\sqrt{x}}$ keys give the reciprocal and the square root, respectively, of the number in the display. The $\boxed{y^x}$ key operates by entering a positive number y, followed by $\boxed{y^x}$, then the number x, followed by $\boxed{=}$. For example, to evaluate 7^3, keys are pressed in the following order: 7 $\boxed{y^x}$ 3 $\boxed{=}$, and the result 343 appears in the display. Similarly, to find $\sqrt[3]{7}$, we evaluate $7^{1/3}$ by pressing the following keys: 7 $\boxed{y^x}$ 3 $\boxed{1/x}$ $\boxed{=}$, which gives $\sqrt[3]{7} = 1.9129$ (to four decimal places).

A lengthy computation frequently involves the evaluation of intermediate numbers that must be recorded and used later to complete the calculation. Scientific calculators allow the user to store a number with the $\boxed{\text{STO}}$ key* and recall it when needed with the $\boxed{\text{RCL}}$ key, thus avoiding the necessity of recording intermediate steps. This feature will be illustrated in examples given in this section.

Example 1   Calculate $\sqrt{3.9^2 + 7.3^2}$

Solution   Press $\boxed{(}$ 3.9 $\boxed{x^2}$ $\boxed{+}$ 7.3 $\boxed{x^2}$ $\boxed{)}$ $\boxed{\sqrt{x}}$. The display shows 8.2764727. Alternative solution:

$$\text{Press } 3.9 \;\boxed{x^2}\; \boxed{+} \; 7.3 \;\boxed{x^2}\; \boxed{=} \;\boxed{\sqrt{x}}.$$

This method uses the $\boxed{=}$ key to calculate the radicand before taking the square root. ∎

Example 2   Calculate $12^3 - 4^5$.

Solution   Press 12 $\boxed{y^x}$ 3 $\boxed{-}$ 4 $\boxed{y^x}$ 5 $\boxed{=}$. The diplay shows 704. ∎

---

*   Some calculators have multiple storage capacity and require a number address to follow the $\boxed{\text{STO}}$ key. The owner's manual that accompanies such a calculator gives details.

Example 3 Calculate $\sqrt[3]{24.3} \cdot \sqrt[5]{32.7}$.

Solution The problem can be rewritten as $(24.3^{1/3}) \cdot (32.7^{1/5})$; then press

$$24.3 \boxed{y^x} \; 3 \boxed{1/x} \; \boxed{\times} \; 32.7 \boxed{y^x} \; 5 \boxed{1/x} \; \boxed{=}.$$

The displays shows 5.8180615. Note that when the $\boxed{\times}$ key is pressed in this sequence, at that point the calculator evaluates $(24.3)^{1/3}$; in this computation it is not necessary to press the $\boxed{=}$ key before the $\boxed{\times}$ key. ∎

Example 4 Calculate $\sqrt{1.3^2 + 2.8^2 - 2(1.3)(2.8)(0.3215)}$.

Solution Press

$$1.3 \boxed{x^2} \; \boxed{+} \; 2.8 \boxed{x^2} \; \boxed{-} \; 2 \boxed{\times} \; 1.3 \boxed{\times} \; 2.8 \boxed{\times} \; 0.3215 \boxed{=} \; \boxed{\sqrt{x}}.$$

The display shows 2.6813206. ∎

Example 5 Calculate $\dfrac{1}{\sqrt{5.61 + 24.93}}$.

Solution Press $\boxed{(}$ 5.61 $\boxed{+}$ 24.93 $\boxed{)}$ $\boxed{\sqrt{x}}$ $\boxed{1/x}$. The display shows 0.18095287. ∎

Example 6 Calculate $\dfrac{1}{5.2^3 + 3.8^4} + \sqrt{4.2^2 + 3.97}$.

Solution Press

$$5.2 \boxed{y^x} \; 3 \boxed{+} \; 3.8 \boxed{y^x} \; 4 \boxed{=} \; \boxed{1/x} \; \boxed{STO} \; 4.2 \boxed{x^2} \; \boxed{+} \; 3.97 \boxed{=}$$
$$\boxed{\sqrt{x}} \; \boxed{+} \; \boxed{RCL} \; \boxed{=}.$$

The display shows 4.6515201. Storage is used to hold the first part while the second part is being calculated. ∎

Example 7 Calculate $(5.873)^3 + 3(5.873)^2 - 9(5.873) + 4$.

Solution Press

$$5.873 \boxed{STO} \; \boxed{y^x} \; 3 \boxed{+} \; 3 \boxed{\times} \; \boxed{RCL} \; \boxed{x^2} \; \boxed{-} \; 9 \boxed{\times} \; \boxed{RCL} \; \boxed{+} \; 4 \boxed{=}.$$

The display shows 257.19166. Use of the $\boxed{STO}$ key eliminates the need to key in the four-digit number 5.873 three separate times.

Note. The $\boxed{y^x}$ key will function only when the base is positive. The calculator will indicate an *Error* if the base is negative. ∎

Example 8 Use the calculator to evaluate the following.

a) $\sqrt{5.3 - 9.7}$ **b)** $\sqrt[3]{-12.97}$ **c)** $(-3.1)^4$ **d)** $(-3.1)^5$

Solution **a)** Press $\boxed{(}$ 5.3 $\boxed{-}$ 9.7 $\boxed{)}$ $\boxed{\sqrt{x}}$. The display will indicate an *Error*. This is predictable since $5.3 - 9.7 = -4.3$, and the square root of a negative number is not a real number.

b) Rewrite $\sqrt[3]{-12.97}$ as $(-12.97)^{1/3}$ and press 12.97 $\boxed{+/-}$ $\boxed{y^x}$ 3 $\boxed{1/x}$ $\boxed{=}$; the result indicates an *Error*. This is because the calculator will not accept a negative base y when the $\boxed{y^x}$ key is used. However $\sqrt[3]{-12.97}$ is a real number equal to $-\sqrt[3]{12.97}$. We therefore calculate $\sqrt[3]{12.97}$ by pressing 12.97 $\boxed{y^x}$ 3 $\boxed{1/x}$ $\boxed{=}$. The display shows 2.3495. Therefore we have $\sqrt[3]{-12.97} = -2.3495$.

c) When evaluating $(-3.1)^4$, the calculator will indicate an *Error* if we press 3.1 $\boxed{+/-}$ $\boxed{y^x}$ 4 $\boxed{=}$, but we know that $(-3.1)^4 = (3.1)^4$, and this can be calculated by using the $\boxed{y^x}$ key. Press 3.1 $\boxed{y^x}$ 4 $\boxed{=}$. The display shows 92.3521. Thus $(-3.1)^4 = 92.3521$.

d) Since $(-3.1)^5 = -(3.1)^5$, we first evaluate $(3.1)^5$ by pressing 3.1 $\boxed{y^x}$ 5 $\boxed{=}$. The display shows 286.2915, so we conclude that $(-3.1)^5 = -286.2915$.

Exercises A.1 (continued)

Use a calculator to solve the following problems. Answers rounded off to three decimal places are given as a check.

1. $\sqrt{47.23 + 52.18}$ 2. $\sqrt{39.4 + (5.8)(7.3)}$ 3. $\sqrt{54.6 - 31.93}$

4. $\sqrt{(9.1)(3.6) - (7.28)(5.97)}$ 5. $\sqrt{9.2^2 + 4.1^2}$ 6. $\sqrt{(3.87 + 9.4) \cdot 4.83^2}$

7. $\sqrt[3]{12.96}$ 8. $\sqrt[3]{-243.78}$ 9. $\sqrt[5]{32.786}$ 10. $\sqrt[4]{17.39}$

11. $\dfrac{1}{2} + \dfrac{1}{3} + \dfrac{1}{4} + \dfrac{1}{5}$ 12. $\dfrac{2}{3} + \dfrac{3}{4} + \dfrac{5}{6}$ 13. $\dfrac{1}{\sqrt{2}} + \dfrac{1}{\sqrt{3}} + \dfrac{1}{\sqrt{4}}$ 14. $\dfrac{5}{\sqrt{12}} + \dfrac{7}{\sqrt{3}}$

15. $\sqrt[3]{3.47^5 + 29.3^3}$ 16. $(-4.3)^2 + (-5.9)^3$ 17. $(-4.1)^3 + (-5.9)^4$

18. $\sqrt{11.9^2 + 13.2^2 - 2(11.9)(13.2)(0.4937)}$ 19. $\sqrt{[11.9^2 + 13.2^2 - 2(11.9)(13.2)](0.4937)}$ 20. $\sqrt{4 - \sqrt{2}}$

Answers to Exercises A.1 (continued)

1. 9.970	**2.** 9.041	**3.** 4.761	**4.** Imaginary number	
5. 10.072	**6.** 17.595	**7.** 2.349	**8.** -6.247	**9.** 2.010
10. 2.042	**11.** 1.283	**12.** 2.250	**13.** 1.784	**14.** 5.485
15. 29.494	**16.** -186.889	**17.** 1142.815	**18.** 12.679	**19.** 0.913	**20.** 1.608

The problems given in Exercises A.2 (pp. 516–517) provide an opportunity for additional practice in using AOS calculators. The student is urged to do most of them.

A.2   RPN CALCULATORS

Calculators using Reverse Polish Notation (RPN) can easily be identified by the presence of the $\boxed{ENT}$ key (and the absence of the $\boxed{=}$ key). A major manufacturer of RPN calculators is Hewlett-Packard (HP). In the following discussion we shall describe the operation of RPN calculators consistent with HP scientific calculators. The student should be able to adapt the treatment found here to other brands quite easily by referring to the owner's manual.

Registers and Use of Stack

The only external means of communication between the calculator and its user is through the keyboard and the numbers appearing in the display. At any time there is only one number in the display; however, the calculator accepts several numbers and stores them for recall on keyboard command. The places used to store the numbers are called registers and may be thought of as physical places inside the machine where a number is kept until needed. HP machines have four such registers. The content of one register is displayed by the machine. This is called the X register. Registers not visible to the user are called Y, Z, and T. These four registers form the *stack* or *automatic memory* of the machine. In order to use RPN calculators efficiently, it is essential to understand the operation of the stack.

If we represent the stack as a mailboxlike set of compartments $\boxed{X \mid Y \mid Z \mid T}$, where X, Y, Z, and T are the addresses for the boxes, then we can visualize what is happening inside the calculator. When a sequence of digit keys is pressed, the corresponding number appears in the X register. Pressing the ⏎(ENT) key shifts the number into the Y register, and the machine is ready to accept a second number. For example, pressing 2 gives $\boxed{2 \mid Y \mid Z \mid T}$; when we follow this with (ENT), we get $\boxed{2 \mid 2 \mid Z \mid T}$. If we now press 3, the 2 in the X register is replaced by 3 and the 2 in the Y register remains. Pressing (ENT) shifts the contents as shown: $X \rightarrow Y \rightarrow Z \rightarrow T \rightarrow$ lost, retaining the number entered in the X register as well as in the Y register. The series of key strokes

$$2 \; (ENT) \; 3 \; (ENT) \; 1 \; 5 \; (ENT) \; 4$$

provides us with this arrangement of numbers in the stack: $\boxed{4 \mid 15 \mid 3 \mid 2}$.

Observe that the 15 was placed in the Y register without pressing key (ENT) between 1 and 5. This feature best describes the purpose of the (ENT) key which is to separate the numbers entered into the machine. Pressing the (ENT) key after 4 will give $\boxed{4 \mid 4 \mid 15 \mid 3}$, losing the 2 (and the calculator is now ready to accept a new number in the X register). It may appear that having only a four-stack capacity is a serious limitation; but this is not so, since we can perform most of our computations without any additional registers, as will be demonstrated in the following examples. In fact, some RPN calculators have only three register stacks, and they perform adequately in most problems.

For arithmetic operations only the numbers in the X and Y registers are used directly. If x is in X and y is in Y, then pressing any one of the keys $(+)$, $(-)$, $(\times)$, or $(\div)$ gives the corresponding result $y + x$, $y - x$, $y \times x$, or $y \div x$ in the display.

For example, to evaluate $2 + 3$, press 2 (ENT) 3 to get $\boxed{3 \mid 2 \mid \mid }$; then pressing the $(+)$ key gives $\boxed{5 \mid \mid \mid }$. To evaluate $15 - 4$, press (1) (5) (ENT) (4) $(-)$; the result will show 11 in the display. Similar steps are followed in the operations of multiplication and division.

In the following examples the grids indicate the content of each register after the key shown in the left column has been pressed. A blank register does not necessarily mean an empty register (contains 0), but rather that we are not concerned with its content in our computation.

Note. Two solutions are given for some of the following problems. It is important for the reader to understand that there are several methods for solving a given problem. After some practice with the calculator, the user will discover efficient keying patterns.

Example 1 Calculate $7 + 6 \cdot 4$.

Solution 1 Press 7 (ENT) 6 (ENT) 4 (×) (+).

Key	X	Y	Z	T
7	7			
ENT	7	7		
6	6	7		
ENT	6	6	7	
4	4	6	7	
×	24	7		
+	31			

Solution 2 We evaluate $6 \cdot 4 + 7$ by pressing 6 (ENT) 4 (×) 7 (+).

Key	X	Y	Z	T
6	6			
ENT	6	6		
4	4	6		
×	24			
7	7	24		
+	31			

Note. In Solution 2, the (ENT) key was not pressed before the 7. The machine knows it is receiving a new number after any operation, and in this example it is not tempted to write 247. Solution 1 is a less natural way to perform the computation, but it illustrates how helpful it is to know the contents of the registers.

Example 2 Calculate $5 \cdot 3 - 4$.

Solution

Key	X	Y	Z	T
5	5			
ENT	5	5		
3	3	5		
×	15			
4	4	15		
−	11			

Example 3   Calculate $7 + 3(4 + 6)$.

Solution 1

Key	X	Y	X	T
7	7			
ENT	7	7		
3	3	7		
ENT	3	3	7	
4	4	3	7	
ENT	4	4	3	7
6	6	4	3	7
+	10	3	7	
×	30	7		
+	37			

Solution 2 Evaluate $(4 + 6) \cdot 3 + 7$.

Key	X	Y	X	T
4	4			
ENT	4	4		
6	6	4		
+	10			
3	3	10		
×	30			
7	7	30		
+	37			

Note. In Solution 1, all the numbers are entered into the stack, and then the operations are performed in the appropriate order. In Solution 2, operations are performed sequentially according to the conventional principle of beginning within the parentheses. This is a more efficient method in terms of number of steps. ∎

Example 4   Calculate $(15 - 4) \cdot 3 + 2$.

Solution 1   (Key 15 means we press the digit keys 1 and 5 in that order.)

Key	X	Y	Z	T
15	15			
ENT	15	15		
4	4	15		
−	11			
3	3	11		
×	33			
2	2	33		
+	35			

Solution 2 $2 + 3(15 - 4)$

Key	X	Y	Z	T
2	2			
ENT	2	2		
3	3	2		
ENT	3	3	2	
15	15	3	2	
ENT	15	15	3	2
4	4	15	3	2
−	11	3	2	2
×	33	2	2	2
+	35	2	2	2

Note. Solution 2 is given to illustrate the contents of the registers when the T register is used. Once a number (2, in this case) is entered into the T register, it remains there and shifts into the Z and then the Y register as the content of the Y register is being used in an operation. This property of the stack is useful in performing some computations (see Example 7 on p. 514).

The contents of the Y, Z, and T registers can be displayed by using the roll key $\boxed{\text{R↓}}$. For example, continuation of Solution 2 by pressing the $\boxed{\text{R↓}}$ key four times would give the results shown below.

	35	2	2	2
$\boxed{\text{R↓}}$	2	2	2	35
$\boxed{\text{R↓}}$	2	2	35	2
$\boxed{\text{R↓}}$	2	35	2	2
$\boxed{\text{R↓}}$	35	2	2	2

Example 5 Given that $f(x) = 5x^2 - 4x + 1$, evaluate $f(3)$.

Solution To evaluate $f(3) = 5 \cdot 3^2 - 4 \cdot 3 + 1$, we proceed in the following way.

Key	X	Y	Z	T
3	3			
$\boxed{\text{ENT}}$	3	3		
$\boxed{\times}$	9			
5	5	9		
$\boxed{\times}$	45			
4	4	45		

Key	X	Y	Z	T
$\boxed{\text{ENT}}$	4	4	45	
3	3	4	45	
$\boxed{\times}$	12	45		
$\boxed{-}$	33			
1	1	33		
$\boxed{+}$	34			

The $\boxed{\text{CHS}}$ and $\boxed{\text{x↔y}}$ Keys

The $\boxed{\text{CHS}}$ key changes the sign of the contents of the X register *only* and must be used to enter a negative number into the machine. The $\boxed{\text{CHS}}$ key does not shift the content of the X register to Y; hence it is necessary to use the $\boxed{\text{ENT}}$ key to separate numbers after the $\boxed{\text{CHS}}$ key is pressed and before a new number is entered. The $\boxed{\text{x↔y}}$ key interchanges the contents of the X and Y registers and leaves the contents of Z and T undisturbed. This key is frequently used in lengthy calculations involving subtraction and/or division.

Example 6   Calculate $-3 + 4 \cdot 5$.

Solution 1   *Solution 2* Treat it as a subtraction.

Key	X	Y	Z	T
3	3			
CHS	-3			
ENT	-3	-3		
4	4	-3		
ENT	4	4	-3	
5	5	4	-3	
×	20	-3		
+	17			

Key	X	Y	Z	T
3	3			
ENT	3	3		
4	4	3		
ENT	4	4	3	
5	5	4	3	
×	20	3		
x↔y	3	20		
−	17			

Note. Solution 2 is given to illustrate the use of the x↔y key. It should be clear that a more efficient sequence of keys is possible by first evaluating $4 \cdot 5$ and then subtracting 3 from the result.

Example 7   Given that $f(x) = 4x^4 + 5x^2$, find $f(-3)$.

Solution   We wish to evaluate $4(-3)^4 + 5(-3)^2$.

Key	X	Y	Z	T
3	3			
CHS	-3			
ENT	-3	-3		
ENT	-3	-3	-3	
ENT	-3	-3	-3	-3
×	9	-3	-3	-3
×	-27	-3	-3	-3
×	81	-3	-3	-3

4	4	81	-3	-3
×	324	-3	-3	-3
x↔y	-3	324	-3	-3
ENT	-3	-3	324	-3
×	9	324	-3	-3
5	5	9	324	-3
×	45	324	-3	-3
+	369	-3	-3	-3

Thus, $f(-3) = 369$.

Overflowing the Stack
Occasionally a given sequence of keying instructions results in overflowing the stack, and an alternative method must be devised to perform the calculations. Obviously, with the numbers used in these examples, one would simply do some

of the calculations mentally; however, if the numbers involved happen to be, say, four-digit numbers, it is helpful to be able to do all of the arithmetic with the calculator.

Example 8   Calculate $\dfrac{25}{2 + 3(4 + 2)}$.

Solution

Key	X	Y	Z	T
25	25			
ENT	25	25		
2	2	25		
ENT	2	2	25	
3	3	2	25	
ENT	3	3	2	25
4	4	3	2	25
ENT	4	4	3	2

At this point the numerator is lost. However, we can continue to evaluate

the denominator, reenter the numerator, and then use the $\boxed{\text{x↔y}}$ key as follows:

2	2	4	3	2
+	6	3	2	
×	18	2		
+	20			
25	25	20		
x↔y	20	25		
÷	1.25			

A more judicious choice of keying the denominator would avoid the overflow problem encountered in the above example. Also, storage registers are available that would alleviate the problem. We shall discuss the use of storage keys later.

Clearing the Calculator

Calculators have various keys for clearing parts of the machine. One key that clears the display only (that is, the X register) is generally labeled $\boxed{\text{CLX}}$ and is especially useful in correcting an error when a wrong number is entered into the display. Some of the more sophisticated calculators have special keys for clearing only the storage registers, or the prefix, or the program in programmable calculators. The owner's manual explains how these keys operate in a particular calculator. In fact, the reader is urged to consult the owner's manual whenever there is a question concerning the operation of any key.

If one wishes to clear the entire machine, turning the calculator off and then on will do it, except for the sophisticated calculators with a continuous memory. It is not always necessary to clear the stack (or even the display) before beginning a new computation, since only the numbers entered for a given calculation are used and the content of the other registers is irrelevant.

Exercises A.2

1. For each indicated calculation two keying methods are given. In each key sequence, fill in a grid giving the contents of the X, Y, Z, and T registers after each command has been executed by the calculator. Determine which method evaluates the given calculation correctly.

a) $8 \cdot 4 - 5$

Key	Key
8	8
ENT	ENT
4	4
ENT	×
5	5
−	−
×	

b) $(7 + 4) \cdot 8$

Key	Key
7	7
ENT	ENT
4	4
+	ENT
8	8
×	×
	+

c) $(9 \cdot 6) \div (4 \cdot 7)$

Key	Key
9	9
ENT	ENT
6	6
×	×
4	4
ENT	+
7	7
×	÷
÷	

d) $10 - 5(7 + 3)$

Key	Key
10	7
ENT	ENT
5	3
ENT	+
7	5
ENT	×
3	10
+	−
×	
−	

2. Determine what numerical expression is being evaluated by each given sequence of keystrokes.

a)

Key
2
ENT
4
ENT
1
−
×
3
+

b)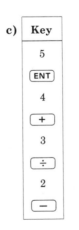

Key
5
ENT
4
×
2
−
3
+

c)

Key
5
ENT
4
+
3
÷
2
−

d)

Key
5
ENT
4
+
3
÷
2
x↔y
−

e)

Key
4
ENT
6
×
3
ENT
6
×
+
7
÷

f)

Key
1
ENT
4
4
÷
1
ENT
5
÷
+

3. Give a sequence of keys that will correctly evaluate each of the given expressions. In each case make a grid showing the contents of all stack registers after each key has been pressed.

a) $2 + 3 + 4 - 6$ **b)** $2 - 4 + 5 \cdot 7$ **c)** $4 \div 2 + 6 \div 3$

d) $\dfrac{4 + 6}{2 + 3}$ **e)** $3(2 - 6) + 4(5 - 2)$

In the following problems, evaluate the given expression using a calculator. Make a grid whenever necessary to get a sequence of keys giving the correct answer. Your computations can be checked with the answers given to four decimal places.

4. $(1.4 + 3.6)(2.1)$ **5.** $(3.8 - 4.3)(6.3)$ **6.** $2.9 + 1.6 \div 3$

7. $\dfrac{1.96 + 2.3}{4.2 - 3.1}$ **8.** $14.98 - \dfrac{4.3 + 2.6}{5.7}$ **9.** $\dfrac{5.4(6.9 - 1.2) + 4}{7 + 4.3}$ **10.** $\dfrac{1}{4} + \dfrac{1}{5} + \dfrac{1}{7}$

11. $\dfrac{3}{4} + \dfrac{4}{5} + \dfrac{2}{7}$ **12.** $5^2 + 7 \cdot 5 - 3$ **13.** $\dfrac{2 \cdot 4^2 - 5 \cdot 4 - 3}{2 \cdot 4 + 1}$ **14.** $\left(\dfrac{3.8}{5.1}\right)^2 + \dfrac{9.6}{4.3}$

15. $5(-1.32)^4 + 4(-1.32)^3$ **16.** $\dfrac{3.48 - (1.23)(4.75)}{8.41 - 2.54(3.57 - 6.75)}$

17. If $f(x) = 1.47x - 5.36$, find $f(3.4)$ **18.** If $f(x) = \dfrac{1.56 - 2.36x}{1.57x}$, find $f(-5.7)$

19. If $f(x) = 7.3x^2 - 4.1x + 3.5$, find $f(3.78)$ **20.** If $f(x) = \dfrac{2.4x^2 - 3.5x - 1.8}{3.2 - 1.5x}$, find $f(-4.3)$.

Answers for Exercises A.2

4. 10.5000	**5.** -3.1500	**6.** 3.4333	**7.** 3.8727
8. 13.7695	**9.** 3.0779	**10.** 0.5929	**11.** 1.8357
12. 57	**13.** 1	**14.** 2.7877	**15.** 5.9799
16. -0.1433	**17.** -0.3620	**18.** -1.6775	**19.** 92.3073

20. 5.9716

The Keys $\boxed{x^2}$ $\boxed{1/x}$ $\boxed{STO}$ $\boxed{RCL}$ $\boxed{y^x}$ $\boxed{\sqrt{x}}$

There is no *one* correct way to perform a given calculation, although some methods of key entry may be more efficient than others. In the preceding section we considered an example in which we evaluated $f(x) = 5x^2 - 4x + 1$ at $x = 3$. A more efficient sequence of keys would include the use of the $\boxed{x^2}$ key. Pressing the $\boxed{x^2}$ key squares the content of the X register, while the contents of the other registers remain unchanged. This is illustrated in the following grid, where we evaluate $f(3)$, given that $f(x) = 5x^2 - 4x + 1$.

The $\boxed{\sqrt{x}}$ and $\boxed{1/x}$ keys operate in a manner similar to that of $\boxed{x^2}$: pressing $\boxed{\sqrt{x}}$ takes the square root of the number in the X register and displays the result, while pressing $\boxed{1/x}$ takes the reciprocal of the number appearing in the X register and displays it. Each of these keys leaves the contents of the Y, Z, and T registers unchanged.

Key	X	Y	Z	T	Remarks
5	5				
ENT	5	5			
3	3	5			
x^2	9	5			3^2 is evaluated
$\times$	45				$5 \cdot 3^2$ in X
4	4	45			
ENT	4	4	45		
3	3	4	45		
$\times$	12	45			$3 \cdot 4$ in X
$-$	33				$5 \cdot 3^2 - 4 \cdot 3$ in X
1	1	33			
$+$	34				$5 \cdot 3^2 - 4 \cdot 3 + 1$ in X

All scientific calculators have at least one memory storage, and some have several. When the STO key is pressed, the content of the X register is placed in a memory storage separate from any of the stack registers. Pressing the recall key RCL will return that number to the X register whenever it is needed and also retain the number in the memory.

If a calculator has more than one memory storage, it is necessary to tell the machine the address of the particular memory to be used. For instance, if the calculator has eight memories numbered 0 through 7, the storage command consists of STO followed by one of the numbers 0 through 7. Similarly for recall, press RCL followed by the number 0 through 7 corresponding to the address where the number is stored.

Example 1   Given that $f(x) = \dfrac{7x^2}{3x - 4}$, find $f\left(\dfrac{1 + \sqrt{5}}{2}\right)$ correct to four decimal places.

Solution   We wish to evaluate

$$7\left(\frac{1 + \sqrt{5}}{2}\right)^2 \div \left[3\left(\frac{1 + \sqrt{5}}{2}\right) - 4\right].$$

We first evaluate $(\sqrt{5} + 1) \div 2$ and store the result for future use. The grid shows decimal values correct to two places.

Key	X	Y	Z	T	Remarks
5	5				
$\boxed{\sqrt{x}}$	2.23				
1	1	2.23			
$\boxed{+}$	3.23				
2	2	3.23			
$\boxed{\div}$	1.61				$(\sqrt{5} + 1) \div 2$ in X
$\boxed{\text{STO}}$ 1	1.61				$(\sqrt{5} + 1) \div 2$ stored in R_1 and still in X
$\boxed{x^2}$	2.61				Square of $(\sqrt{5} + 1) \div 2$
7	7	2.61			
$\boxed{\times}$	18.32				Numerator in X
$\boxed{\text{RCL}}$ 1	1.61	18.32			$(\sqrt{5} + 1) \div 2$ recalled and numerator moved to Y
3	3	1.61	18.32		
$\boxed{\times}$	4.85	18.32			
4	4	4.85	18.32		
$\boxed{-}$	0.85	18.32			Denominator in X and numerator in Y
$\boxed{\div}$	21.4567				Answer

Example 2   Evaluate $\dfrac{1}{\sqrt{2}} - \dfrac{1}{\sqrt{3}}$.

Solution   Here we use the $\boxed{1/x}$ key since this is simpler than using the $\boxed{\div}$ key to evaluate $1 \div \sqrt{2}$ and $1 \div \sqrt{3}$. The grid shows numbers to four decimal places.

Key	X	Y	Z	T	Remarks
2	2				
$\boxed{\sqrt{x}}$	1.4142				$\sqrt{2}$ in X
$\boxed{1/x}$	0.7071				$1/\sqrt{2}$ in X
3	3	0.7071			
$\boxed{\sqrt{x}}$	1.7321	0.7071			$\sqrt{3}$ in X
$\boxed{1/x}$	0.5774	0.7071			$1/\sqrt{3}$ in X and $1/\sqrt{2}$ in Y
$\boxed{-}$	0.1298				Answer (to four places)

Example 3   Evaluate $\dfrac{3.52}{\sqrt{1.63^2 + 3.75^2}}$.

Solution   We begin by evaluating the denominator and then use the $\boxed{1/x}$ key.

Key	X	Y	Z	T	Remarks
1.63	1.63				
$\boxed{x^2}$	2.6569				1.63^2 in X
3.75	3.75	2.6569			
$\boxed{x^2}$	14.0625	2.6569			3.75^2 in X; 1.63^2 in Y
$\boxed{+}$	16.7194				
$\boxed{\sqrt{x}}$	4.0889				Denominator in X
$\boxed{1/x}$	0.2446				Reciprocal of denominator
3.52	3.52	0.2446			
$\boxed{\times}$	0.8609				Answer (to four places)

Another convenience for evaluating polynomial functions of degree greater than 2 and exponential functions in general is the $\boxed{y^x}$ key. This key raises the number in the Y register to the power given in the X register.

We continue with an example where the $\boxed{y^x}$ key is used.

Example 4   Evaluate $f(x) = 4x^3 + 5x^2 - 7$ at $x = 3 - \sqrt{2}$.

Solution   The problem is to evaluate $4(3 - \sqrt{2})^3 + 5(3 - \sqrt{2})^2 - 7$. We first evaluate $3 - \sqrt{2}$, store the result, and recall it when needed.

Key	X	Y	Z	T	Remarks
3	3				
$\boxed{ENT}$	3	3			
2	2	3			
$\boxed{\sqrt{x}}$	1.41...	3			
$\boxed{-}$	1.58...				$3 - \sqrt{2}$ in X
$\boxed{STO}\,4$	1.58...				$3 - \sqrt{2}$ in X and stored in R_4
3	3	1.58...			

Key				Description
y^x	3.98...			$(3 - \sqrt{2})^3$ in X
4	4	3.98...		
$\times$	15.95...			
RCL 4	1.58...	15.95...		$3 - \sqrt{2}$ recalled to X
x^2	2.51...	15.95...		$(3 - \sqrt{2})^2$ in X
5	5	2.51...	15.95...	
$\times$	12.57...	15.95...		$5(3 - \sqrt{2})^2$ in X
$+$	28.52...			$4(3 - \sqrt{2})^3 + 5(3 - \sqrt{2})^2$
7	7	28.52...		
$-$	21.5248			Answer (to four places)

Exercises A.2 (continued)

Evaluate the following expressions to three decimal places. Check your answers with those given at end of this exercise set.

1. Given that $f(x) = 3x^2 - 2x + 1$, find $f(2.13)$.

2. Evaluate $f(x) = 1.6x^2 - 2.4x + 4.1$ at $x = 2.46$.

3. Find the value of $g(x) = 5x^2 + \dfrac{1}{x}$ at $x = -1.57$.

4. Evaluate $\dfrac{1}{2} + \dfrac{1}{3} + \dfrac{1}{4} + \dfrac{1}{5} + \dfrac{1}{6}$.

5. Given that $f(x) = 1 + \dfrac{1}{1 + 1/(1 + 1/x)}$, find the following.

   a) $f(2)$

   b) $f(-1.48)$

6. Evaluate the following expressions by using the $\boxed{\pi}$ key on your calculator.

   a) $(24.67)\left(64 + \dfrac{27}{60}\right)\dfrac{\pi}{180}$

   b) $\dfrac{1}{2}(24.67)^2\left(64 + \dfrac{27}{60}\right)\left(\dfrac{\pi}{180}\right)$

7. Given that $u = 2.21$, $v = \dfrac{7\pi}{10}$, $t = 126.43\left(\dfrac{\pi}{180}\right)$, order these three numbers from smallest to largest.

8. Evaluate.

   a) $(34.63)\left(\dfrac{\pi}{180}\right)\sqrt{\dfrac{2(35.61)(180)}{34.63\pi}}$

   b) $\sqrt{\dfrac{2(35.61)(34.64)\pi}{180}}$

9. Evaluate.

   a) $\dfrac{1 + \sqrt{7}}{3}$

   b) $\left(\dfrac{1 + \sqrt{7}}{3}\right)^2$

   c) $\left(\dfrac{1 + \sqrt{7}}{3}\right)^3$

10. The following numbers may be used as rational approximations of π. Calculate each number and use the $\boxed{\pi}$ key on your calculator to determine the decimal-place accuracy.

a) $\dfrac{22}{7}$ **b)** $\dfrac{333}{106}$ **c)** $\dfrac{355}{113}$ **d)** $\dfrac{208341}{66317}$

11. Evaluate.

     a) $(\sqrt{5.38})^3$ **b)** $\sqrt{5.38^3}$

12. Evaluate.

     a) $\sqrt{24.3 + 36.8}$ **b)** $\sqrt{24.3} + \sqrt{36.8}$

13. Evaluate.

     a) $\dfrac{\sqrt{3} - 1}{\sqrt{3} + 1}$ **b)** $2 - \sqrt{3}$

14. Given that $f(x) = 3x^4 - 8x^2 + 12$, find $f(1.43)$.

15. Given that $f(x) = \dfrac{x^6 - 1}{x - 1}$, find the following.

     a) $f(3)$ **b)** $f(2.3)$ **c)** $f(-1.8)$ **d)** $f(1)$

16. Given that $g(x) = x^5 + x^4 + x^3 + x^2 + x + 1$, find the following.

     a) $g(3)$ **b)** $g(2.3)$ **c)** $g(-1.8)$ **d)** $g(1)$

Compare these results with the answers in problem 15. What conclusions can you draw about the functions f and g?

17. Evaluate.

     a) $\sqrt{24.7} - \sqrt{36.8}$ **b)** $\sqrt{24.7 - 36.8}$

18. Evaluate $\left(\dfrac{1 - \sqrt{5}}{2}\right)^3$.

19. Evaluate $\sqrt{(1 - \sqrt{3})^2 - 1}$.

20. Given that $f(x) = 3x^4 - 4x^3 + x - 5$, find the following.

     a) $f(3)$ **b)** $f(-1.2)$ **c)** $f(\pi)$ **d)** $f\left(\dfrac{1 + \sqrt{5}}{2}\right)$

Answers for Exercises A.2 (continued)

1. 10.351   **2.** 7.879   **3.** 11.688   **4.** 1.450

5. a) 1.600   **b)** 1.243   **6. a)** 27.750   **b)** 342.301

7. $u = 2.210$, $v = 2.199$, $t = 2.207$; $v < t < u$ **8. a)** 6.561   **b)** 6.561

9. a) 1.215   **b)** 1.477   **c)** 1.795

10. Agreement with π: **a)** 2 places   **b)** 4 places   **c)** 6 places   **d)** more than 8 places

11. a) 12.479   **b)** 12.479   **12. a)** 7.817   **b)** 10.996

13. a) 0.268   **b)** 0.268   **14.** 8.186

15. a) 364   **b)** 113.105   **c)** -11.790 **d)** Error, why?

16. a) 364   **b)** 113.105   **c)** -11.790 **d)** 6

17. a) -1.096 **b)** Error, why?   **18.** -0.236 **19.** Error, why?

20. a) 133   **b)** 6.933   **c)** 166.344   **d)** 0.236

Appendix B: Important Properties of Real Numbers

A careful development of the properties of real numbers requires mathematical proofs in which each step is justified by axioms or previously proved theorems. Since many mathematical statements involve the idea of equality, we first state postulates that form the basis for working with the equals sign (=).

Axioms for Equals

Let x, y, and z be elements of a given set S. Then the following properties hold:

Reflexive property: $x = x$.

Symmetric property: If $x = y$, then $y = x$.

Transitive property: If $x = y$ and $y = z$, then $x = z$.

Substitution property: If $x = y$, then we may replace x by y in any expression involving x.

The *system of real numbers* can be characterized as a *complete ordered field*. First we state axioms that describe this system as a field and then give some theorems that follow from these axioms and form a basis for the algebra of real numbers. Consider the set **R** along with the two arithmetic operations of addition and multiplication (subtraction and division are defined later in terms of these). We say that *addition is a binary operation on R*, which means that for each x and y in **R** there is a unique real number denoted by $x + y$. Similarly, *multiplication is a binary operation* denoted by $x \cdot y$.

Many of the fundamental properties of real numbers are introduced in the elementary grades. This is done informally and primarily through examples. For instance, adding 2, 3, and 5, we write $2 + 3 + 5 = 10$ and are not really concerned about whether we arrive at the result by a sequence of binary operations $(2 + 3) + 5$ or by $2 + (3 + 5)$; in either case we arrive at the same answer. Examples of this type lead to generalizations that we accept as axioms satisfied by the system of real numbers. The following list of axioms forms a basis for important properties of real numbers.

Field Axioms for the System of Real Numbers

In dealing with the set of real numbers **R** along with the binary operations of addition and multiplication we assume the truth of the following properties called *postulates,* or *axioms.*

For any x, y, and z in **R**, the following properties hold.

Associativity: $(x + y) + z = x + (y + z)$,
$(x \cdot y) \cdot z = x \cdot (y \cdot z)$.

Commutativity: $x + y = y + x$,
$x \cdot y = y \cdot x$.

Distributivity: $x \cdot (y + z) = (x \cdot y) + (x \cdot z)$.

Identity elements: **R** contains unique numbers 0 and 1 with special properties:

$$x + 0 = x \quad \text{and} \quad x \cdot 1 = x.$$

Inverses: For each x, **R** contains numbers $-x$ and $1/x$ such that

$$x + (-x) = 0 \quad \text{and} \quad x \cdot \frac{1}{x} = 1 \text{ for } x \neq 0.$$

Subtraction and division can now be defined in terms of addition and multiplication.

Definition B.1   Suppose x and y are real numbers. Then,

Subtraction: $x - y$ is given by $x - y = x + (-y)$.

Division: $x \div y$ is given by $x \div y = x \cdot \dfrac{1}{y}$ for $y \neq 0$.

Note that $x \div y$ is also written as x/y.

More Properties of Real Numbers

We now state several important properties that form a basis for the operational rules in the algebra of real numbers. All of these can be proved by using the above definitions and axioms.

In the following theorems, suppose x, y, z, and w are any real numbers.

Theorem 1   If $x = y$, then

$$x + z = y + z, \quad x - z = y - z, \quad x \cdot z = y \cdot z,$$

$$\frac{x}{z} = \frac{y}{z} \quad \text{for} \quad z \neq 0.$$

Theorem 2   If $x + y = 0$, then $x = -y$ and $y = -x$.

If $x \cdot y = 1$, then $x = \dfrac{1}{y}$ and $y = \dfrac{1}{x}$.

Theorem 3

$$\text{If } x + y = x, \text{ then } y = 0.$$
$$\text{If } x \cdot y = x, \text{ then } y = 1.$$

Theorem 4

$$-(-x) = x \quad \text{and} \quad -(x + y) = -x - y.$$

Theorem 5

$$\frac{1}{(1/x)} = x \quad \text{and} \quad \frac{1}{x \cdot y} = \frac{1}{x} \cdot \frac{1}{y}.$$

Theorem 6

$$(-x) \cdot y = -(x \cdot y), \quad (-x) \cdot (-y) = x \cdot y,$$
$$(-1) \cdot x = -x, \quad 0 \cdot x = 0.$$

Theorem 7

$$\text{If } x \cdot y = 0, \text{ then } x = 0 \text{ or } y = 0 \text{ or both.}$$

Theorem 8

$$(x + y) \cdot z = (x \cdot z) + (y \cdot z),$$
$$x \cdot (y - z) = (x \cdot y) - (x \cdot z),$$
$$(y - z) \cdot x = (y \cdot x) - (z \cdot x).$$

Theorem 9

$$\frac{x}{1} = x, \quad \frac{x}{x} = 1 \quad \text{for} \quad x \neq 0.$$

Theorem 10

Equality of fractions

Suppose $y \neq 0$, $w \neq 0$. Then

$$\frac{x}{y} = \frac{z}{w} \quad \text{if and only if} \quad x \cdot w = y \cdot z.$$

Theorem 11

Reducing fractions

$$\frac{z \cdot x}{z \cdot y} = \frac{x}{y}, \quad \text{where} \quad y \neq 0, z \neq 0.$$

Theorem 12 $$\text{If } y \neq 0, \text{ then } \frac{-x}{y} = \frac{x}{-y} = -\frac{x}{y}.$$

Theorem 13

Algebra of fractions

If $z \neq 0$, $w \neq 0$, then

Addition: $$\frac{x}{z} + \frac{y}{z} = \frac{x+y}{z}$$

Subtraction: $$\frac{x}{z} - \frac{y}{z} = \frac{x-y}{z}$$

Multiplication: $$\frac{x}{z} \cdot \frac{y}{w} = \frac{x \cdot y}{z \cdot w}$$

Division: $$\frac{x}{z} \div \frac{y}{w} = \frac{x}{z} \cdot \frac{w}{y} \quad \text{for} \quad y \neq 0.$$

Order Relations

We first assume that **R** contains a subset P, called the *positive numbers,* such that if $x \in P, y \in P$, then $x + y \in P$ and $x \cdot y \in P$. Also, if z is any number in **R**, then exactly one of the following is true: $z \in P$ or $-z \in P$ or $z = 0$. If $x \in P$, we say that $-x$ is a *negative number.*

We now define the *less than* order relation, denoted by $(<)$, as follows:

Suppose x and y are in **R**. Then $x < y$ (read "x is less than y") if and only if there is a positive number z such that $x + z = y$.

Other order relations are defined in terms of $(<)$ as follows:

$x \leq y$ (read "x is less than or equal to y") means $x < y$ or $x = y$;

$x > y$ (read "x is greater than y") means $y < x$;

$x \geq y$ (read "x is greater than or equal to y") means $y < x$ or $y = x$.

Basic properties of the "less than" relation are given in Theorem 14.

Theorem 14   Suppose $x < y$. Then

$$x + u < y + u \text{ for any } u \text{ in } \mathbf{R};$$
$$u \cdot x < u \cdot y \text{ for any positive number } u;$$
$$u \cdot x > u \cdot y \text{ for any negative number } u.$$

System of Real Numbers

It can be shown that the *system consisting of the set of rational numbers* and the usual binary operations of addition and multiplication *forms an ordered field*. However, the property of being an ordered field is *not sufficient* to characterize the system of *real numbers*. At the beginning of Appendix B we referred to a *complete* ordered field. The property of being complete is studied in advanced courses in mathematics.

Exercises B

In problems 1 through 10, justify the given statements on the basis of definition, axioms, or theorems given in the discussion above. Each variable represents a real number.

1. If $3x + 1 = 2x + 1$, then $(3x + 1) - 1 = (2x + 1) - 1$.

2. If $4 + x = 4$, then $x = 0$.

3. $\dfrac{3}{4} = \dfrac{3 \cdot 5}{4 \cdot 5}$

4. $\dfrac{-2}{-(x + 3)} = \dfrac{2}{x + 3}$

5. If $(x + 1)(x - 4) = 0$, then $x + 1 = 0$ or $x - 4 = 0$.

6. If $x(x + 3) = 0$, then $x = 0$ or $x + 3 = 0$.

7. $3x - 2x = (3 - 2)x$

8. $\dfrac{4}{3} - \dfrac{2}{3} = \dfrac{4 - 2}{3}$

9. $\dfrac{4x^2 y}{6xy} = \dfrac{2x}{3}$

10. $\dfrac{2x}{5} + \dfrac{4x}{5} = \dfrac{2x + 4x}{5}$

11. In the following proof of $\dfrac{2}{x} \div \dfrac{6}{y} = \dfrac{y}{3x}$, give a reason for each step.

Step	Reason
$\dfrac{2}{x} \div \dfrac{6}{y} = \dfrac{2}{x} \cdot \dfrac{y}{6}$	
$\dfrac{2}{x} \cdot \dfrac{y}{6} = \dfrac{2y}{x \cdot 6}$	
$\dfrac{2y}{x \cdot 6} = \dfrac{2y}{6x}$	
$\dfrac{2y}{6x} = \dfrac{2y}{(2 \cdot 3)x}$	
$\dfrac{2y}{(2 \cdot 3)x} = \dfrac{2y}{2 \cdot (3x)}$	
$\dfrac{2y}{2 \cdot (3x)} = \dfrac{y}{3x}$	

Therefore

$$\frac{2}{x} \div \frac{6}{y} = \frac{y}{3x}.$$

12. Suppose x and y are real numbers and $x < y$. Prove each of the following.

   a) $x + u < y + u$ for any $u \in \mathbf{R}$.

   b) If u is a positive number, then $u \cdot x < u \cdot y$.

   c) If u is a negative number, then $u \cdot x > u \cdot y$.

13. Given that u is a positive number, prove that $0 < u$.

14. Given that u is a positive number, prove that $-u < 0$.

15. Given that $x < y$, prove that $0 < y - x$.

Appendix C:
Approximate
Numbers

In most applications of mathematics to real-life problems, we encounter two types of numbers: exact and approximate. Examples of exact numbers are 1/2, 4/13, π. However, when these numbers are expressed in decimal form we have

$$\tfrac{1}{2} = 0.5; \qquad \tfrac{4}{13} = 0.307692307 \ldots ; \qquad \pi = 3.141592 \ldots .$$

The decimal representation of 1/2 is finite, whereas for 4/13 and for π it is infinite. There is no problem in replacing 1/2 by 0.5, but when the decimal representation of 4/13 or of π is required, it becomes necessary to round off and use only an approximate decimal value. This is one source of approximate numbers.

Another source of approximate numbers comes from applications involving measurements, and in almost all cases the results are expressed as approximate numbers (limited to the degree of accuracy of the measuring instruments). Approximate numbers are then used in formulas to compute other quantities, and so the final numbers are, of necessity, also approximate. In the following discussion our primary goal is to establish rules that can be used in problems involving computations with approximate numbers. In order to do this, we first discuss significant digits, scientific notation, and rounding off numbers.

Notation. In the main body of the text we used the symbol $=$ to mean both the *exact and the approximate equality,* and its meaning was clear from the context. In this Appendix we wish to emphasize approximate equality, and so we use the symbol $\doteq$ to denote *approximately equal to.*

C.1   SIGNIFICANT DIGITS AND SCIENTIFIC NOTATION

For a better understanding of approximate numbers, it may be helpful to consider some examples first. Suppose that four different objects are measured and their lengths are determined as:

$$a \doteq 24.3 \text{ cm}, \qquad b \doteq 0.00407 \text{ m}, \qquad c \doteq 832.0 \text{ cm}, \qquad d \doteq 34\,700 \text{ cm}.$$

This means that a is an approximate number representing a length that is actually somewhere between 24.25 and 24.35 cm. Similarly, the exact value of b is somewhere between 0.004065 and 0.004075 m, and that of c is between 831.95 and 832.05 cm.

In the case of d it is not clear what accuracy is implied. For example, d might have been measured as 347 meters, in which case the exact value is somewhere between 346.5 and 347.5 m (that is, d is actually between 34\,650 and 34\,750 cm). It is possible that d was measured to the nearest tenth of a meter (nearest 10 cm), in which case we would write $d \doteq 347.0$ m. This implies that d is somewhere between 346.95 and 347.05 m (that is, d is between 34\,695 and 34\,705 cm). Similarly, if d has been measured accurately to the nearest centimeter, then $d \doteq 34\,700$ means that $34\,699.5 < d < 34\,700.5$ cm.

Thus the examples above lead to the following question: When a number is represented in decimal form, which of the digits are significant?

For $a \doteq 24.3$ cm, all three digits 2, 4, 3, are meaningful in expressing accuracy of the measurement; thus we say that a has three significant digits.

For $b \doteq 0.00407$ m, the zero before the decimal and the two zeros after the decimal merely serve the purpose of telling us where the decimal is located, but the remaining digits 4, 0, 7 give information about the accuracy of measurement. If b were expressed in centimeters, then $b \doteq 0.407$ cm, and we would not even encounter the two zeros immediately after the decimal point. Thus b has three significant digits.

In the case of $c \doteq 832.0$ cm, the zero after the decimal tells us that the measurement was made to the nearest tenth of a centimeter, and we do not need to be told where the decimal is located. Therefore all four digits 8, 3, 2, 0 are significant.

In the case of $d \doteq 34\,700$ cm, the two zeros are certainly necessary to locate the decimal point, but it is not clear whether they give us any information about the accuracy of measurement or not. Thus we would say that 3, 4, 7 are significant digits, and an additional statement is required concerning the significance of the two zeros. A convenient way to give this information is to use scientific notation. Thus, if d is accurate to the nearest meter (nearest 100 cm), then we write $d \doteq 3.47 \times 10^4$ cm, and this indicates that only the 3, 4, 7 are significant digits. If d is accurate to the nearest 10 cm, then we write $d \doteq 3.470 \times 10^4$ cm and 3, 4, 7, 0 are significant digits. In a similar way, $d \doteq 3.4700 \times 10^4$ cm implies that d is measured to the nearest centimeter, and so all of the digits 3, 4, 7, 0, 0 are significant.

The above discussion leads us to the following *general statement concerning significant digits:* When a number is written in decimal form, its significant digits begin with the first nonzero digit on the left and end with the last digit on the right that definitely gives information about the accuracy of the number.

That is, all nonzero digits are significant, whereas zeros that merely serve the purpose of locating the decimal point are not, but all other zeros are. In cases when it is not clear whether a zero merely indicates the place of the decimal point (as in d above), scientific notation is useful. To represent a number in *scientific notation,* we write it as a product of a number between 1 and 10 and a power of 10; all digits of the factor between 1 and 10 are significant.

Example 1 Determine which digits are significant in the following numbers.

a) 37.543   b) 136.1030   c) 240.00

d) 0.0048   e) 0.00480   f) 70 400

Solution a) All five digits are significant.

b) All seven digits are significant (including the zero at the end).

c) The three zeros are significant, and so the number has five significant digits.

d) Only the 4 and 8 are significant digits.

e) The 4, 8, and final 0 are significant digits.

f) The digits 7, 0, 4 are significant, but we cannot say without further information whether the last two zeros are significant. ▬

Example 2   Write each of the numbers given in Example 1 in scientific notation.

Solution   **a)** $37.543 = 3.7543 \times 10$

b) $136.1030 = 1.361030 \times 10^2$

c) $240.00 = 2.4000 \times 10^2$

d) $0.0048 = 4.8 \times 10^{-3}$

e) $0.00480 = 4.80 \times 10^{-3}$

f) $70\ 400 = 7.04 \times 10^4$ would indicate that only 7, 0, 4 are significant digits.
$70\ 400 = 7.040 \times 10^4$ would say that 7, 0, 4, 0 are significant digits.
$70\ 400 = 7.0400 \times 10^4$ would tell us that all five digits are significant.   ▪

Example 3   The following numbers are expressed in scientific notation. Write them in ordinary decimal form.

a) 2.78×10^4 **b)** 3.47×10^{-4} **c)** 3.40×10^3 **d)** 4.800×10^{-1}

Solution   **a)** 27 800   **b)** 0.000347   **c)** 3400   **d)** 0.4800   ▪

C.2   **ROUNDING OFF NUMBERS**
When a number is given in decimal form, it is frequently necessary to express it as an approximate number with fewer significant digits. We call this the process of *rounding off a number* and illustrate with the following examples.

Example 4   Round off the following numbers to three significant digits.

a) 3476   **b)** 24.74   **c)** 73.80   **d)** 0.473501

e) 2435   **f)** 69.95   **g)** π **h)** $\pi/2$

Solution   **a)** The number $3480 = 3.48 \times 10^3$ has three significant digits, and it is an approximation to a number between 3475 and 3485. Since the given number 3476 is in this range, we say that 3476 rounded off to three significant digits is 3.48×10^3.

Similarly for (b), (c), (d) we get the following.

b) 24.7   **c)** 73.8   **d)** 0.474

e) Here we encounter a borderline case in which it is not clear whether we should round off to 2430 or 2440. Both appear to be equally good, and so we shall adopt the rule that we round *up* and use $2440 = 2.44 \times 10^3$ as the answer.*

f) This is similar to (e), and so 70.0 is the approximation of 69.95 with three significant digits.

---

*   Some textbooks give a slightly different rule in which the number is sometimes rounded up and other times it is rounded down.

g) Since $\pi = 3.14159\ldots$, we round off to 3.14.

h) $\pi/2 = 1.57079\ldots$ rounded off to three significant digits is 1.57.   ▬

C3   COMPUTATIONS WITH APPROXIMATE NUMBERS

When approximate numbers are used in computations, it is natural to ask: "How many significant digits should we retain in the final result?" To give an answer it is helpful to consider some examples. We first take the problem of multiplying or dividing two approximate numbers, and then we study addition and subtraction of such numbers.*

Multiplication and Division of Approximate Numbers

Suppose the length and width of a rectangular object are measured with a ruler marked in millimeters and are found to be $l \doteq 16.4$ cm, $w \doteq 8.6$ cm. We wish to find the area of the rectangle. Since Area $= l \times w$, we get

$$\text{Area} \doteq (16.4 \times 8.6) \text{ cm}^2 \doteq 141.04 \text{ cm}^2.$$

This is a computed value based on the measurements of l and w expressed as approximate numbers. How many of the five digits in 141.04 are really meaningful and not misleading in terms of stating the actual area of the object?

On the basis of the given information about l and w, all we can say is that

$$16.35 < l < 16.45 \text{ cm} \qquad \text{and} \qquad 8.55 < w < 8.65 \text{ cm}.$$

This implies that

$$16.35 \times 8.55 < A < 16.45 \times 8.65 \text{ cm}^2.$$

That is, all we can really say about the actual area is

$$139.7925 < A < 142.2925 \text{ cm}^2. \tag{C.1}$$

This is the best claim we can make about the area on the basis of the given measurements.

Our computed value of $A \doteq 141.04$ cm^2 is certainly in the range given by expression (C.1), but stating that $A \doteq 141.04$ cm^2 implies that we know $141.035 < A < 141.045$ cm^2. This says considerably more than we actually do know.

Suppose we round off the computed value to three significant digits: $A \doteq 141$ cm^2. This implies that $140.5 < A < 141.5$ cm^2, and clearly this still claims more than the inequality given in (C.1). Therefore we try rounding off to two significant digits: $A \doteq 140$ cm$^2 = 1.4 \times 10^2$ cm^2. This means that $135 < A < 145$ cm^2, and making such a statement is consistent with the inequality given by (C.1).

---

*   The general problem of accuracy in computations involving other operations (such as square root, logarithm, etc.) is a topic for numerical-analysis courses.

In conclusion, rounding off the computed value of the area to two significant digits results in the best statement we can make that is consistent with what the given measurements tell us about the actual area. Since l was measured to three significant digits and w to two significant digits, this suggests that we should round off the product to the smaller number of significant digits of the measured values.

The problem of dividing two approximate numbers is similar. Suppose $a \doteq 34.6$ and $b \doteq 8.4$ are approximate numbers, and we wish to determine $c = a \div b$. Using a calculator to evaluate c, we get

$$c = \frac{34.6}{8.4} = 4.1190 \ldots .$$

How many digits should we retain in the answer? Since $34.55 < a < 34.65$ and $8.35 < b < 8.45$, we obtain

$$\frac{34.55}{8.45} < \frac{a}{b} < \frac{34.65}{8.35}.$$

Thus, all we know about c is that

$$4.0888 < c < 4.1497 \text{ (to four decimal places).} \qquad (C.2)$$

If we round off c to three significant digits ($c \doteq 4.12$), then we are saying that $4.115 < c < 4.125$, and this is not consistent with what we know about c, as given by (C.2). If we round off to two significant digits ($c \doteq 4.1$), then we imply that $4.05 < c < 4.15$, which is in agreement with statement (C.2). Since $a = 34.6$ has three significant digits and $b = 8.4$ has two significant digits, this example suggests that the quotient of two approximate numbers should be rounded off to the smaller number of significant digits of the two measured values.

The above examples suggest the following.

Rule for multiplying and dividing approximate numbers

In the multiplication and division of approximate numbers, the result should be rounded off to the smallest number of significant digits in the data used.

For example, suppose $x \doteq 47.36$, $y \doteq 17.5$, $z \doteq 5.2$, and we wish to evaluate $u = (xy) \div z$. Since the numbers of significant digits in x, y, z are four, three, two, respectively, we should retain two significant digits for u. Thus

$$u \doteq (47.36 \times 17.5) \div 5.2 = 159.3846 \ldots ,$$

and so we have $u \doteq 160 = 1.6 \times 10^2$. If this value is to be used in subsequent computations, then we should use one more significant digit ($u \doteq 159$) for that purpose, but we must remember that in the final round off, u is accurate to only two significant digits.

Addition and Subtraction of Approximate Numbers

When one is adding or subtracting approximate numbers, the situation is a little different from that of multiplying or dividing. For example, suppose a bank reports that a certain fund has $248 000 in it, where this is accurate to the nearest thousand dollars. Now suppose that $72.35 is added to this fund. It would be misleading to say that the fund now has $248 072.35 in it. We would say that the fund still has $248 000 in it to the nearest thousand dollars (based on the given information). That is, we would write $248\,000 + 72.35 \doteq 248\,000$.

It is clear from this example that when we add two approximate numbers, we are not interested in the number of significant digits each has, but we are primarily interested in the *level of precision* of each number. We say that the level of precision of 248 000 is the nearest thousand, and that of 72.35 is the nearest hundredth; thus the level of precision of 72.35 is greater than that of 248 000.

As another example, suppose x, y, and z are approximate numbers given by $x \doteq 24.65$, $y \doteq 0.036$, $z \doteq 132.4$. The levels of precision of x, y, z are hundredths, thousandths, tenths, respectively. Common sense would suggest that the sum

$$x + y + z \doteq 24.65 + 0.036 + 132.4 = 157.086$$

should be rounded off to the nearest tenth, since z is no more accurate than the nearest tenth, and we cannot expect $x + y + z$ to be more accurate. Thus

$$x + y + z \doteq 157.1.$$

The above examples lead us to the following common-sense rule.

Rule for adding and subtracting approximate numbers

In the addition and subtraction of approximate numbers, the result should be rounded off to the lowest level of precision in the data used.

Linear and Angle Measurements

When one is solving triangles, the angle and length measurements are usually given as approximate numbers. Therefore it is desirable to have a guide that can be used to determine the angle measurements with an accuracy corresponding to that of the length measurements. For angles that are not too close to $0°$ or $90°$, the following table provides a satisfactory rule.

Lengths accurate to	Corresponding angles accurate to
Two significant digits	Nearest degree
Three significant digits	Nearest 10′
Four significant digits	Nearest minute
Five significant digits	Nearest tenth of a minute

In the following examples, suppose x, y, z, u, v, t are approximate numbers given by

$$x \doteq 3.48, \quad y \doteq 0.0360, \quad z \doteq 3251, \quad u \doteq 5.004,$$
$$v \doteq 84\,000 \quad \text{(only 8 and 4 are significant)},$$
$$t \doteq 24\,800 \quad \text{(the tens 0 is significant)}.$$

Example 5   Write the above numbers in scientific notation.

Solution
$$x \doteq 3.48 \times 10^0, \quad y \doteq 3.60 \times 10^{-2}, \quad z \doteq 3.251 \times 10^3,$$
$$u \doteq 5.004 \times 10^0, \quad v \doteq 8.4 \times 10^4, \quad t \doteq 2.480 \times 10^4,$$ ■

Example 6   Give the number of significant digits in each of the numbers above.

Solution   In x there are three; y has three; z has four; u has four; v has two; t has four.   ■

Example 7   State the level of precision of the given numbers.

Solution   The level of precision of x is hundredths, of y is ten thousandths, of z is units, of u is thousandths, of v is thousands, and of t is tens.   ■

Example 8   Using the rule for multiplication and division of approximate numbers, evaluate the following.

a) $x \cdot z$ **b)** $\dfrac{yv}{x}$ **c)** $u \cdot t$

Solution   **a)** $x \cdot z \doteq (3.48)(3251) = 11\,313.48$. Since x has three and z has four significant digits, the result should be rounded off to three significant digits. Thus $x \cdot z \doteq 11\,300 = 1.13 \times 10^4$.

b) $\dfrac{y \cdot v}{x} \doteq \dfrac{(0.0360)(84\,000)}{3.48} = 868.9655 \ldots$.

The smallest number of significant digits of x, y, and v is two, and so the answer should be rounded off to two significant digits. That is,

$$\frac{y \cdot v}{x} \doteq 870 = 8.7 \times 10^2.$$

c) Both u and t have four significant digits, and so $u \cdot t$ should be rounded off to four significant digits: $u \cdot t \doteq (5.004)(24\ 800) \doteq 1.241 \times 10^5$.　■

Example 9　Using the rule for addition and subtraction of approximate numbers, evaluate the following.

a) $x + y$　　　**b)** $z + t$　　　**c)** $u - x$　　　**d)** $v + t$　　　**e)** $x + z - u$

Solution　**a)** $x + y \doteq 3.48 + 0.0360 = 3.516$. Since the level of precision of x is hundredths and that of y is ten thousandths, we round off the sum to hundredths: $x + y \doteq 3.52$.

b) $z + t \doteq 3251 + 24\ 800 = 28\ 051$. The level of precision of z is units and that of t is tens, and so we round off the sum to tens:

$$z + t \doteq 28\ 050 = 2.805 \times 10^4.$$

c) $u - x \doteq 5.004 - 3.48 = 1.524$. The result should be rounded off to the nearest hundredth, and so we have $u - x \doteq 1.52$.

d) $v + t \doteq 84\ 000 + 24\ 800 = 108\ 800$. Since v is correct to the nearest thousand and t is accurate to the nearest tens, we round off the sum to the nearest thousand:

$$v + t \doteq 109\ 000 = 1.09 \times 10^5.$$

e) $x + z - u \doteq 3.48 + 3251 - 5.004 = 3249.476$. Since the least precise of x, z, u is z (to the nearest unit), we round off the result to the nearest unit: $x + z - u \doteq 3249$.　■

Example 10　Using the rules for computation with approximate numbers, evaluate the following.

a) $z - xu$　　　　　　　　**b)** $\dfrac{v - t}{x}$

Solution　**a)** We first evaluate xu:

$$xu \doteq (3.48)(5.004) = 17.41392 \doteq 17.41.$$

Therefore

$$z - xu \doteq 3251 - 17.41 = 3233.59 \doteq 3234.$$

Note that in the final computation we used an extra digit for xu.

b) We first evaluate $v - t$:

$$v - t \doteq 84\ 000 - 24\ 800 = 59\ 200 \doteq 59\ 000 = 5.9 \times 10^4.$$

Thus

$$\frac{v - t}{x} \doteq \frac{59\ 200}{3.48} = 17\ 011.494 \ldots \doteq 17\ 000 = 1.7 \times 10^4.$$

Note that in the final computation we used $v - t = 59\,200$ (an extra significant digit), but we rounded off the final result to two significant digits. ▪

Example 11   The radius of a circle is measured as $r \doteq 6.41$ cm. Find the area of the circle.

Solution   We use the formula Area $= \pi r^2$. Since r is measured to three significant digits, the result should be rounded off to three significant digits. We use π as given by the calculator and find that

$$\text{Area} \doteq \pi(6.41)^2 = 129.082 \ldots \text{ cm}^2 \doteq 129 \text{ cm}^2. \qquad ▪$$

Exercises C

In problems 1 through 7, suppose x, y, z, u, v, t are approximate numbers given by

$$x \doteq 64.75, \qquad y \doteq 4830, \qquad z \doteq 0.0045, \qquad u \doteq 0.0370, \qquad v \doteq 3005.2,$$
$$t \doteq 3100 \quad \text{(the tens 0 is significant and the units 0 is not)}.$$

1. Write each of the above numbers in scientific notation.

2. Determine the number of significant digits in each of the above numbers.

3. State the level of precision of each of the above numbers.

4. Round off the above numbers to two significant digits.

Using the rules for computing with approximate numbers, evaluate the expressions given in problems 5 through 7.

5. **a)** xu **b)** vz **c)** $t \div y$ **d)** $(uy) \div z$

6. **a)** $x + y$ **b)** $u - z$ **c)** $y - t$ **d)** $y - x - v$

7. **a)** $xz - u$ **b)** $\dfrac{y - v}{t}$ **c)** $y + ut$

8. The radius of a circle (measured accurately to the nearest millimeter) is found to be $r \doteq 2.476$ m. Find the circumference and area of the circle.

9. The radius of a sphere is measured as $r \doteq 3.47$ cm. Find the surface area and volume of the sphere.

10. The lengths of the edges of a rectangular box are measured to the nearest millimeter and found to be

$$a \doteq 23.4 \text{ cm}, \qquad b \doteq 12.8 \text{ cm}, \qquad c \doteq 8.4 \text{ cm}.$$

Determine the volume and the total surface area of the box.

11. The speed of light is approximately 3×10^5 km/sec. A light-year is defined as the distance travelled by light in one year. Assuming 365 days in a year, find the number of kilometers in a light-year. Express your answer in scientific notation.

12. The hypotenuse and an angle of a right triangle are measured and found to be 32.4 cm and $23°40'$, respectively. Calculate the area and the perimeter of the triangle.

Answers for Exercises C

1. $x \doteq 6.475 \times 10$ $y \doteq 4.83 \times 10^3$ $z \doteq 4.5 \times 10^{-3}$
     $u \doteq 3.70 \times 10^{-2}$ $v \doteq 3.0052 \times 10^3$ $t \doteq 3.10 \times 10^3$

3. x, hundredths; y, tens; z, ten thousandths; u, ten thousandths; v, tenths; t, tens

5. a) 2.40   b) 14   c) 0.642   d) 4.0×10^5

7. a) 0.25   b) 0.589   c) 4.94×10^3

9. 3.78 cm^2; 175 cm^3 **11.** $9 \times 10^{12} \text{ km}$

Appendix D:
Table
of
Prime
Numbers

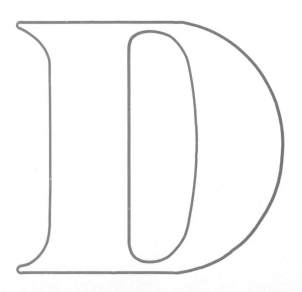

2	151	353	577	811	1049	1297	1559
3	157	359	587	821	1051	1301	1567
5	163	367	593	823	1061	1303	1571
7	167	373	599	827	1063	1307	1579
11	173	379	601	829	1069	1319	1583
13	179	383	607	839	1087	1321	1597
17	181	389	613	853	1091	1327	1601
19	191	397	617	857	1093	1361	1607
23	193	401	619	859	1097	1367	1609
29	197	409	631	863	1103	1373	1613
31	199	419	641	877	1109	1381	1619
37	211	421	643	881	1117	1399	1621
41	223	431	647	883	1123	1409	1627
43	227	433	653	887	1129	1423	1637
47	229	439	659	907	1151	1427	1657
53	233	443	661	911	1153	1429	1663
59	239	449	673	919	1163	1433	1667
61	241	457	677	929	1171	1439	1669
67	251	461	683	937	1181	1447	1693
71	257	463	691	941	1187	1451	1697
73	263	467	701	947	1193	1453	1699
79	269	479	709	953	1201	1459	1709
83	271	487	719	967	1213	1471	1721
89	277	491	727	971	1217	1481	1723
97	281	499	733	977	1223	1483	1733
101	283	503	739	983	1229	1487	1741
103	293	509	743	991	1231	1489	1747
107	307	521	751	997	1237	1493	1753
109	311	523	757	1009	1249	1499	1759
113	313	541	761	1013	1259	1511	1777
127	317	547	769	1019	1277	1523	1783
131	331	557	773	1021	1279	1531	1787
137	337	563	787	1031	1283	1543	1789
139	347	569	797	1033	1289	1549	1801
149	349	571	809	1039	1291	1553	1811

Answers to Odd-Numbered Exercises

Exercises 1.0, pages 13–15

1. $\{4, 5, 6\}$ **3.** $\{1, 2, 3, 4, 5\}$ **5.** $\{0.1, 1\}$

7. False   **9.** True   **11.** True

13. False   **15.** True   **17.** $\{2, 3\}$

19. $\{5, 7\}$ **21. a)** True   **b)** False   **c)** True   **d)** True

23. a) True   **b)** True   **c)** False   **d)** True

25. a) True   **b)** False   **c)** True   **d)** True

27. a)

29. a)

   b)

   c)

31. A in I, B in II, C in IV, D in III   **35. a)** $\sqrt{29}$; 5.39   **b)** $\sqrt{13}$; 3.61

37. 9.43   **39.** 42.20   **41.** 52

43. Yes   **45.** Yes

47. a) $-2 + 3i$ **b)** $-4 + 5i$ **c)** $1 + 7i$ **d)** $-3.5 + 0.5i$

49. a) $-i$ **b)** $-0.2 + 0.4i$ **c)** $-1 - 2i$ **d)** $-2 + i$ **51. a)** 3   **b)** 3.5

53. a) 2   **b)** 1   **55. a)** $x \geq -1$ **b)** $x < -0.5$

57. $x - y$ **59.** $x - 2$ **61.** $x^2 - 5x + 7$

63. $3x^2 + 11x - 4$ **65.** $2x^3 - 6x^2 + x - 3$ **67.** $x^2 + x - 3$

69. $x^2 + 3x + 11 + \dfrac{32}{x - 3}$ **71.** $3(x - 2)(x + 2)$ **73.** $(2x - 1)(x - 2)$

75. Does not factor   **77.** $(x - 2)(x + 10)$ **79.** $8x^2(1 - 2x)(1 + 2x)$

Exercises 1.1, pages 24–28

1. a) $f = \{(-1, -2), (0, 1), (1, 4), (2, 7), (3, 10)\}$ **b)** Yes; Yes

   c) $\mathcal{R}(f) = \{-2, 1, 4, 7, 10\}$ **d)** $\begin{array}{ccccc} -1 & 0 & 1 & 2 & 3 \\ \downarrow & \downarrow & \downarrow & \downarrow & \downarrow \\ -2 & 1 & 4 & 7 & 10 \end{array}$

3. a) $f = \{(-4, 3), (-1, 1), (2, -1), (5, -3)\}$ **b)** Yes; Yes   **c)** $\mathcal{R}(f) = \{-3, -1, 1, 3\}$

   d) $\begin{array}{cccc} -4 & -1 & 2 & 5 \\ \downarrow & \downarrow & \downarrow & \downarrow \\ 3 & 1 & -1 & -3 \end{array}$

5. a) $f = \{(-1, -1), (0, 0.5), (1, -1), (3, -13)\}$ **b)** Yes; No

   c) $\mathcal{R}(f) = \{-13, -1, 0.5\}$ **d)** $\begin{array}{cccc} -1 & 0 & 1 & 3 \\ \downarrow & \downarrow & \downarrow & \downarrow \\ -1 & 0.5 & -1 & -13 \end{array}$

7. a) $f = \{(0, 0), (1, 0), (2, 2), (3, 6)\}$ **b)** Yes; No   **c)** $\mathcal{R}(f) = \{0, 2, 6\}$

d) 0 1 2 3
↓ ↓ ↓ ↓
0 0 2 6

9. a) Yes; No   **b)** $\mathcal{D}(f) = \{-1, 0, 1, 2, 3, 4\}$; $\mathcal{R}(f) = \{0, 1, 4, 9, 16\}$
c) $f(x)$ equals the square of x

11. a) No; No   **b)** $\mathcal{D}(h) = \{0, 1, 4, 9\}$; $\mathcal{R}(h) = \{-3, -2, -1, 0, 1, 2, 3\}$
c) $h(x) = \sqrt{x}$ or $h(x) = -\sqrt{x}$

13. a) $f = \{(-1, 1), (0, 0), (1, 1), (2, 4)\}$ **b)** Yes
c) $\mathcal{D}(f) = \{-1, 0, 1, 2\}$ $\mathcal{R}(f) = \{0, 1, 4\}$

15. a) $\{(1, 0), (3, 2), (5, 4)\}$ **b)** Yes

17. a) $\mathcal{D} = \{x \mid x \geq 0\}$ **b)** $\{(1, 15), (2, 30), (3.6, 54), (10, 150)\}$

19. a) $\mathcal{D}(f) = \mathcal{R}$ **b)** $f(-2) = 11, f(0) = 3, f(\sqrt{5}) = 3.53$

21. a) $\mathcal{D}(h) = \mathcal{R}$ **b)** $h(-2) = 11, h(1) = -1, h(1 + \sqrt{5}) = -9.94$

23. a) $\mathcal{D}(g) = \{x \mid x \neq 2\}$ **b)** $g(-2) = -1, g(0) = 0, g(\sqrt{5}) = 21.18$

25. a) $\mathcal{D}(g) = \{t \mid t \neq -1, t \neq 0\}$ **b)** $g(-2) = -(\frac{1}{2}), g(0.5) = -(\frac{4}{3}), g(\sqrt{3}) = -0.21$

27. a) $\mathcal{D}(f) = \{t \mid t \leq 1\}$ **b)** $f(1) = 0, f(-3) = 2, f(-\sqrt{7}) = 1.91$

29. a) 4   **b)** 4   **c)** 1.76   **31. a)** 1   **b)** 1.65   **c)** 2

33. a) 0   **b)** 1   **c)** 1.21

35. a) $3u^2 - 8u + 2$ **b)** $3x^2 + 20x + 30$

37. a) $\dfrac{u^2 - 2u + 1}{u}$ **b)** $\dfrac{4x^2 - 4x + 1}{2x}$ **39. a)** 1   **b)** 1   **c)** 0

41.

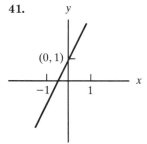

43.

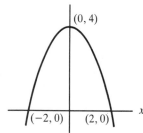

45.

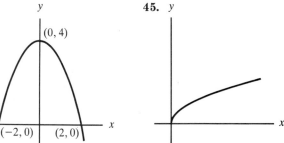

47.

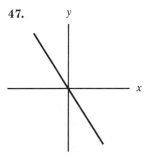

49.

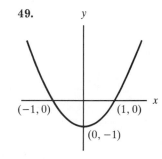

51.

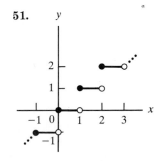

53.

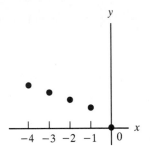

55. a) Yes   **b) R**

57. a) No   **b)** $\{x \mid -4 \le x \le 4\}$

59. a) Yes   **b) R**

61. a) 0   **b)** -3 **c)** 2

63. a) 3   **b)** 2   **c)** -2

Exercises 1.2, pages 40-42

1. 21

3. 280

5. -2.75

7. Undefined

9. -7

11. -3

13. 6.5

15. $\frac{16}{7}$

17. 3

19. Undefined

21. 0

23. $\dfrac{x^2 + 4}{x - 3}$; $\{x \mid x \ne 3\}$

25. $\sqrt{x}(x - 3)$; $\{x \mid x \ge 0\}$

27. $\sqrt{x - 3}$; $\{x \mid x \ge 3\}$

29. $x^2 - 6x + 13$; **R**

31. $x + 4$; $\{x \mid x \ge 0\}$

33. $\sqrt{x^2 + 4}$; **R**

35. -2

37. No solution

39. No solution

41. Yes

43. $\dfrac{x - 3}{2}$

45. $\dfrac{4 - x}{2}$

47. $\dfrac{2(6 - x)}{3}$

49. 6.04

51. 1.47

53. 2.37

55. 0.18

57. No, $\mathfrak{D}(f \circ g) \ne \mathfrak{D}(g \circ f)$

59. $f(x) = x^2$, $g(x) = x + 1$; other possible answers   **61.** 41; -4; 16; 121

63. 4; -6; 2; -2; 5   **65. a)** 1   **b)** 1   **c)** 2   **d)** 2

Exercises 1.3, pages 47-49

1. a) $-\frac{3}{2}$ **b)** 0

3. a) $-\frac{2}{3}$ **b)** Undefined

5. a) $x - 2y + 5 = 0$ **b)** $y - 4 = 0$

7. a) $2x + y = 0$ **b)** $x + 2 = 0$

9. $2x - y + 8 = 0$

11. $3x + 5y = 15$

13. $-\frac{3}{2}$; $(-2, 0)$, $(0, -3)$

15. $\frac{3}{4}$; $(2, 0)$, $(0, -\frac{3}{2})$

17.

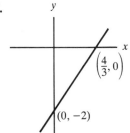

19.

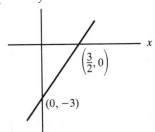

21.

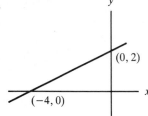

23.

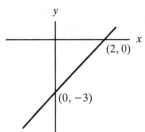

25. a) $2x - 3y = -7$ **b)** $3x + 2y = -4$ **27. a)** $x + 1 = 0$ **b)** $y - 3 = 0$

29. No $\qquad$ **31.** Yes $\qquad$ **33.** $x - 2y = 0$

35. $16x + 6y = 25$ $\qquad$ **37.** $(1, 3), (4, 1)$ $\qquad$ **39.** $(2, 1)$

41. $x + 2$; Yes $\qquad$ **43.** $\dfrac{2x - 1}{-x + 3}$; No $\qquad$ **45.** $-2x + 4$; Yes

Exercises 1.4, pages 59-61

1. a) $-0.5, 2$ **b)** Rational $\qquad$ **3. a)** $\dfrac{2 \pm \sqrt{10}}{3}$ **b)** Irrational; $-0.39, 1.72$

5. a) $1 \pm \sqrt{3}$ **b)** Irrational; $-0.73, 2.73$ **7. a)** $1 \pm i$ **b)** Complex

9. a) $\sqrt{3} \pm \sqrt{5}$ **b)** Irrational; $-0.50, 3.97$

11. a) $\dfrac{2.3 \pm \sqrt{11.69}}{2}$ **b)** Irrational; $-0.56, 2.86$

13.

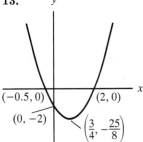

15.

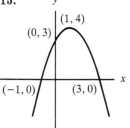

17.

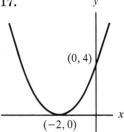

19.

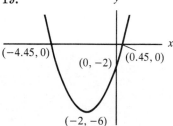

21. Min of $f(x)$ is -1; no max

23. Max of $f(x)$ is -3; no min $\qquad$ **25.** Min of $f(x)$ is -4.84; no max

27. Max of $f(x)$ is 4; min is -6 **29.** Min of $f(x)$ is -7; no max

31. Max of $f(x)$ is 1; min is -15 **33.** $S = \{x \mid x < 1 \text{ or } x > 3\}$

35. $S = \{x \mid -1 \leq x \leq 1.5\}$ **37.** $S = \{x \mid x < 1 - \sqrt{5} \text{ or } x > 1 + \sqrt{5}\}$

39. $S = \{2\}$ **41. a)** $K = (\sqrt{15}x/2)(2 - x), 0 < x < 2$ **b)** 1 by $\sqrt{15}/2$ **c)** $\sqrt{15}/2$

43. a) 90 m by 180 m   **b)** 63 800 m²

45. a) $T = \begin{cases} 1400x & \text{if } 0 \leq x \leq 120 \\ 2600x - 10x^2 & \text{if } 120 < x \leq 150 \end{cases}$ **b)** $x = 130$ gives max of \$169 000

Exercises 1.5, pages 68–69

1. a) 4   **b)** 5   **c)** $\sqrt{3}$ **3. a)** 12   **b)** 4   **c)** $\dfrac{3 + \sqrt{5}}{2}$ **5. a)** 3   **b)** 1   **c)** 3

7. a) $(f \circ g)(x) = \begin{cases} x^2 - 3x + 2 & \text{if } x \geq 1 \\ x^2 - x & \text{if } x < 1 \end{cases}$ **b)** $\left(\dfrac{f}{g}\right)(x) = \begin{cases} x & \text{if } x > 1 \\ -x & \text{if } x < 1 \end{cases}$

    c) $\mathcal{D}(f \circ g) = \mathbf{R}; \ \mathcal{D}\left(\dfrac{f}{g}\right) = \{x \mid x \neq 1\}$

9. $S = \{-1, 3\}$ **11.** $S = \{-1, 3\}$ **13.** $S = \{-1\}$

15. $S = \{x \mid x \leq 1\}$ **17.** $S = \{x \mid x \geq 2\}$ **19.** $S = \emptyset$

21. $S = \{-4, 4\}$ **23.** $S = \left\{2, \dfrac{-1 - \sqrt{17}}{2}\right\}$ **25.** $S = \{x \mid 0 < x < 2\}$

27. $S = \{x \mid x \geq 2\}$ **29.** $S = \{x \mid x \leq -\frac{1}{2}\}$

31. $f(x) = \begin{cases} 2x - 2 & \text{if } x \geq 2 \\ 2 & \text{if } x < 2 \end{cases}$

33. $f(x) = \begin{cases} x^2 + x + 4 & \text{if } x \geq -1 \\ x^2 - x + 2 & \text{if } x < -1 \end{cases}$ **35.**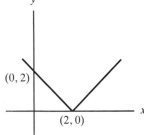

37. 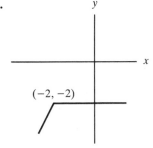   **39.**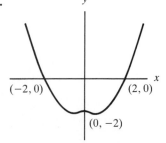

41. a) $\mathcal{D}(f) = \mathbf{R}, \mathcal{R}(f) = \{y \mid y \geq 0\}$ **b)** $\mathcal{D}(g) = \mathbf{R}, \mathcal{R}(g) = \{y \mid y \leq 1\}$

Exercises 1.6, page 77

1. a) Circle   **b)** $y = \sqrt{9 - x^2}$; $y = -\sqrt{9 - x^2}$

c)

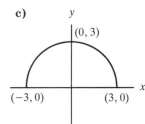
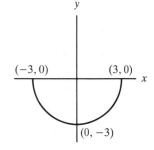
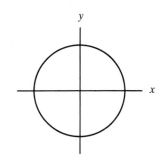

3. a) Ellipse   **b)** $y = \frac{1}{2}\sqrt{12 - x^2}$; $y = -\frac{1}{2}\sqrt{12 - x^2}$

c)

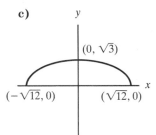
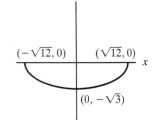
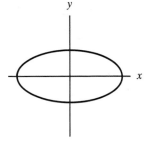

5. a) Parabola   **b)** $y = 4x^2$ **c)**

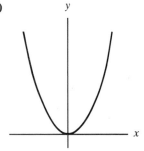

7. a) Parabola   **b)** $y = \frac{4}{3}\sqrt{-x}$; $y = -\frac{4}{3}\sqrt{-x}$

c)

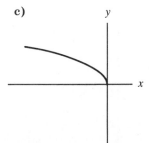
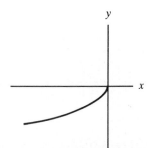
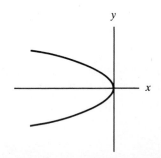

9. a) Parabola   **b)** $y = 3\sqrt{x};\ y = -3\sqrt{x}$

c)

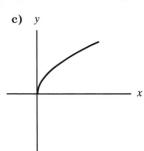 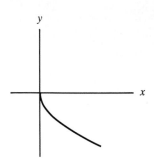 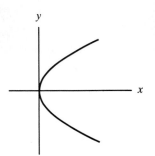

11. a) Circle   **b)** $y = 0.5\sqrt{9 - 4x^2};\ y = -0.5\sqrt{9 - 4x^2}$

c)

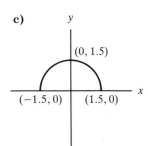 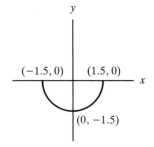 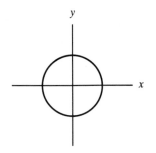

13. Graph is the point $(0, 0)$

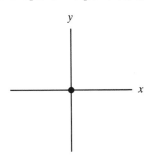

15. a) Hyperbola   **b)** $y = \sqrt{x^2 - 9};\ y = -\sqrt{x^2 - 9}$

c)

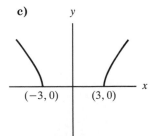 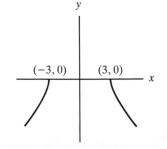 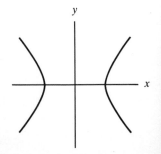

17. a) Hyperbola   **b)** $y = \frac{2}{3}\sqrt{x^2 - 9}$; $y = -\frac{2}{3}\sqrt{x^2 - 9}$

c)

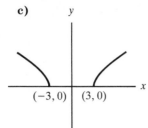

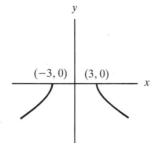

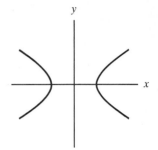

19. a) Two lines   **b)** $y = 0.4x$; $y = -0.4x$

c)

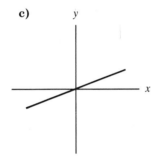

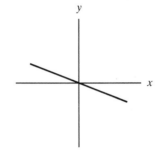

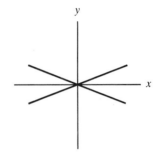

21. a) Hyperbola   **b)** $y = \frac{3}{2}\sqrt{x^2 - 4}$; $y = -\frac{3}{2}\sqrt{x^2 - 4}$

c)

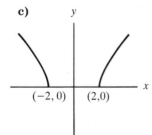

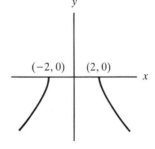

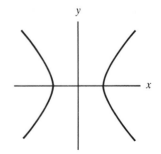

23. No graph

Exercises 1.7, pages 90–91

   1. Even   **3.** Odd   **5.** Neither   **7.** Even

   9. Odd   **11.** Even   **13.** Even

15. Neither

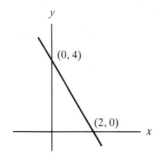

17. Origin

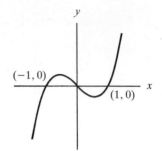

19. Origin

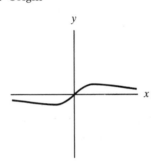

21. y-axis

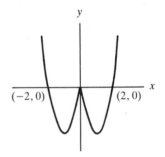

23. Neither

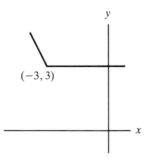

25. Neither

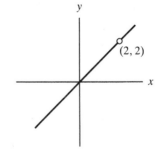

27. $\mathfrak{D}(f) = \mathbf{R}$, $\mathfrak{R}(f) = \mathbf{R}$ **29.** $\mathfrak{D}(g) = \mathbf{R}$, $\mathfrak{R}(g) = \mathbf{R}$

31. $\mathfrak{D}(g) = \mathbf{R}$; $\mathfrak{R}(g) = \{y \mid y \geq -4\}$ **33.** $\mathfrak{D}(f) = \{x \mid \leq 0\}$; $\mathfrak{R}(f) = \{y \mid y \leq 0\}$

35. $\mathfrak{D}(f) = \{x \mid -4 \leq x \leq 4\}$; $\mathfrak{R}(f) = \{y \mid 0 \leq y \leq 2\}$

37. $\mathfrak{D}(f) = \{x \mid x \neq -2\}$; $\mathfrak{R}(f) = \{y \mid y \neq -2\}$

39. a) Increasing   **b)** One to one   **41. a)** Decreasing   **b)** One to one

43. a) Increasing   **b)** One to one   **45. a)** Neither   **b)** Not one to one

47. a) Increasing   **b)** One to one   **49. a)** Neither   **b)** One to one

51.

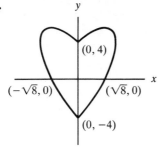

Exercises 1.8, pages 97–99

1. $f^{-1} = \{(0, -2), (1, 0), (3, 2), (4, 4)\}$ **3.** $f^{-1} = \{(\sqrt{5}, 3), (\sqrt{6}, 4), (3, 1), (4, 2)\}$

5. a) One to one   **b)** $f^{-1}(x) = \dfrac{3 - x}{2}$; $\mathcal{D}(f^{-1}) = \mathbf{R}$ **7. a)** Not one to one

9. a) One to one   **b)** $f^{-1}(x) = -x^2$; $\mathcal{D}(f^{-1}) = \{x \mid x \geq 0\}$

11. a) One to one   **b)** $f^{-1}(x) = x^2 + 1$, $\mathcal{D}(f^{-1}) = \{x \mid x \geq 0\}$

13. a) One to one   **b)** $f^{-1}(x) = \sqrt[3]{x}$; $\mathcal{D}(f^{-1}) = \mathbf{R}$

15. a) One to one   **b)** $f^{-1}(x) = \begin{cases} -x & \text{if } x < 0, \\ -\frac{1}{3}x & \text{if } x \geq 0; \end{cases}$ $\mathcal{D}(f^{-1}) = \mathbf{R}$

17. a) One to one   **b)** $f^{-1}(x) = \sqrt{\dfrac{1 - x}{x}}$; $\mathcal{D}(f^{-1}) = \{x \mid 0 < x \leq 1\}$

19. a) One to one   **b)** $f^{-1}(x) = \dfrac{1}{x}$; $\mathcal{D}(f^{-1}) = \{x \mid 0 < x \leq 1\}$

21. a) One to one   **b)** $f^{-1}(x) = 1 + \sqrt{1 + x}$; $\mathcal{D}(f^{-1}) = \{x \mid x \geq 0\}$

23. a) One to one   **b)** $f^{-1}(x) = 1 + \sqrt{x}$; $\mathcal{D}(f^{-1}) = \{x \mid x \geq 0\}$

25. $f^{-1}(x) = \dfrac{4 - x}{2}$; $\mathcal{D}(f^{-1}) = \mathbf{R}$ **27.** $f^{-1}(x) = \sqrt{x + 1}$; $\mathcal{D}(f^{-1}) = \{x \mid x \geq -1\}$

29. $f^{-1}(x) = 1 - x^2$; $\mathcal{D}(f^{-1}) = \{x \mid x \geq 0\}$

31. $f^{-1}(x) = \dfrac{x}{2}$ **33.** $f^{-1}(x) = \sqrt{x}$, $x \geq 0$

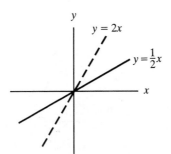

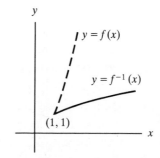

35. $f^{-1}(x) = x^2 + 1, x \geq 0$

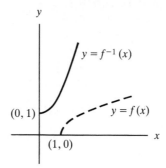

37. $f^{-1}(v) = \sqrt[3]{\dfrac{3v}{4\pi}}$

v	1	2	3.72	5.64
r	0.62	0.78	0.96	1.10

39. a) $\mathcal{D}(f) = \{t \mid 0 \leq t \leq 4\}$ **b)**

s	10	48	60	75
t	0.26	1.51	2.06	3.17

Review Exercises, pages 99–101

1. Yes; $\mathcal{D}(g) = \{-1, 0, 1, 2\}$; $\mathcal{R}(g) = \{2, 4, 6, 8\}$

3. $\mathcal{D}(f) = \mathbf{R}$; $\mathcal{D}(g) = \{x \mid -1 \leq x \leq 1\}$; $\mathcal{D}(h) = \mathbf{R}$

5. $(g \circ f)(x) = 2\sqrt{x - x^2}$; $\mathcal{D}(g \circ f) = \{x \mid 0 \leq x \leq 1\}$

7. $\left(\dfrac{f}{g}\right)(x) = \dfrac{2x - 1}{\sqrt{1 - x^2}}$; $\mathcal{D}\left(\dfrac{f}{g}\right) = \{x \mid -1 < x < 1\}$ **9. a)** 0   **b)** -3

11. a) 5   **b)** Undefined   **13.** 0.876   **15.** 0.967

17. $-1, 1$ **19.** 1, 3   **21.** $5x + 2y + 1 = 0$

23. $x - 2y + 7 = 0$ **25.** Yes; $x + 4y - 11 = 0$

27. f is not an increasing function; f is one to one

29.

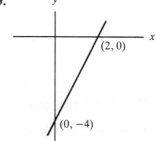

31.

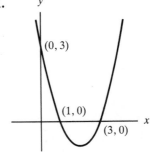

33. a) $(1, 0)$ **b)** $(0, 3)$ **c)** Lowest point $(1, 0)$

35. a) $(\frac{3}{2}, 0)$, $(1, 0)$ **b)** $(0, -3)$ **c)** Highest point $(\frac{5}{4}, \frac{1}{8})$

37. $\left\{ \dfrac{-3 + \sqrt{13}}{2}, \dfrac{-3 - \sqrt{13}}{2} \right\}$ **39.** $\{-7, 1\}$

41. $\{x \mid x < -1 \text{ or } x > 4\}$ **43.** Empty set

45. Ellipse   **47.** Parabola

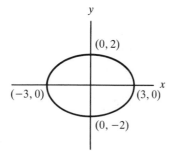   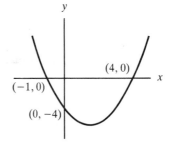

49. Semicircle   **51.** Line with point $(-2, 4)$ missing

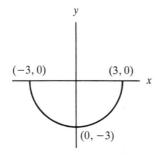   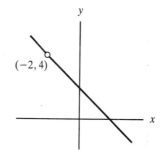

53. $f^{-1}(x) = 5 - x$; $\mathfrak{D}(f^{-1}) = \mathbf{R}$; $\mathfrak{R}(f^{-1}) = \mathbf{R}$

55. $f^{-1}(x) = -x^2$; $\mathfrak{D}(f^{-1}) = \{x \mid x \le 0\}$; $\mathfrak{R}(f^{-1}) = \{y \mid y \le 0\}$

57. $f^{-1}(x) = 1 + \sqrt{x}$; $\mathfrak{D}(f^{-1}) = \{x \mid x \ge 1\}$; $\mathfrak{R}(f^{-1}) = \{y \mid y \ge 2\}$

59. $f^{-1}(x) = \dfrac{6 - x}{3}$; $\mathfrak{D}(f^{-1}) = \{x \mid x \le 6\}$; $\mathfrak{R}(f^{-1}) = \{y \mid y \le 0\}$

Exercises 2.0, pages 108–109

1. $(x + 3)(x^2 - 3x + 9)$ **3.** $x(x^2 - 6x + 12)$

5. $(x - 2)(x^2 + 1)$ **7.** $x(x + 1)(x - 1)^2$

9. $(x + 1)(x - 2)(x^2 + x + 2)$ **11.** $5x^2 - x$

13. $3x^3 - 3x^2 + 6$ **15.** $x^4 - 4x^2 + 16x - 16$

17. 3   **19.** $q(x) = x^2 + x + 5$; $r(x) = 8x + 9$

21. -2

23. $\dfrac{x-1}{x}$

25. $\dfrac{1-x}{2x-1}$

27. $\dfrac{4x^2 - 10x + 25}{x}$

29. $\dfrac{x^2 - 5x + 13}{4 - x}$

31. $\dfrac{x^2 + 2x + 6}{x^2 + x - 2}$

33. $\dfrac{5x^2 + x - 1}{x^2 + x}$

35. $\dfrac{x^2 - x}{x + 1}$

37. $\dfrac{2x^3 + 2x^2 + 2x}{x + 1}$

39. $\dfrac{x^2 - 4x + 4}{2x + 1}$

41. $\dfrac{1}{x^3 - x^2 - 2x - 12}$

43. $\dfrac{1}{3x^2 - 3x - 2}$

45. 1

47. $\dfrac{3x - 1}{3}$

Exercises 2.1, pages 113–114
1. a) Yes   **b)** No   **c)** Yes

Problem	Degree	Leading coefficient	Constant term	Over the rational numbers?	Standard form
3	4	3	3	Yes	$3x^4 - 2x^3 + x^2 - x + 3$
5	3	-1	0	Yes	$-x^3 + x^2 + x$
7	3	2	-2	Yes	$2x^3 - 2x^2 + 2x - 2$
9	3	$3\sqrt{3}$	$\sqrt{3}$	No	$3\sqrt{3}x^3 - 2\sqrt{3}x + \sqrt{3}$

11. Yes; 2, 4

13. No

15. Yes; 4, 3

17. a) 13   **b)** -40.12

19. a) 0.96   **b)** -7.83

21. 23.80

23. 0.80

25.

x	-1.48	-0.43	0	0.83	1.64
y	-5.0	-3.0	-4.0	-6.6	-7.2

27.

x	-2.3	-2.4	-2.33	-2.34	-2.331
y	0.33	-0.82	0.00066	-0.11	-0.011

29. $P(x) = 4x + 1$

31. Any polynomial of the type $P(x) = 2x^3 + bx^2 + cx + d$, where
$4b + 2c + d = -17$

Exercises 2.2, page 119

1.

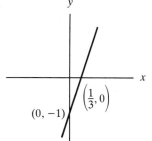

3.

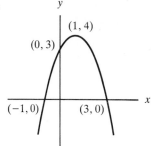

5.

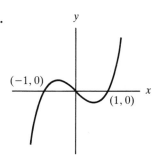

7.

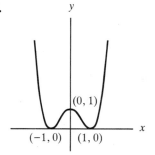

9.

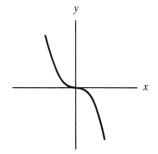

11.

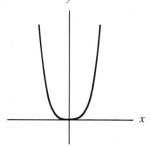

13.

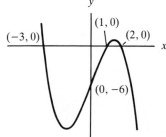

15.

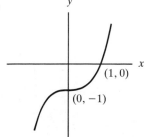

17.

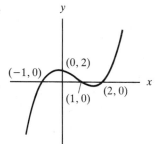

19.

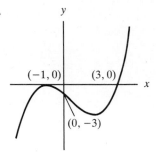

21.

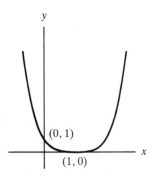

23.

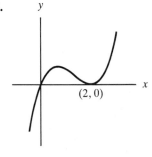

25.

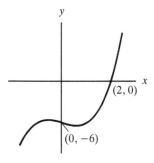

27.

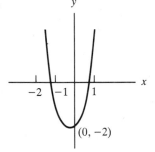

29. $\mathcal{R}(f) = \mathbf{R}$; x intercept at about $x = 2.4$

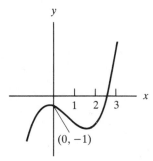

Exercises 2.3, page 126

1. $(x - 2)(3x + 4) + 9$

3. $(x + 2)(3x^3 - 6x^2 + 10x - 17) + 29$

5. $(x - 3)(-4x^2 - 7x - 23) - 66$

7. $q(x) = 2x^2 + 3x + 13,\ r = 40$

9. $q(x) = x^4 + 2x^3 + 4x^2 + 8x + 16,\ r = 64$

11. $q(x) = 3x^2 - 9.2x + 25.08,\ r = -64.192$

13. $q(x) = x^2 + 1.2x + 3.44,\ r = 1.128$ **15.** $r = 1$ **17.** $p(-3) = -100$

19. $p(-1) = 0$ and so $(x + 1)$ is a factor; $p(1) = 0$ and so $(x - 1)$ is a factor

21. $p(-2) = 0$ and so $(x + 2)$ is a factor

23. $x + 1$ is a factor if n is an odd positive integer   **25.** $-3/2$ **27.** 2

29. $p(x) = x^3 + 2x^2 - x - 2$ **31.** $p(x) = x^3 - 1.4x^2 - x + 1.4$

Exercises 2.4, page 132

1. $p(x) = x^3 - 2x^2 - 5x + 6$

3. $p(x) = x^3 + x^2 - 16x - 16$

5. $p(x) = x^4 - 6x^2 + 8x - 3$

7. $p(x) = x^3 - 3x^2 - 2x + 4$

9. $p(x) = x^3 + (1 - \sqrt{3})x^2 - \sqrt{3}x$

11. $0,\ -2,\ \sqrt{5}$, each of multiplicity one

13. 0 and 1 of multiplicity two; -2 of multiplicity one   **15.** $\sqrt{2}$ is a zero

17. -2 is a zero   **19. a)** No   **b)** Yes   **21. a), b)** $1, 2, 4, 8$ **c)** 4 is a root

23. a), b) $\pm 1,\ \pm 3,\ \pm\frac{1}{2},\ \pm\frac{3}{2}$ **c)** $-\frac{3}{2}$ is a root

25. a), b) $\pm 1,\ \pm 3,\ \pm\frac{1}{3}$ **c)** $-1,\ -\frac{1}{3},\ 3$ are roots

27. a), b) $\pm 1,\ \pm 3,\ \pm\frac{1}{5},\ \pm\frac{3}{5}$ **c)** $-\frac{1}{5}$ is a root

29. a), b) $-1,\ -2,\ -\frac{1}{2},\ -\frac{1}{3},\ -\frac{2}{3},\ -\frac{1}{6},\ -\frac{1}{9},\ -\frac{2}{9},\ -\frac{1}{18}$ **c)** $-\frac{2}{3},\ -\frac{1}{2},\ -\frac{1}{3}$ are roots

31. a), b) $\pm 1,\ \pm 2,\ \pm 3,\ \pm 4,\ \pm 6,\ \pm 9,\ \pm 18,\ \pm 36$ **c)** $-3,\ -2,\ -2,\ 3$ are roots

33. $p(x) = (2x + 1)(x^2 + x + 1)$ **35.** $p(x) = (x + 1)(2x - 1)(x^2 + 1)$

37. $-1 - \sqrt{2}$ is a root of $x^2 + 2x - 1 = 0$, an equation with integer coefficients

39. $\dfrac{1 + \sqrt{5}}{2}$ is a root of $x^2 - x - 1 = 0$, an equation with integer coefficients

Exercises 2.5, page 138

1. a) 3 or 1   **b)**
c) 3

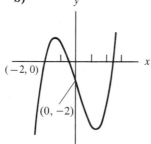

(−2, 0)

(0, −2)

3. a) 3 or 1   **b)**
c) 3

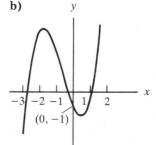

(0, −1)

5. a) 4, 2 or 0   **b)**
  c) 2

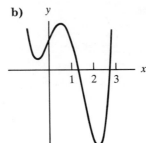

7. a) 4, 2, or 0   **b)**
  c) No zeros

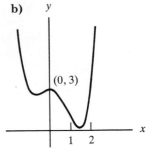

9. Only one zero; 2.1 to one decimal place

11. Only one zero; -3.2 to one decimal place

13. Four zeros: between $-2, -1$; $-1, 0$; 2, 3; 3, 4; smallest is -1.6 to one decimal place

15. Only one zero; 3.28 to two decimal places

17. Three zeros: between $-3, -2$; $-1, 0$; 1, 2; largest is 1.34 to two decimal places

19. Two zeros: between $-2, -1$; 1, 2; largest is 1.91 to two decimal places

Exercises 2.6, pages 145–146

1. 4, 4   **3.** 0, 0   **5.** $\infty, -\infty$

7. a) $x = -2$ **b)** $y = \frac{3}{2}$ **9. a)** $x = 1, x = -1$ **b)** $y = 3$

11. a) $x = -2, x = 1$ **b)** $y = 0$ **13. a)** $x = -2$ **b)** $y = 1$

15. a) $x = 0$ **b)** $y = 0$

17.

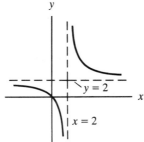

19.

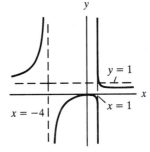

21.

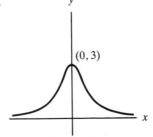

23.

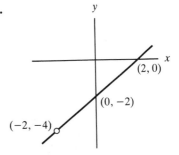

25.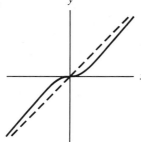

27.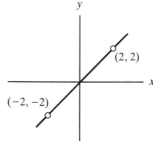

29.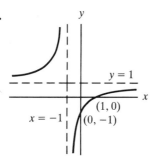

31.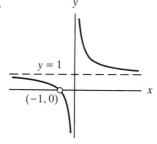

33.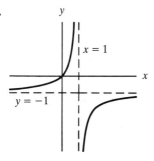

35.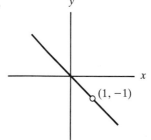

Exercises 2.7, pages 151–152

1. Slope at $(-1, -2)$ is -1

3. Slope at $(\frac{3}{2}, \frac{21}{8})$ is -2.75

5. Slope at $(5, 2)$ is 0.25

7. Slope at $(-2, \sqrt{5})$ is 0.894

9. Slope at $(0, 1)$ is -0.5

11. Slope at $(4, 3)$ is 0.25

13. $y = 2x - 2$

15. $y = 0.25x + 2$

17. $y = 0.25x + 2.75$

19. $y = -0.354x - 3.182$

21. $y = 0.75x - 0.50$

23. $y = -0.25x + 1.25$

Review Exercises, pages 152–153

1. a) -12 **b)** -2.784

3. a) -79.30 **b)** -40.08

5.

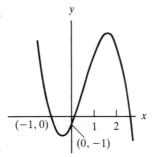

7.

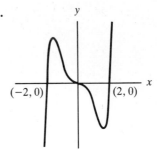

9. $f(x) = x^3 - 7x - 6$

11. 0, 1, 4

13. One zero, between -1 and 0

15. $q(x) = 3x^2 - 7x + 2, r = -4$

17. $r = 12$

19. $-\frac{1}{2}$ is the only rational zero

21. Irrational roots between: $-3, -2; 0, 1; 1, 2$; largest is 1.8 to one decimal place

23. $\mathcal{D}(f) = \{x \mid x \neq -1, x \neq 2\}$

25.

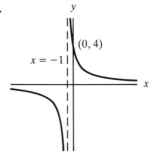

27. a) $(-3,0); (0, \frac{3}{2})$
  b) $x = 2; y = -1$
  c)

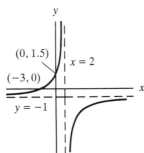

29. a) $(3, 0), (-1, 0); (0, 3)$
  b) $x = 1; y = x - 1$ is an oblique asymptote
  c)

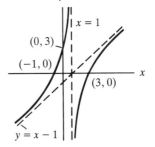

Exercises 3.0, pages 160–162

1. 2   **3.** 1   **5.** $\frac{512}{27}$ **7.** -39 **9.** 7

11. 1   **13.** 1   **15.** 0   **17.** 2   **19.** 0

21. 2865 **23.** 19.37 **25.** 18.68 **27.** 7.240 **29.** 3.828

31. x^6; **R** **33.** $1/x$; $\{x \mid x \neq 0\}$ **35.** 1; $\{x \mid x \neq -1, x \neq 1\}$

37. $2 + \sqrt{x} - x$; $\{x \mid x \geq 0\}$ **39.** $|x|$; **R**

41. 1; $\{x \mid x \neq 0, x \neq -2, x \neq -3\}$ **43.** $\sqrt{2} - 1$

45. $-1 - 2\sqrt{7}$ **47.** $\dfrac{1}{2\sqrt{3}}$ **49.** $\dfrac{1}{5 + 2\sqrt{6}}$ **51.** $S = \{\frac{1}{2}\}$

53. $S = \{1\}$ **55.** $S = \{1\}$ **57.** $S = \emptyset$ **59.** $S = \{-1, 1\}$

61. True **63.** True **65.** True **67.** False **69.** False

Exercises 3.1, pages 168–170

1. 3 **3.** 27 **5.** 81 **7.** 45 **9.** 1

11. 54 **13.** -140 **15.** 1585 **17.** -1.428 **19.** 0.215

21. 1.293 **23.** 1.341 **35.** 0.523 **27.** -1.106

29. $x + 2 + \dfrac{1}{x}$

31. 2 **33.** $x - 2$ **35. a)** $-\dfrac{1}{1 + \sqrt{3}}$ **b)** $\dfrac{1}{4 - 2\sqrt{3}}$ **c)** $\dfrac{1}{\sqrt{x} + 3}$

37. 3.31 **39.** $\frac{10}{3}$ **41.** $\frac{109}{30}$

43. $\sqrt{3} + 2$ is a root of $x^2 - 4x + 1 = 0$; an equation with integer coefficients and no rational roots

45. $\sqrt[3]{2} - 1$ is a root of $x^3 + 3x^2 + 3x - 1 = 0$; an equation with integer coefficients and no rational roots

47. a) $V = \frac{4}{3}\pi(1 + 2\sqrt{t})^3$; domain is $\{t \mid 0 \leq t \leq 49\}$

b) $t = 4$ sec, $V = 523.60$ cm; $t = 60$ sec, V is not defined

49. a) Yes **b)** No **c)** No

Exercises 3.2, pages 174–175

1. 4.729 **3.** 0.062 **5.** 0.177 **7.** 2.952 **9.** 1.435

11. a) $3^{\sqrt{3}}$ is greater **b)** e^3 is greater

13.

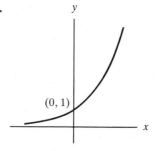

15.

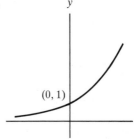

17.

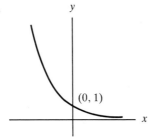

$(0, 1)$

19.

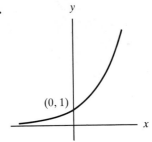

$(0, 1)$

21.

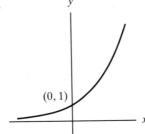

$(0, 1)$

23.

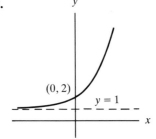

$(0, 2)$ $y = 1$

25.

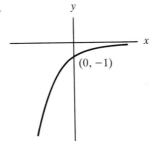

$(0, -1)$

27.

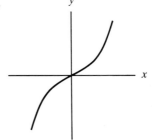

29. a) 0.50   **b)** 0.27   **c)** 0.12   **d)** 0.95   **e)** 0.65

31.

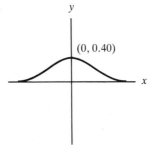

$(0, 0.40)$

33. -0.282 **35.** 80.012   **37.** 3.333

39. Three solutions: 2, 4, -0.8 (to one decimal place)

Exercises 3.3, pages 182–183

1. 5   **3.** $\frac{5}{2}$ **5.** 2   **7.** -2

9. -1 **11.** Undefined   **13.** Undefined   **15.** $\frac{5}{2}$

17. $\frac{3}{2}\log_b p + 2\log_b q$ **19.** $\frac{3}{2}\log_b p + \frac{4}{3}\log_b q$

21. $\log_b p + \log_b q$ **23.** $\log_b p + \log_b q - 1$

25. 1.1133   **27.** 2.6826   **29.** 3.8136   **31.** 1.8155

33. 0.4729   **35.** 1.4610   **37.** 0.4307   **39.** 1.1656

41. $\log_3 100$ **43.** $\log_7 9$ **45.** $\log_2 \frac{3}{2}$ **47.** $b = 4$

49. $x = 2$ **51.** $x = 1$ **53.** $x = 2$ **55.** $x = 2$

57. True   **59.** Meaningless   **61.** False   **63.** Meaningless

65. False   **67.** False   **69.** $\mathcal{D}(f) = \{x \mid -5 < x < 5\}$

71. $\mathcal{D}(h) = \{x \mid 3 < x < 5\}$ **73.** $\mathcal{D}(g) = \{x \mid x < 0 \text{ or } x > 4\}$

75. $\mathcal{D}(h) = \{x \mid x > 0\}$

Exercises 3.4, pages 190–191

1. 1.6094   **3.** 0.2718   **5.** 2.2912   **7.** Undefined

9. -0.4666 **11.** 0.7730   **13.** 1.43   **15.** -0.42

17. $\frac{1}{2}$ **19.** 3.0061   **21.** 4.6274   **23.** 0.4246

25. No solution   **27.** -0.2135 **29.** -0.1392 **31.** -0.5815

33. 0.4916   **35.** 0.6545   **37.** $-\frac{1}{2}, 2$ **39.** 6.9584

41. -1 **43.** True   **45.** Meaningless   **47.** True

49. False   **51. a)** $\mathcal{D}(f) = \{x \mid x > 2\}$ **b)** $\mathcal{D}(g) = \{x \mid x > 2\}$

53.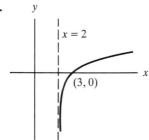

55.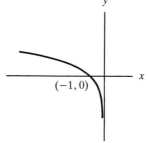

57.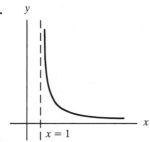

59.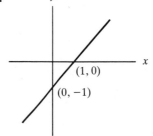

61. 0.86   **63.** 0.43   **65.** 1.284   **67.** 0.149

69. $(g \circ f)(x) = e^{\ln x} = x,\ x > 0$

Exercises 3.5, pages 197–198

1. b) $N = 3\,174\,802$ **c)** 3 hours 59 minutes

3. a) 30%   **b)** approximately 19000 years   **5.** 7.96 days   **7.** 62%

9. a) 9 times   **b)** 20.5 times   **11.** $2163.57   **13.** 8.75% earns $85.52 more

15. Needs to wait 8 months   **17.** $6711.29

19.

r	4	6	8	9	12	18	24
N	18	12	9	8	6	4	3
rN	72	72	72	72	72	72	72

Exercises 3.6, page 202

1. 1.39   **3.** 1.00   **5.** 0.92   **7.** -0.50

9. 0   **11.** 2.72   **13.** 1.00   **15.** 1.00

17. $y = 0.37x$ **19.** $y = 5.44x + 2.72$

Review Exercises, pages 202–203

1. 0.903   **3.** 3.135   **5.** 10.751   **7.** 0.520

9. 1.292   **11.** 0.166   **13.** Undefined   **15.** 3.5

17. Undefined   **19.** 12.265   **21.** 24.799   **23.** 6.167

25. 3   **27.** -0.432 **29.** 522.735   **31.** 0.434

33. 2.262   **35.** No solution

37.

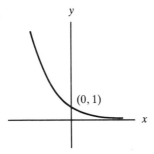

39.

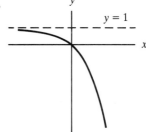

41. a) 2.223   **b)** -1.223, 2.223   **43.**

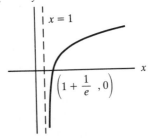

45. Maximum value of y is 0.5307 to four decimal places   **47.** 11460 years
49. $25 563.23

 Exercises 4.1, pages 211–213

1. a)   **b)**   **c)**   **d)**

$A = 135°$

 $B = 720°$

 $C = -60°$

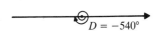 $D = -540°$

e)   **f)**   **g)**   **h)**

$E = 210°$

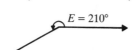 $F = 10°$

 $G = -300°$

 $H = 22°30'$

3. a)   **b)**   **c)**   **d)**

$A = 2\pi$

 $B = \dfrac{17\pi}{6}$

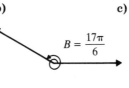 $C = \dfrac{\pi}{2}$

 $D = \dfrac{\pi}{4}$

e)   **f)**   **g)**   **h)**

$E = -\dfrac{7\pi}{2}$

 $F = -\dfrac{3\pi}{2}$

 $G = \dfrac{9\pi}{4}$

 $H = \dfrac{\pi}{3}$

5. a)   **b)**   **c)**

 θ

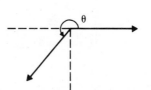 θ

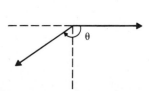 θ

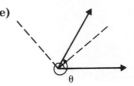
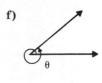

7. a) 48.6617°   **b)** −75.2114°   **9. a)** 37°34′59″   **b)** 321°34′35″

11. a) $\dfrac{2\pi}{3}$; 2.094   **b)** $\dfrac{7\pi}{4}$; 5.498   **c)** $\dfrac{\pi}{8}$; 0.393   **d)** $-\dfrac{11\pi}{6}$; −5.760

13. a) 1.12   **b)** 4.01   **c)** −0.64   **d)** 2.58   **e)** 8.24

15. a) 135°   **b)** −630°   **c)** 110°   **d)** −3060°   **e)** 675°

17. a) 78.495°; 78°29′43″   **b)** 0.195°; 0°11′41″   **c)** 92.707°; 92°42′23″
    d) −197.670°; −197°40′14″   **e)** 1718.873°; 1718°52′24″

Exercises 4.2, pages 217–221

1. a) 23.52 cm   **b)** 47.94 cm   **c)** 134.00 cm   **3. a)** 0.49   **b)** 2.39   **c)** 1.16

5. a) 16.01 m   **b)** 89.72 m   **c)** 50.47 m   **d)** 392.54 m

7. a) 131.60 cm/sec   **b)** 30389.39 cm/min   **c)** 9723.20 cm/min   **d)** 52.74 cm/sec

9. a) 1 rev/hr   **b)** 1/60 rev/min   **c)** 6 deg/min   **d)** $\pi/30$ rad/min

11. a) 40.84 cm/hr   **b)** 0.68 cm/min   **c)** 0.011 cm/sec

13. 9972.67 m/min   **15.** 1675.52 km/hr   **17.** 10109 km/hr

19. a) 1.1041 rad   **b)** 63.26°   **21. a)** 202.2756 rad   **b)** 11589.54°

23. a) $\dfrac{2\pi}{365.25}$; 0.0172 rad/day   **b)** $\dfrac{\pi}{4383}$; 0.000717 rad/hr   **c)** 106798 km/hr

25. 136.35 cm/sec; 1090.76 cm   **27.** 7.52 m   **29.** Time is 1:17

Exercises 4.3, pages 229–231

1. a)   **b)**   **c)**

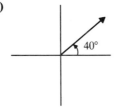
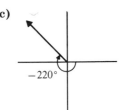

d)   **e)**

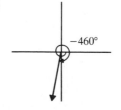

3. a) I **b)** IV **c)** III **d)** III **e)** III

5. a) $260°$ **b)** $180°$ **c)** $180°$ **d)** $5\pi/4$ **e)** 1.96

7. Any angle of the type $90° + k \cdot 360°$, $k \in \mathbf{J}$ **9.** $\{\theta \mid \theta = -\dfrac{2\pi}{3} + k \cdot 2\pi, k \in \mathbf{J}\}$

11. a) $\{\theta \mid \theta = 45° + k \cdot 360°\}$ **b)** $\{\theta \mid \theta = 225° + k \cdot 360°\}$
   c) $\{\theta \mid \theta = 120° + k \cdot 360°\}$

13.

	$\sin\theta$	$\cos\theta$	$\tan\theta$	$\cot\theta$	$\sec\theta$	$\csc\theta$
a)	$-3/5$	$4/5$	$-3/4$	$-4/3$	$5/4$	$-5/3$
b)	-0.60	0.80	-0.75	-1.33	1.25	-1.67

Problem	a)	b)	c)	d)
15	$\sqrt{3}/2$	$1/2$	$-1/2$	$-\sqrt{3}/2$
17	-1	$-\sqrt{2}$	1	$\sqrt{2}$
19	$1/2$	$-1/\sqrt{3}$	$1/2$	$-1/2$
21	1	1	undef	-1
23	2	-1	$1/\sqrt{3}$	$-1/2$

25.

	$\sin\theta$	$\tan\theta$	$\cot\theta$	$\sec\theta$	$\csc\theta$
a)	$4/5$	$-4/3$	$-3/4$	$-5/3$	$5/4$
b)	0.800	-1.333	-0.750	-1.667	1.250

27.

	$\sin\beta$	$\cos\beta$	$\tan\beta$	$\sec\beta$	$\csc\beta$
a)	$-4/5$	$-3/5$	$4/3$	$-5/3$	$-5/4$
b)	-0.800	-0.600	1.333	-1.667	-1.250

29.

	$\cos\theta$	$\tan\theta$	$\cot\theta$	$\sec\theta$	$\csc\theta$
a)	$\sqrt{15}/4$	$-1/\sqrt{15}$	$-\sqrt{15}$	$4/\sqrt{15}$	-4
b)	0.968	-0.258	-3.873	1.033	-4.000

31. $\dfrac{2 + \sqrt{3}}{2}$; 1.866

Exercises 4.4, pages 233–234

1. 0.4695	**3.** 1.2208	**5.** 0.6865	**7.** −0.6123	**9.** −0.3190
11. 0.3894	**13.** Undefined	**15.** 0.8502	**17.** −2.8267	**25.** 1.9882
21. 2.3750	**23.** 2.8083	**25.** 0.8859	**27.** 0.2737	**29.** 1.0000
31. 0.3199	**33.** 0.4643	**35.** 1.0419	**37.** 3.57	**39.** Undefined
41. 2.38	**43.** 0	**45.** 0.84	**47.** 0.17	

Exercises 4.5, pages 246–247

In problems 1 through 8, let $P:(a, b)$ correspond to s and draw diagrams similar to that in problem 1 below. In each case use congruent triangles to determine the coordinates of point Q. For example, in problem 1, point $Q:(-a, b)$ corresponds to $(\pi - s)$.

1. $Q:(-a, b)$

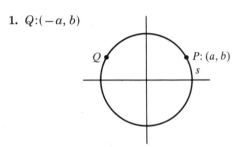

3. $Q:(b, -a)$ **5.** $Q:(-a, -b)$ **7.** $Q:(-b, a)$

9. See Fig. 4.34   **11.** See Fig. 4.36   **13.** See Fig. 4.38

15. Yes   **17.** $\{x \mid x = (2k + 1)\dfrac{\pi}{2}, k \in \mathbf{J}\}$ **19.** $\{1\}$

Exercises 4.6, pages 252–256

1. $b = 4.60$ cm; $c = 5.64$ cm; $\beta = 54°36'$ **3.** $b = 288$ cm; $\alpha = 31°17'$; $\beta = 58°43'$

5. $\beta = 62°43'$; $a = 25.90$ cm; $b = 50.21$ cm; Area $= 650$ cm²

7. $c = 26417$ m; $\alpha = 66°24'$; $\beta = 23°36'$

9. $a = 108.90$ cm; $\alpha = 55°58'$; $\beta = 34°02'$; Area $= 4005$ cm²

11. $77°00'$ **13.** 3.03 m²

15. a) 6498.0 cm²;   **b)** 8441.1 cm²;   **c)** 9189.6 cm²;   **d)** $3249n \sin \dfrac{180°}{n} \cos \dfrac{180°}{n}$

17. 43.47 m   **19.** 5.00 m   **21.** 2.47 cm²   **23.** $62 029.49   **25.** 25.3 m

29. 10.07 cm; 4.91 cm   **31.** $a = 1.47$ m; $b = 3.39$ m

Exercises 4.7, pages 264–269

1. $c = 48$; $\beta = 105°53'$; $\alpha = 31°07'$ **3.** $b = 90$; $\alpha = 69°06'$; $\gamma = 27°30'$

5. $c = 1.03$; $\beta = 35°48'$; $\gamma = 20°12'$ **7.** $\alpha = 25°35'$; $\beta = 119°39'$; $\gamma = 34°46'$

9. $\gamma = 80°$; $b = 34$; $c = 35$ **11.** $\beta = 21°$; $a = 64$; $b = 31$

13. Two solutions: $b_1 = 5114$, $\beta_1 = 83°05'$, $\gamma_1 = 64°55'$;
$b_2 = 2800$, $\beta_2 = 32°55'$, $\gamma_2 = 115°05'$

15. $\gamma = 82°47'$; $a = 4.711$; $b = 1.979$

19. 140.1 m   **21.** 149 m   **23.** 23.69   **25.** 22.9 m   **27.** 388 m

29. 52°   **31.** (107, 42)   **35.** 23 m   **37.** 63°42'   **39.** $5 + 4\sqrt{2}$

Review Exercises, pages 269–272

1. a)   **b)**   **c)**

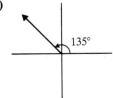   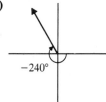   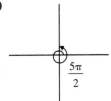

d)   **e)**   **f)**

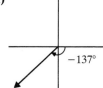   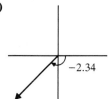   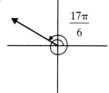

3. a) 1   **b)** $\dfrac{1}{\sqrt{3}}$ **c)** $-\dfrac{2}{\sqrt{3}}$ **d)** $-\dfrac{1}{2}$ **e)** 0   **f)** 1   **g)** 1   **h)** -1

5. a) $-\dfrac{4}{5}$ **b)** $-\dfrac{5}{3}$ **c)** $\dfrac{3}{5}$ **d)** $\dfrac{4}{3}$ **e)** $\dfrac{5}{3}$ **f)** $\dfrac{4}{5}$

7. a) 270°   **b)** 30°   **c)** 135°   **d)** $-45°$

9. a) $-\dfrac{\sqrt{3}}{2}$ **b)** 0   **c)** $\dfrac{1}{2}$ **d)** $\dfrac{\sqrt{3}}{2}$ **e)** $\sqrt{3}$ **f)** $\dfrac{2}{\sqrt{3}}$

11. a) 0.6820   **b)** -0.4877 **c)** 0.5407   **d)** 0.9004   **e)** 1.1897   **f)** 0.7771

13. a) 0.7880   **b)** 1.7646   **15. a)** 1   **b)** 1

17. a) True   **b)** False   **c)** True   **d)** False

19.   **21.** $a = 27.90$ cm; $b = 24.94$ cm

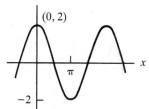

23. $b = 18.8$ cm; $\alpha = 39°20'$; $\beta = 50°40'$ **25.** No solution

27. $\alpha = 46°$; $\beta = 76°$; $\gamma = 58°$ **29.** Infinitely many solutions   **31.** No solution

33. 53 m; 197 m

Exercises 5.1, pages 278–279

1. a) Yes   **b)** Yes   **c)** Yes

43. a) Even   **b)** Neither

47. $\mathfrak{D}(f) = \mathbf{R}$, $\mathfrak{R}(f) = \{1\}$

41. a) Odd   **b)** Neither

45. a) Even   **b)** Odd

49. $\mathfrak{D}(f) = \mathbf{R}$, $\mathfrak{R}(f) = \{y \mid 0 \le y \le 1\}$

Exercises 5.2, pages 283–285

1. $\cot(\alpha + \beta) = \dfrac{\cot\alpha \cot\beta - 1}{\cot\alpha + \cot\beta}$

5. a) $\dfrac{\sqrt{6} - \sqrt{2}}{4}$ **b)** $\dfrac{\sqrt{2} - \sqrt{6}}{4}$ **c)** $-2 - \sqrt{3}$ **d)** $2 + \sqrt{3}$ **e)** $-\sqrt{2} - \sqrt{6}$

f) $\sqrt{2} - \sqrt{6}$

7. $\frac{1}{7}$ **9.** $\frac{1}{2}$ **11.** Identity   **13.** Identity

15. Identity   **17.** Identity   **19.** $\dfrac{\sqrt{10 + 2\sqrt{5}}}{4}$

23. a) $2\sin 4\alpha \cos\alpha$ **b)** $2\cos 4\alpha \cos\alpha$ **c)** $-2\sin 2\alpha \sin\alpha$

25. a) $\dfrac{\sqrt{3}}{2}$ **b)** $-\dfrac{\sqrt{2}}{2}$ **c)** 1 **d)** $-\dfrac{\sqrt{3}}{3}$

Exercises 5.3, pages 290–292

3. a) $-\frac{120}{169}$ **b)** $\frac{119}{169}$ **c)** $-\frac{120}{119}$ **5. a)** 0.7882 **b)** 0.2075 **c)** 0.2121

7. a) $\sqrt{3} - 2$ **b)** $-\frac{1}{2}\sqrt{2 - \sqrt{2}}$ **c)** $\sqrt{3} - 2$ **d)** $\frac{1}{2}\sqrt{2 - \sqrt{3}}$

9. a) $\frac{1}{2}\sqrt{2 - \sqrt{2}}$ **b)** $-\frac{1}{2}\sqrt{2 - \sqrt{2}}$ **c)** $\frac{1}{2}\sqrt{2 + \sqrt{2}}$ **d)** $-2 - \sqrt{3}$

11. a) 0.96126 **b)** 0.74314 **c)** 0.26750 **13.** $-\dfrac{5\sqrt{10} + 48}{50}$ **35.** Identity

37. Not an identity   **39.** Identity   **41.** $-\frac{1}{8}$ **43.** $\dfrac{7}{\sqrt{65}}$ **45.** $-\frac{116}{845}$

Exercises 5.4, pages 302–304

1.

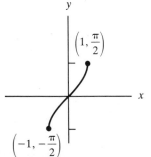

3.

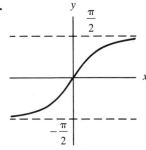

5. π　　　　**7.** $-\pi/4$　　**9.** $-\pi/6$　　**11.** Undefined　**13.** $\frac{3}{4}$

15. π　　　　**17.** Undefined　**19.** $-\frac{3}{4}$　　**21.** $\frac{7}{24}$　　**23.** $-\frac{4}{5}$

25. 0.39　　　**27.** -0.98　　**29.** 1.30　　**31.** 0.73　　**33.** Undefined

35. -0.18　　**37.** Yes　　　**39.** Yes　　**41. a)** -0.48　**b)** 0.07

43. a) Solution set is $\emptyset$　**b)** Solution set $= \{x \mid x \le -1.56\}$

45. a) Solution set $= \{-1, 1\}$　**b)** Solution set $= \{x \mid -1 < x < 0\}$

47. a) Solution set $= \{x \mid -1 \le x \le 1\}$　**b)** Solution set $= \{x \mid 0 \le x \le \pi\}$

49. c) 3.16 m

Exercises 5.5, pages 309–310

1. $S = \left\{ x \mid x = \dfrac{2\pi}{3} + k \cdot 2\pi \text{ or } x = \dfrac{4\pi}{3} + k \cdot 2\pi \right\}$　　**3.** $S = \left\{ x \mid x = \dfrac{\pi}{6} + k\pi \right\}$

5. $S = \left\{ x \mid x = \dfrac{\pi}{4} + \dfrac{k\pi}{2} \right\}$　　**7.** $S = \{ x \mid x = (2k + 1)\pi \}$

9. $S = \{ x \mid x = -30° + k \cdot 360° \text{ or } x = 210° + k \cdot 360° \}$

11. $S = \{ x \mid x = 120° + k \cdot 180° \}$　　**13.** $S = \emptyset$

15. $S = \{ x \mid x = 60° + k \cdot 360° \text{ or } x = 180° + k \cdot 360° \}$　　**17.** $S = \left\{ \dfrac{\pi}{3}, \dfrac{2\pi}{3} \right\}$

19. $S = \emptyset$　　**21.** $S = \{0.76, 2.39\}$　　**23.** $S = \{1.03, 4.17\}$

25. $S = \{1.34, 2.91, 4.48, 6.05\}$　　**27.** $S = \left\{ \dfrac{3\pi}{2} \right\}$　　**29.** $S = \left\{ \dfrac{\pi}{3}, \dfrac{5\pi}{3} \right\}$　　**31.** $S = \emptyset$

33. $S = \{1.85, 4.44\}$　　**35.** $S = \{0.34, 2.80\}$　　**37.** $S = \left\{ \dfrac{\pi}{3} \right\}$

39. $S = \left\{ \dfrac{\pi}{2}, \dfrac{3\pi}{2}, 1.11, 4.25 \right\}$　　**41.** $S = \left\{ \dfrac{\pi}{2}, \dfrac{3\pi}{2} \right\}$　　**43.** $\dfrac{15\pi}{8}; \dfrac{23\pi}{8}$

45. $\dfrac{3\pi}{8}; \dfrac{7\pi}{8}; \dfrac{11\pi}{8}; \dfrac{15\pi}{8}$

Exercises 5.6, page 315

1. $p = 2\pi, A = 2$　　　　　　　　　　　**3.** $p = 2\pi, A = 2$

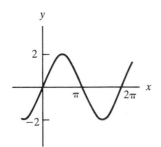　　　　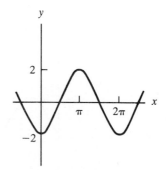

5. $p = \pi$, $A = 3$

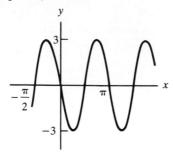

7. $p = 2$, $A = 3$

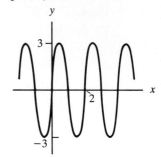

9. $p = 4$, $A = 2$

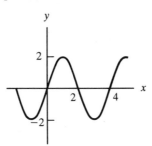

11. $p = 2\pi$, $A = 2$, phase shift $\dfrac{\pi}{2}$ to right

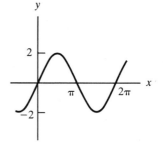

13. $p = \pi$, $A = 3$, phase shift $\dfrac{\pi}{6}$ to right

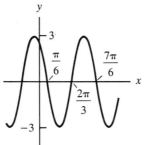

15. $p = 1$, $A = 4$, phase shift $\frac{1}{4}$ to right

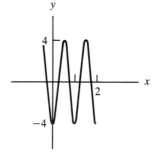

17. $y = \sqrt{2}\sin\left(x - \dfrac{\pi}{4}\right)$; $p = 2\pi$; $A = \sqrt{2}$; phase shift $\dfrac{\pi}{4}$ to right

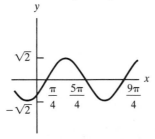

19. $y = 2 \sin\left(x - \dfrac{\pi}{6}\right)$; $p = 2\pi$; $A = 2$; phase shift $\dfrac{\pi}{6}$ to right

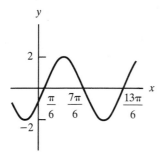

Exercises 5.7, pages 317–318

1. 2   **3.** 0   **5.** -0.5 **7.** 1   **9.** -0.58 **11.** 0.87   **13.** 2

15. a)

c	0	1	1.5	3.4	-0.8
$m(c)$	1	0.540	0.071	-0.967	0.697

b)

c	0	1	1.5	3.4	-0.8
$\cos c$	1	0.540	0.071	-0.967	0.697

c) $m(c) = \cos c$; $m(2.5) = -0.801$; $m(-\sqrt{5}) = -0.617$

Review Exercises, pages 318–320

9. $-\frac{24}{25}$ **11.** $\frac{56}{65}$ **13.** $\frac{119}{169}$ **15.** $\frac{3713}{4225}$ **17.** $-\frac{24}{7}$

19. $-\dfrac{4 + 3\sqrt{3}}{10}$ **21.** $-\dfrac{\pi}{3}$ **23.** $\dfrac{5\pi}{4}$ **25.** $\dfrac{\pi}{6}$ **27.** $\frac{63}{65}$

29. $\frac{1}{5}$ **31.** $\frac{3}{5}$ **33.** 0.436   **35.** Undefined   **37.** 0.935

39. -0.990 **41.** $\dfrac{\pi}{3}$; $\dfrac{5\pi}{3}$ **43.** 1.030; 4.172   **45.** 2.498

47. $\dfrac{\pi}{2}$; $\dfrac{3\pi}{2}$ **49.** No solution   **51.** $\dfrac{\pi}{6}$ **53.** 0.84

55. 2.91   **57.** False   **59.** True   **61.** False   **63.** False

Exercises 6.1, pages 332–335

1. $x = 1$, $y = 3$

3. Dependent; solutions are given by $x = t$, $y = \dfrac{3 - 2t}{4}$, where t is any number

5. Inconsistent   **7.** $x = 3$, $y = -1$ **9.** Inconsistent   **11.** $x = -2$, $y = -1$

13. Dependent; solutions are given by $z = t$, $x = \dfrac{3 + 2t}{3}$, $y = \dfrac{13t + 3}{3}$, where t is any number

15. Dependent; solutions are given by $z = t$, $x = \dfrac{5 - 4t}{3}$, $y = \dfrac{4t - 11}{3}$, where t is any number

17. $x = 2$, $y = 1$ **19.** Inconsistent

21. Dependent; solutions are given by $x = t$, $y = \dfrac{4t + 5}{8}$, where t is any number

23. $x = \frac{22}{3}$, $y = \frac{22}{7}$ **25.** $x = 4$, $y = 1$, $z = -2$

27. Dependent; solutions are given by $z = t$, $x = \dfrac{4t - 4}{3}$, $y = \dfrac{5t + 1}{3}$, where t is any number

29. $x = -3$, $y = -2$, $z = 2$

31. A produces 5 items per hour; B produces 15 items per hour

33. Sandwich $1.60, drink $0.30, pie $0.60   **35.** 400 g of A, 1600 g of B

37. 300 g of A, 1200 g of B, 900 g of C **39.** 6 hours for A, 12 hours for B

41. Area of first is 64π, area of second is 16π

43. a) $x = \frac{424}{117}$, $y = \frac{116}{39}$, $z = 2$ b) $x = -\frac{328}{117}$, $y = -\frac{128}{39}$, $z = -3$

Exercises 6.2, pages 345–347

1. 11   **3.** 1   **5.** 0.60

7. $x = \frac{5}{7}$, $y = -\frac{1}{7}$ **9.** Inconsistent   **11.** $x = -\frac{7}{17}$, $y = -\frac{7}{13}$

13. 9   **15.** 75   **17.** $x = 1$, $y = -1$, $z = 1$

19. Inconsistent   **21.** 7   **23.** 1   **25.** 0, -2, and 50

27. One cup of pudding contains 360 calories; one tablespoon of cream contains 50 calories

29. 14 hours and 24 minutes   **31.** $\dfrac{1}{x - 2} + \dfrac{3}{x + 3}$

33. $\dfrac{3}{x + 2} - \dfrac{1}{x - 2}$ **35.** $\dfrac{1}{x} + \dfrac{2}{x + 2} - \dfrac{3}{x - 2}$

37. $\dfrac{3}{x + 1} - \dfrac{1}{x - 2} + \dfrac{1}{(x - 2)^2}$

Exercises 6.3, pages 355–357

1.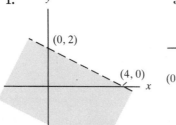

3.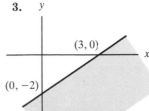

5.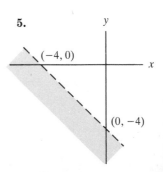

7.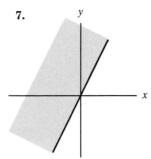

9.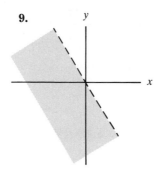

11.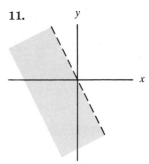

13.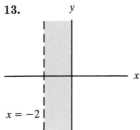

15.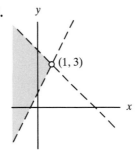

17.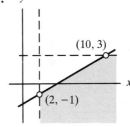

19.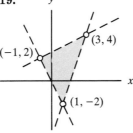

21.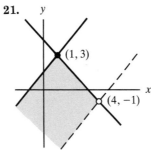

23.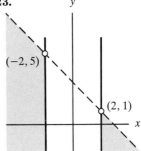

25. Empty set

27. a) Yes **b)** No

29. a) No **b)** No

31.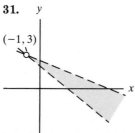

33.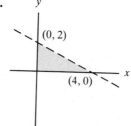

35.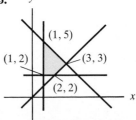

37.

Lambs	5	5	5	5	5	6	6	6	7
Goats	4	5	6	7	8	4	5	6	4

39. x: units of A; y: units of B; $2x + 3y \geq 8$, $5x + 2y \geq 9$, $2x + y \leq 8$, $x \geq 0$, $y \geq 0$

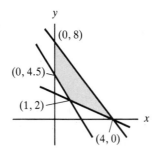

41. x: units of C; y: units of D; $0 \leq x \leq 3$, $3x + y \geq 5$, $3x + 4y \leq 21$, $y \geq 2$

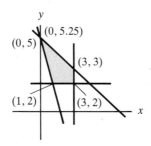

Exercises 6.4, pages 362–363

1. $x = 4$, $y = 16$; $x = -1$, $y = 1$

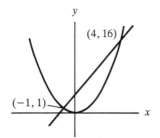

3. $x = 0$, $y = 0$; $x = -\frac{1}{2}$, $y = \frac{3}{2}$

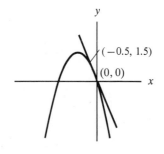

5. $x = \frac{3}{2}, y = -2; x = -3, y = 1$

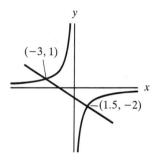

7. No solution

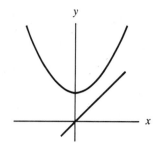

9. $x = -\frac{4}{3}, y = \frac{100}{9}; x = 3, y = 1$

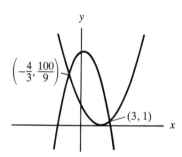

11.

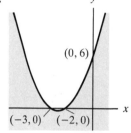

13.

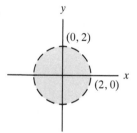

15.

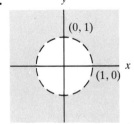

17.

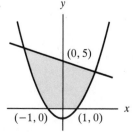

19.

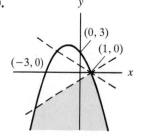

21.

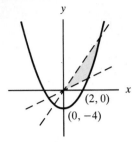

23. 5 cm by 12 cm

25. Base is 6 cm, altitude is 12 cm

Review Exercises, pages 363–364

1. $x = 7, y = 8$ **3.** $x = 4, y = -6$

5. $x = -2, y = -1, z = 3$

7. Dependent; solutions are given by $z = t, x = -3t - 1.5, y = 4t + 1.25$, where t is any number

9. $x = \frac{15}{37}, y = \frac{15}{16}$ **11.** 23   **13.** 30

15.

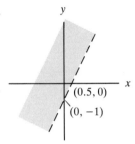

17.

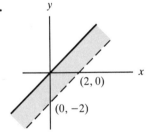

19.

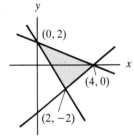

21.

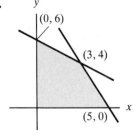

23. $\dfrac{2}{x - 1} - \dfrac{1}{x + 2}$ **25.** $\dfrac{5}{x + 1} - \dfrac{3}{x}$ **27.** 4 of A and 10 of B

7

Exercises 7.1, pages 271–273

1. a) 1, 4, 9, 16   **b)** 2, 4, 8, 16   **3. a)** $-1, 1, -1, 1$ **b)** $\frac{1}{2}, -\frac{1}{4}, \frac{1}{8}, -\frac{1}{16}$

5. 1, $\sqrt{2}$, $\sqrt{3}$, 2   **7.** 2, $\frac{9}{4}, \frac{64}{27}, \frac{625}{256}$ **9.** 24

11. 18   **13.** $a_n = 4n - 1$ **15.** $a_n = -(-\frac{2}{3})^n$

17. $\displaystyle\sum_{k=1}^{4} k^3$ **19.** $\displaystyle\sum_{k=1}^{n} \left(-\frac{1}{2}\right)^{k-1}$

21. a) $1, 4, 2, 8, 5, 7, 1;\ a_{10} = 8,\ a_{12} = 7,\ a_{20} = 4,\ a_{6k} = 7$
   b) $0, 1, 1, 1, 0, 0, 0;\ b_8 = 1,\ b_{17} = 0,\ b_{24} = 0$

23. a) $3, 1, 4, 1, 5, 9;\ a_{16} = 3,\ \displaystyle\sum_{k=1}^{16} a_k = 80;$ no general formula for a_n

   b) $0, 0, 1, 0, 0, 0;\ b_{12} = 1,\ \displaystyle\sum_{k=1}^{12} b_k = 4;$ no general formula for b_n

25. $2, \sqrt{6}, \sqrt{6}, 2, 0$ **27.** $\dfrac{1}{2}, \dfrac{2}{3}, \dfrac{3}{4}, \dfrac{4}{5}, \dfrac{5}{6}, \dfrac{6}{7};\ S_n = \dfrac{n}{n+1}$

29. a) $3, 6, 12, 24, 48$ **b)** $a_n = 3 \cdot (2^{n-1})$

31. Let P_n be the perimeter of an inscribed regular polygon of n sides.
$P_n = 2n \sin \dfrac{180°}{n}$. Show that $P_n \to 2\pi$ as $n \to \infty$.

33.

c \ n	1	2	3	4	5	6	7	8
1	0.37	0.69	0.50	0.61	0.55	0.58	0.56	0.57
0.5	0.61	0.55	0.58	0.56	0.57	0.56	0.57	0.57
2	0.14	0.87	0.42	0.66	0.52	0.60	0.55	0.58

In each case, a_n is approaching a fixed number $0.57\ldots$ as n becomes large.

Exercises 7.2, pages 376–378

1. a) 6   **b)** 21   **3. a)** 53   **b)** 375   **5. a)** -344 **b)** -408
7. a) 3   **b)** 2540   **9. a)** 16   **b)** -528 **11. a)** -5 **b)** 7
13. a) $\frac{36}{5}$ **b)** $\frac{204}{5}$ **15. a)** -6 **b)** $-\frac{27}{4}$ **17.** 34950
19. 576   **21.** $\frac{140}{3}$ **23.** 780
25. -3 **27.** $x = -17,\ a_8 = -156$
29. a) 191.1 m   **b)** 1960 m   **31.** Yes; $d = 8$
33. a) b_n: $3, 7, 11, 15, 19;\ b_n^2$: $9, 49, 121, 225, 361$ **b)** No
35. a) \$32 400   **b)** \$225 600   **c)** Almost 49 years old

Exercises 7.3, pages 382–383

1. a) Yes   **b)** Yes   **c)** Yes   **d)** No   **3.** $a_8 = -243,\ S_8 = -\frac{1640}{9}$
5. $a_8 = \frac{5}{128},\ S_8 = \frac{1275}{128}$ **7.** $a_4 = 64,\ a_5 = 128$
9. $r = \frac{1}{2},\ a_8 = \frac{1}{192}$ **11.** $r = -\frac{1}{2},\ a_1 = -\frac{32}{5},\ S_{10} = -\frac{341}{80}$
13. $n = 5$ **15.** $n = 8$ **17.** 62   **19.** 0.333333   **21.** $\frac{93}{16}$
23. 0.33333   **25.** $\dfrac{-5 \pm \sqrt{65}}{4}$ **27.** No solution in $\mathbf{R}$
29. 16   **31.** $\frac{20}{11}$ **33.** Converges to $-\frac{12}{7}$

35. Diverges   **37.** $0.41\overline{6}$ **39.** $0.\overline{142857}$ **41.** $0.\overline{076923}$ **43.** $\frac{37}{30}$

45. $\frac{87}{55}$ **47.** $\frac{4}{11}$ **49.** $\frac{1}{7}$ **51.** $n = 4$ **53.** 100 cm

Exercises 7.4, pages 388–390

1. $P(1)$ is false, $P(2)$, $P(3)$, $P(4)$ are true   **3.** $P(1)$, $P(2)$, $P(3)$, $P(4)$ are true

5. 4   **7.** 2   **23.** True   **25.** True

29. a)

n	1	2	3	4	5	6	7	8
$f(n)$	383	347	313	281	251	223	197	173

b) Yes

c) $n = 60$

Exercises 7.5, pages 396–397

1. 120   **3.** 132   **5.** 15   **7.** 1   **9.** 124

11. 56   **13.** 1   **15.** 1

21. $16x^4 + 32x^3y + 24x^2y^2 + 8xy^3 + y^4$

23. $x^{12} - 6x^{10} + 15x^8 - 20x^6 + 15x^4 - 6x^2 + 1$

25. $x^{10} - 5x^8y + 10x^6y^2 - 10x^4y^3 + 5x^2y^4 - y^5$

27. $x^4 - 8x^2 + 24 - \dfrac{32}{x^2} + \dfrac{16}{x^4}$ **29.** $-792x^7y^5$ **31.** $-280x^5$ **33.** $-4480x^7$

35. 90720

Exercises 7.6, pages 405–406

1. \$840   **3.** \$5550   **5. a)** \$4317.85   **b)** \$4416.08   **c)** \$4439.28   **d)** \$4451.08

7. a) \$8253.09   **b)** \$8537.18   **c)** \$8604.66   **d)** \$8639.03

9. \$1500 now is worth \$2146.15 in three years   **11.** 8.33%

13. a) \$468.63   **b)** \$442.48   **15.** \$38992.73

Exercises 7.7, pages 408–409

Sequences converge to the following limits:

1. 1   **3.** 0.3679   **5.** 0   **7.** 6.2832   **9.** 0.5   **11.** 2.3028

Review Exercises, pages 409–410

1. a) $\frac{1}{2}, \frac{3}{4}, \frac{7}{8}, \frac{15}{16}$ **b)** $\frac{49}{16}$ **3. a)** 2, 5, 8, 11   **b)** 26   **5. a)** 118   **b)** 1452

7. -1 or 3   **9. a)** $\frac{3}{2}, \frac{5}{4}, \frac{9}{8}, \frac{17}{16}$ **b)** No   **c)** $\frac{79}{16}$

11. a) $0.2\overline{6}$ **b)** $1.\overline{63}$ **c)** $0.2\overline{142857}$ **15.** No, not true for $n = 4$

17. a) 15   **b)** 455   **c)** 84   **19.** $6435x^6$

21. a) 0.34868   **b)** 0.36603   **c)** 0.36770   **d)** 0.36786

23. a) \$5877.85   **b)** \$6028.96   **c)** \$6082.82   **25.** \$514.31   **27.** 3925   **29.** 1

Exercises 8.1, pages 423–424

1. Circle, center $(0, 0)$ and radius 3

3. Parabola, vertex $(0, 0)$, focus $(2, 0)$, directrix $x + 2 = 0$

5. Circle, center $(1, -3)$, radius 2   **7.** Circle, center $(\frac{3}{2}, 2)$, radius $\dfrac{3\sqrt{2}}{2}$

9. Parabola, vertex $(-2, 1)$, focus $(-3, 1)$, directrix $x + 1 = 0$

11. Parabola, vertex $(\frac{1}{2}, 1)$, focus $(\frac{1}{2}, \frac{7}{12})$, directrix $y = \frac{17}{12}$

13. $(x - 2)^2 + (y - 4)^2 = 2^2$, $x^2 + y^2 - 4x - 8y + 16 = 0$

15. $(x + 2)^2 + y^2 = 2^2$, $x^2 + y^2 + 4x = 0$ **17.** $y^2 = 8x$ **19.** $x^2 = -4y$

21. $y^2 - 4y + 4x = 4$ **23.** $x^2 - 6x - 16y - 55 = 0$

25. $y = \sqrt{3 + 2x - x^2}$, $y = -\sqrt{3 + 2x - x^2}$; each is a function

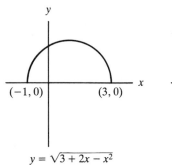

$y = \sqrt{3 + 2x - x^2}$

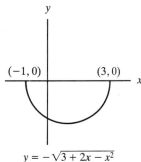

$y = -\sqrt{3 + 2x - x^2}$

27. $y = 2\sqrt{-x}$ or $y = -2\sqrt{-x}$ **29.** $x^2 - 200x + 200y = 0$, $0 \le x \le 200$

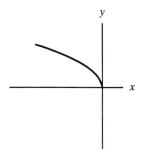

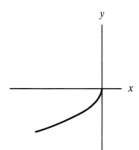

Exercises 8.2, page 432

1. Vertices $(3, 0)$ and $(-3, 0)$; foci $(\sqrt{5}, 0)$ and $(-\sqrt{5}, 0)$

3. Vertices $(0, 4)$ and $(0, -4)$; foci: $\left(0, \dfrac{8\sqrt{2}}{3}\right)$ and $\left(0, -\dfrac{8\sqrt{2}}{3}\right)$

5. Vertices $(1, 3)$, $(1, -5)$; foci $(1, 2\sqrt{3} - 1)$, $(1, -2\sqrt{3} - 1)$

7. Vertices $(2.5, 1.5)$ and $(-5.5, 1.5)$; foci: $(\sqrt{15} - 1.5, 1.5)$ and $(-\sqrt{15} - 1.5, 1.5)$

9. Vertices $(1, 3)$ and $(1, -3)$; foci $(1, \sqrt{5})$ and $(1, -\sqrt{5})$

11. Vertices $(1, 2)$ and $(-3, 2)$; foci $(\sqrt{2} - 1, 2)$ and $(-\sqrt{2} - 1, 2)$

13. $16x^2 + 25y^2 = 400$ **15.** $9(x - 3)^2 + 5y^2 = 45$

17. $(x - 1)^2 + 4(y + 1)^2 = 4$ **19.** $7(x - 3)^2 + 16(y + 1)^2 = 28$

Exercises 8.3, page 440

1. a) Center $(0, 0)$; vertices $(3, 0)$ and $(-3, 0)$; foci $(\sqrt{13}, 0)$ and $(-\sqrt{13}, 0)$
b) $2x - 3y = 0$, $2x + 3y = 0$

3. a) Center $(0, 0)$; vertices $(0, 4)$ and $(0, -4)$; foci $(0, 3\sqrt{2})$ and $(0, -3\sqrt{2})$
b) $2\sqrt{2}x - y = 0$, $2\sqrt{2}x + y = 0$

5. a) Center $(1, -1)$; vertices $(3, -1)$ and $(-1, -1)$; foci $(1 + \sqrt{13}, -1)$ and
$(1 - \sqrt{13}, -1)$ **b)** $3x - 2y = 5$, $3x + 2y = 1$

7. a) Center $(-2, 1)$; vertices $(-2, 3)$ and $(-2, -1)$; foci $(-2, 1 + 2\sqrt{2})$ and
$(-2, 1 - 2\sqrt{2})$ **b)** $x - y + 3 = 0$, $x + y + 1 = 0$

9. a) Center $(-1, 2)$; vertices $(1, 2)$ and $(-3, 2)$; foci $(-1 + \sqrt{5}, 2)$ and
$(-1 - \sqrt{5}, 2)$ **b)** $2y = x + 5$, $2y = -x + 3$

11. a) Center $(-1, \frac{3}{2})$; vertices $(-1, \frac{5}{2})$ and $(-1, \frac{1}{2})$; foci $\left(-1, \dfrac{3 + \sqrt{5}}{2}\right)$ and

$\left(-1, \dfrac{3 - \sqrt{5}}{2}\right)$ **b)** $4x - 2y + 7 = 0$, $4x + 2y + 1 = 0$

13. $5x^2 - 4y^2 = 20$ **15.** $y^2 - 12(x - 3)^2 = 4$

17. $5(x - 3)^2 - 4(y + 1)^2 = 20$ **19.** $x^2 - y^2 = 9$

Exercises 8.4, pages 448–449

1. a)

b) $\begin{cases} X = x - 1 \\ Y = y - 3 \end{cases} \begin{cases} x = X + 1 \\ y = Y + 3 \end{cases}$ **c)** A:$(3, 4)$, $[2, 1]$
B:$[3, -2]$, $(4, 1)$

3. a)

b) $\begin{cases} X = x - 2 \\ Y = y + 3 \end{cases} \begin{cases} x = X + 2 \\ y = Y - 3 \end{cases}$ **c)** A:$(3, 4)$, $[1, 7]$
B:$[3, -2]$, $(5, -5)$

5. $\begin{cases} X = x + 2 \\ Y = y - 2 \end{cases}$
$X^2 + 2Y^2 = 16$

7. $\begin{cases} X = x + 1, \\ Y = y + 2 \end{cases}$
$Y^2 = 3X$

9. $\begin{cases} X = x + 3 \\ Y = y - 1 \end{cases}$
$X^2 + Y^2 = 4$

11. a) $\langle 0, 0 \rangle$ **b)** $(0, 0)$ **13. a)** $\left(-\dfrac{\sqrt{2}}{2}, \dfrac{5\sqrt{2}}{2} \right)$ **b)** $\left(\dfrac{5\sqrt{2}}{2}, \dfrac{3\sqrt{2}}{2} \right)$

15. $(x')^2 - (y')^2 = 8$ **17.** $5(x')^2 + (y')^2 = 10$ **19.** $(x')^2 + 2(y')^2 = 8$

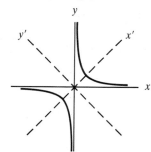   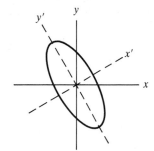   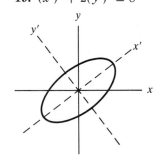

Exercises 8.5, page 453

1.
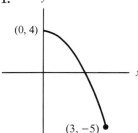

3.
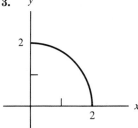

5.
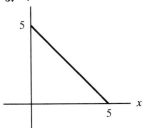

7. Particle starts at $(4, 0)$, moves along elliptic path to point $(-4, 0)$ at the end of π seconds.

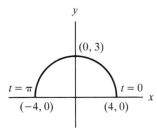

9. Particle starts at $(1, 0)$ moves along line segment to $(0, 1)$ at $t = \frac{1}{2}$, then back to $(1, 0)$ at $t = 1$; repeats this motion for each of the next 3 seconds.

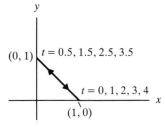

11. $x^2 + y^2 = 1$

13. $2x + y = 3$, $-3 \le x \le 1$; line segment between $(-3, 9)$ and $(1, 1)$

15. $y = x^2 + 1$, $x \ge 0$

17. $bx - ay = bx_0 - ay_0$ represents a line through (x_0, y_0) with slope b/a.

Review Exercises, pages 453–454

1. a) $(x + 2)^2 + (y - 1)^2 = 16$ **b)** $x^2 + y^2 + 4x - 2y = 11$ **3.** $y^2 = 12x$

5. a) $\dfrac{(x - 1)^2}{12} + \dfrac{(y - 4)^2}{16} = 1$ **b)** $4x^2 + 3y^2 - 8x - 24y + 4 = 0$

7. a) $\dfrac{(x - 1)^2}{4} - \dfrac{(y + 1)^2}{5} = 1$ **b)** $5x^2 - 4y^2 - 10x - 8y - 19 = 0$

9. Center $(-1, 2)$, radius 2

11. a) Vertex $(1, -1)$ **b)** Focus $(1, -\frac{3}{4})$ **c)**

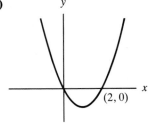

13. a) Center $(0, 1)$ **b)** Foci $(0, 1 + \sqrt{5})$ and $(0, 1 - \sqrt{5})$

15. Semicircle

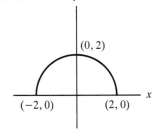

17. a) Parabola   **b)** Ellipse   **c)** Hyperbola   **d)** Circle

19. a) $[4, 2]$ **b)** $(-2, 4)$ **c)** $(-1, 5)$

21. Circular arc between $(0, 2)$ and $(2, 0)$

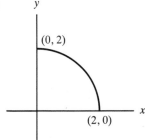

Exercises 9.1, page 460

1. a) P:[3, 50°], [−3, 230°], [−3, −130°], [3, 410°]

   b) Q:[4, −60°], [−4, 120°], [4, 300°], [4, 660°]

   c) T:[2, 540°], [2, 180°], [2, −180°], [−2, 0°]

3. P_2:[3, 310°], Q_2:[4, 60°], T_2: [2, 180°]   5. P_1:$\left[2, \frac{5\pi}{3}\right]$; Q_1:$\left[3, \frac{\pi}{12}\right]$; T_1:$\left[4, \frac{11\pi}{6}\right]$

7. a)

   b)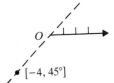

   c)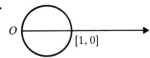

   d)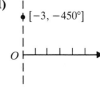

9. a) $\left[3, \frac{4\pi}{3}\right]$ b) $\left[4, \frac{\pi}{4}\right]$ c) [2, π]   d) $\left[3, \frac{3\pi}{2}\right]$

Exercises 9.2, page 464

1.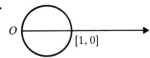

3.

5.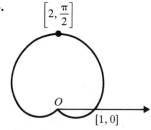

7.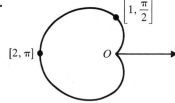

9.

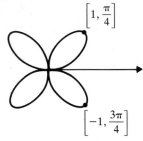

$\left[1, \dfrac{\pi}{4}\right]$

$\left[-1, \dfrac{3\pi}{4}\right]$

11.

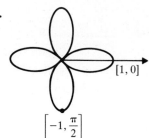

$[1, 0]$

$\left[-1, \dfrac{\pi}{2}\right]$

13.

$\left[1, \dfrac{\pi}{2}\right]$

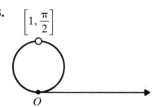

O

15.

$\left[1, \dfrac{\pi}{4}\right]$

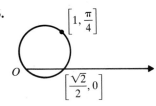

O

$\left[\dfrac{\sqrt{2}}{2}, 0\right]$

17.

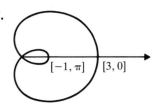

$[-1, \pi]$ $[3, 0]$

19.

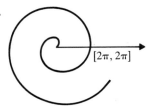

$[2\pi, 2\pi]$

Exercises 9.3, pages 468–469

1. a) $[\sqrt{2}, 135°]$ **b)** $[2, 240°]$ **c)** $[5.09, 51.85°]$ **d)** $[2.89, 122.87°]$

3. a) $\left[3\sqrt{2}, \dfrac{3\pi}{4}\right]$ **b)** $[\sqrt{10}, 5.03]$ **c)** $[3.53, 0.48]$

5. a) $(0, 2)$ **b)** $\left(\dfrac{3\sqrt{2}}{2}, \dfrac{3\sqrt{2}}{2}\right)$ **c)** $(2.09, -0.81)$

7. a) Yes   **b)** No   **c)** Yes   **d)** No   **e)** No

11. $x^2 + y^2 - 2y = 0$ **13.** $y = (\tan\frac{4}{3})x$ or $y = 4.13x$

15. $y = 0.25x^2 - 1$ **17.** $x^2 + y^2 + 2x = 0$

19. $r = 1$, circle   **21.** $\theta = \text{Tan}^{-1}3 = 1.25$; line through the origin with slope 3

23. No, the origin is a point on $r = \sin\theta$ but not on $r\csc\theta = 1$ **25.** Yes

Review Exercises, page 470

1. a) $[1, 0]$ **b)** $[3, \pi]$ **c)** $\left[4\sqrt{2}, \dfrac{\pi}{4}\right]$ **d)** $\left[2\sqrt{2}, \dfrac{3\pi}{4}\right]$

   e) $\left[2, \dfrac{7\pi}{6}\right]$ **f)** $\left[2, -\dfrac{\pi}{4}\right]$ **g)** $\left[4, \dfrac{\pi}{2}\right]$ **h)** $\left[3, \dfrac{3\pi}{2}\right]$

3. a) $(2, 2\sqrt{3})$ **b)** $(\sqrt{3}, -1)$ **c)** $(-4, 0)$ **d)** $\left(-\dfrac{1}{\sqrt{2}}, -\dfrac{1}{\sqrt{2}}\right)$ **e)** $\left(\dfrac{3}{\sqrt{2}}, \dfrac{3}{\sqrt{2}}\right)$

5. Graph is a circle of radius $\frac{1}{2}$ **7.** Graph is a circle of radius 1

9. Graph is a vertical line three units to the right of the origin

11. Graph is a spiral   **13.** $r^2 = 4$; circle with center at the origin and radius 2

15. $y^2 + 2x - 1 = 0$; parabola that opens to the left

Exercises 10.1, pages 476–478

1. a) $-i$ **b)** -1 **c)** 1 **d)** i **e)** i **f)** $-i$ **g)** 1 **h)** $-i$

3. a) 12 **b)** $12i$ **c)** -12 **d)** $-\frac{3}{4}i$ **e)** $\frac{3}{4}i$ **f)** $\frac{3}{4}$

5. a) 4 **b)** $-1 - 5i$ **c)** $\dfrac{4 - 3\sqrt{2}}{2} - \dfrac{2 + 3\sqrt{2}}{2}i$

9. a) $3 - 8i$ **b)** $2 + \sqrt{3}i$ **c)** $-4 + 4\sqrt{2}i$ **d)** 9 **e)** $\frac{1}{10} + \frac{3}{10}i$ **f)** $\dfrac{4}{17} + \dfrac{3\sqrt{2}}{17}i$

11. a) $2i$; $-\frac{1}{2}i$ **b)** i; $-3i$ **c)** $\dfrac{\sqrt{13} - 3}{2}i$; $-\dfrac{\sqrt{13} + 3}{2}i$ **d)** i; $-\frac{1}{2}i$

13. $x = -1, y = 2$ **15.** $x = -5, y = 13$ or $x = 2, y = 6$ **17.** Yes

19. a) No   **b)** No   **21.** $2 - 5i$ **23.** $1.5 - 0.5i$ **25.** $-2 + 6i$

Exercises 10.2, pages 480–481

1. $(3, 5)$ **3.** $(0, 4)$ **5.** $(0, 0)$ **7.** $(2, 1)$ **9.** $-4i$ **11.** $-4 - 3i$

13.

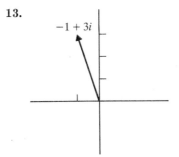

15.

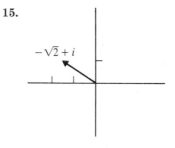

17.

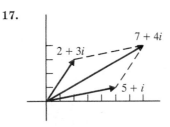

19.

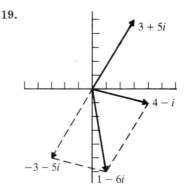

21.

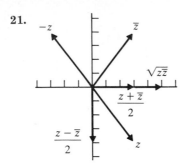

23. a) $-\dfrac{1}{2}+\dfrac{\sqrt{3}}{2}i$ **b)** $-\dfrac{1}{2}-\dfrac{\sqrt{3}}{2}i$ **c)** $-\dfrac{1}{2}-\dfrac{\sqrt{3}}{2}i$ **d)** 1

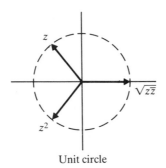

Unit circle

25. a) $x>0$ and $y=0$ **b)** $x=0$ **c)** $x>0$ and $y>0$ **d)** $x>0$ **e)** $y<0$

Exercises 10.3, pages 487–488

3. a) $\pi(\cos 0° + i\sin 0°)$ **b)** $5(\cos 306.87° + i\sin 306.87°)$
  c) $\sqrt{2}(\cos 135° + i\sin 135°)$ **d)** $13(\cos 337.38° + i\sin 337.38°)$

5. a) $\dfrac{3\sqrt{2}}{2}+\dfrac{3\sqrt{2}}{2}i$ **b)** -5 **c)** $-\dfrac{1}{2}-\dfrac{\sqrt{3}}{2}i$

7. a) $4(\cos 315° + i\sin 315°)$ **b)** $3(\cos 120° + i\sin 120°)$ **c)** $\cos\dfrac{\pi}{6}+i\sin\dfrac{\pi}{6}$

9. a) $\cos 45° + i\sin 45°$ **b)** $\dfrac{\sqrt{2}}{2}+\dfrac{\sqrt{2}}{2}i$

11. a) $2(\cos 120° + i\sin 120°)$ **b)** $-1+\sqrt{3}i$

13. $18(\cos 180° + i\sin 180°) = -18$ **15.** $\dfrac{1}{6}\left[\cos(-30°)+i\sin(-30°)\right]=\dfrac{\sqrt{3}}{12}-\dfrac{1}{12}i$

17. a) $2[\cos(-30°)+i\sin(-30°)]$ **b)** $2\sqrt{2}[\cos(-135°)+i\sin(-135°)]$

19. a) $\dfrac{\sqrt{2}}{2}[\cos(-105°)+i\sin(-105°)]$ **b)** $\dfrac{\sqrt{2}}{2}(\cos 105° + i\sin 105°)$

23. a) $-4i$ **b)** -8

Exercises 10.4, pages 492–493

1. a) $\cos 150° + i \sin 150° = -\dfrac{\sqrt{3}}{2} + \dfrac{1}{2}i$ **b)** $16[\cos(-180°) + i \sin(-180°)] = -16$

   c) $\cos 240° + i \sin 240° = -\dfrac{1}{2} - \dfrac{\sqrt{3}}{2}i$

3. a) $8[\cos(-90°) + i \sin(-90°)] = -8i$ **b)** $\cos 180° + i \sin 180° = -1$

5. a) $16(\cos 0° + i \sin 0°) = 16$ **b)** $16(\cos 240° + i \sin 240°) = -8 - 8\sqrt{3}i$

   c) $\dfrac{\sqrt{2}}{4}(\cos 225° + i \sin 225°) = -\dfrac{1}{4} - \dfrac{1}{4}i$

7. a) $256(\cos 180° + i \sin 180°) = -256$ **b)** $8(\cos 90° + i \sin 90°) = 8i$

9. $-5 + i$ **11.** $4 + 8\sqrt{3}i$ **13. a)** $1 + 3i$ **b)** $-9 - 3i$

15. $\sin 3\theta = 3 \sin \theta \cos^2\theta - \sin^3\theta = 3 \sin \theta - 4 \sin^3\theta$;
   $\cos 3\theta = \cos^3\theta - 3 \sin^2\theta \cos \theta = 4 \cos^3\theta - 3 \cos \theta$

Exercises 10.5, page 498

1. $1; -\dfrac{1}{2} + \dfrac{\sqrt{3}}{2}i; -\dfrac{1}{2} - \dfrac{\sqrt{3}}{2}i$

3. $1.12 - 0.24i; 0.57 + 0.99i; -0.77 - 0.66i; -1.05 - 0.47i; 0.12 - 1.14i$

5. $\dfrac{\sqrt{3}}{2} + \dfrac{1}{2}i; i; -\dfrac{\sqrt{3}}{2} + \dfrac{1}{2}i; -\dfrac{\sqrt{3}}{2} - \dfrac{1}{2}i; -i; \dfrac{\sqrt{3}}{2} - \dfrac{1}{2}i$

7. $2.36 + 0.31i; -0.31 + 2.36i; -2.36 - 0.31i; 0.31 - 2.36i$

9. $\frac{1}{2}(\sqrt{3} + i); \frac{1}{2}(-\sqrt{3} + i); -i$ **11.** $2 + i; 1 - i$ **13.** $i; -1 - i$

15. $-1; \dfrac{\sqrt{2}}{2}(-1 + i); \dfrac{\sqrt{2}}{2}(1 - i)$ **19.** $2 + 3i; -2 - 3i$

Review Exercises, page 499

1. $-2 + 2i$ **3.** $-7 - 24i$ **5.** $-\frac{1}{2}i$

7. $\dfrac{\sqrt{3} - 1}{32} - \dfrac{\sqrt{3} + 1}{32}i$ **9.** $-2.65 - 69.83i$ **11.** $2 - 2i$

13. 24 **15.** $-3 + 2\sqrt{3}i$

17. $0.98 - 0.17i; -0.34 + 0.94i; -0.64 - 0.77i$

19. $i; -i; -\dfrac{\sqrt{2}}{2} + \dfrac{\sqrt{2}}{2}i; \dfrac{\sqrt{2}}{2} - \dfrac{\sqrt{2}}{2}i$

Index

TRIGONOMETRY FORMULAS

Angular Measure

Convert degrees to radian measure: Multiply by $\pi/180$.
Convert radians to degree measure: Multiply by $180/\pi$.

Triangles

Law of Cosines:

$$a^2 = b^2 + c^2 - 2bc \cos \alpha$$

$$b^2 = a^2 + c^2 - 2ac \cos \beta$$

$$c^2 = a^2 + b^2 - 2ab \cos \gamma$$

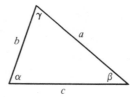

Law of Sines:

$$\frac{\sin \alpha}{a} = \frac{\sin \beta}{b} = \frac{\sin \gamma}{c}$$

Area:

$$\text{Area} = \tfrac{1}{2} ab \sin \gamma = \tfrac{1}{2} bc \sin \alpha = \tfrac{1}{2} ac \sin \beta$$

$$\text{Area} = \sqrt{s(s - a)(s - b)(s - c)}, \quad \text{where } s = \tfrac{1}{2} (a + b + c)$$

Complex Numbers

DeMoivre's Formula:

$$(\cos \theta + i \sin \theta)^n = \cos n\theta + i \sin n\theta, \quad \text{where } i^2 = -1 \text{ and } n \text{ is any integer.}$$